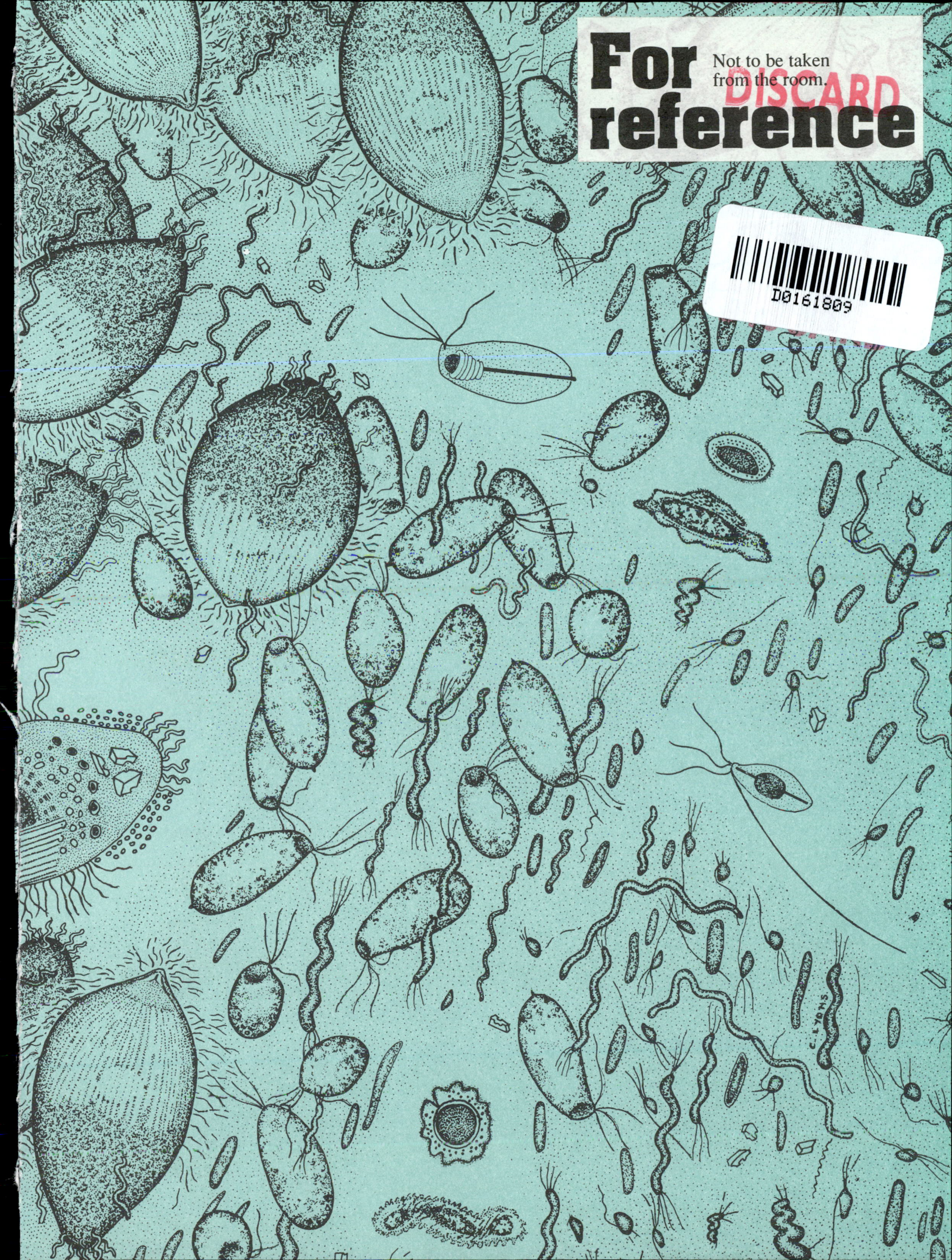

ILLUSTRATED GLOSSARY OF PROTOCTISTA

1

A. A. van Leeuwenhoek 1632-1723

B. C.A. Agardh 1785-1859

C. C.G. Ehrenberg 1795-1876

D. F. Dujardin 1801-1860

E. J. Leidy 1823-1891

2

A. E. Haeckel 1834-1919

B. M. Woronin 1838-1903

C. W.G. Farlow 1844-1919

D. O. Bütschli 1848-1920

E. W. Schewiakoff 1859-1930

3

A. C.A. Kofoid 1865-1947

B. E.A. Minchin 1866-1915

C. H.S. Jennings 1868-1947

D. A. Kahl 1877-1946

E. F.E. Fritsch (*left*) 1879-1954

G.M. Smith

1

A. A. Pascher 1881-1945

B. V.A. Dogiel 1882-1955

C. E. Chatton 1883-1947

D. E. Fauré-Fremiet 1883-1971

E. C. Dobell 1886-1949

2

A. L.R. Cleveland 1892-1969

B. H. Skuja 1892-1972

C. K. Bělař 1895-1931

D. P.-P. Grassé 1895-1985

E. W.R. Taylor 1895-1990

3

A. G.W. Prescott 1899-1988

B. G.F. Papenfuss 1903-1981

C. I. Manton 1904-1988

D. T.M. Sonneborn 1905-1981

E. H.C. Bold 1909-1987

B.M. Honigberg
1920-1992

Leading Deceased Contributors to Protoctist Research

"About two hours distant from this Town [Delft, Holland] there lies an inland lake, called the Berkelse Mere ... whose bottom in many places is marshy, or boggy," wrote Antony van Leeuwenhoek in 1674 (see Dobell, 1958, in Bibliography, p. 192). In describing, with astonishment, the inhabitants of the whitish water, this Dutchman founded, single-handedly, at least two sciences: **protoctistology,** the subject of this Glossary, and **bacteriology.** As the first person to see and write about microbes, not surprisingly he categorized them as either animals or plants. He saw *"an abundance of very little and odd animalcules, whereof some were incredibly small"* (Dobell, 1958, p. 111). He wrote in a letter to the Royal Society:

> *A third sort of little animals that were about twice as long as broad, and to my eye quite eight times smaller than the animalcules first mentioned; and I imagined, although they were so small, that I could yet make out their legs, or little fins. Their motion was very quick, both roundabout and in a straight line.*

Since the 17th century the lively objects of van Leeuwenhoek's studies have been parceled into many fragmented "fields" of science; none of the eminent scholars who in fact studied these protoctists called themselves **protoctistologists.** Indeed, each who thought he studied algae (literally water plants) called himself a botanist or phycologist. Those working primarily on slime nets or water molds identified themselves as students of fungi, mycologists, the kind of botanists who study mushrooms. Each of the others, self-categorized as a student of little animals (with legs or fins), referred to his work in "invertebrate zoology." In the three centuries since the revelation of the swimming marvels of the microcosm, such scientists have called themselves protozoologists, parasitologists, or invertebrate zoologists. Even today, more than 130 years after the Englishman, J. Hogg, coined the term **Protoctista** for the kingdom, there are still no self-identified protoctistologists! Because of lack of space, we picture here only a few of these superb scientists, who, in retrospect—because they loved the frisky ciliates, glimmering diatoms, and other inhabitants of the microscopic world as explained in this Glossary—were all practicing protoctistologists.

C.A. Agardh was followed, at Lund, by his even more prolific son, J.G. Agardh, who by rights should be in this gallery of protoctist researchers. Professor Robert Wilce tells us that Agardh the younger, who developed benthic algal systematics on the basis of their reproductive patterns, was probably more important to the development of the science than his father. The omission of Jacob Whitman Bailey, the first North American phycologist who, along with the German, C.G. Ehrenberg, founded the study of diatoms, has been noted by B.C. Parker, Professor at Virginia Polytechnic Institute. We, of course, never predicted that on May 1, 1992, our University of Massachusetts colleague, expert in trichomonads and founder of the taxon **Kinetoplastida,** Bronislaw M. Honigberg, would die. Had we known, his photograph certainly would have been included among these distinguished contributors to the newly consolidated field of protoctistology.

ILLUSTRATED GLOSSARY OF PROTOCTISTA

Vocabulary of the algae, apicomplexa, ciliates, foraminifera, microspora, water molds, slime molds, and the other protoctists

This glossary, for students, teachers, and researchers, is an illustrated, abbreviated version of the

HANDBOOK OF PROTOCTISTA

The structure, cultivation, habitats, and life histories of the eukaryotic microorganisms and their descendants exclusive of animals, plants, and fungi

EDITORS

Lynn Margulis
University of Massachusetts - Amherst

Heather I. McKhann
University of California - Los Angeles

Lorraine Olendzenski
University of Massachusetts - Amherst

EDITORIAL COORDINATOR

Stephanie Hiebert

JONES AND BARTLETT PUBLISHERS
BOSTON **LONDON**

Editorial, Sales, and Customer Service Offices

Jones and Bartlett Publishers
One Exeter Plaza
Boston, MA 02116

Jones and Bartlett Publishers International
PO Box 1498
London W6 7RS
England

Library of Congress Cataloging-in-Publication Data

Illustrated glossary of protoctista : the structure, cultivation, habitats, and life histories of the eukaryotic microorganisms and their descendants exclusive of animals, plants, and fungi : vocabulary of the algae, apicomplexa, ciliates, foraminifera, microspora, water molds, slime molds, and other protoctists / editors, Lynn Margulis, Heather I. McKhann, Lorraine Olendzenski; editorial coordinator, Stephanie Hiebert.

p. cm.

Abbreviated version of: Handbook of protoctista. c1990.

Includes bibliographical references (p.)

ISBN 0-86720-081-2

1. Protista. 2. Protozoa. 3. Algae. 4. Myxomycetes. 5. Eukaryotic cells. I. Margulis, Lynn, 1938- . II. McKhann, Heather I. III. Olendzenski, Lorraine. IV. Handbook of protoctista.

QR74.5.I44 1992
576--dc20 92-40468
CIP

Editorial Production: Peter H. Neumann Associates

Cover illustration: *Pterotermes occidentis* hindgut microbial community (Phylum Zoomastigina) drawn by Christie Lyons based on electron micrographs by David Chase.

Printed in the United States of America
97 96 95 94 93 10 9 8 7 6 5 4 3 2 1

Table of Contents

Frontispiece: Leading Deceased Contributors to Protoctist Research ii
Foreword to *Handbook of Protoctista* (Lewis Thomas) ix
Editors of *Handbook of Protoctista* xi
Editors of *Illustrated Glossary of Protoctista* xii
Contributors to *Handbook of Protoctista* and Other Protoctist Experts xiii

Introduction to *Illustrated Glossary of Protoctista* xvii

1. For whom this book is intended 2. How many protoctists are there? 3. There are no single-celled animals or plants 4. Like lichens, all protoctists are co-evolved microbial symbionts 5. Terminology: Flagella/undulipodia; flagellate/mastigote 6. Undulipodia, sex and the four groups of phyla 7. Kinetids, sexual life cycles and systematic chapter order 8. Life histories 9. Kingdom Protoctista 10. How many kingdoms of organisms should be recognized? 11. Kingdom criteria 12. Invalidity of nutritional criteria 13. Size and multicellularity: Protoctist vs. protist 14. Organization of this book 15. Polyphyly and anastomosing branches on family trees 16. International recognition of protoctists 17. Acknowledgments and apologies References Table 1. Modes of Nutrition for Life on Earth Table 2. Modes of Nutrition in Protoctista Table 3. Units and Dimensions

Toward a Nomenclatural Protist Perspective (John O. Corliss) xxvii

1. Neo-Haeckelian protistology 2. Protists and codes of nomenclature 3. Protist macrosystematics and nomenclature 4. Where do we go from here? References

Sources of Living Protoctists and Their Culture xxxiii

Access to Protoctist Information: Toward a Database (Kenneth W. Estep) xxxix

1. Type specimens and traditional collections 2. Video information 3. A Protoctist Knowledge Base 4. Information retrieval from the Protoctist Knowledge Base Further Information Protoctist collections Video microscopy and image analysis Linnaeus databases Expert Systems, HyperText, and Neural Networks References

Molecular Biology and Protoctist Phylogeny (Mitchell L. Sogin) xlv

1. Ribosomal RNA phylogenies 2. Molecular analysis of ribosomal RNAs 3. Data analysis References Computer software for analysis of molecular data

Protoctist Glossary

Etymological Roots and Examples of Use 3

General Glossary 5

Organism Glossary 137

Bibliography of Glossaries 192

Classification 193

Table 4. Classes of the Phyla of the Kingdom Protoctista 197 Table 5. Informal Names of Protoctistan Phyla and Classes 201 Table 6A. Phyla of Protoctists 204 Table 6B. Classes of Phylum Zoomastigina 236 Table 6C. Classes of Phylum Chlorophyta 244 Table 7. Classification: Summary of Phyla and Lower Taxa 251

Figure Credits 287

Foreword
to *Handbook of Protoctista*

Just a few years back, this book would have seemed to most readers trained in either medicine or biology nothing more than an arcane compendium of nature's most outlandish oddities, queer single cells of primary interest only to the most specialized of specialists.

Indeed, such a book as this could not have been put together a few years back because most of the details contained here have emerged so recently. But, more than this, the intellectual reasons for deep interest in these organisms did not exist in many minds until the revolution in cell biology was well under way.

Now, it is an obvious though startling fact of life, accepted all round, that the Kingdom of Protoctista represents our distant parents and now awaits deep exploration, with irresistible opportunities at every hand for good reductionist research. The appearance of these cells a billion-odd years ago was the second major event in planetary evolution and led directly, lineage by lineage, to our own complex selves, brain and all.

The first, of course, was the appearance on the planet of our Ur-ancestors, the prokaryotes, 3.5 or more billion years ago. *Bergey's Manual* became this century's comprehensive encyclopedia for the investigators of the Kingdom of Bacteria and remains an indispensable tool in every laboratory engaged in microbiology.

The *Handbook of Protoctista* will surely become a volume of similar scientific indispensability, not just for students of cellular evolution and symbiosis but for all sorts of scientists engrossed by the facts of life itself. Many, maybe most of the creatures described here are still total mysteries, but biological science has now evolved to a stage when it is the likeliest of prospects that they will turn soon into everyday objects of bench research.

It is a safe prediction that the present book will have its own kind of evolution in the decades ahead as the science itself evolves. I only wish I could guess at what its editions will look like twenty years from now. But this depends on the science that the book will foster and stimulate, and if the science is as good as it should be, the outcome will have to be unpredictable and always a surprise.

Lewis Thomas, M.D.
Scholar-in-Residence
Cornell University Medical College

President Emeritus
Memorial Sloan-Kettering Cancer Center

Editors
of Handbook of Protoctista

Lynn Margulis
Department of Biology
University of Massachusetts
Amherst, MA 01003 USA

John O. Corliss
Department of Zoology
University of Maryland
College Park, MD 20742 USA

Michael Melkonian
Universität zu Köln
Botanisches Institut
Gyrhofstrasse 15
D-5000 Köln 41 GERMANY

David J. Chapman
Department of Biology
University of California at Los Angeles
Los Angeles, CA 90024 USA

EDITORIAL COORDINATOR
Heather I. McKhann
Department of Biology
University of California at Los Angeles
Los Angeles, CA 90024 USA

Editors
of *Illustrated Glossary of Protoctista*

Lynn Margulis
Department of Biology
University of Massachusetts
Amherst, MA 01003 USA

Heather I. McKhann
Department of Biology
University of California at Los Angeles
Los Angeles, CA 90024 USA

Lorraine Olendzenski
Department of Biology
University of Massachusetts
Amherst, MA 01003 USA

EDITORIAL COORDINATOR
Stephanie Hiebert

Contributors

to *Handbook of Protoctista* and Other Protoctist Experts*

Christian F. Bardele (Actinopoda)
Institut für Biologie III
Auf der Morgenstelle 28
D-7400 Tübingen GERMANY

Donald J.S. Barr (Chytridiomycota)
Biosystematics Research Centre
Central Experimental Farm
Ottawa K1A 0C6 CANADA

Richard L. Blanton (Acrasea)
Biological Sciences
Texas Tech University
P. O. Box 4149
Lubbock, TX 79409 USA

Guy Brugerolle (Zoomastigina: Parabasalia, Proteromonadida, Pyrsonymphida, Retortamonadida)
Groupe de Zoologie et Protistologie
Université de Clermont-Ferrand II
63177 Aubière FRANCE

Kurt Randall Buck (Choanomastigotes)
Monterrey Bay Aquarium Research Institute
160 Central Avenue
Pacific Grove, CA 93950 USA

Monique Cachon (Phaeodaria and Polycystina)
Laboratoire de Protistologie Marine
UA 671, CNRS
06230 Villefranche-sur-Mer FRANCE

Elizabeth U. Canning (Microspora)
Department of Pure and Applied Biology
Imperial College of Science, Technology and Medicine
London SW7 2AZ UNITED KINGDOM

*Authors of *Handbook of Protoctista* chapters and other researchers who generously provided information are listed with the groups that they study. Regrettably, since the *Handbook of Protoctista* project began two contributors, Jean Cachon and Charles Miller, have died. Dr. Miller's former associate, Daniel Dylewski, has taken over the phylum Plasmodiophoromycota.

James Cavender (Dictyostelida)
Department of Botany
Porter Hall
Ohio University
Athens, OH 45701 USA

David J. Chapman (Chlorophyta, other algae)
Department of Biology
University of California at Los Angeles
Los Angeles, CA 90024 USA

Cicily Chapman-Andresen (Karyoblastea and other large free-living amebas)
Institute of Cell Biology
The Zoological Institutes
University of Copenhagen
15 Universitetsparken
DK-2100 Copenhagen O DENMARK

Margaret N. Clayton (Phaeophyta)
Department of Ecology and Evolutionary Biology
Monash University
Clayton, Victoria 3168 AUSTRALIA

John O. Corliss (Ciliophora, Opalinata, Pseudociliata)
P.O. Box 53008
Albuquerque, NM 87153 USA

Richard M. Crawford (Bacillariophyta)
Department of Botany
University of Bristol
Bristol, B58 1UG UNITED KINGDOM

Johan F. De Jonckheere (Rhizopoda, Amebomastigota)
Instituut voor Hygiëne en Epidemiologie
Juliette Wytsmanstraat 14
B-1050 Brussels BELGIUM

Isabelle Desportes (Apicomplexa, Paramyxea)
Laboratoire de Physiologie Cellulaire
Université Pierre et Marie Curie
4 place Jussieu
75005 Paris FRANCE

Michael W. Dick (Oomycota)
Department of Botany
School of Plant Sciences
The University of Reading
Building 2, Earley Gate
Whiteknights
Reading RG6 2AU UNITED KINGDOM

Betsey D. Dyer (Amebomastigota, Bicosoecids, Parabasalia, Pyrsonymphida)
Department of Biology
Wheaton College
Norton, MA 02766 USA

Daniel P. Dylewski (Plasmodiophoromycota)
Kraft, Inc.
Microscopy and Rheology
Technology Center
801 Waukegan Road
Glenview, IL 60025 USA

Kenneth W. Estep (Actinopoda, Prymnesiophyta)
Institute for Taxonomic Zoology
University of Amsterdam
P.O. Box 4766
N-1009 Amsterdam THE NETHERLANDS

Jean Febvre (Acantharia)
Laboratoire de Biologie Cellulaire
Marine Station Zoologique
CEROV
06230 Villefranche-sur-Mer FRANCE

Colette Febvre-Chevalier (Heliozoa)
Laboratoire de Biologie Cellulaire
Marine Station Zoologique
CEROV
06230 Villefranche-sur-Mer FRANCE

Tom Fenchel (Ciliophora, free-living mastigotes)
Marine Biological Laboratory
University of Copenhagen
DK-3000 Helsingør DENMARK

Gary L. Floyd (Prasiolales, Trentepohliales, Ulvophyceae)
Department of Plant Biology
Ohio State University
Columbus, OH 43210 USA

Wilhelm Foissner (Ciliophora, Rhizopoda)
Institute for Zoology
University of Salzburg
Heilbrunnerstrasse 34
A-5020 Salzburg AUSTRIA

Lafayette Frederick (Myxomycota)
Department of Botany
Howard University
Washington, DC 20059 USA

Melvin S. Fuller (Hyphochytriomycota, "Fungi")
Department of Botany
University of Georgia
Athens, GA 30602 USA

Paul W. Gabrielson (Rhodophyta)
Department of Biology
William Jewell College
Liberty, MO 64068-1896 USA

David J. Garbary (Rhodophyta)
Department of Biology
St. Francis Xavier University
Antigonish, NS B26 1C0 CANADA

Marcelle A. Gillott (Cryptophyta)
Department of Biological Sciences
University of Wisconsin - Milwaukee
P.O. Box 413
Milwaukee, WI 53211 USA

Linda Graham (Charophyceae)
132 Birge Hall
Department of Botany
430 Lincoln Drive
University of Wisconsin
Madison, WI 53706 USA

Michael C. Grant [Charophyceae (Order Charales)]
Department of Environmental, Population and Organismic Biology
University of Colorado
Boulder, CO 80309 USA

J.C. Green (Phytoflagellates, esp. Prymnesiophyta)
Plymouth Marine Laboratory
The Laboratory
Citadel Hill
Plymouth PLI 2PB UNITED KINGDOM

Karl G. Grell (Foraminifera, Rhizopoda)
Friedlandstrasse 27
D-7404 Rotenburg GERMANY

I. Brent Heath (Chytridiomycota, Oomycota)
Department of Biology
York University
4700 Keele Street
North York, ON M3J 1P3 CANADA

Peter Heywood (Raphidophyta)
Division of Biology and Medicine
Brown University
Providence, RI 02912 USA

David J. Hibberd (Chlorarachnida, Eustigmatophyta, Xanthophyta)
Axle Tree Cottage
Starvecrow Lane
Peasmarsh, Rye
East Sussex TN31 6XL UNITED KINGDOM

Robert W. Hoshaw (Conjugaphyta)
Department of Ecology and Evolutionary Biology
University of Arizona
Tucson, AZ 85721 USA

Ludwig Kies (Glaucocystophyta)
Institut für Allgemeine Botanik und Botanischer Garten
Ohnhorststrasse 18
Universität Hamburg
D-2000 Hamburg 52 GERMANY

Peter A. Kivic (Euglenida)
Fay County Road
North Bangor, NY 12966 USA

Bruno P. Kremer (Glaucocystophyta)
Institut für Naturwissenschaften und ihre Didaktik
Abteilung für Biologie
Universität zu Köln
Gronewaldstrasse 2
D-5000 Köln 41 GERMANY

Jørgen Kristiansen (Chrysophyta)
Institut for Sporeplanter
University of Copenhagen
Øster Farimagsgade 2D
DK-1353 Copenhagen K DENMARK

Mark Leckie (Fossil Foraminifera)
Department of Geology
University of Massachusetts
Amherst, MA 01003 USA

John J. Lee (Granuloreticulosa)
Biology Department
City University of New York
Convent Ave. at 138th St.
New York, NY 10031 USA

Jiří Lom (Microspora, Myxozoa)
Institute of Parasitology
Czechoslovak Academy of Science
Branišovská 31
370/05, České Budějovice CZECHOSLOVAKIA

Denis H. Lynn (Ciliophora)
Department of Zoology
University of Guelph
Guelph, ON N1G 2W1 CANADA

Richard M. McCourt (Conjugaphyta)
Department of Biological Sciences
De Paul University
Chicago, IL 60614 USA

Michael Melkonian (Algal cell biology, Chlorophyceae, Microthamniales, Pedinomonadales, Prasinophyceae)
Universität zu Köln
Botanisches Institut
Lehrstuhl 1
Gyrhofstrasse 15
D-5000 Köln 41 GERMANY

Jean-Pierre Mignot (Bicosoecids, Euglenida)
Laboratoire de Biologie des Protistes
Université Blaise-Pascal de Clermont-Ferrand
Complexe Scientifique des Cézeaux
63177 Aubière FRANCE

Øjvind Moestrup (Algae)
Institut for Sporeplanter
Øster Farimagsgade 2D
DK-1353 Copenhagen K DENMARK

Miklos Muller (Kinetoplastida, Trichomonads)
The Rockefeller University
New York, NY 10021 USA

Charles J. O'Kelly (Prasiolales, Trentepohliales, Ulvophyceae)
Department of Botany and Zoology
Massey University
Palmerston North NEW ZEALAND

David Patterson (Zoomastigina, other free-living protists)
Department of Zoology
Bristol University
Woodland Road
Bristol B58 1UG UNITED KINGDOM

K. Perch-Nielsen (Prymnesiophyta)
Geological Institute ETH-Z
CH-8092 Zürich SWITZERLAND

Frank O. Perkins (Haplosporidia, Paramyxea)
Virginia Institute of Marine Science
College of William and Mary
Gloucester Point, VA 23062 USA

David Porter (Labyrinthulomycota)
Department of Botany
University of Georgia
Athens, GA 30602 USA

Igor B. Raikov (Ciliophora, mastigotes)
Institute of Cytology
4, Trikhoretsky Avenue
194064 St. Petersburg, RUSSIA

Frank E. Round (Bacillariophyta)
Department of Botany
University of Bristol
Bristol BS8 1UG UNITED KINGDOM

Frederick L. Schuster (Rhizopoda)
Department of Biology
Brooklyn College
City University of New York
Brooklyn, NY 11210 USA

Eugene B. Small (Ciliophora)
Department of Zoology
University of Maryland
College Park, MD 20742 USA

Mitchell L. Sogin (Molecular evolution)
Marine Biological Laboratory
Woods Hole, MA 02543 USA

Milton Sommerfeld (Rhodophyta)
Department of Botany
Arizona State University
Tempe, AZ 85287 USA

Frederick W. Spiegel (Acrasea, Cellular slime molds, Dictyostelida, Myxomycota, Protostelida)
Department of Biological Sciences
SE 632
University of Arkansas
Fayetteville, AR 72701 USA

F.J.R. Taylor (Dinomastigota, Ebridians)
Institute of Oceanography
University of British Columbia
Vancouver, BC V6T 1W5 CANADA

Øle Secher Tendal (Xenophyophora)
Zoological Museum
University of Copenhagen
Universitetsparken 15
DK-2100 Copenhagen O DENMARK

Roberta A. Townsend (Rhodophyta)
University of Sydney
School of Biological Sciences
Sydney, NSW 2006 AUSTRALIA

William Trager (Apicomplexa)
The Rockefeller University
New York, NY 10021 USA

Patti L. Tyler (Rhodophyta)
Department of Botany and Microbiology
Arizona State University
Tempe, AZ 85287 USA

Keith Vickerman, F.R.S. (Diplomonadida, Kinetoplastida)
Department of Zoology
University of Glasgow
Glasgow, G12 8QQ SCOTLAND

Emile Vivier (Apicomplexa)
Service de Biologie Animale
Université des Sciences et Techniques de Lille
59655 Villeneuve d'Ascq, Cedex FRANCE

Patricia L. Walne (Euglenida)
Department of Botany
University of Tennessee
Knoxville, TN 37996-1100 USA

Jen-Chyong Wang (Conjugaphyta)
Department of Ecology and Evolutionary Biology
University of Arizona
Tucson, AZ 85721 USA

Peter Westbroek (Fossil protists, Prymnesiophyta)
Department of Chemistry
Leiden University
P.O. Box 9502
2300 RA Leiden THE NETHERLANDS

Jean M. Whatley (Karyoblastea)
Department of Plant Sciences
University of Oxford
South Parks Road
Oxford OX1 3RA UNITED KINGDOM

Howard C. Whisler (Ellobiopsida, Oomycota)
Department of Botany AJ-10
University of Washington
Seattle, WA 98195 USA

Introduction
to *Illustrated Glossary of Protoctista*

1. For whom this book is intended

This reference book is an abbreviated form of the *Handbook of Protoctista* (Margulis *et al.*, 1990). Designed for all investigators, instructors and students who deal with protoctists, the eukaryotic microorganisms and their descendants (exclusive of the animals, fungi and plants), it contains the latest information. Understanding the relations among live organisms is essential for biochemists, botanists, ecologists, cell and molecular biologists, medical researchers, microbiologists, mycologists, parasitologists, phycologists, protozoologists, and zoologists.

2. How many protoctists are there?

We estimate that there are more than 100,000 species of described, extant protoctists and that many thousands more await discovery. Probably the number in each category is even greater for extinct forms. On encountering the hypertrophied intestines of the East African rhinoceros, W. Van Hoven (1987) discovered a new world of symbiotic eukaryotes using scanning electron microscopy. A wood-ingesting termite may contain as many as thirty different protist species. Beavers and cervids enjoy diets extremely rich in cellulose; who can even predict the protoctistan populations residing in these and so many other animals?

3. There are no single-celled animals or plants

The bewildering diversity of protoctists, so much of it unknown, must be organized on a rational basis. All earlier schemes conceived of members of the protoctists as tiny animals, tiny plants ... and later fungi (water molds or aquatic fungi). Even today, many scientists (e.g., especially cell biologists, plankton ecologists and geologists) routinely write about Protozoa and Algae as if they were phyla in the Animal and Plant kingdoms, respectively. These organisms are no more "one-celled animals and one-celled plants" than people are shell-less multicellular amebas. Indeed, since animals and plants always develop from embryos, neither one-celled animals nor one-celled plants even exist! Unlike all previous works on protoctists (including the *Illustrated Guide to the Protozoa*, Lee *et al.*, 1985), the *Handbook of Protoctista* does not operate from "the top down," imposing an obsolete two-kingdom view on the unaccommodating Protoctista. Rather, we have attempted to respect this great realm in its own right, conscious of its legacy from the prokaryotes. All scientists agree that protoctists originated by symbiotic mergers of bacteria.

4. Like lichens, all protoctists are co-evolved microbial symbionts

All protoctists are composites; chimeras with multiple ancestry. They have all evolved from more than a single type of microbial symbiont. Within the perspective of formal divisions of the biological sciences: botany, zoology and mycology, we find ourselves in a period comparable to that of lichenology in the late nineteenth century. The realization that all lichens (superficially "primitive plants") are symbionts of algae or cyanobacteria with fungi was jarring; the implications of the symbiotic nature of lichens for their systematics and taxonomy were profound.[1]

Like lichens, all algae have secondarily and, in some cases, independently, acquired photosynthetic symbionts. Analogous to the fungi of lichens, the heterotrophic components of algae, rather than the phototrophic plastids, tend to be diverse. In all cases plastids in whatever their glorious color (green - chloroplasts, red - rhodoplasts, blue-green - cyanelles, etc.) are not directly related to their "hosts" (the rest of the heterotrophic cytoplasm in which the plastids reside). Indeed, in many groups of algae (e.g., euglenids, prasinophytes, chlorophytes and conjugating green algae) it is questionable that the heterotrophic hosts are directly related to each other.

Given their symbiotic nature, in spite of botanical tradition, we can no longer tolerate classification of protoctists on the basis of the colors of their co-evolved phototrophic symbionts, organelles derived from undigested food. Mitochondria, like plastids, originated from respiring bacteria by several independent acquisitions. Therefore, mitochondrial characteristics in

[1] Serious scholars such as Dr. W. Nylander (1867 cited in Abbayes, 1954) denounced derisively the lichen symbiosis concept, which he named "Schwendenerisme" after Schwendener, who articulated the "theorie algo-lichénique" of the symbiotic nature of all lichens. Lichenologists all accept "Schwendenerisme"; most now agree that lichens need to be named and classified with their heterotrophic fungal partners. Whereas in the 25,000 or so species of lichens the diversity in the phycobiont (algal or cyanobacterial symbiont) is relatively limited, lichen fungi are profoundly diverse; the lichen symbiosis is highly polyphyletic.

different protoctist lineages cannot be used as the basis for classification until details of mitochondrial polyphyly are available.

Taking our cues from lichenologists (Hawksworth and Hill, 1984), in this book we consider the ultrastructure and sexual patterns of the cytoplasmic (heterotrophic) components of protoctist cells of paramount importance for determination of phyla, classes and the other higher taxa. Cell structure and developmental patterns, exclusive of the mitochondria, plastids and other xenosomal organelles (Corliss, 1987) provide the primary basis of our classification.

5. Terminology: Flagella/undulipodia; flagellate/mastigote

The senior editors and contributors of the *Handbook of Protoctista* nearly came to blows concerning aspects of terminology. We are dealing with the collapse of the walled structures of academic disciplines such as protozoology. As Fleck said in 1935, "This social character inherent in the very nature of scientific activity is not without its substantive consequences. Words that were formerly simple terms become slogans; sentences that were once simple statements become calls to battle" (Fleck, 1979, p. 43. See Margulis and Sagan, 1986, for further discussion of the Fleckian notion of scientific thought-styles).

The *Illustrated Glossary of Protoctista* is one result of this scientific tension: we have attempted to include all the defined uses of words our authors and editors sent us. Words such as "spore" may have several entirely different meanings. Furthermore, "spores" of some organisms (e.g., slime molds) may be identical to "cysts" of others (e.g., amebas). As senior editor I have taken a heavy hand concerning the use of the terms flagella/ undulipodia in attempting to obtain consistent usage amongst at least my younger colleagues.

The term "flagellum" is ambiguous: it has been used to refer to both prokaryotic and eukaryotic motility organelles, nonhomologous structures that differ in organization and composition. The bacteriologists have claimed as their own the term "flagellum" to refer to the rigid, extracellular rotary structure composed of flagellin protein (Margulis and Sagan, 1985; Sieburth and Estep, 1985). Eukaryotic "flagella," and the nearly identical cilia, are, in contrast, intrinsically motile structures consisting of tubulin and more than one hundred other proteins with a characteristic 9-fold symmetrical arrangement. The term "flagella" thus obscures the fact that these are completely different structures. It has therefore been proposed (Margulis, 1980; 1985) that the term "undulipodium," already in general use in the Russian and other literature early in this century (Shmagina, 1948), be used for eukaryotic flagella and cilia, and that "mastigote" be used to refer to the traditional "flagellates," that is, undulipodiated cells.

Undulipodia consist of microtubules arranged in a "9+2" or, as here designated, [9(2)+2] array, forming a shaft called the axoneme. That is, the axoneme consists of an outer ring of 9 groups of doublets of microtubules, with two singlet central microtubules. Each undulipodium develops from its kinetosome (also called a basal body), a cylinder of triplet microtubules lacking the central pair, i.e., the "9+0," here more accurately designated the [9(3)+0] array. When the kinetosome lacks an axoneme, it is called a centriole. Flagellar insertions, designated kinetids, are kinetosomes and associated fibrils and tubules that form specific unit patterns in all undulipodiated cells. Both traditionally and in this book undulipodia and their kinetids are of great taxonomic significance, as discussed below.

Flagella and undulipodia are therefore defined as follows (Margulis, 1980):

Undulipodia (Latin: *undula*, a little wave; Greek: *podos*, foot. Little waving feet) s. undulipodium

Cilia and [9(2)+2] flagella: Long slender tubulin-containing intracellular organelles of motility of eukaryotes, intrinsically motile throughout their length, capable of movement when severed from the cell.

Diameter: 0.25 micrometers (250 nanometers)

The axoneme is the [9(2)+2] shaft of the undulipodium; that is, it is the undulipodium lacking its surrounding membrane.

Flagella (Greek: whip) s. flagellum

Solid bacterial organelles of motility composed of flagellin, intrinsically nonmotile. Rotary locomotion is generated at the points of insertion of organelle into the cell (Berg, 1975). Flagella are extracellular in that they always extend externally beyond the plasma membrane of the prokaryotic cell.

Diameter: 15-30 nanometers

Includes the axial filaments or axial fibrils of spirochetes.

In these latter microbes the flagella are situated in the periplasm, the space between the outer and the inner (plasma) lipoprotein membranes of the gram negative cell wall.

Although many eminent scholars retain "flagellum" and "basal body" for the ninefold symmetrical microtubular motility organelles of eukaryotes, we feel that they are reluctant to change because, in the confines of their restricted subdisciplines, they confront little ambiguity. Because they do not daily inhabit the world of bacteriologists, the fact that "flagellum" refers both to cilium and to the rotating, bacterial organelle is not a constant source of confusion. The terminological conflicts will resolve with more knowledge of the protoctists, their structures, life cycles, biochemistry, and evolutionary history.

6. Undulipodia, sex and the four groups of phyla

Our perspective leads us to distinguish four groups of phyla (I-IV) based on absence or presence of undulipodia and absence or presence of complex sexual life cycles. By "complex" we refer to morphologically distinct life cycle stages generated by meiotic or fertilization-mediated changes in genetic organization such as ploidy. Within these four groups, details of classes, orders and families are primarily established from ultrastructural details, exclusive of plastids and mitochondria since these cell organelles tend to be uniform within phyla. The lower taxa (*genus, species*) are unchanged from the "protozoological," "algal" and "mycological" traditions. A summary of the four groups, thirty-five phyla and their classes is shown in Table 4 (p. 197).

7. Kinetids, sexual life cycles and systematic order

Variations on the standard [9(2)+2] organization of undulipodia are rare. All axonemes are underlain by a [9(3)+0] kinetosome. Extreme variation (e.g., [9(1)+0]; [10(1)+1]; [14(1)+3], has been reported in *Pelomyxa palustris* (Seravin and Goodkov, 1987); indeed these structures may not be axonemes at all! Whether *Pelomyxa* was an anaerobic consortium that early in cell evolution gained undulipodia prior to the origin of mitosis and meiotic sex (as argued in Margulis and Sagan, 1986 and Margulis, 1988b) or *P. palustris* was derived from a standard [9(2)+2] mastigote (as suggested by Seravin and Goodkov) cannot yet be determined. In any case we are especially conscious of the ultrastructure of kinetids (undulipodial bases), undulipodia, mitotic apparatus and other relatively conserved cell features. (See Margulis and Sagan, 1985 and the Glossary for explanations of kinetids and related microtubular organellar systems.) Kinetids (e.g., under various names such as "flagellar basal apparatus") have been used for decades as phylogenetic markers.

No kinetids, undulipodia or any other evidence for undulipodia-based transformation or sexual system has ever been presented for the conjugating green algae. Therefore, admittedly against the wishes of Michael Melkonian and most of the chlorophyte authors, we have placed the desmids and "Conjugales" with rhodophytes and cellular slime molds as major lineages that lack undulipodia at all stages and yet display complex sexual life cycles. Within each of the four phyla groupings we attempt to order the phyla themselves by morphological complexity: from least to most. While we are painfully aware of our ignorance concerning protoctist phylogeny, we believe this is the most effective way to align these groups according to the evolutionary trends they display. No doubt the organization of this book, like that of *Bergey's Manual* (Krieg and Holt, 1984; Sneath *et al.*, 1986), itself will evolve. Thus we are satisfied, in the first edition of the *Handbook of Protoctista,* to have purged the essentialist animal/plant dichotomy and to have raised the level of consciousness toward these extraordinarily diverse eukaryotic organisms. Perusal of this Glossary makes it evident that protoctists are not "lower plants" nor "lower animals" nor "lower fungi."

8. Life histories

A universal feature of sexuality in eukaryotes is gender and the ability of complementary cells to recognize each other's gender. Recognition of gender is followed by nuclear exchange or cell fusion. Although gender-determining mechanisms differ widely, mature cells differing in gender are attracted to each other, their membranes fuse, and they transfer either nuclei alone or nuclei accompanied by cytoplasm. These sexual acts are followed eventually by karyogamy. The karyogamy process recombines, in a single nucleus, genes from different parents. The reiterative, cyclical nature of cell fusion (usually fertilization) and its relief (usually meiosis) is a major characteristic that distinguishes all sexual protoctists from both their prokaryotic antecedents and nonsexual relatives. Often the fusion events are followed by programmed death, i.e., pycnosis (chromosome degeneration) leading to nuclear or whole-cell disintegration. Although bacteria readily transfer small replicons (plasmids, viruses, etc.) and even large ones (e.g., genophores in conjugation), the ritualized cell membrane fusion events required for meiotic sexuality are a distinguishing feature of the Protoctista and their descendants.

Certain amebas, euglenids and other protists showing no propensity whatsoever for sexual fusion probably evolved prior to the origins of meiotic sexuality; it is likely that, in principle, such organisms are incapable of tissue differentiation (Margulis and Sagan, 1986). Many species in the Kingdom Protoctista display idiosyncratic yet recognizable meiotic sexual life cycles; the life cycles of most species are unknown.

The generalized life cycle drawing represents an attempt to include the fundamental features, in principle, of any meiotic sexual system and its relation to the visible processes of cell or multicellular organismic reproduction (see Life Cycle diagram, p. 65). The external features of the development of any protoctist constitute its life history; life histories can be documented by microscopic observation of the entire cycle of events in development. Life cycles, which require ploidy analyses, are far more difficult to establish. Not surprisingly, the best life cycle descriptions are available from the most devastating parasites and the largest, most conspicuous algae. The number, organization, and interaction of the nuclear genomes and their gene products must be known before life cycles of protoctist taxa are established with confidence. Because all protoctists are products of stable associations between exogenously derived microbial genomes (former bacterial symbionts), the difficulties of life cycle data and phylogenetic reconstruction are exacerbated. With the advent of molecular data, the situation is improving.

If karyokinesis is uncoupled to cytokinesis either before (haploid) or after karyogamy (diploid), plasmodia (also called syncytia, coenobia, coenocytes, etc.) result. When these multinucleated plasmodia reproduce as entities by cytokinesis, the reproductive process is called plasmotomy. Such tendencies toward plasmotomy occur in several protoctist groups.

If karyokinesis is followed by differential changes in the offspring nuclei relative to each other, "nuclear dimorphism" occurs; this tendency is highly marked both in Granuloreticulosa and Ciliophora. Foraminifera have generative and somatic nuclei, ciliates have macro- and micronuclei. In both groups the nuclei are conspicuously dimorphic.

A goal of protoctistology is understanding the genetic basis of life histories, i.e., documentation of karyological and other aspects of the life cycles of its members. Here an attempt is made to place the members of the two major sexual groups of phyla (those with and those without undulipodia) on the generalized life cycle drawing according to the most probable observed stages, i.e., those that dominate the life histories of each phylum. This tentative placement is merely preliminary, and inevitably because of the marked, often unstudied diversity in the group, it is grossly oversimplified. The analysis of the genetic and karyological bases of differentiation reflected on this life cycle diagram (p. 65) is clearly irrelevant to the phyla in groups I and III that lack all sexual stages. Like prokaryotes, they cannot, of course, be placed anywhere on the diagram. An awareness of our ignorance of the detailed karyological and genetic features of these organisms is an absolute prerequisite for the eventual solution of the life cycle problem.

9. Kingdom Protoctista

The "Protoctista" comprises the entire motley and unruly group of non-plant, non-animal, non-fungal organisms representative of lineages of the earliest descendants of eukaryotes. Protoctista are accepted here as one of the five kingdoms of life.

Unfortunately no neat definition encompasses all the protoctistan diversity, except a definition by exclusion. The term Protoctista, invented by the English biologist John Hogg (1860) for organisms neither animal nor plant, although not euphonious, is apt.

Although the Kingdom Protoctista, as "organisms neither animal nor plant," was invented by John Hogg, our debt is to H. F. Copeland of Sacramento, California, who resuscitated the protoctist kingdom in modern form. Copeland's contribution differed from ours in two major ways. Copeland (1956), in his idiosyncratic masterpiece buried in the arcane biological literature, recognized only four kingdoms. He included fungi (as "phylum Inophyta") in Kingdom Protoctista and he excluded all green algae, which he placed in the Kingdom Plantae. We agree with Kendrick (1985) that the zygospore-, ascospore- and basidiospore-forming fungi deserve their own kingdom status. Whittaker (1959), like us heavily indebted to Copeland, was first to recognize five kingdoms of life. He delimited a kingdom of single-cell organisms, Kingdom Protista. Although we agree fundamentally with Whittaker's five-kingdom formulation, we reject his formal Kingdom Protista because it is so heavily dependent on the false dichotomy of single vs. multicellularity as explained below.

Our five-kingdom system is modified slightly from both Copeland's and Whittaker's original work. In both Whittaker and here, the other four kingdoms are Monera (Prokaryotae), Fungi, Animalia and Plantae (Margulis and Schwartz, 1988). Protoctista are, of course, excluded from the Prokaryotae (the kingdom of organisms with bacterial cell organization) on the basis of their nucleated cells.[2] Whereas two comprehensive treatises dealing with prokaryotes already are available: *Bergey's Manual* (Krieg and Holt, 1984; Sneath *et al.*, 1986) and *The Prokaryotes*, Second Edition (Balows *et al.*, 1992), none but this extensively treats all the protoctists.

10. How many kingdoms of organisms should be recognized?

A flurry of kingdom schemes embellishes the modern biological literature. A perusal of current textbooks at university and secondary school levels indicates that most authors and biologists who work with live organisms use one or another of the variations on Whittaker's five-kingdom scheme as presented here. However, in a burst of activity generated primarily by Carl Woese and his colleagues (Fox *et al.*, 1980; Woese, 1987), molecular biologists have claimed first three ("*Archaebacteria*," "*Eubacteria*" and "*Eukaryotes*") "primary kingdoms" and now two ["*Parkaryotae*": eubacteria, halobacteria and methanogenic bacteria and "*Karyotae*": eocytes, certain genera of thermoacidophilic bacteria and all eukaryotes (Lake, 1988)]. Based on calculations derived from nucleic acid sequence data primarily derived from the genes of ribosomal RNA (rRNA), these schemes make several questionable assumptions that render them far less valid than work based on live organisms correlated with information taken directly from the fossil record.

Molecular biologists assume the complete correlation of the sequence of nucleotide bases in RNA with the phylogenetic history of the organisms in question. Thus they consistently use partial, rather than total, phylogenies. They fail to take into account rampant horizontal transfer of genomes which leads to the concept of the absence of meaningful species in the bacterial world (Sonea and Panisset, 1983). They ignore definitive research on the molecular recombinant nature of all eukaryotes (Margulis, 1992) and thus the consequences of anastomosing (as opposed to branching) phylogenetic trees (see Section 15). Although molecular biology provides an extraordinary tool for the comparative study of extant bacterial groups, its claims for reconstruction of evolutionary history during the Archean and Proterozoic eons (from 3500 until 570 million years ago) must be cautiously evaluated. This is especially true in light of the contradictory claims of the molecular evolutionists (Lake, 1988; Lazcano, 1993).

Our five-part scheme, in which the protoctists are raised to kingdom status, makes no claim to monophyly. Rather the groupings reflect insofar as possible the entire biology of their members. We have attempted to use ultrastructure, genetics, physiology, development, behavior and ecology of these vast assemblages of life all as part of our systematic scheme. We hope we have provided the reader with an unambiguous and accessible systematic classification. The powerful combination of molecu-

[2] E. Chatton was the first to recognize the prokaryote-eukaryote distinction (in 1925). In fact, he handled protoctists in an extremely similar way to ours considering the "lowest common denominator" (see p. 385 in his paper "*Pansporella perplexa: Reflexions sur la biologie et la phylogenie des protozoaires.*" In: *Annales des Sciences Naturelles Zoologie: L'Anatomie, le physiologie, la classification et l'histoire naturelle des animaux,* Bouvier, M.E.-L., ed., 10th Series. Vol. 8. Masson et Companie, Publishers, Paris, 1925).

lar biological detail, ultrastructural and genetic analysis and the revolution in paleobiology (Schopf, 1992) augur well for a reunification of the biological sciences and an eventual international consensus for biological classification. We assert that this reunification heavily depends on the recognition of the key importance of the protoctists in the evolution and present-day distribution of life on Earth.

11. Kingdom criteria

Biologists agree that all protoctists have prokaryotic ancestors. The prokaryotes first appear in the fossil record approximately 3500 million years ago. We all agree that from some protoctists emerged members of the three other eukaryotic kingdoms: Animalia and Fungi from unknown heterotrophs and Plantae from green algae. These three kingdoms, which in general appear later than protoctists in the fossil record, can be precisely defined and their approximate dates of appearance recorded. Animals, diploid organisms developing from blastular embryos, as members of the Ediacara fauna, are found in the fossil record over 700 million years ago (Glaessner, 1984). The fungi, haploid and dikaryotic organisms developing from desiccation-tolerant spores and lacking undulipodia at all stages of their life cycle, appeared, primarily in association with plant roots, in the late Silurian or lower Devonian some 400 million years ago (Pirozynski and Malloch, 1975). Plants, organisms that develop from embryos surrounded by maternal tissue, and in which the haploid alternates with the diploid generation, also appeared in the lower Paleozoic era (Richardson, 1992). Fossils interpreted to be robust-walled cysts of protoctists are recorded in the fossil record well over a billion years ago (Vidal, 1984).

12. Invalidity of nutritional criteria

Phototrophic animals (e.g., *Convoluta roscoffensis*) and heterotrophic plants (e.g., *Monotropa*) are known: nutritional criteria do not suffice in the definition of highest taxa.

Indeed, protoctists display a remarkable diversity of nutritional modes: many are listed in Table 1 and Table 2 (pp. xxv-xxvi). Every nutritional mode is represented in protoctists with the exceptions of photoorganoheterotrophy and chemolithotrophy. The photoorganoheterotrophs use organic compounds as food sources while simultaneously employing visible light to generate ATP directly whereas chemoautotrophs can exclusively use carbon dioxide and other inorganic compounds as sources of carbon and energy. In strict chemoautotrophy neither the source of carbon nor the source of energy is from carbon-hydrogen (organic) compounds (e.g., methanogenesis, methylotrophy, ammonia oxidation, sulfide oxidation and the like). These two metabolic virtuosities are nutritional modes entirely limited to bacteria. Thus, although the protoctists display a greater range of nutritional types than do plants or fungi or animals, they are far more limited in energy and nutrient-gathering capability than are bacteria.

13. Size and multicellularity: Protoctist vs. protist

As a group, the Protoctista range enormously in size. From the smallest micromonads, chlorellas and *Nanochlorum* (Zahn, 1984), which measure about a single micrometer in diameter, to the giant kelps, we encompass here organisms that extend in size over seven orders of magnitude (see Table 3, p. xxvi). Botanists and marine biologists are loath to call the gigantic members of the kingdom "protists," a term with connotations of the very small. Indeed, in the history of biology the term protist has included even bacteria (Poindexter, 1971). Thus we restrict the term "protist" to an informal usage. Protist, in this volume, refers to the protoctistan members of the microcosm that require use of microscopes for their visualization. Whereas the term "protoctist" includes all members of the kingdom, "protist" refers to only the smaller organisms, generally composed of a single or only a few cells.

Multicellular, even differentiated, organisms are known in all five kingdoms, therefore the dichotomy "unicellular-multicellular" does not help define protoctists. Many lineages of single-celled prokaryotes and nearly all lineages of protists, independently from each other, reached multicellular status (e.g., cyanobacteria, myxobacteria, actinobacteria, ameba-slime molds, diatoms, chrysomonads and so forth). Although members of the kingdom Fungi may be unicellular (e.g., the yeasts), animals and plants, because they grow from multicellular embryos, are always multicellular. Organisms traditionally labeled "unicellular animals" or "unicellular plants" are protoctists and therefore presented in this book.

14. Organization of this book

With the publication of the *Handbook of Protoctista* the protoctists seceded from the plant, animal, and fungal kingdoms (Margulis, 1992). This event heralds a new phase in the study of protoctists. Yet, as discussed by Corliss in the next section, no international guidelines for naming and banking these organisms yet exist. New and expanding technology, applied to the study of protoctists, will allow the compilation of diverse information that will end the traditional separation between the phycologists, zoologists, and mycologists. The accessibility of computer data banks and new information permit the development of Linnaeus and other protoctist computer information systems, as discussed here by Estep (p. xxxix). The generation of molecular sequence data from proteins, ribosomal and other RNAs yields essential information for the development of protoctist phylogenies from detailed information about their representative macromolecules, as Sogin discusses (p. xlv). Terminology pertaining to these many varied organisms is often complex and confusing. Here we present a General Glossary (p. 5), expanded and illustrated from the one in the *Handbook of Protoctista* from which, for clarity, all taxonomic entries have been removed to the Organism Glossary (p. 137). Only phyla illustrations are in the Organism Glossary; all others are with the general entries. Species have been omitted from the Organism Glossary except those belonging to monotypic genera. The *Handbook of Protoctista* organism index and original literature must be consulted for species names.

Additional information contained in this book will permit the reader an overview of the kingdom. General features of protoctist life histories and life cycles are discussed and diagrammed (p. xix), whereas sources for obtaining live cultures of these organisms are listed on page xxxiii. Table 4 (p. 197) lists current classes of the phyla of protoctists, Table 5 (p. 201) lists common names, Table 6 (p. 204) compares the currently recognized 35 phyla, and Table 7 (p. 251) presents the lower taxa of the various phyla insofar as they have been created. The references used to compile them are found after the tables.

15. Polyphyly and anastomosing branches on family trees

Like lichenologists, protoctistologists deal with chimeras (Schwemmler, 1989). Given the bacterial ancestry of both mitochondria and plastids (Gray, 1984; Margulis and Bermudes, 1985; Bermudes and Margulis, 1987), it is clear that *all* protoctists descend from tightly knit bacterial communities. Indeed, it is now accepted that all photosynthetic protoctists have at least three different types of bacterial ancestors (host, mitochondria, plastid) whereas all heterotrophic protoctists have at least two (host, mitochondria). If the intracellular motility system of eukaryotes is also derived from symbiotic bacteria then photosynthetic protoctists descended from at least four and heterotrophic protoctists from at least three different bacterial lineages (Schwemmler, 1984; Margulis, 1992). This book has grappled with the fact that the formal systematics of the eukaryotic microorganisms has never included consideration of their multiple microbial ancestry. The heterotrophic host portion of the protist cell has traditionally been of concern to zoologically-oriented scientists while colors and pigments have preoccupied the botanically inclined. This, in part, has resulted in the chaos of nomenclature described so aptly here (p. xxvii) by Corliss. The fact that the protoctists evolved from microbial communities is both our strength and our weakness: we can see how evolutionary innovation has appeared, but we deal with extreme polyphyly that is an anathema to streamlined taxonomic schemes. Our strategy in the *Handbook* has been one of "biological common denominator": to erect a chapter for each group that everyone agrees contains only closely related members. Thus obscure organisms like *Stephanopogon*, pyrsonymphids (metamonads), bicosoecids, *Chlorarachnion*, ellobiopsids, ebridians, haplosporidians and xenophyophorans have merited their own short chapters.[3] Future editions of this work must determine how these groups should be best classified.

16. International recognition of protoctists

Methods for preserving and naming protoctists may fundamentally differ from those of the other forms of life. Certainly (like animals, plants and fungi) some protists can be pickled or dried: herbarium and museum specimens may be used for identification. Those protoctists that are amenable to such methods, e.g., brown and red algae, tend to be the best known and the most traditionally accommodated in the other kingdoms. On the other hand, the small amebas, the slime molds and chytridiomycotes and many other protoctists are handled with sterile technique and other methods of microbiology. What is desperately needed, in addition to the code-of-nomenclature resolution, is general agreement on introducing new protoctists to the literature and banking the equivalent of the "type specimen."

In some cases stained microscopical slide preparations are most appropriate and adequate. Microcinematography using film or video of live material may tell us far more about the nature of a protoctist than any preserved specimen. Many protoctists are amenable to placement in standard microbiological culture collections. Others can be preserved in collections as standards for identification by histological sections of host animal tissue accompanied by appropriate color projection slides or electron micrographs. In many cases the only acceptable substitutes for live material are photographs of live material. International cooperation amongst "protoctistologists" to develop standardized methods for handling "type specimens" appropriate to the organisms in question is strongly recommended. In short, protoctistologists must formally remove themselves from zoological, mycological, microbiological and botanical unions. In addition to the solution of issues of nomenclature advocated by Corliss, a new international "protoctistological type collection" should be instituted. The problems of banking type specimens perhaps appropriately fall to the International Society of Evolutionary Protistology (ISEP). This young organization, which meets biennially, has published the proceedings of the 7th meeting as a Special Issue of *BioSystems*, Volume 21, numbers 3 and 4, 1988, Elsevier Scientific Publishers, Ltd., Ireland, edited by Ø. Moestrup, S.T. Moss, D.J. Patterson, P.J. Rizzo and M.A. Sleigh (pages 177-425). This publication records the scientific results reported at the meeting, which took place at Royal Holloway and Bedford New College, Egham, England, July 19-24, 1987.[4] If protoctistologists have had any voice in the form of a scientific journal, until now *BioSystems* has been that sound. With the reconstitution of the French journal *Protistologica* as *The European Journal of Protistology* and with eventual ISEP publication in *Symbiosis*, we anticipate new music from our professional colleagues.

Integration of the algae, water molds and other nonprotozoans into *An Illustrated Guide to the Protozoa* (Lee et al., 1985) is an essential next step in the establishment of a rational biological science.

---

[3] The "incertae sedis" groups, the ellobiopsids are thought to be related to the intracellular parasites lacking undulipodia and complex life cycles, whereas the ebridians have been aligned with dinomastigotes. Both require further characterization before they can be placed within existing phyla or raised to independent phylum status.

[4] The subsequent ISEP meeting (July 2-9, 1992) took place at Orsay, France. The publication of the invited and contributed papers is planned for *BioSystems*, 1993, Elsevier, The Netherlands. (Subsequent volumes are planned to be published by Balaban Publishers, Rehoboth, Israel.)

17. Acknowledgments and apologies

The preparation of this *Illustrated Glossary of Protoctista* was particularly aided by J. Steven Alexander, Jon Ashen, David Bermudes, Stuart Brown, Paula Carroll, Theresa Chan, Eileen Crist, Maureen Cunningham Neumann, Kathryn R. Delisle, Brian Dempsey, René Fester, Gail Fleischaker, Rodney Fujita, Ricardo Guerrero, Terry Hill, Gregory Hinkle, R. W. Hyde, John Kearney, Sally Klingener, Tom Lang, Jane Leighton, Ellen Leotsakos, Richard Lounsbery Foundation, Caroline Lupfer, Christie Lyons, Sheila Manion-Artz, Jennifer Margulis, Zachary Margulis, Lynn Massie, Alan McHenry, Kelly McKinney, Donna Mehos, Rafael Millán, Jeris Miller, Karen Nelson, Carl Pisaturo, Laurie Read, Donna Reppard, Dorion Sagan, Jeremy Sagan, Stephanie Seber, Jacob Seeler, Landi Stone, Isobel Taylor, Elizabeth Thomson, Joseph Volosin, Rae Wallhausser, and Oona West. The contribution to the glossary by Greg Hinkle, Laurie Read and Dorion Sagan, and to the illustrations by Kathy Delisle requires special mention.

Our publishers, Donald Jones and Arthur Bartlett, in contracting the *Handbook* and the *Glossary* have graciously, if inadvertently, supported scientific research. They and their dedicated employees, Ellen Leotsakos, Terry Hill, Elizabeth Thomson, Maureen Cunningham Neumann, Paula Carroll, and Rafael Millán must take credit for being benefactors of original scientific findings. Alan McHenry and his board of directors at the Richard Lounsbery Foundation of New York City provided funds for the completion of this project twice when its future was bleak. The Life Sciences Office of the National Aeronautics and Space Agency, under grant NGR-004-025 to Lynn Margulis, has supported more protoctistological science than they realize.

Since these eukaryotic microorganisms do not categorize appropriately, standard sources of research funding for studying them do not exist. Although botanists might support work on algae and zoologists agree to fund work on trypanosomes or apicomplexans, no government or other scientific funding agency is equipped to cross discipline boundaries such as those perceived to divide, e.g., taxonomy from cell biology, plant pathology from plankton ecology or parasitology from phycology. Unlike government granting agencies with their rigid categories, the Richard Lounsbery Foundation responded flexibly to our scientific needs. The role of Mr. McHenry and his board of directors in bringing the *Protoctista* project to completion cannot be underestimated. Lounsbery funding was unique in that it permitted new integrative intellectual work that otherwise, unfunded, is not undertaken. Lounsbery supported manuscript preparation such as communication with scientists from different backgrounds, drawing of new illustrations under the direction of the investigators, and library research. Without the generosity of the Lounsbery grant, quite simply, neither the *Handbook of Protoctista* nor the *Illustrated Glossary of Protoctista* would exist. The Lounsbery Foundation has been very sensitive in the production of this reference work to the needs of those of modest means - the students, instructors, and specialists - as has Arthur Bartlett, who approved publication of this comprehensive reference book priced within their means both in paperbound and hardbound editions.

We are, as Corliss mentions, in the midst of a protoctistological revolution. More than most books (and much like the Bergey's trust, Krieg and Holt, 1984; Sneath *et al.*, 1986) we are dependent on our readers. With admiration for the diversity of the living world and especially the protoctistan microbes that are our ancestors, your editors invite you to send us corrections, comments, opinions and suggestions of any kind.

Lynn Margulis
Spring 1993

REFERENCES

Abbayes, H. Des: In: *Histoire de la Botanique en France* (D. de Virille, ed.), pp. 235-241. Paris: Eighth International Botanical Congress Paris-Nice, 1954.

Balows, A., Trüper, H.G., Dworkin, M., Harder, W., Schleifer, K.-H.: *The Prokaryotes*, 2nd ed., Vols. I-IV. New York: Springer-Verlag, 1992.

Berg, H.C.: Bacterial behavior. *Nature* 254, 389-391 (1975).

Bermudes, D., Margulis, L.: Symbiont acquisition as neoseme: Origin of species and higher taxa. *Symbiosis* 4, 185-198 (1987).

Copeland, H.F.: *Classification of the Lower Organisms.* Palo Alto: Pacific Books, 1956.

Corliss, J.O.: Protistan phylogeny and eukaryogenesis. *International Review of Cytology* 100, 319-370 (1987).

Fleck, L.: *Genesis and Development of a Scientific Fact.* Chicago and London: University of Chicago Press, 1979.

Fox, G.E., Stackebrandt, R.B., Hespell, R.B., Gibson, J., Maniloff, J., Dyer, T.A., Wolfe, R.S., Balch, W.E., Tanner, R., Magrum, L., Zablen, L.B., Blakemore, R., Gupta, R., Bonen, L., Lewis, B.J., Stahl, D.A., Luehrsen, K.R., Chen, K.N., Woese, C.R.: The phylogeny of prokaryotes. *Science* 209, 457-463 (1980).

Glaessner, M.F.: *Dawn of Animal Life.* Cambridge, England: Cambridge University Press, 1984.

Gray, M.: The bacterial ancestry of mitochondria and plastids. *BioScience* 33, 693-699 (1984).

Hawksworth, D.L., Hill, D.J.V.: *The Lichen-forming Fungi.* New York: Chapman and Hall, 1984.

Hogg, J.: On the distinctions between a plant and an animal, and on a fourth kingdom of nature. *The Edinburgh New Philosophical Journal* (new series) 12, 216-225 (1860).

Kendrick, B.: *The Fifth Kingdom.* Waterloo, Ontario: Mycologue Publications, 1985.

Krieg, N.R., Holt, J.G., eds.: *Bergey's Manual of Systematic Bacteriology*, Vol I. Baltimore: Williams and Wilkins, 1984.

Lake, J.A.: Origin of the eukaryotic nucleus determined by rate-invariant analysis of rRNA sequences. *Nature* 331, 184-186 (1988).

Lazcano, A.: RNA world and molecular phylogeny. In: *Evolution on the Early Earth* (S. Bengtson, ed.). New York: Columbia University Press, 1993 (in press).

Lee, J.J., Hutner, S.H., Bovee, E.C., eds.: *An Illustrated Guide to the Protozoa*, Lawrence, KS: Society of Protozoologists, 1985.

Margulis, L.: Undulipodia, flagella and cilia. *BioSystems* 12, 105-108 (1980).

Margulis, L.: Undulipodiated cells. *BioScience* 35, 333 (1985).

Margulis, L.: Systematics: The view from the origin and early evolution of life. Secession of the protoctista from the animal and plant kingdoms. In.: *Prospects in Systematics* (D. Hawksworth and R.G. Davies, eds.), pp. 430-443. Oxford, UK: Clarendon Press, 1988.

Margulis, L.: *Symbiosis in Cell Evolution: Microbial Communities in the Archean and Proterozoic Eons*, 2nd ed. New York: W.H. Freeman Co., 1992.

Margulis, L., Bermudes, D.: Symbiosis as a mechanism of evolution: Status of cell symbiosis theory. *Symbiosis* 1, 101-124 (1985).

Margulis, L., Corliss, J.O., Melkonian, M., Chapman, D.J.: *Handbook of Protoctista*. Boston: Jones and Bartlett Publishers, 1990.

Margulis, L., Sagan, D.: Order amongst animalcules: The protoctista kingdom and its undulipodiated cells. *BioSystems* 18, 141-147 (1985).

Margulis, L., Sagan, D.: *Origins of Sex*. New Haven, CT: Yale University Press, 1986.

Margulis, L., Schwartz, K.: *Five Kingdoms: An Illustrated Guide to the Phyla of Life on Earth*, 2nd ed. New York: W.H. Freeman Co., 1988.

Pirozynski, K., Malloch, D.: The origin of land plants: A matter of mycotrophism. *BioSystems* 6, 153-164 (1975).

Poindexter, J.: *Microbiology: An Introduction to Protists*. New York: Macmillan Publishers, 1971.

Richardson, T.B.: Origins and evolution of the earliest land plants. In: *Major Events in the History of Life* (J.W. Schopf, ed.), pp. 95-118. Boston: Jones and Bartlett Publishers, 1992.

Schopf, J.W., ed.: *Major Events in the History of Life*. Boston: Jones and Bartlett Publishers, 1992.

Schwemmler, W.: *Reconstruction of Cell Evolution: A Periodic System*. Boca Raton: CRC Press, 1984.

Schwemmler, W.: *Symbiogenesis: A Macro-mechanism of Evolution*. Berlin: Walter de Gruyter, 1989.

Seravin, L. N., Goodkov, A. V.: The flagella of the freshwater amoeba *Pelomyxa palustris*. *Tsitoligya* 29, 721-724 (1987) (in Russian).

Sieburth, J. McN., Estep, K.: Precise and meaningful terminology in marine microbial ecology. *Marine Microbial Food Webs* 1, 1-16 (1985).

Shmagina, A.P.: Mertsatel' Noe Dvizhenie. *(Ciliary Movement.)* Moscow: Medgiz, 1948 (in Russian).

Sneath, P.H.A., Mair, N.S., Sharpe, M.E., Holt, J.G., eds.: *Bergey's Manual of Systematic Bacteriology*, Volume 2. Baltimore: Williams and Wilkins, 1986.

Sonea, S., Panisset, M.: *A New Bacteriology*, Boston: Jones and Bartlett Publishers, 1983.

Van Hoven, W. Isolated cilioprotistan evolution in African rhino intestines. International Society for Evolutionary Protistology. Royal Holloway and Bedford New College, Egham, England. Oral Communication, July 1987.

Vidal, G.: The oldest eukaryotic cells. *Scientific American* 250, 48-57 (1984).

Whittaker, R.H.: On the broad classification of organisms. *Quarterly Review of Biology* 34, 210-226 (1959).

Woese, C.R.: Bacterial evolution. *Microbial Reviews* 51, 221-271 (1987).

Zahn, R.K.: A green alga with minimal eukaryotic features: *Nanochlorum eukaryotum*. *Origins of Life* 13, 289-303 (1984).

TABLE 1. MODES OF NUTRITION FOR LIFE ON EARTH[1]

A list of the sources of energy, electrons and carbon for metabolism; name of each mode with examples of growth of organisms to which names apply. Names constructed by addition of suffix "-troph", e.g., photolithoautotroph (plants).

ENERGY (light or chemical compounds)	ELECTRONS (or hydrogen donors)	CARBON sources	ORGANISMS and their hydrogen or electron donors
Photo- (light)	Litho- (inorganic compounds and C_1)	Auto- (CO_2)	Prokaryotes: Chlorobiaceae, H_2S, S Chromatiaceae, H_2S, S Rhodospirillaceae, H_2 Cyanobacteria, H_2O Chloroxybacteria, H_2O Protoctista (algae), H_2O Plants, H_2O
		Hetero- $(CH_2O)_n$	None
	Organo- (organic compounds)	Auto-	None
		Hetero-	Prokaryotes: Chromatiaceae, org. comp.[2] Chloroflexaceae, org. comp.[2] Rhodospirillaceae[2] *Rhodomicrobium*, C_2, C_3 comp. Heliobacteriaceae, org. comp.[2] Halobacteria
Chemo- (chemical compounds)	Litho-	Auto-	Prokaryotes: methanogens, H_2 hydrogen oxidizers, H_2 methylotrophs, CH_4, CHOH, etc. ammonia, nitrite oxidizers, NH_3, NO_2^-
		Hetero-	Prokaryotes: "sulfur bacteria," S manganese oxidizers, Mn^{++} iron bacteria, Fe^{++} sulfide oxidizers, e.g., *Beggiatoa*; sulfate reducers, e.g., *Desulfovibrio*
	Organo-	Auto-	Prokaryotes: clostridia, etc., grown on CO_2 as sole source of carbon (H_2, $-CH_2$)
		Hetero-	Prokaryotes (most) (including nitrate, sulfate, oxygen and phosphate[3] as terminal electron acceptors) Protoctista[4] (most) Fungi[4] Plants[4] (achlorophyllous) Animals[4]

[1] Tables 1 to 3 devised in collaboration with R. Guerrero.

[2] Organic compounds e.g., acetate, proprionate, pyruvate. Comp.–compounds

[3] Detection of phosphine: I. Dévai, L. Felföldy, I. Wittner, S. Plósz. 1988. New aspects of the phosphorus cycle in the hydrosphere. *Nature* 333:343-345.

[4] Oxygen as terminal electron acceptor.

TABLE 2. MODES OF NUTRITION IN PROTOCTISTA

Photolithoautotrophy	Chemoorganoheterotrophy	
Photosynthetic capability[1]	**Osmotrophy** (traditional protozoa)	
Photoautotrophy (all algae)	*Mode*	*Adjective/synonyms*
	saprotrophy	(saprophytous, saprophyte)
	histotrophy	(histophagous, tissue eater)
	detritrophy	(detrivorous, detrivore)
Mixotrophy conditionally chemoorgano-heterotrophic and photolitho-autotrophic modes (mastigote algae)	**Phagotrophy/Ingestive Nutrition**	
	Mode	*Adjective/synonyms*
	bactivory	(bactivorous)
	herbivory	(herbivorous, primary consumer, plant eater)
	algivory	(algivorous)
	fructivory	(fructivorous)
	insectivory	(insectivorous)
	carnivory	(carnivorous, meat eater)
	predation	(predaceous, predatory)
	Biotrophy	
	Mode	*Adjective/synonyms*
	symbiotrophy	(symbiosis, histophagous, parasitic)
	necrotrophy	(parasitic, pathogenic)

[1] Always associated with the presence of chlorophyll-containing plastids. No strict chemolithoautotrophs known that lack all organic requirements in the protoctists, e.g., many algae have vitamin requirements.

TABLE 3. UNITS AND DIMENSIONS

ABBREVIATION	UNIT	NAME	USED TO MEASURE
m	1	Meter; standard unit; (=39.4 inches)	Organisms (animals, plants)
Larger than meter			
Mm	10^6 m	Megameter	Biomes, ecosystems
km	10^3 m	Kilometer	Ecosystems, habitats
Smaller than meter			
dm	10^{-1} m	Decimeter	Individual organisms, animals, plants, fungi
cm	10^{-2} m	Centimeter	Plankton, protoctists, animals
mm	10^{-3} m	Millimeter	Plankton, protoctists, fungal sporocarps
μm	10^{-6} m	Micrometer (often called micron)	Protists, individual cells, nanoplankton, bacteria, organelles
nm	10^{-9} m	Nanometer (formerly called mμ or millimicron)	Subcellular structures, wavelengths of visible light
Å	10^{-10} m	Angstrom	Macromolecules, (proteins, nucleic acids), organic compounds

Toward a Nomenclatural Protist Perspective

1. Neo-Haeckelian protistology

The emergence of a neo-Haeckelian protistology a few years ago (Corliss, 1986) has had a heuristic effect on interest in these "lower eukaryotes" (some 200,000 species strong, and still counting!). Biologists of many persuasions are carrying out exciting research on algae, protozoa, and various "lower" fungi - and their findings have caught the attention of other workers without regard for the conventional fields of or loyalties to botany, zoology, and mycology. This is progress: I have called it "the Protist Revolution" (Corliss, 1983).

Unfortunately, as is inevitable, with progress come problems. In the case of protistology (or "protoctistology," see section 13, p. xxi), such challenges are of diverse and myriad sorts. I wish to focus here attention on only one serious problem: that of nomenclature. Biologists cannot function, or at least cannot communicate very well nor carry out meaningful comparative studies, without a common language. Under "nomenclature," I shall not include discussion of biological terminology, although that also represents a troublesome area of potentially high controversy: for example, different names for the same organelles and the same names for nonhomologous organelles or cell structures (Corliss, 1986; Margulis, 1992; Margulis and Sagan, 1985). What remains is complicated enough, and the subject has a number of more or less interrelated components outlined briefly below.

Although nomenclature serves only as a handmaiden to taxonomy (not interfering with the freedom, etc., of the latter), no system of classification and no "phylogenetic tree" can really be proposed and discussed without the use of labels for the groups under consideration. Thus we have nomenclatural problems to be dealt with at both lower and higher levels in the taxonomic hierarchy of the protists. While true for all organisms, the problems here are exacerbated by a number of major factors. One is that species of protists, to date, have been divided between the plant and animal kingdoms (by botanists and zoologists, of course). To add further confusion, some groups have been claimed *simultaneously* by both (taxonomic) kinds of biologists. Finally, even protistologists interested in macrosystematics are not in agreement as to how *many* (let alone how to *name* them) phyla or kingdoms should be recognized for the diverse assemblages of these heretofore neglected eukaryotes. To date, from two to 17 kingdoms have been proposed or endorsed and from 12-45 phyla! Such confusion leads to chaos and loss of credibility: protistologists *must* come to some agreement.

Fortunately, biologists do not have to be in perfect agreement to recognize the value of a degree of stability in nomenclatural matters affecting the protists. To attempt to await harmony would be futile. But, before anyone can propose any logical solutions, the problems themselves need to be better understood. The whole biological community should appreciate why the protist situation is unique.

Discussion below is continued under two separate headings: the area affecting infraordinal taxonomic levels, putatively covered by exisiting codes of nomenclature; and the suprafamilial area, where both ranks and names are legally left to the creativity and imagination of the investigators involved.

2. Protists and codes of nomenclature

As long as eukaryotic biologists were presumed to be working with organisms assignable to only the two well established kingdoms, Animalia and Plantae, there appeared to be no irresolvable nomenclatural problems. Mini-plants fell under the jurisdiction of provisions of the International Code of Botanical Nomenclature (latest edition, 1983); mini-animals, under the International Code of Zoological Nomenclature (latest, 1985). Codes of nomenclature guarantee a high level of stability and a universality of usage (Latin is the common language); and they are essential for assuring the uniqueness of every taxonomic name (family level and below). But international bickering has slowed progress in solving cases involving "lower eukaryotes" or protists.

Problems, generally ignored, have long existed, as mentioned briefly above, because of the *taxonomic* assignment of some protists to *both* kingdoms. Now, with the recognition of multiple kingdoms of eukaryotes (four being a conservative number: plants, animals, fungi, protists: see such works as Barnes, 1984; Corliss, 1983, 1984, 1987; Margulis, 1974a, b; Margulis and Schwartz, 1982, 1988; Sleigh, 1979; Taylor, 1978; Whittaker, 1969, 1977; Whittaker and Margulis, 1978), the matter is made only more complicated.

What are the possible solutions, and which ones seem most reasonable for adoption? Here we are concerned with nomenclatural questions at the family-level and below (for the higher categories, see pages xix, xxi, Tables 4 and 7, pp. 197 and 251).

Elsewhere, I have briefly listed some seven ways to resolve difficulties at this taxonomic level (Corliss, 1986). Here I shall offer some discussion, pro and con, for each of these, indicating my own preferences. [Comments on some of these options have been offered earlier by Corliss (1983, 1984), Jeffrey (1982), Patterson (1986b), and Ride (1982; Ride and Younès, 1986).]

(1) Amalgamation of all existing codes into a single, unified new code for all organisms (including, in theory, the prokaryotic bacterial groups and possibly even the viruses). This ecumenical approach is too ideal, too utopian to be realized. In view of the general lack of cooperation to date between "code-workers" for the different groups, some of it petty (factors of tradition and authoritarianism intervene that are not apt to evaporate totally in the near future), one must conclude that this purported solution, a very long-range alternative at best, is impractical.

(2) Establish new codes for every new kingdom that is created. But probably great confusion would result from endorsement of this suggestion, for at least three reasons: workers cannot agree on the *number* of "new" kingdoms (they range from one to 15 at the moment); too many sets of (differing) rules would, understandably, be unacceptable to the biological community as a whole and even to the specialists; and most new codes would only "steal" rules from the two exisiting ones (botanical and zoological), with "improvements" over which disputation might be endless. Admittedly, at first blush a number of us saw this alternative as an excellent one. That it might (still) be *if* the only new kingdom were to be the Protista (or Protoctista). But, today (see Corliss, 1986, 1987), there is not only a considerable tendency to multiple kingdoms of eukaryotes (e.g., see Cavalier-Smith, 1981, 1983; Möhn, 1984), with an isolated "Protista" *not* even among them, but also one of Leedale's (1974) notions - that the protists are simply a "level of organization," not a discrete taxonomic entity - still has a lot of appeal among practicing taxonomic protistologists (though the feeling is not often expressed in print). Also, for the sake of parallel treatment, the Fungi (in the popular four-eukaryotic-kingdom system) presumably would need a completely separate set of codified provisions, too; and there has been no mention of this in the taxonomic mycological literature to date, to my knowledge.

(3) Harmonize, on a case-by-case basis, provisions of the present zoological and botanical codes that are relevant to the protists. Some such action is already taking place (e.g., see Ride and Younès, 1986; Silva, 1980), and it needs to continue. The problem remains, however, that taxonomic protistologists would have to know/consult *both* codes and would have to, in effect, continue supporting the highly artificial taxonomy of their particular forms. That is, some species would have to be recognized, identified, and treated as "plants" and some as "animals" in order to apply the harmonizing provisions of the "correct" code in any given case. Nevertheless, this solution seems to appeal to many workers, perhaps principally phycologists. Its helpfulness, even now, cannot be denied; and I hope that positive actions continue along such approaches.

(4) Allocate some protist taxa, *arbitrarily*, to treatment under one code; others, under the other. But who would make such jurisprudential decisions? And how would/could enforcement be applied? More importantly, what happens as/if the *composition* of the affected protist groups is altered by subsequent taxonomic workers? It seems a pity to (re)invoke the arbitrary division of protists along the "pigmented-non-pigmented" line, using that outworn characteristic to separate "plants" from "animals." But that is what supporters of this "solution" would probably cause to happen, while lamenting their own actions. Admittedly, this seems to be a popular option.

(5) Designate an international committee or commission to serve as "arbitrator" in any disputes that arise because of the current "overlapping" nomenclatural situation (e.g., dinomastigote species are treatable differently under botanical and zoological codes). The problems here are similar to those mentioned above. Who would designate the "arbitration board"? What would happen if its decisions were ignored or, more likely, if disputes were not brought to its attention? Once again, too, resolutions would likely be made along conventional botanical/zoological lines: this is not progress, in my opinion.

(6) Require persons working on taxonomic/nomenclature problems involving protists currently coming under provisions of *both* codes to meet the most stringent conditions imposed by the rules set forth in *either* code. The principal drawback to this solution is not unlike ones discussed in the other cases presented above. Is it fair to oblige a large group of systematic protistologists to become familiar with both complex codes? Indeed, today, many such workers are not even sufficiently informed about *either* one of the sets of regulations. Again, the emphasis on "plant" and "animal" rules for protists seems patently unjust. If anything, this proposal drives one back to supporting resolution number 2 above, viz., a separate code for the "lower eukaryotes."

(7) Allow the worker him/herself to have "freedom of choice" with respect to which code to apply in any given case [see Patterson's (1986b) interesting paper on "ambiregnal taxonomy"]. There are two obvious problems with this, however, it seems to me. Would followers of the *unused* code recognize/accept such choices, which *they* would not have favored? And are we not once again assigning a "plant-like" nature to the involved protists if a worker selects the botanical rules, and "animal-like" in the other case? What has happened to the "*protist*-like" nature of protists?

A relevant important booklet (edited by Ride and Younès, 1986), reviews the present state and current issues of biological nomenclature. Including contributions from some nine specialists in taxonomic/nomenclatural matters from diverse fields (botany, zoology, bacteriology, virology, etc.), some of its discussion is relevant to the "problem of codes" mentioned here. Also, Taylor *et al.* (1986, 1987) published a work on nomenclature in mastigote groups currently treated under both botanical and zoological codes. Apparently independently of Corliss (1983, 1984, 1986), Jeffrey (1982), Patterson (1986b), and Ride (1982; and see Ride and Younès, 1986), whose papers are not cited, these workers have proposed, in effect, "solutions" that I have numbered 2, 3, and 4, above. They offer some concrete suggestions regarding modification of provisions now in force in the codes involved.

3. Protist macrosystematics and nomenclature

Above the taxonomic level of the family, currect codes of nomenclature make no real attempt at controlling the names and naming of such high-level groups (see the codes themselves, cited above, and Corliss, 1983, 1984; Jeffrey, 1977). This has generally been accepted by biologists of all persuasions, including the writer. The overall stability of high-level taxa in conventional systems of classification has rendered largely unnecessary any special "legalistic" attention to the matter.

In recent years, however, there has occurred a sudden bustle of activity in the areas of class/phylum/kingdom creation and subsequent naming (see Corliss, 1983, 1986 and Leedale, 1974). The principal cause has been the renewed interest in protists. In attempting to accommodate multiple high-level protist groups, workers have found it advisable to reconsider all eukaryotic assemblages. The most prolific worker to date (16 kingdoms!) has been Möhn (1984).

Such researchers or macrosystematists are clearly in need of some guidelines for handling the nomenclature of suprafamilial taxa. The only alternative is chaos, as each worker ignores or treats differently the variety of problems involved: choice of names, determination of ranks, significance of implicated concepts, and even assignment of appropriate authorships and dates to the names used. Proliferation of unnecessary names must come to a halt.

By no means am I suggesting that only a *conservative* approach would be acceptable. The ultimate in conservation, of course, would return us to the two-kingdom system, relegating the highly diverse assemblages of protists to their former few phyla or classes of plants (algae and "lower" fungi) or animals (protozoa). To me, this is unthinkable! As our information becomes more precise and constellations of data support the distinctiveness of carefully studied organisms, we need to reflect such uniqueness in our classification schemes. A single species can, of course, serve as the basis for creation of a separate high-level taxon. Witness the recent erection of a new phylum (division) by Hibberd and Norris (1984) for the single "algal" protist *Chlorarachnion reptans*.

What should the guidelines for treating suprafamilial nomenclatural problems include? Should they comprise a whole new code, a code usable by botanists and zoologists as well as protistologists? Who should propose them, and how, and when? We should avoid the pitfalls discussed in the preceding section: that is, to have an international body arbitrarily decide on a set of rather rigid guidelines and then rapidly impose it on the unsuspecting broad community of biologists would never do. Guidelines should evolve, and they should arise from the grass-roots practitioners, in my opinion, rather than being legislated from above. It will take years to develop a reasonable, agreeable set of provisions, as I see it. But there is already a growing and serious interest in the subject (e.g., as displayed at several international congresses, especially ISEP, in discussions offered by such workers as Andersen, Corliss, Margulis, Merinfeld, Patterson, and Silva, protistologists of taxonomically diverse backgrounds and training).

In the present essay, it is premature to propose any specific nomenclatural guidelines for the modern macrosystematist. Rather, I wish to stress the absolute *need* for such in due time, if chaos is to be averted and protistologists are to be enabled to maintain their credibility, and to review briefly the areas to which they should relate. The guidelines may - unlike the codes controlling infrafamilial *nomenclature* only - have to include reference to certain *taxonomic* matters as well, without, however, attempting to actually dictate decisions that must remain the prerogative of the investigator.

Especially in times of systematic instability—unfortunately, this is the situation that obtains today—names and ranks and concepts are inextricably intertwined at the highest levels. Once a rank is determined, a taxonomic judgment decision, the choice of a name for the taxon involved is not a simple matter. Although priority is not demanded for the higher taxa, the fact that a number of more or less appropriate names may already be available certainly exacerbates the problem. Considerations run the gamut from "common sense and courtesy" (Corliss, 1984) to conventional practice, authoritarianism, and the like (Corliss, 1983, 1984; Silva, 1984). Concepts and diagnoses concerning the taxonomic limits of the involved groups are very important, too. To what extent or degree do changes in the boundaries of a group (taxon) influence the retention (*or* the dropping) of names formerly employed for the taxon in question?

Some biologists consider questions of authorships and dates of names of high-level taxa as self-serving at best ("ego-trips" for the coiners of the names). Nevertheless, such problems cannot be ignored. Newly proposed names inevitably, and quite properly, are "credited" to the proposer, with the date being that of the publication containing them. But, if the taxon under concern is *essentially* identical with one named by another worker at an earlier date, are we not (morally, at least) obliged to record such an act by citation of the first namer and date? However, how is one to define or interpret "essentially" here?

Additional minor matters exist with respect to names. One deserving at least brief mention is the perennial controversial topic of whether or not to designate uniform prefixes or suffixes for all names at the same rank (e.g., -ida as a suffix for ordinal names, -ina for subordinal, and the like). Proponents claim, quite correctly, that such identification tags remind the student, at a glance, what rank is involved. Two pedagogical problems arise, however: the names sometimes become unwieldy with the addition of several letters to them; and, more grave, if a group's place in the hierarchy in the classification scheme is subsequently changed, even repeatedly up and down, the student must use a possibly familiar name in a constantly *altered* state, taxing his/her memory and patience!

Some protozoologists, in fact, insert letters before an added suffix, in order to indicate taxonomic rank. For example, Levine (1985, and earlier papers) inserts an "-as-" (before the terminations of "-ida" and "-ina") in the case of class and subclass names, and an "-or-" for orders and suborders. Thus, the familiar ciliate class Kinetofragminophora becomes "Kinetofragminophorasida," the trypanosome order Kinetoplastida becomes "Kinetoplastorida," etc. Should a subsequent worker reduce or raise the rank of such groups, a still more unfamiliar and equally lengthy name would have to be used under the Levine convention. I consider this to be adding still further to the student's confusion.

4. Where do we go from here?

Some system had to be adopted for use in the *Handbook.* It seemed unwise to the editors to allow individual contributors to use their own macrosystems for the protists, particularly since they are treating only segments of the vast whole (although they have been allowed considerable freedom in choice of terminology: e.g., whether to call protist locomotory organelles cilia, flagella, or undulipodia). But it is also clear to the editors, when plans for this compendium were conceived four or five years ago, that no single satisfactory taxonomic-nomenclatural solution for the array of complex problems treated in the preceding sections could be devised in a short period of time and without the help of a large number of knowledgeable and motivated persons. Nevertheless, some system had to be employed: it was natural, at the time the decision needed to be made, to choose the most authoritative "overview" taxonomic approach then available, that of Margulis and Schwartz (1982). Only a few modifications, generally expansions (e.g., incorporation of the phylum Chlorarachnida), have had to be made subsequently. Some refinements have derived from Corliss (1984, 1987) and, more recently, from Margulis and Sagan (1985) and Margulis and Schwartz (1988). [The arrangement of the three dozen phyla described in the *Handbook of Protoctista* - into four groups based essentially on the presence or absence of non-pseudopodial locomotory organelles (i.e., cilia or flagella) combined with degree of complexity in sexual life cycles - is the brainchild of senior editor Margulis.]

Still within the framework of the "four-eukaryotic-kingdoms" hypothesis, the most recent detailed overall treatment of high-level *taxonomic-nomenclatural* matters - in which some 45 phyla (grouped into 18 or 19 supraphyletic assemblages) are recognized - is that of the author (Corliss, 1984; and see slight refinements in Corliss, 1987). But I admit that I have only scratched the surface and exposed the problems, not solved them.

With respect to composition of appropriate, often meaning "neutral" with respect to "plant" or "animal" nature, names for phyletic groups, the very recent (at the original time of this writing) and very similar, yet independently derived, proposals of two teams merit mention here. Rothschild and Heywood (1987), while confining their attention to the vernacular level of nomenclature, have suggested that major protist assemblages be given names made distinctive by wide usage of the suffix "-protist." For example, the 14 largely pigmented groups treated in their long paper bear these labels (with conventionally used common names appearing in parentheses): chloroprotists (green algae), euglenoprotists (euglenoids), chlorarachnioprotists (chlorarachniophytes), rhodoprotists (red algae), cryptoprotists (cryptomonads), chrysoprotists (golden-brown algae), xanthoprotists (yellow-green algae), haptoprotists (haptophytes or prymnesiophytes), bacillarioprotists (diatoms), raphidoprotists (raphidophytes), phaeoprotists (brown algae), dinoprotists (dinoflagellates), eustigmatoprotists (eustigmatophytes), and glaucoprotists (glaucophytes). [See also, now, Heywood and Rothschild (1987).]

Margulis and Sagan (1985), in a much shorter paper but one devoted primarily to matters of nomenclature and terminology, rename 42 protist ("protoctist") phyla of which 17 (again, largely taxa of pigmented forms, except for the last five) are given a "-protista" suffix, often resulting in names essentially identical to those of Rothschild and Heywood (1987)! In this work, however, the names are formally presented in latinized style with capital initial letters and a terminal "-a", as follows: Haptoprotista, Xanthoprotista, Eustigmatoprotista, Chrysoprotista, Cryptoprotista, Raphidoprotista, Bacillarioprotista, Chloroprotista, Ulvaprotista [*should* be *Ulvo*protista], Conjugaprotista [Conjug*o*- or Conjugato], Glaucoprotista, Chlorarachnioprotista, Labyrinthulaprotista [Labyrinthul*o*-], Acrasioprotista, Mycetoprotista, Ooprotista, and Chytridioprotista. An additional 25 phyla were left with their generally more conventional names, frequently identical with, or derived from, those endorsed by Corliss (1984): Prasinophyta, Phaeophyta, Rhodophyta, Dinomastigota, Euglenida, Caryoblastea (I suggest that "K" replace the "C"), Rhizopoda, Actinopoda, Foraminifera, Choanomastigota, Bicoecida [Bicosoecida], Kinetoplastida, Ciliophora, Amoebomastigota, Opalinida, Parabasalida, Metamonadida, Apicomplexa, Microsporidia, Myxosporidia, Paramyxida, Actinomyxida, Haplosporea, Plasmodiophorida, and Hyphochytrida.

In the same year, Sieburth and Estep (1985) offered a classification of some 36 protist phyla in which they propose an original and interesting synthesis of trophic and taxonomic information/observations. They also included a short glossary of terms meaningful, especially, for marine ecologists working with mixtures of prokaryotic and protistan organisms. Nomenclaturally, however, this third "concomitant" team presents nothing new with respect to high-level taxonomic names or their suffixes, content with the standard phyletic/divisional terminology long in use by phycologists and protozoologists.

For the sake of comparison, the total list of names employed by Corliss (from Corliss, 1987), is mentioned here. Some 19 supraphyletic assemblages, given vernacular names and divided into two groupings, are recognized. One is composed of those representing major and often better known evolutionary lines; and the other, of assemblages that appear to be more isolated from a phylogenetic point of view. These assemblages are postulated to contain phyletic (divisional) groups that are presented below with latinized names.

First Grouping. Assemblage of the chlorobionts or chloroprotists (latter after Rothschild and Heywood, 1987): phlya Chlorophyta, Prasinophyta, Conjugatophyta, Charophyta, Ulvophyta. The chromobionts (essentially "heterokonts" of the literature): Chrysophyta, Haptophyta, Bacillariophyta, Xanthophyta, Eustigmatophyta, Phaeophyta, Raphidophyta, possibly Proteromonadea. The rhizopods: Karyoblastea [perhaps not independent at this high a level?], Amoebozoa, Acrasia, Eumycetozoa, Plasmodiophorea, Granuloreticulosa (essentially the Foraminifera): and the enigmatic Xenophyophora might belong here. The mastigomycetes (phycomycetes): Hyphochytridiomycota, Oomycota, Chytridiomycota. The actinopods: Heliozoa (admittedly highly polyphyletic in its conventional composition: e.g., see especially Smith and Patterson, 1986), Taxopoda, Acantharia, Polycystina, Phaeodaria. The polymastigotes (formerly "higher zooflagellates"): Metamonadea, Parabasalia; with Proteromonadea possibly better here than in the first assemblage, the chlorobionts, above? The sporozoa (apicomplexans): single phylum, the Sporozoa, in my view, leaving aside the enigmatic and highly controversial genus *Perkinsus*. The ciliates (heterokaryotes): single well-defined phylum, the Ciliophora.

Second Grouping. Assemblage of the euglenozoa: phyla Euglenophyta, Kinetoplastidea, and *perhaps* Pseudociliata (for the single genus *Stephanopogon* (Lipscomb and Corliss, 1982). The dinoflagellates (dinozoa, mesokaryotes): Peridinea, Syndinea, plus some "Anhangen" groups at subphyletic rankings. The rhodophytes: single phylum, Rhodophyta (but the Glaucophyta are still considered closely allied to the red algae by some workers, who would include them here as a second phylum). The cryptomonads (cryptophytes); Cryptophyta. The chlorarachniophytes: Chlorarachniophyta. The choanoflagellates: Choanoflagellata. The paraflagellates ("opalinids"): Opalinata (Patterson, 1986a, would incorporate the enigmatic proteromonads here in some way, calling the combined group the Slopalinida). The labyrinthomorphs: Labyrinthulea, Thraustochytriacea. The microsporidia (microsporans): Microsporidia. The haplosporidia (ascetosporans): Haplosporidia. The myxosporidia (myxosporans, myxozoa): Myxosporidia, with the Actinomyxidea possibly appended at a subphyletic level.

Rationale for my albeit somewhat tentative or "exploratory" choices of the names listed above is given in Corliss (1984).

An encouraging sign, as protistologists seem to be on the brink of chaos, nomenclaturally speaking, is the deep interest taken by various international bodies of biologists in the problems exposed in this paper. For example, the International Union of Biological Sciences has established a committee charged with a thorough investigation of the conflicts arising, particularly, from a "dual system" of nomenclature of some protist groups (Ride and Younès, 1986). ISEP, i.e., the International Society for Evolutionary Protistology, is forming its own groups to study the issue on a broader scale. The International Congress of Systematic and Evolutionary Biology holds special symposia on these continuing problems. Other societies - protozoological, phycological, and mycological - are similarly beginning to show interest and concern.

So, a "protist perspective" or outlook at the nomenclatural level, both infra- and suprafamiliar, may emerge before long, to the relief of taxonomic protistologists everywhere who are beginning to view with dismay the dilemma in which we currently seem to find ourselves.

While there is thus hope for the future, production of the *Handbook* could not be delayed awaiting the eventual resolution of nomenclatural problems, particularly those at the higher levels in the classification of protist groups. The other editors and contributors, as well as some readers, may favor alternative approaches to protist systematics, but the phyletic macroscheme adopted for this book, as pointed out above, is basically the one found in Copeland, 1956 (ref. p. xxiii) as improved by Corliss, Margulis and others, a system that is probably about as useful as any other at this stage in the evolution of taxonomic protistology.

John O. Corliss

REFERENCES

Barnes, R.S.K., ed.: *A Synoptic Classification of Living Organisms*. Oxford: Blackwell Scientific Publications and Sunderland, MA: Sinauer Associates, 1984.

Cavalier-Smith, T.: Eukaryote kingdoms: seven or nine? *BioSystems* 14, 461-481 (1981).

Cavalier-Smith, T.: A 6-kingdom classification and a unified phylogeny. In: *Endocytobiology II: Intracellular Space as Oligogenetic Ecosystem*, (Schenk, H.E.A., Schwemmler, W., eds.), Vol. 2, pp. 1027-1034. Berlin-New York: Walter de Gruyter, 1983.

Corliss, J.O.: Consequences of creating new kingdoms of organisms. *BioScience* 33, 314-318 (1983).

Corliss, J.O.: The kingdom Protista and its 45 phyla. *BioSystems* 17, 87-126 (1984).

Corliss, J.O.: Progress in protistology during the first decade following reemergence of the field as a respectable interdisciplinary area in modern biological research. *Progress in Protistology* 1, 11-63 (1986).

Corliss, J.O.: Protistan phylogeny and eukaryogenesis. *International Review of Cytology* 100, 319-370 (1987).

Heywood, P., Rothschild, L.J.: Reconciliation of evolution and nomenclature among the higher taxa of protists. *Biological Journal of the Linnean Society* 30, 91-98 (1987).

Hibberd, D.J., Norris, R.E.: Cytology and ultrastructure of *Chlorarachnion reptans* (Chlorarachniophyta divisio nova, Chlorarachniophyceae classis nova). *Journal of Phycology* 20, 310-330 (1984).

International Code of Botanical Nomenclature, adopted by the XIII International Botanical Congress, Sydney, 1981. Utrecht, The Netherlands: International Association for Plant Taxonomy, 1983.

International Code of Zoological Nomenclature, adopted by the XX General Assembly of the International Union of Biological Sciences, Helsinki, 1979. London: International Trust for Zoological Nomenclature, 1985.

Jeffrey, C.: *Biological Nomenclature*, 2nd ed. New York: Crane, Russak, 1977.

Jeffrey, C.: Kingdoms, codes and classification. *Kew Bulletin* 37, 403-416 (1982).

Leedale, G.F.: How many are the kingdoms of organisms? *Taxon* 23, 261-270 (1974).

Levine, N.D.: *Veterinary Protozoology*. Ames: Iowa State University Press, 1985.

Lipscomb, D.L., Corliss, J.O.: *Stephanopogon*, a phylogenetically important "ciliate," shown by ultrastructural studies to be a flagellate. *Science* 215, 303-304 (1982).

Margulis, L.: Five-kingdom classification and the origin and evolution of cells. *Evolutionary Biology* 7, 45-78 (1974a).

Margulis, L.: The classification and evolution of prokaryotes and eukaryotes. In: *Handbook of Genetics* (King, R.C., ed.), Vol. 1, pp. 1-41. New York: Plenum Press, 1974b.

Margulis, L.: *Symbiosis in Cell Evolution: Microbial Communities in the Archean and Proterozoic Eons*, 2nd ed. New York: W.H. Freeman Co., 1992.

Margulis, L., Sagan, D.: Order amidst animalcules: The Protoctista kingdom and its undulipodiated cells. *BioSystems* 18, 141-147 (1985).

Margulis, L., Schwartz, K.V.: *Five Kingdoms: An Illustrated Guide to the Phyla of Life on Earth*. San Francisco: W.H. Freeman, 1982; 2nd ed., 1988.

Möhn, E.: *System und Phylogenie der Lebewesen*, Vol. 1. *Physikalische, chemische und biologische Evolution. Prokaryonta, Eukaryonta (bis Ctenophora)*. Stuttgart: E. Schweizerbart'sche Verlagsbuchhandlung (Nagele u. Obermiller), 1984.

Patterson, D.J.: The fine structure of *Opalina ranarum* (family Opalinidae): Opalinid phylogeny and classification. *Protistologica* 21, 413-428 (1986a).

Patterson, D.J.: Some problems of ambiregnal taxonomy, and a possible solution. *Symposia Biologica Hungarica* 33, 87-93 (1986b).

Ride, W.D.L.: Nomenclature of organisms treated as both plants and animals. *Biology International* 6, 15-16 (1982).

Ride, W.D.L., Younès, T., eds.: *Biological Nomenclature Today*. (IUBS Monograph Series, 2) Oxford: IRL Press, 1986.

Rothschild, L.J., Heywood, P.: Protistan phylogeny and chloroplast evolution: conflicts and congruence. *Progress in Protistology* 2, 1-68 (1987).

Sieburth, J. McN., Estep, K.W.: Precise and meaningful terminology in marine microbial ecology. *Marine Microbial Food Webs* 1, 1-15 (1985).

Silva, P.C.: Remarks on algal nomenclature VI. *Taxon* 29, 121-145 (1980).

Silva, P.C.: The role of extrinsic factors in the past and future of green algae systematics. In: *Systematics of the Green Algae* (Irvine, D.E.G., John, D.M., eds.), Systematics Association Special Vol. No. 27, pp. 419-433. New York-London: Academic Press, 1984.

Sleigh, M.A.: Radiation of the eukaryote Protista. In: *The Origin of Major Invertebrate Groups* (House, M.R., ed.), Systematics Association Special Vol. No. 12, pp. 23-54. New York-London: Academic Press, 1979.

Smith, R. McK., Patterson, D.J.: Analyses of heliozoan interrelationships: an example of the potentials and limitations of ultrastructural approaches to the study of protistan phylogeny. *Proceedings of the Royal Society of London* 227, 325-366. (1986).

Taylor, F.J.R.: Problems in the development of an explicit hypothetical phylogeny of the lower eukaryotes. *BioSystems* 10, 67-89 (1978).

Taylor, F.J.R., Sarjeant, W.A.S., Fensome, R.A., Williams, G.L.: Proposals to standardize the nomenclature in flagellate groups currently treated by both the botanical and zoological codes of nomenclature. *Taxon* 35, 890-896 (1986).

Taylor, F.J.R., Sarjeant, W.A.S., Fensome, R.A., Williams, G.L.: Standardization of nomenclature in flagellate groups treated by both the botanical and zoological codes of nomenclature. *Systematic Zoology* 36, 79-85 (1987).

Whittaker, R.H.: New concepts of kingdoms of organisms. *Science* 163, 150-160 (1969).

Whittaker, R.H.: Broad classification: the kingdoms and the protozoans. In: *Parasitic Protozoa* (Kreier, J.P., ed.), Vol. 1, *Taxonomy, Kinetoplastids, and Flagellates of Fish*, pp. 1-34. New York-London: Academic Press, 1977.

Whittaker, R. H., Margulis, L.: Protist classification and the kingdoms of organisms. *BioSystems* 10, 3-18 (1978).

Sources of Living Protoctists and Their Culture[1]

ABBREVIATION	ADDRESS
ATCC [2]	American Type Culture Collection 12301 Parklawn Drive Rockville, MD 20852 USA
Carolina [2]	Carolina Biological Supply Company 2700 York Road Burlington, NC 27215 USA
CBS	Centraalbureau voor Schimmelcultures Baarn NETHERLANDS
CCAP1 [2]	Culture Collection of Algae and Protozoa Institute of Terrestrial Ecology 36 Storeys Way Cambridge CB3 ODT UNITED KINGDOM
CCAP2	Culture Collection of Algae and Protozoa Freshwater Biological Association The Ferry House Ambleside, Cumbria LA22 OLP UNITED KINGDOM
CCAP3	Scottish Marine Biological Association Laboratory Oban SCOTLAND
CCMP [2]	Culture Collection for Marine Phytoplankton Bigelow Laboratory for Ocean Sciences McKown Point West Boothbay Harbor, ME 04575 USA
CMI	Commonwealth Mycological Institute Kew UNITED KINGDOM
CSIRO [2]	CSIRO Culture Collection of Unicellular Marine Algae CSIRO Marine Laboratories Box 21 Cronulla, N.S.W. 2230 AUSTRALIA
GUE	University of Guelph Dr. Joseph Gerrath Botany and Genetics Department University of Guelph Guelph, Ontario N1G 2W1 CANADA
IABH	Institut für Allgemeine Botanik der Universität Ohnhorststrasse 18 Universität Hamburg D-2000 Hamburg 52 GERMANY (Collection of Dr. L. Kies)
MBAUK [2]	Marine Biological Association of the U.K. Dr. J.C. Green The Laboratory Citadel Hill Plymouth PL1 1PB UNITED KINGDOM
NEPCC [2]	Northeast Pacific Culture Collection Department of Oceanography University of British Columbia Vancouver, British Columbia V6T 1W5 CANADA

[1] Much more detailed information on sources and culture methods can be found in the *Handbook of Protoctista*.
[2] Indicates catalog or computer printout of cultures available.

ABBREVIATION	ADDRESS
NIES	National Institute for Environmental Studies JAPAN
OSU	Ohio State University Algae Collection Gary L. Floyd 1735 Neil Avenue Columbus, OH 43210 USA
PRA [2]	Culture Collection of Autotrophic Organisms Dr. Hans Ettl Institute of Botany Czechoslovak Academy of Sciences Dukelska 145 CS-379 83 Trebon CZECHOSLOVAKIA
SAG [2]	Sammlung von Algenkulturen Pflanzenphysiologisches Institut der Universität Nikolausberger Weg 18 D-3400 Göttingen GERMANY
SWC	Laboratory of Sea Water Chemistry Kagawa University Prof. T. Okaichi Faculty of Agriculture Miki-cho, Kida-gun Kagawa Pref. 761-07 JAPAN
UGA	Department of Botany Culture Collection University of Georgia Athens, GA 30602 USA
UTEX [2]	Culture Collection of Algae at University of Texas Dr. Richard C. Starr Department of Botany University of Texas at Austin Austin, TX 78712 USA
UWA	University of Washington Oceanography Collection School of Oceanography University of Washington Seattle, WA 98195 USA

PHYLUM	CULTURE COLLECTION	SOURCES IN NATURE[3]
RHIZOPODA	ATCC, Carolina, CCAP2	*Entamoeba histolytica*, *E. gingivalis*: humans; *E.invadens*: reptiles
HAPLOSPORIDIA	NA[4]	*Haplosporidium*, *Minchinia*: molluscs, polychaetes, decapod crustaceans, or echinoderms of marine or freshwater environments; *Urosporidium*: in trematodes and nematodes parasitizing invertebrates
PARAMYXEA	NA	Parasites of digestive organs of marine invertebrates
MYXOZOA	NA	Myxosporea: Parasites of marine and freshwater fish, reptiles; Actinosporea: parasites of marine sipunculid worms and freshwater oligochaetes
MICROSPORA	NA	Parasites of insects, fish
ACRASEA	ATCC, CCAP1, CBS	Dead plant parts, tree bark, dung, and soil
DICTYOSTELIDA	ATCC, CBS	Animal dung, soil, decaying plants and mushrooms

[3] See Tables 6 and 7, pp. 204 and 251.
[4] NA=Not available from culture collections

PHYLUM	CULTURE COLLECTION	SOURCES IN NATURE
RHODOPHYTA	Carolina, CCAP1, OSU, SAG, UTEX	Florideophycidae: Primarily marine: rocky intertidal communities, temperate subtidal communities, and coral reefs; Bangiophycidae: Oceans, lakes, streams, hot springs, and soil, but primarily marine intertidal, growing as epiliths or epiphytes
CONJUGAPHYTA	CCAP1, GUE, UTEX	Almost entirely freshwater: Ponds, lakes, streams, rivers, marshes, and bogs
XENOPHYOPHORA	NA	Deep sea
CRYPTOPHYTA	CCAP1, CCMP, CSIRO, MBAUK, NEPCC, SAG	Oligotrophic temperate and northern marine and freshwater environments
GLAUCOCYSTOPHYTA	IABH, SAG, UTEX	Freshwater: Planktonic or benthic in lakes, ponds, or ditches
KARYOBLASTEA	NA	Freshwater: Mud or sand in stagnant or near-stagnant water; Northern hemisphere
ZOOMASTIGINA Amebomastigota	ATCC	Soil, freshwater, and rarely, marine sediments and water
ZOOMASTIGINA Bicosoecids (Bicoecids)	NA	Marine or freshwater plankton
ZOOMASTIGINA Choanomastigota	NA	Primarily marine: Neustonic, epibiotic on bryozoa, phaeophytes, rhodophytes, and diatoms, or planktonic
ZOOMASTIGINA Diplomonadida	*Giardia intestinalis*: ATCC	*Trepomonas, Hexamitus*: Freeliving in mesosaprobic or polysaprobic freshwater habitats and the sea; rest are commensals or parasites in a variety of animals, often in alimentary tract
ZOOMASTIGINA Pseudociliata	NA	Marine benthic environments
ZOOMASTIGINA Kinetoplastida	ATCC	Bodonids: Freeliving or epizoic; some parasites of fish and other aquatic organisms. Trypanosomatids: All parasites; found in insects, plants, mammals, and dipterans
ZOOMASTIGINA Opalinata	NA	Endosymbionts in large intestine or rectum of poikilothermic vertebrates, mainly anuran amphibians
ZOOMASTIGINA Proteromonadida	NA	Posterior intestinal tract of many amphibians, reptiles, and mammals
ZOOMASTIGINA Parabasalia	*Trichomonas*: ATCC	Digestive system of termites and wood-eating cockroaches and respiratory, digestive, and reproductive systems of mammals and birds

PHYLUM	CULTURAL COLLECTION	SOURCES IN NATURE
ZOOMASTIGINA Retortamonadida	NA	Intestines of animals, including insects, reptiles, fish, and vertebrates
ZOOMASTIGINA Pyrsonymphida	NA	Hindguts of termites and wood-eating cockroaches
EUGLENIDA	Carolina, CCAP1, NEPCC, SAG, UTEX	Freshwater habitats, esp. those rich in decaying organic matter; others in marine/brackish areas
CHLORARACHNIDA	NA	Marine
PRYMNESIOPHYTA	Carolina, CCAP1, CCAP3, CCMP, CSIRO, MBAUK, NEPCC, OSU, UTEX, UWA	Marine and rarely freshwater habitats; planktonic
RAPHIDOPHYTA	CSIRO, NIES, OSU, SWC, UTEX	Freshwater species: Acidic or neutral environments where vegetation abundant; marine species: open sea or brackish water
EUSTIGMATOPHYTA	CCAP1, CCMP, UTEX	Marine and freshwater
ACTINOPODA Polycystina and Phaeodaria	NA	Marine pelagic
ACTINOPODA Heliozoa	*Actinosphaerium (Echinosphaerium) nucleofilum*: Carolina	Marine or freshwater in shallow water, just above the sediment-water interface
ACTINOPODA Acantharia	NA	Marine planktonic, temperate or tropical waters
HYPHOCHYTRIOMYCOTA	ATCC, UGA	Soil, freshwater or marine habitats; usu. found saprobic on pollen, carcasses, or plant debris
LABYRINTHULOMYCOTA	NA	Marine and estuarine environments, usu. associated with benthic algae, marine vascular plants, and detrital sediments
PLASMODIOPHOROMYCOTA	J.P. Braselton, Ohio U., Dept. of Botany, Athens, OH 45701 USA; J.S. Karling, Purdue U., Dept. of Biology, W. Lafayette, IN 47906 USA	Found in soil or fresh water in their protist or plant hosts
DINOMASTIGOTA	ATCC, Carolina, CCAP1, CCMP, CSIRO, MBAUK, NEPCC, SAG, SWC, UTEX	Marine and freshwater; symbiotic in protist or invertebrate hosts as zooxanthellae; parasitic on protists, plants and animals
CHRYSOPHYTA	CCAP1, SAG, UTEX	Freshwater plankton; some epibiotic, neustonic, or benthic; rarely marine
CHYTRIDIOMYCOTA	ATCC, CMI	Aquatic environments and soil

PHYLUM	CULTURE COLLECTION	SOURCES IN NATURE
PLASMODIAL SLIME MOLDS Myxomycota	Carolina	Forest environments, typically in forest floor litter, on fallen tree trunks, and decaying stumps
PLASMODIAL SLIME MOLDS Protostelida	F.W. Spiegel, Dept. of Botany and Microbiology, U. of Arkansas, Fayetteville, AR 72701 USA	Aerial portions of dead plants or from bark of living or dead trees; soil
CILIOPHORA	ATCC, Carolina, CCAP1	Aquatic environments including ponds, lakes, estuaries, saltmarshes, and oceans
GRANULORETICULOSA	*Allogromia laticollaris*: ATCC, J.J. Lee, Biology Dept., City Univ. of New York, Convent Ave. at 138 St., New York, NY 10031 USA	Marine habitats; planktonic, littoral
APICOMPLEXA	NA	Parasites of invertebrates and vertebrates
BACILLARIOPHYTA	CCAP1, CCMP, CSIRO, NEPCC, SAG, UTEX, UWA	Marine and freshwater habitats; planktonic, benthic, epipelic, epipsammic, epilithic, epizoic
CHLOROPHYTA	ATCC, Carolina, CCAP1,CCMP, CSIRO, MBAUK, NEPCC, OSU, PRA, SAG, UTEX, UWA	Marine and fresh water, brackish water, soil, terrestrial and subaerial, epiphytic, endophytic, epizoic, endozoic; extreme environments
OOMYCOTA	ATCC, CBS, CMI	Mostly in freshwater or terrestrial habitats
XANTHOPHYTA	ATCC, Carolina, CCAP1, OSU, PRA, SAG, UTEX	Majority in freshwater habitats and soil, saltmarshes; *Vaucheria*: freshwater, marine, and brackish habitats
PHAEOPHYTA	Carolina, CCAP1, OSU, SAG, UTEX	Exclusively marine: intertidal and subtidal zones of coastal regions throughout the world
ELLOBIOPSIDS	NA	Parasites of euphasids, mysids, amphipods, shrimp, and copepods
EBRIDIANS	NA	Coastal marine habitats

REFERENCES

Haines, K.C., Hoagland, K.D., Fryxell, G.A.: A preliminary list of algal culture collections of the world. In: *Selected Papers in Phycology II*. (J.R. Rosowski and B.C. Parker, eds.), pp. 820-826. Lawrence, KS: Phycological Society of America, 1982. (Provides list of algal culture collections with addresses and telephone numbers, fee and restriction information, cross-referenced taxonomically.)

Hindák, Frantisek: Culture collection of algae at Laboratory of Algology in Trebon. *Archiv für Hydrobiologie Supplement 39, Algological Studies* 2/3, 86-126 (1970).

Schlösser, U.G.: Sammlung von Algenkulturen. *Berichte der Deutschen Botanischen Gesellschaft* 95, 181-287 (1982).

Starr, R.C.: The culture collection of algae at the University of Texas at Austin. *Journal of Phycology* 14 (suppl. 47), 47-100 (1978). (Provides listing of cultures held, cross-referenced to CCAP1, along with media recipes for growth of algae and information about ordering.)

Access to Protoctist Information: Toward a Database

The protoctist kingdom encompasses an incredibly diverse group of organisms, including species with and without skeletal structures, with varying tolerance to fixation and preservation techniques, and with dimensions that span many orders of magnitude. This diversity makes difficult the collection of information into a central archive. Further complicating the problem is the current system of collecting disparate information by differently trained scientists. Traditional techniques for collection of information on protoctists are described here along with new methods using computer systems, all of which contribute to a "Protoctist Knowledge Base."

1. Type specimens and traditional collections

The only collection information that currently follows "rules" is the cataloging of type specimens. Type specimens are original materials from which a species is described. Although often specified with the original published description, a type specimen also may be designated later. Many type specimens reside in botanical and zoological collections in Europe, for example, the seaweed collection begun in 1820 at the University of Leiden, The Netherlands. More than 100,000 identified seaweeds, of which over 5,000 are type specimens mounted dry on paper sheets, are housed in the Leiden herbarium.

Such cataloging on herbarium sheets of specimens of larger algae (confusingly referred to by phycologists as "macrophytes," which means "large plants") can be done in a consistent manner. However, type material for other protoctists is recorded in a large variety of ways: live cultures, preserved whole organisms, whole mounts and sections on glass microscopy slides, mounted material for transmission (TEM) and scanning (SEM) electron microscopy, published photographs (light micrographs, scanning and transmission electron micrographs), and drawings. Many of these kinds of information can be stored in computer databases. The diversity of techniques leads to inconsistencies in practice and problems in the recording of results and retrieval of information especially for the smaller protoctists. The larger protoctists are preserved whole using a variety of preservation techniques. Permanent preparations on glass microscope slides archive whole specimens, sections, or skeletal parts, of which the diatom herbarium at the Academy of Natural Sciences in Philadelphia is an excellent example. The Academy includes 334,000 specimens of cleaned, mounted diatom frustules in its collection only a third of which are cataloged. Since many of the smaller protoctists are unidentifiable with light microscopy alone, the cataloging of images derived from scanning or transmission electron microscopy is indispensable.

Live cultures serve as an important source of information. Many large collections include clones derived from the original specimen from which species were described (see page xxxiii for sources of protoctist live material). Finally, some species lack type material: neither the original specimen nor cultures derived are available. For these, published photographs or drawings must suffice. These species include those for which type specimens were lost or never deposited in an herbarium or type culture collection, particularly uncultivable species or species documented early in the history of biology. Type material is often missing for those species that have never been observed alive.

2. Video information

Video microscopy, which allows recording of motility and behavior, significantly augments our ability to identify and catalog information about protoctists. Such information about live organisms was formerly available only from texts written by the observing scientist. For small mastigotes in particular, motility is often of vital importance for species determination. Enhanced video microscopy and image analysis may be used to extend the human eye, allowing observation of structures otherwise difficult to observe or invisible (Inoué, 1986).

Professional video systems (e.g., 3/4" U-matic) are required when preparing video for copying and distribution; VHS video equipment should be used only for personal archiving. Video disks, which can be used to store enhanced images and which also may be prepared from original videotape material, ease the access and retrieval of computer information. This technology promises to become an important supplement to tapes for recording and distribution of critical information on behavior, sexuality, and reproduction of protoctists. Video playback systems for PAL, SECAM, or NTSC television standards are required to view video archived information from investigators around the world and to translate from one television standard to another.

3. A Protoctist Knowledge Base

Traditional illustrated publications for collecting protoctist information will remain important, but computer technologies will create a new unified structure for storing and retrieving results in a

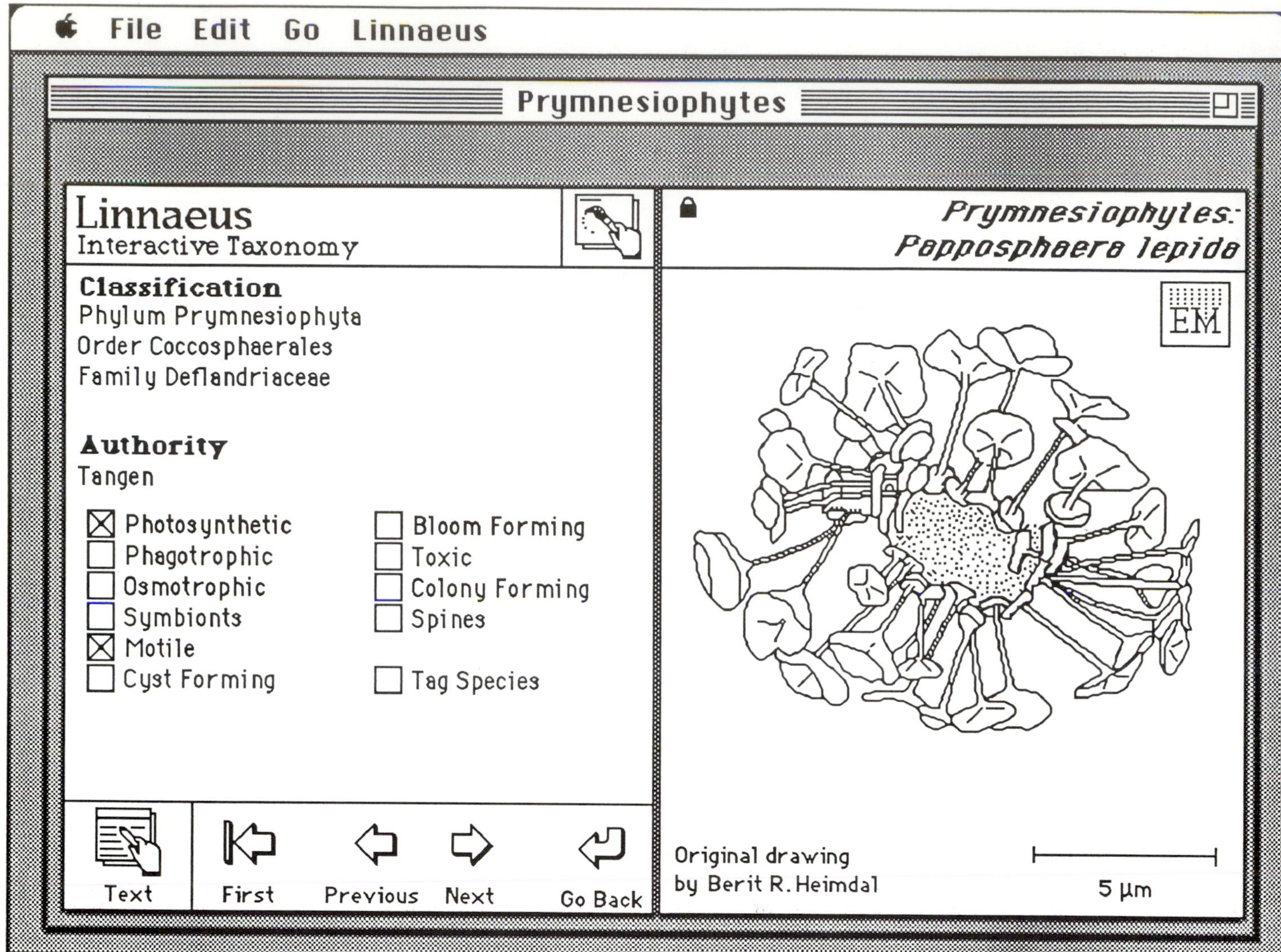

Figure 1. **Linnaeus Protist**, now containing 300 species, mainly those of economic importance from the North Sea, collects descriptions, literature references, line drawings, and micrographs into a single computer program.

single, easily accessible source. Protoctist information has only begun to find its way into an electronic format. The forthcoming Apple/Macintosh computer program, the **Linnaeus Protist** (a development of its predecessor, **Linnaeus Zooplankton**, Estep *et al.*, 1989), is an example.

Linnaeus Protist collects species descriptions, literature references, line drawings, and micrographs into a single program (Figure 1). This type of software now allows full-color pictures to be displayed on relatively inexpensive computers (Figure 2). Line drawings and micrographs for **Linnaeus** are entered into the computer by scanner; this device, which digitizes information in the original drawings and micrographs, resembles a small copy machine. Though limited to approximately 300 protist species from coastal Scandinavia, the system provides an example of software expansible for construction of a Protoctist Knowledge Base.

Any Protoctist Knowledge Base is anticipated to be an electronic archive containing species descriptions, digitized drawings and micrographs, and literature references. The base could consist of "small" modules pressed onto single CD-ROMs (Compact Disc Read Only Memory). CD-ROM technology allows 400-700 Megabytes (equivalent to 10-100 thousand illustrations and pages of text) to be stored on a single disk. A single group, such as most protoctist phyla, fits comfortably into this size. If the system were modular problems associated with large databases, i.e., protracted search time and very large computer requirement can be avoided. Such disks, designed only to be read and not altered by the user, represent an inexpensive way to store and distribute huge amounts of information.

An additional feature of a Protoctist Knowledge Base is digital video. Recent chip technology allows video to be recorded, compressed, and played back in real time. This technology makes possible the storage of 1-2 hours of video on a CD, rather than a large video disk, allowing all data, including video, to be stored in a single disk format and permitting video to play back in a window

Developments in computer technologies allow full-color pictures and digital sound to be presented on relatively inexpensive (i.e., Macintosh II) computers.

on the computer screen, rather than on a separate monitor. Digital video also increases the possibilities for the use of sound; e.g., **Linnaeus Protist** stores the pronunciation of the species names in the voices of the original authors. Digital sound video material unobtainable elsewhere may be extended by experts.

4. Information retrieval from the Protoctist Knowledge Base

Recent developments in computer software suggest several possible approaches to the retrieval of information in a prospective Protoctist Knowledge Base. Software approaches include: HyperText Systems, Guided Searches, Expert Systems, and Neural Networks to impart information most easily to the user.

A HyperText System, like that for **Linnaeus**, is a computer system that allows multimedia information to be entered into a single computer file and retrieved as in a database program. Its main advantage is the capacity to link pieces of information together: for example, an algal species description linked to information about toxicity, geographical distribution, and appropriate literature references. This designation of links makes HyperText Systems ideally suited to knowledge base construction; it permits users to move from one kind of information to another by clicking with the mouse.

Guided Searches, like ordinary database searches, seek specific criteria upon which the computer performs a search. The difference in a guided search is in the information that the expert attaches to the search criteria. Search criteria may be sorted by the expert, and text or pictures may be attached to particular criteria to guide the user in making choices.

Expert Systems encode knowledge in rule-based systems. In this case a rule is nothing more than the logical structure first stated concerning Aristotle:

All men are mortal.
Aristotle is a man.
Therefore Aristotle is mortal.

The Expert System consists of rules of this type that define the knowledge base and an "inference engine" that uses the rules to ask questions and draw conclusions. Expert Systems have been applied to zoological and botanical taxonomic problems by Gautier and Pavé (1990).

Neural Networks, like Expert Systems, provide a logic structure for drawing conclusions about information. The main difference for Neural Networks is in the method of data entry. Expert Systems assume that the expert knows the rules exactly, and they make logical inferences using these entered rules. Neural Networks permit more "fuzzy" data to be entered into the system; logical weights are constructed by the software for further inferences.

The optimal software for a potentially comprehensive Protoctist Knowledge Base is still in question. It is clear, however, that a system like that in the figure on the next page could be constructed with separate CD-ROM databases connected via several CD players to a single personal computer (PC). If additional access were

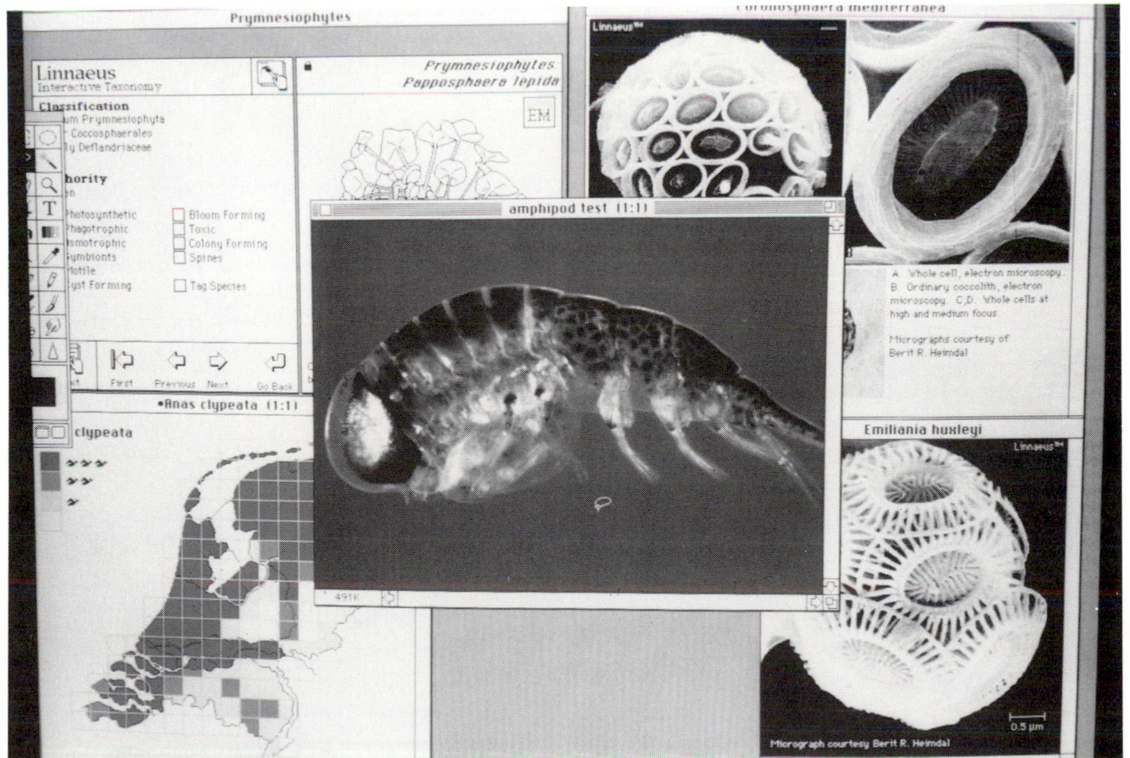

Figure 2. Sample screens from **Linnaeus Protist**: classification, distribution, morphology, consumer and other information on prymnesiophytes.

A Prototype for the Protoctist Knowledge Base

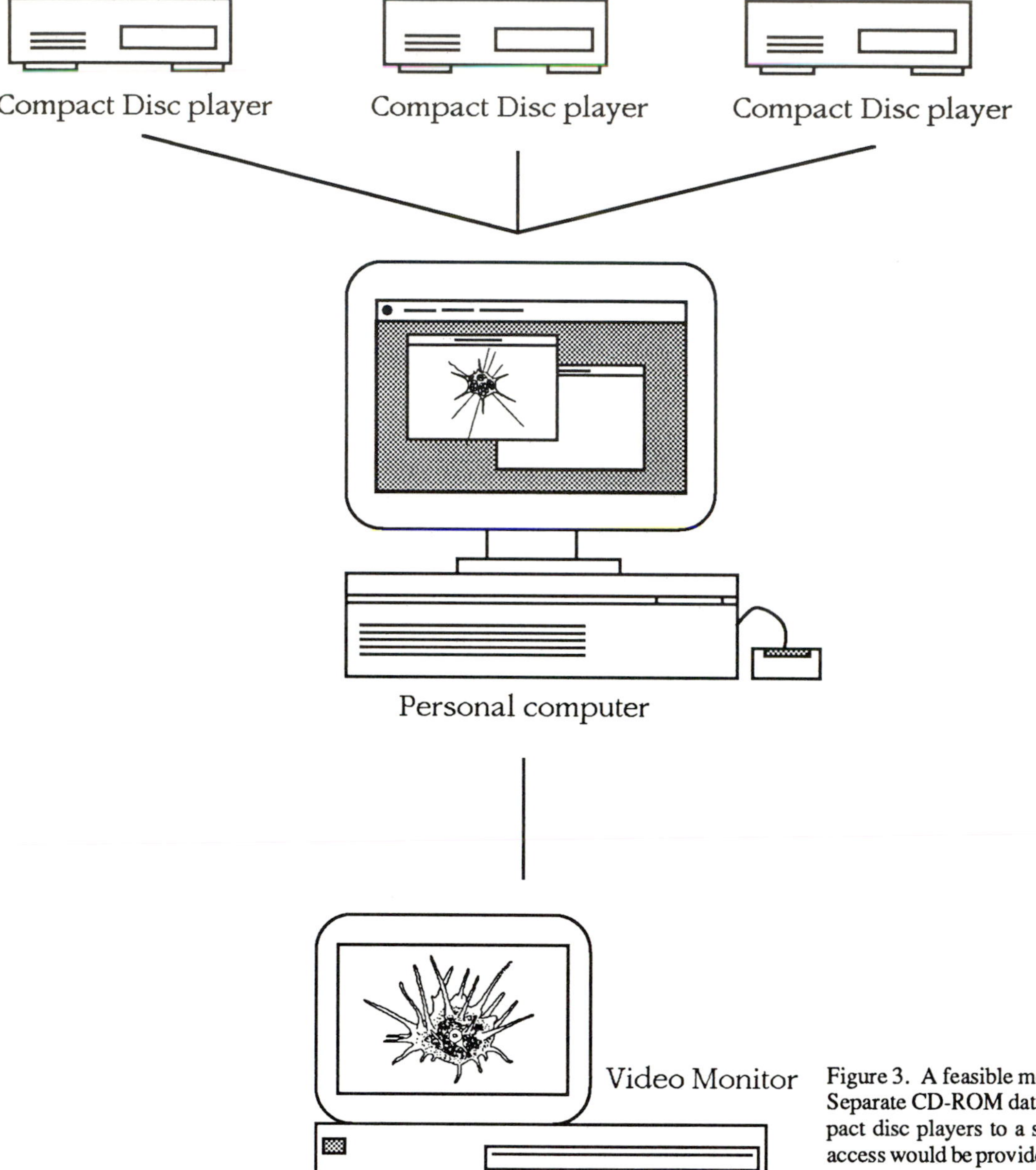

Figure 3. A feasible model for a Protoctist Knowledge Base. Separate CD-ROM databases are connected via several compact disc players to a single personal computer. Additional access would be provided via modem to users outside the local area.

provided via modem to users outside the local area of this system, it could serve as an invaluable rapid source of information on protoctist species.

FURTHER INFORMATION

Further information on existing type collections, techniques for establishing new collections of preserved materials, or representations of life material are obtainable from:

Protoctist collections:

Academy of Natural Sciences
19th and the Parkway
Philadelphia, PA 19103 USA

Institut for Sporeplanter
University of Copenhagen
Øster Farimagsgade 2D
DK-1353 Copenhagen K DENMARK

Dr. W.F. Prud'homme van Reine
Rijksherbarium, Algologie
Schelpenkade 6
Postbox 9514
2300 RA Leiden THE NETHERLANDS

Video microscopy and image analysis:

Inoué, Shinyo. 1986. *Video Microscopy.* Plenum Press, New York.

Linnaeus databases:

Francisco Rey
Institute of Marine Research
Postbox 1870
5013 Bergen NORWAY

Expert Systems, HyperText, and Neural Networks:

University of Amsterdam
Expert Center for Taxonomic Identification
Postbox 4766
1009 AT Amsterdam THE NETHERLANDS

Aleksander, I., Burnett, P.: *Thinking Machines.* Oxford, UK: Oxford University Press, 1987.

Fraase, M.: *Macintosh Hypermedia. Vol. II: Uses and Implementations.* Glenview, IL: Scott, Foresman and Company, 1990.

McAleese, R.: *Hypertext: Theory Into Practice.* Oxford, UK: Blackwell Scientific Publications, 1989.

For collections of live material, see list beginning on page xxxiii.

Kenneth W. Estep

REFERENCES

Estep, K.W., Hassel, A., Omli, L., MacIntyre, F.: **Linnaeus**: Interactive taxonomy using the Macintosh computer and HyperCard. *BioScience* 39, 635-638 (1989).

Gautier, N., Pavé, A.: Object-centered representation for species systematics and identification in living systems in nature. *CABIOS* 6, 383-386 (1990).

Inoué, S.: *Video Microscopy.* New York: Plenum Press, 1986.

Molecular Biology and Protoctist Phylogeny

The evolutionary history of eukaryotes has been dominated by members of the Protoctista, the most diverse of the nucleated organisms. The earliest eukaryotes and the last common ancestor to plants, animals, and fungi were protoctists, but this taxonomic group is not a cohesive evolutionary assemblage. Rather "protoctist" describes organisms unified by levels of organization represented by paraphyletic lines of descent; they cannot be definitively placed in a phylogenetic context using traditional methods of comparative morphology, physiology, or biochemistry. Molecular comparisons offer new and powerful quantitative methods for the development of phylogenies for protoctists; these frameworks can be used to reevaluate phylogenetic inferences based upon nonmolecular markers.

1. Ribosomal RNA phylogenies

Similarities in the sequences of monomers of macromolecules (and/or their coding regions) can be used to infer phylogenetic relationships between organisms if the following criteria are satisfied:* 1) The coding regions must be present in all compared organisms, and they should be subject to similar or identical functional constraints. 2) They should contain a statistically significant number of nucleotide or amino acid positions (greater than 1000 sites) that can vary independently. 3) The coding regions must not be capable of undergoing horizontal gene flow between distinct evolutionary lineages. 4) The sequences to be compared must represent orthologous (directly related) genomes, i.e., nuclear genes should be compared to other genes from nuclear, not plastid, sources. The sequence to be compared is extracted from the organelle to be compared (e.g., nucleocytoplasm, chloroplast) and is not from a different organellar lineage or unidentified endosymbiont.

The small subunit (16S-like) and large subunit (23S-like) ribosomal RNAs (rRNAs) are particularly well suited for molecular systematic studies of divergent protist lineages (Sogin, 1989). Sequence domains that are invariant in all eukaryotes or all organisms are interspersed among partially conserved and rapidly evolving regions. Comparisons of sequence data from the conserved and partially conserved regions are used to estimate extent of genetic relatedness between the most divergent taxa, while the rapidly evolving domains are useful in defining relationships of members of lower taxa, e.g., a single genus.

2. Molecular analysis of ribosomal RNAs

Comparative sequence analysis of ribosomal RNAs and the inference of molecular phylogenetic relationships requires the growth of axenic or monoprotist cultures, extraction of nucleic acids, isolation and preparation of sequencing template(s), acquisition of the sequence data, alignment of the sequences to existing databases, and phylogenetic inference. Sources and growth conditions for protoctists are described elsewhere in this volume. Whenever possible, nucleic acids should be isolated from exponentially growing cultures, however, sufficient amounts of DNA can be produced from preserved specimens for polymerase chain reaction experiments. Total nucleic acids are extracted with phenol from disrupted cells (Sambrook *et al.*, 1989). Cell disruption protocols vary from simple sodium dodecyl sulfate (SDS) detergent lysis of amebas or mechanical disruption of centric diatoms by vortex treatment to more sophisticated methods including sonication, glass bead homogenization, or passage through a French pressure cell. A variety of procedures, including density gradient centrifugation and chromatography (Sambrook *et al.*, 1989; Hillis *et al.*, 1990), can be used to fractionate bulk nucleotide preparations into DNA and RNA components. The preparation of good sequencing templates requires high-quality DNA and RNA preparations.

Methods for characterizing nucleic acid sequences require the preparation and screening of genomic libraries for recombinant clones. This rate-limiting step now can be circumvented by direct sequence analysis of rRNAs (Qu *et al.*, 1983) and/or sequence analysis of rRNA coding regions that have been amplified using "polymerase chain reaction" (PCR) procedures (Mullis and Faloona, 1987). Strategies for rapidly sequencing ribosomal RNAs or their corresponding DNA coding regions are based on the interspersion pattern of conserved sequences among different, more unique domains. Oligonucleotide "primers," complementary to conserved domains, are used to initiate synthesis in dideoxynucleotide chain termination DNA sequencing reactions (Elwood *et al.*, 1985). The distribution of primer sites throughout the length of the genes for the rRNAs permits complete and rapid sequence analysis (at least one week for a full sequence) on both strands of a 16S-like or 23S-like rRNA coding region.

* The organism from which the sequence has been determined is representative of the named group, and the sequence is constant relative to ontogeny of the organism in question. This is especially important since it has been recognized that the same organism (i.e., *Plasmodium berghei*) can have different sequences in the ribosomal DNA genes depending on the life cycle stage sampled (Gunderson, *et al.*, *Science* 238, 933-937, 1987). (Editor's note)

Direct sequence analysis of rRNAs in total cell RNA can be achieved using conserved primers to initiate synthesis in reverse transcriptase-mediated sequencing reactions (Lane *et al.*, 1985). This method permits sequence determinations 300 nucleotides in length and is commonly used to sample regions in 23S rRNAs that include both partially conserved and rapidly evolving domains. The major attribute of this strategy is speed and assurance that the analysis corresponds to a transcribed rRNA gene (and not a similar, nontranscribed "pseudogene"). Unfortunately, sequences determined by this technique are rarely longer than a few hundred nucleotides, and the ambiguity and error rate may exceed five percent. This error/ambiguity rate profoundly affects fine resolution branching topologies where evolutionary distances between nodes correspond to as few as one or two nucleotides per hundred positions.

Polymerase chain reaction techniques, which are both rapid and very accurate, provide an alternative approach to direct sequence analysis of rRNAs. Oligonucleotides complementary to conserved sequences near the 5' and 3' regions of 16S-18S rRNAs are used to initiate DNA synthesis in PCR experiments. The primers permit the specific amplification of 16S-18S rRNA genes defined by the nuclear genome, even in the presence of prokaryotic DNA (Medlin *et al.*, 1988). Sequences between the conserved elements are exponentially amplified by repetitive cycles of denaturing duplex DNA, annealing primers complementary to the conserved sequence elements, and primer extension using DNA polymerase. The products of the primer extension reaction and the original duplex DNA serve as templates in successive amplification cycles. Within a few hours, several micrograms of DNA coding for ribosomal RNAs can be obtained from less than 0.1 nanograms of bulk genome DNA. The resulting product is cloned into the single-stranded phage M13 or characterized by modifications of the dideoxynucleotide sequencing protocols for analyzing double-stranded DNA templates. This rapid strategy requires a limited number of organisms for DNA extraction and, of great significance, permits the simultaneous analysis of several representatives from the multi-copy rDNA genes of a single species. By direct sequence analysis of PCR products or concurrent analysis of multiple recombinant M13 clones containing representative amplification products, the extent of sequence heterogeneity in the rDNA gene family can be assessed.

3. Data analysis

Ribosomal RNAs from eukaryotes vary in length from 1250 nucleotides to as many as 2300. Homologous positions in rRNA coding regions must be aligned before the extent of structural similarity between rRNA genes can be determined. Multiple sequence alignment algorithms are available for a variety of computers, but none is capable of considering sequence patterns in more than a few sequences simultaneously. Approximate alignments can be obtained using automated procedures that consider pairs of sequences. The most closely related sequences are placed in an optimal alignment using algorithms that consider the merits of inserting alignment gaps versus mismatches in the compared sequences. The order of addition for new sequences in the aligned data set is based upon their similarity to the most closely related sequence pairs. Since this is a clustering procedure, the alignments must be further refined to reflect conservation patterns that can be found in more distantly related sequences. The alignments can be dramatically improved by juxtaposition of regions that define conserved secondary structures. Computer programs capable of aligning sequences on the basis of conserved higher order structures are under development. Software packages for the analysis of molecular data, including multiple sequence alignment algorithms and computer-assisted sequence alignment editors, are available on a variety of computers. Many investigators employ commercial programs, including the GCG (Genetics Computer Group, Inc.) or the IG suite package (IntelliGenetics) for VMS and UNIX machines. The most sophisticated multiple sequence alignment editors are available from university sources, including SeqEdit for VMS/VAX machines, MASE for UNIX-based machines, and a new generation editor, GDE, for Xwindows.

Nearly identical sequences (i.e., greater than 95% similar) can be unambiguously aligned over their entire length. Uncertainties will occur in rapidly evolving regions when overall sequence similarities fall below 95%. Regions that cannot be unambiguously aligned should not be included in an analysis. For example, in alignments between *Ochromonas danica* and *Skeletonema costata* 16S-like rRNAs, 14% of the positions can differ and approximately 100 of their 1840 nucleotides cannot be aligned with certainty. When sequences are 60%-65% similar, only 1000 sites can be unambiguously aligned. The number of positions that can be included in an analysis depends on phylogenetic context; in general, it is inversely proportional to the maximum evolutionary distances separating sequences in the data set. The establishment of the Ribosomal RNA Data Base Project, funded by the National Science Foundation, has aided in the resolution of issues surrounding alignments and selection of aligned positions. The project maintains a database of aligned ribosomal RNA sequences that can be accessed by E-mail or electronic data transfer protocols. The goal of those constructing the database is to provide uniform alignments based upon large data sets and more sophisticated algorithms not routinely available to independent investigators.

Many techniques have been described for inferring phylogenetic relationships from molecular data. The most common are maximum parsimony and distance matrix approaches (Felsenstein, 1988). The utility of one method over that of the other sparks vigorous debate among proponents of each and can lead to major conflicts between practitioners using similar techniques. Many molecular systematists employ specific "black box approaches" determined in large part by the tools available in their local computing environment. The advantages and disadvantages of various techniques under specific circumstances have been reviewed by Felsenstein (1988) and Swofford and Olsen (1990). Numerous computer packages have been developed for phylogenetic inference. PHYLIP provides more than 30 programs for the inference of phylogenetic relationships using a variety of techniques, and it can be installed on most computers. PAUP and MacClade for the Macintosh and Hennig86 for IBM compatible PCs are excellent packages for inference of maximum parsimony. Most of the multiple sequence analysis editors can generate data output formats compatible with PAUP, PHYLIP, MacClade, and Hennig86. The GDE and SeqEdit programs provide interfaces to distance matrix method based algorithms, including Fitch and Margoliash and/or Desote Matrix tree inferences. The methods, conclusions, and current state of phylogenetic inferences of protoctists will be reviewed by Sogin and Patterson (in preparation).

Mitchell L. Sogin

REFERENCES

Elwood, H.J., Olsen, G.J., Sogin, M.L.: The small subunit ribosomal RNA gene sequences from the hypotrichous ciliates *Oxytricha nova* and *Stylonychia pustulata*. *Molecular Biology and Evolution* 2, 399-410 (1985).

Felsenstein, J.: Phylogenies from molecular sequences: Inference and reliability. *Annual Review of Genetics* 22, 521-565 (1988).

Hillis, D.M., Larson, A., Davis, S.K., Zimmer, E.: Nucleic acids III: Sequencing. In: *Molecular Systematics*. (D. M. Hillis and C. Moritz, eds.). Sunderland, MA: Sinauer Associates, 1990.

Lane, D.J., Pace, B., Olsen, G.J., Stahl, D.A., Sogin, M.L., Pace, N.R.: Rapid determination of 16S ribosomal RNA sequences for phylogenetic analyses. *Proceedings of the National Academy of Sciences, USA* 82, 6955-6959 (1985).

Medlin, L., Elwood, H.J., Stickel, S., Sogin, M.L.: The characterization of enzymatically amplified eukaryotic 16S-like rRNA coding regions. *Gene* 71, 491-499 (1988).

Mullis, K.B., Faloona, F.A.: Specific synthesis of DNA *in vitro* via a polymerase-catalysed chain reaction. *Methods in Enzymology* 155, 335-350 (1987).

Qu, L.H., Michot, B., Bachellerie, J-P.: Improved methods for structure probing in large RNAs: A rapid "heterologous" sequencing approach is coupled to the direct mapping of nuclease accessible sites. Applications to the 5' terminal domain of eukaryotic 28S rRNA. *Nucleic Acids Research* 11, 5903-5920 (1983).

Sambrook, J., Maniatis, T., Fritsch, E.F.: *Molecular Cloning: A Laboratory Manual*, 2nd ed. Cold Spring Harbor, NY: Cold Spring Harbor Laboratory, 1989.

Sogin, M.L. Evolution of eukaryotic microorganisms and their small subunit ribosomal RNAs. *American Zoologist* 29, 487-499 (1989).

Sogin, M.L., Patterson, D.: Protist phylogeny from 16S RNA data. *Microbiological Reviews* (in preparation).

Swofford, D.L., Olsen, G.O.: Phylogeny reconstruction. In: *Molecular Systematics*. (D.M. Hillis, and C. Moritz, eds.). Sunderland, MA: Sinauer Associates, 1990.

COMPUTER SOFTWARE FOR ANALYSIS OF MOLECULAR DATA

GCG — **Genetics Computer Group Package**
Genetics Computer Group, Inc.
University Research Park
575 Science Drive, Suite B
Madison, WI 53711 USA

IG — **IntelliGenetics Suite**
IntelliGenetics, Inc.
700 East El Camino Real
Mountain View, CA 94040 USA
(Available for VAX, SUN, PC, and Macintosh.)

MASE — **Multiple Alignment Sequence Editor**
Professor Temple Smith
Dana Farber Cancer Institute
44 Binney Street
Boston, MA 02155 USA
(Source code for SUN workstations.)

SEQEDIT — **Sequence Editor**
Gary Olsen
Department of Microbiology
131 Burril Hall
407 South Goodwin
Urbana, IL 61801 USA
(Multiple sequence alignment editor distributed as precompiled code for VAX/VMS machines.)

GDE — **Genetic Data Environment**
Steve Smith
Harvard Genome Laboratory
16 Divinity Avenue
Cambridge, MA 02138 USA
(Available as source code for GDE editor.)

PHYLIP **Phylogeny Inference Package**
Joseph Felsenstein
Department of Genetics SK-50
University of Washington
Seattle, WA 98195 USA
(Pascal source code that can be compiled on most computer systems.)

PAUP **Phylogenetic Analysis Using Parsimony**
David L. Swofford
Illinois Natural History Survey
607 East Peabody Drive
Champaign, IL 61820 USA
(Precompiled code is distributed for Macintosh and IBM-PC computers and as C source code for minicomputers and mainframes.)

Henig86 **James S. Farris**
41 Admiral Street
Port Jefferson Station
New York, NY 11776 USA
(Executable code is distributed for IBM-compatible computers only.)

MacClade **Wayne P. Maddison and D.R. Maddison**
Distributed by Sinauer Associates
Sunderland, MA 01375 USA.
(Distributed as precompiled code for Macintosh only.)

RDP **Ribosomal RNA Database**
Carl Woese and Gary Olsen
Department of Microbiology
131 Burril Hall
407 South Goodwin
Urbana, IL 61801 USA
(Aligned database for ribosomal RNA sequences. Information may be obtained by sending E-mail inquiry "help" to "rdp@antares.mcs.anl.gov.")

PROTOCTIST GLOSSARY

Etymological Roots and Examples of Use

General Glossary

Organism Glossary

Bibliography of Glossaries

Etymological Roots and Examples of Use

a-, an- Gr. not, without (amastigote, amitosis, anaerobe)
acro- <Gr. *akron,** top, peak; *akros*, at the end, top, first, highest (acrocentric)
adelpho- Gr. *adelphos*, brother, twin, near kinsman (adelphoparasite)
aero- <L. *aer*, the lower atmosphere, air (aerobe, aerotolerant)
agnoto- Gr. *agnostos*, unknown, ignorant of (agnotobiotic)
allo- <Gr. *allos*, other (alloparasite)
amphi- Gr. around, on both sides, double (amphiesma)
ana- Gr. up, back, again (anabiosis)
aniso- Gr. *anisos*, unequal (anisogametes)
anti-, anta- <Gr. *anti*, against, opposed to (antapical)
apo- Gr. from, off, away, after, without, separate (aposymbiotic)
archae-, arche-, archi-, archo- <Gr. *arche*, beginning, first cause, chief; *archaios*, from the beginning, old (archegonium)
auto- Gr. *autos*, self (autogamy)
auxo- Gr. increase, grow (auxospore)
axo- L. *axis*; Gr. *axonium*, *axoniskos*, axle, pole (axopod)

base, basi- <L., Gr. *basis*, *basilaris*, at the base (basipetal)
bi-, bin-, bis-, <L. *bis*, twice, in a twofold manner (bimastigote, binary fission)
bios- <Gr. *biosis*, life, manner of life (anabiosis)
brady- Gr. slow (bradyzoites)

calyx L. (Gr. *kalyx*) cup, cover, outer envelope (glycocalyx)
carpo-, -carpic Gr. *karpos*, fruit (eucarpic, carpospore)
center, centri-, centro- <L. *centrum*, midpoint of a circle (centripetal cleavage)
chasmo- <Gr. *chasma*, hole, gulf (chasmolithic)
chemo- <Ar. *alchimia*, chemist; Gr. pertaining to chemistry and chemical terms (chemotaxonomy)
chlamydo- <Gr. *chlamys*, cloak (chlamydospore)
chloro- <Gr. *chloros*, green (chloroplast)
choano- <Gr. *choane*, funnel (choanomastigote)
chromato- <Gr. *chroma*, color (chromatophore)
chryso- <Gr. *chrysos*, gold (chrysoplast)
cline <L. *clino*, slant, slope, tend (geosyncline)
cocco- L. *coccum*, berry (coccolith)
coelo- <Gr. *koilos*, hollow; *koiloma*, a cavity (coelozoic)
coeno- <Gr. *koinos*, common (coenocytic)
con- <L. *cum*, together, with (conspecific)
copro- <Gr. *kopros*, dung (coprophile)
cryo- <Gr. *kryos*, *krymos*, icy, cold, frost (cryophile)
crypto- <Gr. *krypto*, hide, conceal (cryptostomata)
cyano- <Gr. *kyanos*, *kyaneos*, dark blue (cyanobacteria)
cyclo- <Gr. *kyklos*, circle (cyclosis)

Key
< from -word prefix
Ar. Arabic word- suffix
Gr. Greek -word- either prefix or suffix
L. Latin
* Italics indicate the original Latin word or the transliteration of the original word from Greek or Arabic.

cyrto- <Gr. *kyrtos*, curved; *kyrte*, fish-basket (cyrtos)
cysti-, cysto- <Gr. *kystis*, bladder, sac, cell (cystosorus)
cyto- <Gr. *kytos*, hollow place, cell (cytoplasm)

desmo- <Gr. *desmos*, bond, fetter, chain (desmosome)
di-, dif-, dir-, dis- <L. *dis*, in two, apart, away from, without, not; <Gr. *dis*, twice, double (dimorphic)
dicho- <Gr. *dicha*, in two (dichotomous)
dictyo- <Gr. *diktyon*, net (dictyosome)
diplo- <Gr. *diploos*, twofold (diplokaryon)
dys- Gr. bad, ill, with difficulty, hard (dyskinetoplastic)

e- <L. *ex*, out of, from (enucleation)
eco- <Gr. *oikos*, house, home (ecology)
ecto- <Gr. *ek*, out of, from (ectosymbiont)
endo- <Gr. *endon*, within, inside (endosymbiont)
ento- <Gr. *entos*, within, inside (entozoic)
epi- Gr. upon, on; over (epilith)
eu- Gr. good, true, original, primitive (euploid, eucarpic)
eurys-, eury- Gr. broad, wide, widespread (euryhaline)
exo- <Gr. *ex*, out of, without (exocytosis)

gameto- <Gr. *gamete*, wife; *gametes*, husband (gametes)
gamos-, -gamy Gr. *gamos*, marriage, union (gamont)
gen-, -gen <L. *gigno*, *genitus*, be born, causing, producing, forming; *genea*, race, stock, family (biogenic)
-genesis Gr. beginning, birth, origin (oogenesis)
geo- <Gr. *ge*, *gaia*, Earth (geosyncline)
germen L. bud, seed, sprout (germinate)
glyco-, glycy- <Gr. *glykys*, sweet (glycocalyx)
gnoto-, gnosis Gr. wisdom (gnotobiotic)
-gono- <Gr. *gonos*, seed, offspring, product (gonomere)
gymno- <Gr. *gymnos*, naked (gymnosperm)
-gyne- Gr. woman (trichogyne)

hali-, halo- <Gr. *hals*, *halos*, sea, salt; *halios*, of the sea (stenohaline)
haplo- <Gr. *haploos*, single, simple (haploid)
hapto- Gr. join, fasten to, fix upon (haptonema)
hema-, hemato- <Gr. *haima*, blood (hematozoic)
hetero- <Gr. *heteros*, other, different (heterokaryotic)
histo- Gr. *histos*, tissue (histozoic)
holo- <Gr. *holos*, whole, entire, all (holozoic)
homo- <Gr. *homos*, same, uniform, similar (homodynamic)
hyalo- <Gr. *hyalos*, glass; *hyaleos*, *hyalinos*, of glass, glassy, transparent (hyaloplasm)
hyper-, hypero- <Gr. *hyper*, beyond, over, above, very (hypertrophy)
hypno- Gr. *hypnos*, sleep (hypnozygote)
hypo- Gr. under, beneath, less than (hypovalve)

ichno- Gr. footprint, track (ichnofossil)
inter- L. between, among (intercellular)
intra-, intro- L. within (intracellular)

iso- <Gr. *isos*, equal, like (isogametes)

-karyo- Gr. seed, kernel, nut (eukaryotes)

kineto-, -kinetico <Gr. pertaining to motion (kinetosome)

-kont <Gr. movement (heterokont)

lamella- L. dim. of lamina, plate (plastid lamella)

leuco- <Gr. *leukos*, white (leucoplast)

limno- <Gr. *leimon*, marsh, lake, pool (epilimnion)

litho- <Gr. *lithos*, stone (lithology)

lobe-, lobo- <L. *lobos*, a rounded projection or protuberance (lobopodium)

lysis, lyso-, -lytic <Gr. *lyo*, loose, dissolve, break up (lysosome)

macro- <Gr. *makros*, long, large (macrospore)

mastigo- <Gr. *mastix, -igos*, whip (mastigoneme)

mega- <Gr. *megas*, large, great (megaspore)

meio- <Gr. *meiom*, to make smaller (meiosis)

mero- Gr. *meris, meros*, part, portion, share (merogony)

meso- <Gr. *mesos*, middle (mesotrophic)

meta- Gr. between, among, near (metacentric)

micro- <Gr. *mikros*, small, little (microspore)

micto-, -mixis <Gr. *miktos*, mixed (apomixis)

mitos- Gr. thread (mitosis)

morpho- <Gr. *morphe*, shape, form (morphology)

myco-, myceto- <Gr. *mykes, etos*, fungus (mycophagy)

nano- L. *nanus*; Gr. *nanos*, a dwarf (nanoplankton)

necro- <Gr. *nekros*, dead body, corpse (necrotrophic)

nemato- <Gr. *nema, -tos*, thread (pantacronematic)

nucleus L. *nux*, kernel, nut (macronucleus)

oligo- <Gr. *oligos*, few, scanty (oligotrophic)

-oma Gr. denoting a tumor or morbid condition (xenoma)

omni- L. *omnis*, all (omnivorous)

oo-, -oon, Gr. egg (oosphere)

opistho- <Gr. *opisthen*, behind (opisthokont)

osmo- <Gr. *osmos*, a pushing (osmoregulation)

paleo- <Gr. *palaios*, ancient, old (paleontology)

palin-, palim- <Gr. *palin*, again, back, repetition (palintomy)

pan-, panto- <Gr. *pas, pantos*, all, the whole, every (pantacronematic)

para- Gr. beside, near, by (parasome)

partheno- Gr. *parthenos*, virgin (parthenospore)

pedo- <Gr. *pais, paidos*, child (pedogamy)

pellic L. *pellis*, skin (pellicle)

peri- Gr. around, near (periphyton)

phaeo- <Gr. *phaios*, dusky, brown (phaeoplast)

-phage, phago- <Gr. *phagein*, to eat (phagocytosis)

phil-, philo- <Gr. *phileo*, love (coprophile, lithophile)

-phore, phoro- <Gr. *phoreus*, carrier, bearer (chromatophore)

photic, photo- <Gr. *phos, photos*, light (photosynthesis)

phyco- <Gr. *phykos*, seaweed, alga (phycoma)

phyl- <Gr. *phyle*, tribe, race (phylogeny)

phyto- <Gr. *phyton*, plant (phytoalexin)

pico- Gr. tiny (picoplankton)

placo- <Gr. *plax, -akos*, anything flat and wide (placolith)

plasma Gr. that which is formed or molded, substance (nucleoplasm)

pleuro- <Gr. *pleura*, side (cryptopleuromitosis)

pluri- <L. *pluralis* , from *plus, plur*, more (pluriseriate)

podo-, podi-, -poda <Gr. *podus, podos*, foot (pseudopod)

proto- Gr. *protos*, first (protozoa)

pseudo- <Gr. *pseudos*, lie (pseudopod)

reticulo- L. *reticulatus*, netlike; *rete, retis*, net (reticulopodia)

rhabdo- <Gr. *rhabdos*, rod, stick (rhabde)

rheo- Gr. *rheos*, stream, current (rheoplasm)

rhizo- <Gr. *rhiza*, root (rhizoplast)

rhodo- <Gr. *rhodon*, rose, red (rhodoplast)

sapro- Gr. *sapros*, rotten (saprotrophy)

schiza-, schizo- Gr. splinter or chip of wood (schizogony)

seme Gr. sign, signal (hyposeme)

siphon <L. *sipho*, pipe, bent tube (siphonous)

-some, soma- Gr. *soma*, body or flesh (xenosome, somatic)

spore, sporo- <Gr. *spora*, a sowing, seed (sporophorous vesicle)

steno- Gr. *stenos*, narrow (stenohaline)

sticho- Gr. *stichos*, line, row (stichonematic)

stoma, -stome Gr. *stoma*, mouth (pseudostome)

stria L. groove or ridge (striated)

stroma- Gr. *stroma*, bed, mattress (stromatolite)

sub- L. under (sublittoral)

supra- L. above, over (supralittoral)

sym- Gr. together, with (symbiosis)

syn- Gr. together, with (syngamy)

tax-, taxia-, taxis, taxi- <Gr. *tasso*, arrange, classify, place (magnetotaxis)

telo- Gr. *telos*, end (telomere)

tetra- <Gr. *tessares*, four (tetrapore)

theca L. (Gr. *theke)* case, container, envelope, sheath (epitheca)

thermo- <Gr. *therme*, heat (thermophilic)

thigmo- Gr. *thigma*, touch (thigmotaxis)

-tomy, -tome <Gr. *temno*, dissection, excision, cut (plasmotomy)

tricho- <Gr. *thrix, trichos*, hair (trichogyne)

-troph Gr. *trophe*, food, nourishment (heterotrophic)

vor- L. *vorax*, gluttonous, greedy (omnivorous)

xantho- Gr. *xanthos*, yellow (xanthophyll)

xeno- Gr. *xenos*, stranger, guest (xenophyae, xenosome)

zoo- Gr. *zoön*, living animal (zoonosis)

zygo- Gr. *zygon, zygos*, yolk (zygote)

-zyme, zymo- <Gr. yeast (enzyme, zymogram)

General Glossary

Every effort has been made to make this glossary as comprehensive as possible. We have tried to include definitions for words that appear in the glossary as part of another entry. Examples are given of organisms bearing unique structures. For illustrations of representative organisms from each phylum and some classes, see Organism Glossary beginning on page 137. We have included definitions submitted to us for the same word with different meanings (see, for example, the discussion of the use of "cyst" and "spore" in the Introduction). For each entry, synonyms, if any, appear first. If they are true synonyms, a definition appears only once. Definitions appear for each "synonym" with slightly different meaning. Different forms of the same word are listed as synonyms (e.g., pseudopod, pseudopodium; monothalamic, monothalamous). Throughout the figure legends we have used the informal names of phyla (see Table 5) to designate the organisms illustrated. Abbreviations used in the figure legends are: DIC=Nomarski differential interference contrast light microscopy; PC=phase contrast light microscopy; SEM=scanning electron microscopy; TEM=transmission electron microscopy.

A

A-axis

Morphological term describing foraminiferan shells; the A-axis is the shortest axis of the hexagonal pattern. The C-axis is the longest axis. See *C-axis*.

A form

In foraminifera with alternation of generations, the sexual organism, the gamont. In dinomastigotes, arrangement of thecal plates. Illustration: Plate formula.

A-tubule

Tubule (subfiber) of doublet of axoneme of undulipodium; the innermost tubule of axoneme doublets to which dynein arms are attached or the microtubule comprising the wall of the complete tubule of centriolar triplets is called the A-tubule. Illustrations: Kinetid, Kinetosomes.

AAA pathway

See *Alpha aminoadipic acid pathway.*

Aboral (adj.)

Away from the oral opening.

Abyssal (adj.)

Hadal. Pertaining to the ocean environment or depth zone of 4,000 meters or deeper; also pertaining to the organisms of that environment. Illustration: Habitat.

Acanthopod

See *Acanthopodium.*

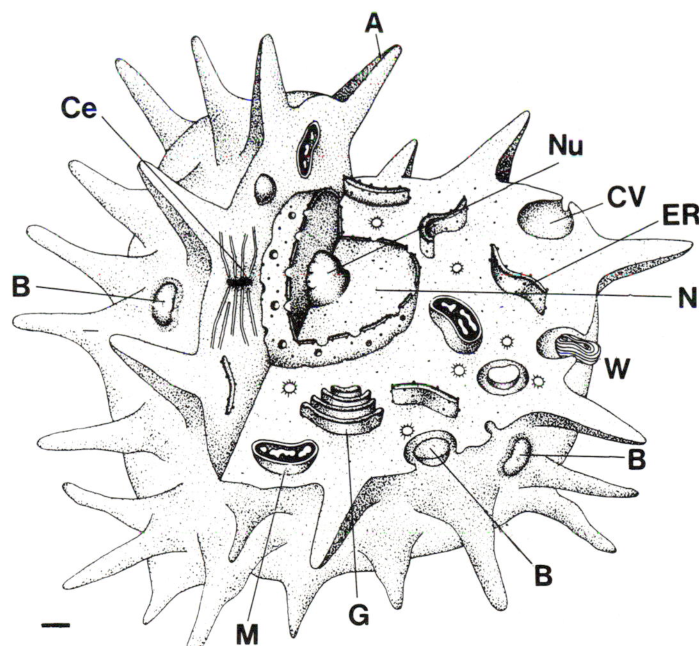

Acanthopodium

Acanthamoeba-like trophic ameba. A=acanthopodium; B=bacteria; Ce=centriolelike organelle; CV=contractile vacuole; ER=endoplasmic reticulum; G=golgi apparatus; M=mitochondrion; N=nucleus; Nu=nucleolus; W=whorl. Bar=1 µm.

Acanthopodium (pl. acanthopodia)

Acanthopod; fine-tapering pseudopod of *Acanthamoeba.*

Key to abbreviations

adj.	adjective
cf.	compare
e.g.	for example (Latin *exempli gratia*)
esp.	especially
i.e.	that is (Latin *id est*)
kb	kilobases
n.	noun
pl.	plural
sing.	singular
usu.	usually
v.	verb

Acellular slime molds
Plasmodial slime molds; myxomycotes (i.e., myxomycotes and protostelids). Illustration: Plasmodial slime molds.

Acentric chromosome
Chromosome that lacks a centromere.

Acentric mitosis
Mitosis that occurs in the absence of centrioles, centriolar plaques, kinetosomes, or other microtubule-organizing centers at the poles of the cell. Illustration: Mitosis.

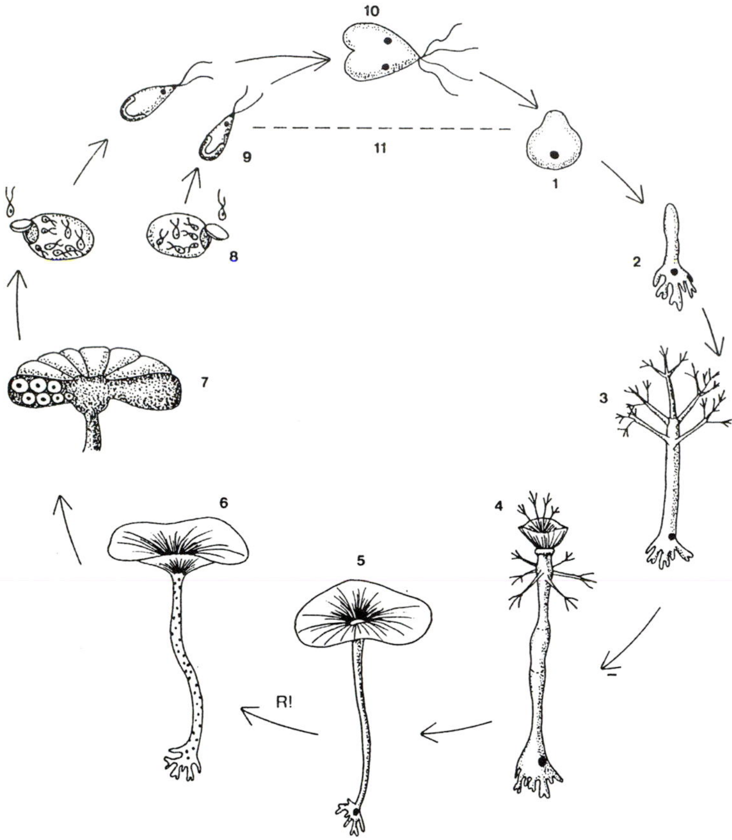

Acetabularian life history
The life history of the ulvophycean chlorophyte *Acetabularia acetabulum* (L.) Silva (= *A. mediterranea* Lamouroux). 1. germinating zygote or parthenogenetic gamete. 2. development of stalk and rhizoids. 3. whorl formation. 4. initiation of cap. 5. growth of cap and meiotic division (R!) of primary nucleus. 6. migration of secondary nuclei into cap rays. 7. development of cysts in rays. 8. release of gametes from cysts (gametocysts). 9. isogametes. 10. syngamy. 11. parthenogenesis.

Acetabularian life history
Stages in development of chlorophytes belonging to the genus *Acetabularia* and their class relatives.

Acrasin
Chemical attractant secreted by dictyostelid amebas that signals aggregation into a multicellular structure (e.g., cyclic adenosine 3', 5'-monophosphate (cAMP) or glorin, a dipeptide).

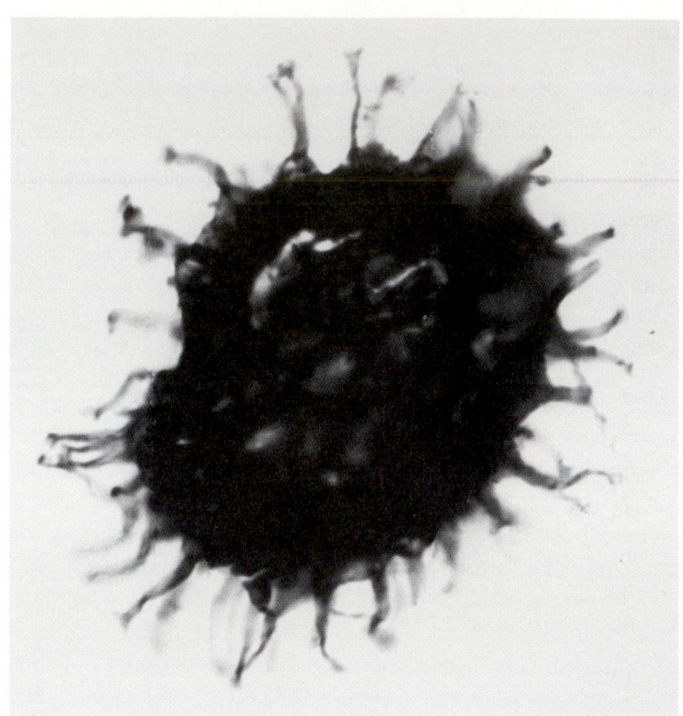

Acritarch
Skiagia ornata.

Acritarch
Hollow, organic unidentified microfossil; may be spherical, ellipsoidal, or polygonal, smooth or granulated, or with spinose projections. Proterozoic to Recent. Probably eggs or protoctist cysts, but lack sufficient morphological detail to be classified.

Acrobase
Groove or surface marking that extends anteriorly from the sulcus onto the epicone of unarmored dinomastigotes. Illustration: Dinokont.

Acrocentric chromosome
Chromosome with a terminal or nearly terminal centromere (kinetochore). Illustration: Chromosome.

Acroneme (adj. acronematic)
Mastigoneme at terminus of undulipodium or other mastigoneme. Smooth undulipodium with a fine fibril at the distal end. Illustrations: Mastigoneme, Phaeophyta.

Actines
Main branches of an ebridian skeleton, arising from the rhabde (longitudinal rod). Illustration: Ebridians.

Actinopod
Spine, or thin cell process, characteristic of heliozoans, acantharians, phaeodarians, and polycystines; underlain by microtubules. Used in feeding, locomotion, etc. Informal name of organism in phylum Actinopoda. See also illustration: Heliozoa.

Adelphoparasite
Parasite closely related to its host; shares family or lower taxon with its host. See *Alloparasite*.

Adhesive disc
Basal disc; striated disc; sucking disc; ventral disc; ventral sucker. Cup-shaped attachment of some protists; e.g., in mastigotes and some spirotrichous ciliates, a thigmotactic cup-shaped organelle at the aboral end of the protist used for attachment to its substratum (usu. the surface of a host organism). Organelle of the diplomonad *Giardia* (phylum Zoomastigina) that attaches *Giardia* to animal epithelium. Supported by complex cytoskeleton and delimited by a ridge, the lateral crest, it is composed of tubulin and a 30-kilodalton protein, giardin. Illustration: Karyomastigont system.

Adhesorium (pl. adhesoria)
Adhesive organelle (e.g., of plasmodiophorid plant parasites).

Adoral (adj.)
Toward the oral opening.

Aerobe
Organism active and capable of completing its life cycle only in the presence of gaseous oxygen, O_2.

Aerophyte (adj. aerophytic)
Air dweller; refers to algae and plants.

Aerotolerant (adj.)
Anaerobic, but not inhibited by low concentrations of O_2.

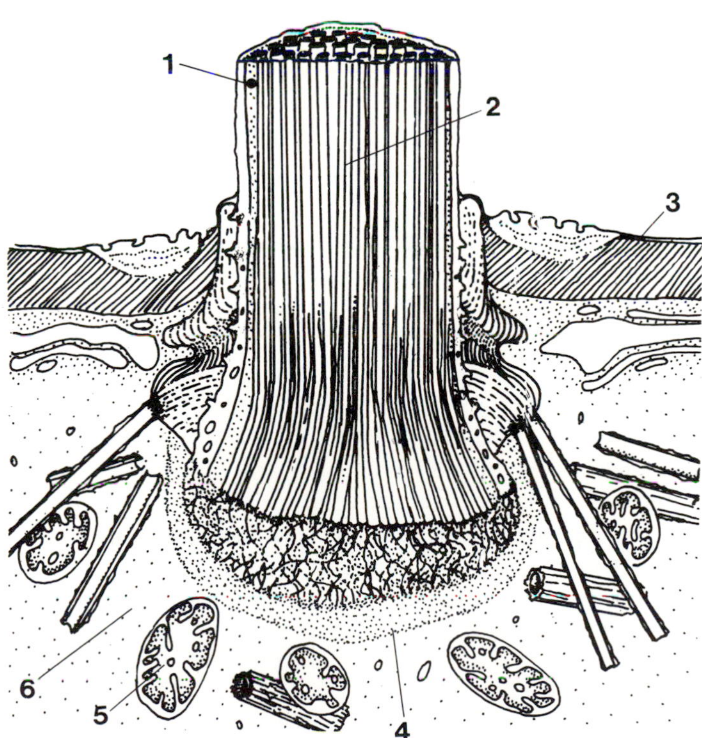

Actinopod
1=stereoplasm; 2=microtubular bundles; 3=capsular wall; 4=axoplast; 5=mitochondrion; 6= endoplasm.

Aethalia
The myxomycote *Lycogala epidendrum*.

Aethalia (sing. aethalium)
One of three types of sporophores in myxomycotes; large syncytial sporocarp of certain plasmodial slime molds. See *Plasmodiocarp, Sporangium.*

Aflagellate
See *Amastigote*.

Agamogony
Series of nuclear or cell divisions producing individuals that are neither gametes nor capable of forming gametes. Illustrations: Foraminifera, Life cycle.

Agamont
Reproducing organism at a stage in its life cycle during which it lacks gametes or other sexual structures (e.g., foraminifera, schizonts of apicomplexans). Illustrations: Foraminifera, Life cycle.

Agar
Type of phycocolloid; α-sulfated carbohydrate composed of ß-1,3 linked D-galactose and α-1,4 linked anhydro-L-galactose extractable from cell walls and intercellular spaces of the rhodophytes *Gelidium* and *Gracilaria*. Resistance to digestion and transparency make agar (to which nutrients are added) an ideal matrix upon which to grow microbes.

Agglutinated test
Glued test; covering or shell produced by protoctist from sediment particles including tests of other organisms, usu. with organic lining (e.g., foraminifera). Illustration: Rhizopoda.

Aggregation center
Structure of dictyostelid cellular slime molds formed by the coming together of hundreds of amebas. Illustration next page. See also illustration: Dictyostelida.

Agnotobiotic culture
Crude culture. Mixed culture; pertaining to a heterogeneous culture in which the microbiota is unidentified. See *Gnotobiotic*.

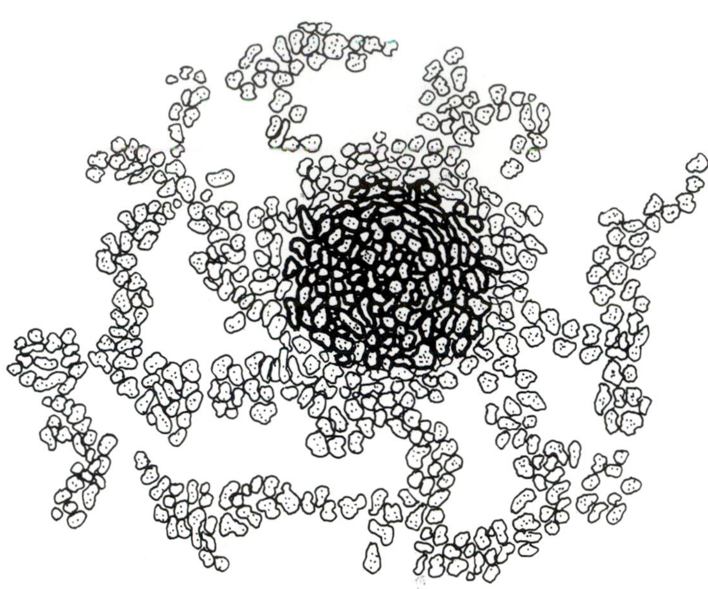

Aggregation center
The acrasid *Copromyxella spicata*.

Agnotoxenic (adj.)
Pertaining to a culture medium contaminated by one or more unknown organisms. See *Monoxenic, Polyxenic.*

Akinete
Type of propagule; nonmotile single- or few-celled structure formed by thickening of the cell wall of a growing cell; capable of passive propagation and germination, usu. of cyanobacteria or algae (e.g., conjugating green algae, xanthophytes).

Akont
See *Amastigote.*

Algae (sing. alga)
Photoautotrophic protoctists; all oxygenic phototrophs exclusive of cyanobacteria, chloroxybacteria, and plants; ecological term for aquatic oxygenic phototroph.

Alginate
Salt of alginic acid, produced in walls of phaeophytes; a polysaccharide with ß-1,4 linked D-mannuronic acid and α- 1,4 linked L-guluronic acid in varying ratios.

Algivory (n. algivore; adj. algivorous)
Mode of nutrition; referring to organisms that feed on algae (see Table 2).

Algology
See *Phycology.*

Allelochemic (adj.; n. allelochemicals)
Ecological term referring to chemical substances (secondary metabolites) which, when released into the surroundings of organisms, influence the behavior or development of other individuals of different species. See *Antibiotic, Pheromone, Secondary metabolite.*

Allometric transformation
Growth in three dimensions that can be described by simple rules (e.g., in protostelids).

Alloparasite
Parasite not closely related to its host; i.e., does not share family or lower taxon with its host. See *Adelphoparasite.*

Allophycocyanin
Type of phycobiliprotein; blue water-soluble extract; found in cyanobacteria, rhodophytes, and cryptomonads. See *Phycocyanin, Phycoerythrin.*

Allosteric (adj.)
Referring to the change in some enzymes whereby a small molecule combines with a site on the protein (other than the active site) resulting in a change in catalytic activity via a change in protein conformation.

Allozymes
Alternative enzyme forms encoded by different alleles at the same genetic locus.

Aloricate (adj.)
Lacking a lorica. See *Lorica.*

Alpha aminoadipic acid pathway
AAA pathway. Biosynthetic metabolic pathway forming the amino acid lysine. This pathway is characteristic of some protoctists and fungi and is entirely different from the diaminopimelic acid pathway of lysine biosynthesis, found in bacteria, other protoctists, and plants. See *Diaminopimelic acid pathway.*

Alpha spore
See *Carpospore.*

Alpine (adj.)
Characteristic or descriptive of the mountainous regions lying between timber line and snow line. Illustration: Habitat.

Alternation of generations
Life cycles of plants and protoctists usu. in which haploid (1N, gametophyte) generation alternates with diploid (2N, sporophyte) generation; the haploid and diploid organisms may be morphologically identical or extremely different. Also refers to alternation of morphological types in a given life cycle even when there is no change in ploidy (e.g., hydroid-medusoid transformation in coelenterate animals). Illustration: Life cycle.

Alveolus (pl. alveoli)
Small cavity or pit (e.g., bubblelike cytoplasmic compartments filled with either fluid or gas and often forming a soap bubble-like frothy layer around certain spumellarian radiolaria; cavities in the cortex of ciliates or valves of diatoms; pellicular alveoli enclose the thecal plates in the armored dinomastigotes). Illustrations: Cortex, Glaucocystophyta, Kinetid, Kinetosomes, Oral region.

Amastigote
Aflagellate; akont. Protoctist cell lacking undulipodia, either for the entire life cycle or for stages of the life cycle; nonmotile cell; also, a specific morphological stage in the life cycle of Trypanosomatidae which is rounded, lacks external undulipodia, but has a prominent kinetoplast and a short, internal undulipodium. Illustration: Kinetoplastida.

Ameba (adj. ameboid)
Amoeba. Member of the phylum Rhizopoda; also refers to stages in the life cycles of other organisms that move by means of pseudopods; descriptive term for habit, i.e., movement by pseudopod formation and protoplasmic streaming (e.g., in dictyostelids). Illustrations: Dictyostelida, Rhizopoda.

Amebomastigote
Amebas that undergo transformation to mastigote stage. Informal name of members of the class Amebomastigota. Illustration: Protostelida.

Amicronucleate (adj.)
Lacking micronuclei (e.g., in ciliates).

Amitosis (adj. amitotic)
Cell divisions of eukaryotes that lack chromosome changes typical of mitosis.

Amoeba (pl. amoebae)
See *Ameba.*

Amphibious (adj.)
Referring to the ability to live both on land and in water.

Amphiesma
Outer, peripheral complex of dinomastigotes, consisting of the cell membrane, a single layer of (amphiesmal) vesicles, trichocysts, and sometimes a pellicle. Illustration: Dinomastigota.

Amphiesmal vesicles
Vesicles directly under the amphiesma of dinomastigotes, thought to be responsible for test production. Illustrations: Dinomastigota, Extrusome.

Ampulla (pl. ampullae)
Accessory branch systems, usually congested in appearance (e.g., in rhodophytes, ciliates).

Amylopectin
Storage polysaccharide of algae composed of α-1,4 glucoside linkages, with α-1,6 linked side chains. Can be detected by its red-purple staining when treated with iodide-potassium iodine (Lugol's) solution. Illustration: Merogony.

Anabiosis
Reviving; restoring to active metabolism and growth from a deathlike or suspended condition; resuscitation from dry or frozen state. See *Cryptobiosis.*

Anadromous (adj.)
Referring to organisms that normally live in a marine environment but mate in fresh water (e.g., salmon).

Anaerobe (adj. anaerobic)
Organism active and capable of completing its life cycle in the absence of gaseous O_2.

Analogous (adj.)
Of macromolecules, structures, or behaviors that have evolved convergently; similar in function but different in evolutionary origin.

Anano
See *Aplastidic nanoplankton.*

Anaphase
Stage in mitosis in which chromatids separate (segregate) at their kinetochores and move toward opposite poles. See *Mitosis.*

Anastomosis (v. anastomose)
The process of linking branches, filaments, or tubes by fusion to form networks.

Anchoring disc
See *Polar sac.*

Aneuploid (adj.; n. aneuploidy)
Possessing a number of chromosomes that is not an exact multiple of the typical haploid set for the species; bearing translocations or other chromosome abnormalities. See *Euploid.*

Animal
Multicellular, diploid organism that develops from a blastula, product of fertilization of eggs and sperm, generally heterotrophic.

Anisofilar (adj.)
Referring to a filamentous structure which, in the everted state, is composed of stretches of markedly unequal width along its length (e.g., microsporan polar tube, myxozoan polar filaments). See *Isofilar.*

Anisogametes (adj. anisogametous)
Gametes of a given species that differ in size or form. See *Isogametes.*

Anisogamonts (adj. anisogamontous)
Gamonts of a given species that differ in size or form. See *Isogamonts.*

Anisogamy (adj. anisogamous)
Pairing of gametes that differ in size or form (anisogametes). See *Isogamy.*

Anisokont (adj.)
Heterokont. Refers to cell with undulipodia (or other motile organelles) unequal in length or unlike in movement or form (e.g., usu. one with short mastigonemes and the other smooth, lacking mastigonemes). See *Isokont, Heterodynamic undulipodia.* Illustration next page.

Anisoplanogametes
Motile gametes ("swarmers") of different size.

Anlage (pl. anlagen)
Primordium; the first recognizable part of a developing organ in an embryo or organelle in a cell (e.g., immature ciliate macronucleus or suctorian tentacle).

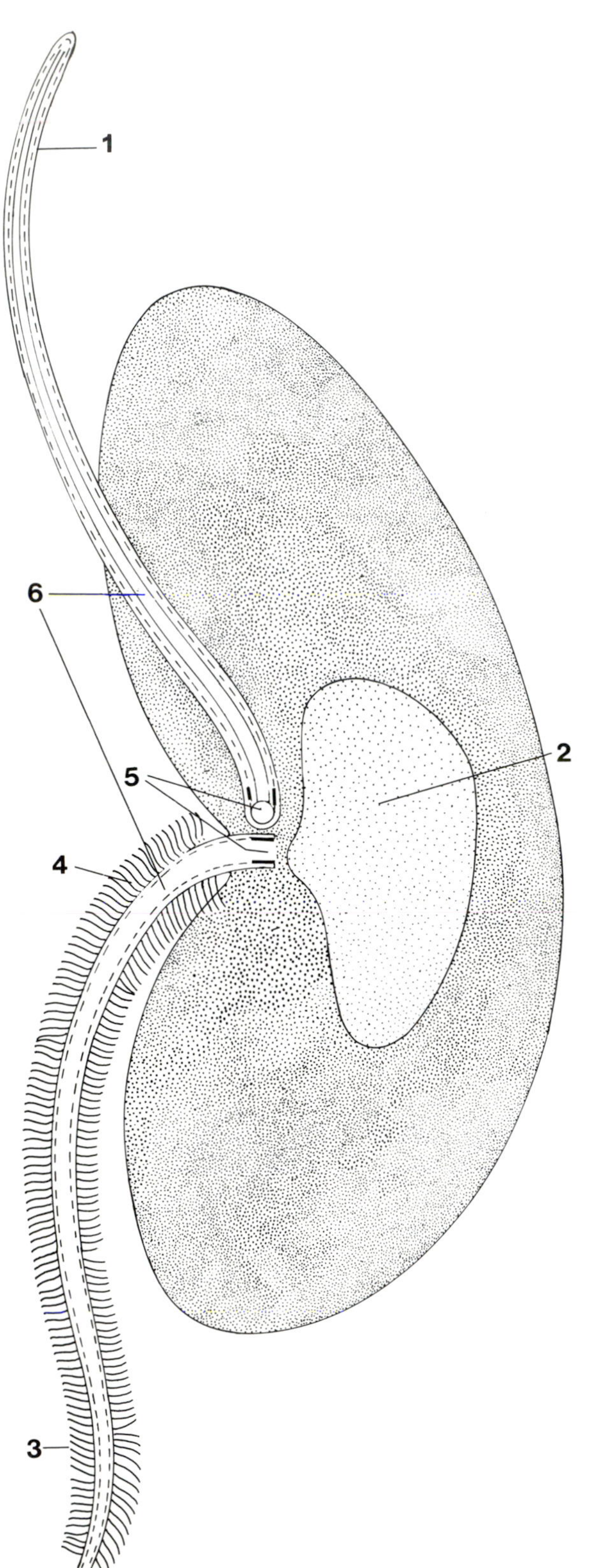

Anisokont
Heterokont structure of a cell. 1=forwardly directed undulipodium; 2=nucleus; 3=trailing undulipodium; 4=mastigonemes (hairs, tinsel, flimmers); 5=kinetosomes; 6=[9(2)+2] axonemes.

Annulus (pl. annuli; adj. annular)
Structure or part resembling a ring (e.g., the central area on the valve face of some centric diatoms); general term for a girdle or an equatorial belt, band, or groove.

Anoxia (adj. anoxic)
Oxygen deficiency; lack of gaseous oxygen.

Anoxic layer
Layer of water or air in which molecular oxygen (O_2) is absent. Illustration: Habitat.

Antapical (adj.)
Opposite to or on the other side of the apex or tip. Illustration: Thecal plate.

Anterior flagellum (pl. anterior flagella)
See *Anterior undulipodium.*

Anterior undulipodium (pl. anterior undulipodia)
Forwardly directed undulipodium.

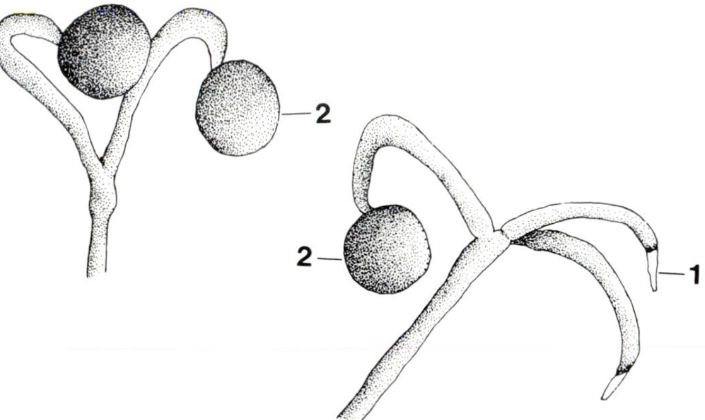

Antheridium
Antheridia (1) and oogonia (2) of the ulvophycean chlorophyte *Dichotomosiphon tuberosus* (Caulerpales).

Antheridium (pl. antheridia)
Male sex organ; sperm-producing gametangium. See also illustrations: Life cycle, Oomycota.

Antherozoid
Male motile gamete; undulipodiated sperm.

Antibiotic
Substance produced and released typically by bacteria or fungi that injures or prevents growth of organisms belonging to a different species. Type of allelochemical.

Antibody
Protein produced by vertebrate blood cells capable of defending animal against a specific foreign substance (antigen).

Antigen
Foreign material that, upon introduction into a vertebrate animal, stimulates antibody production.

Antigenic determinant
See *Epitope.*

Antigenic variation
Change in the surface antigen type expressed. A process that enables parasites to evade the host's immune response (e.g., in trypanosomes).

Antiplectic (adj.)
See *Metachronal waves.*

APC
See *Apical pore complex.*

Aperture
Opening; e.g., the major openings to the exterior through which cytoplasm extends in foraminiferan tests; usu. larger than the pores.

Aphanoplasmodium (pl. aphanoplasmodia)
One of three types of plasmodium formed by myxomycotes; intermediate between phaneroplasmodia and protoplasmodia in size and complexity. Formed by most members of the subclass Stemonitomycetidae. See *Phaneroplasmodium, Protoplasmodium.*

Aphotic zone
That region of the ocean or a body of fresh water in which too little sunlight penetrates for any kind of photosynthesis to occur. Illustration: Habitat.

Aphytal (adj.)
Ecological term for nonphotic, benthic zones in aquatic environments.

Apical (adj.; n. apex)
Pertaining to top, tip. Illustration: Thecal plate.

Apical complex
Structure at the apex of members of the phylum Apicomplexa generally consisting of two apical conoidal rings, a conoid, and a polar ring, to which subpellicular microtubules and electron-dense membrane-bounded organelles composed of rhoptries and micronemes are attached. The name of the phylum Apicomplexa is derived from this structure, which facilitates attachment and penetration of the protoctist to its host cell. Illustrations: Apicomplexa, Conoid.

Apical conoidal ring
Cone-shaped structure at anterior end of apicomplexan cell that is part of the apical complex. Illustration: Conoid.

Apical depression
Depression in the anterior portion of a cell (e.g., the epicone of a dinomastigote). Illustration: Glaucocystophyta.

Apical growth
Growth at the tip or apex.

Apical pore complex
APC. Complex of openings at the apex of a structure or organism (e.g., in dinomastigotes). Illustrations: Epicone, Thecal plate.

Aplanospore
Hemiautospore. Nonundulipodiated spore; a nonsexual, nonmotile propagule.

Aplastidic nanoplankton
Anano. Planktonic protists in the 2-20 µm size range which lack plastids.

Aplerotic (adj.)
Descriptive of oogenesis in oomycote oospores in which the oospore clearly does not fill the oogonial cavity. See *Plerotic.* Illustration: Oospore.

Apoagamy
See *Apogamy.*

Apochlorotic (adj.)
Referring to the lack of photosynthetic pigments in organisms or cells that once contained them or whose ancestors once contained them; usu. refers to algae.

Apogamy (adj. apogamous, apogamic)
Apoagamy. Development of organism without fusion of gametes; development of a diploid phase from a haploid phase without fertilization in organisms with sexual ancestors.

Apomeiosis
Nuclear division without meiosis in a cell that usu. divides by meiosis.

Apomixis (adj. apomictic)
Altered meiosis or fertilization such that mixis is bypassed (e.g., parthenogenesis); condition of being formerly sexual. See *Mixis.*

Apomorphy (adj. apomorphic)
Advanced, specialized, derived taxonomic character (seme) that was absent in the ancestor at the bifurcation of the lineage. See *Plesiomorphy, Symplesiomorphy, Synapomorphy.*

Apophysis
Swollen region.

Aposeme
Seme identified as an altered form of an earlier seme. See *Seme.*

Aposymbiotic (adj.)
Condition of lacking symbionts in formerly symbiotic organisms.

Appressorium (pl. appressoria)
Swollen hyphal tip used to penetrate other organisms (usu. plants) by means of attachment, building of turgor pressure, and growth of a thin hyphal peg into the host organism.

Approsorium (pl. approsoria)
Specialized cell structure that functions in the penetration of a host cell wall and presumably in the uptake of nutrients in the host. In plasmodiophorids, arises at the end of a short germ tube that is formed by an encysted zoospore.

Aragonite (adj. aragonitic)
Mineral, like calcite, composed of calcium carbonate ($CaCO_3$), but differing from calcite in having orthorhombic crystallization, greater density, and less distinct cleavage.

Arbusculate (adj.)
Having the form of a bush or tree (e.g., some protoctists, fungi in mycorrhizal associations).

Archegonium (pl. archegonia; adj. archegoniate)
Female sex organ; the egg- or oogonium-producing gametangium characteristic of some plants and protoctists.

Archeoplasmic spheres
Proteinaceous structures visible as dots by light microscopy and resolvable by electron microscopy as spherical organelles from which spindle microtubules seem to emerge. Associated with the rostrum of hypermastigotes.

Archeopyle
Opening or rupture commonly observed in cysts of dinomastigotes and their microfossils; its position is of taxonomic significance. Illustration: Dinomastigote life history.

Arenaceous test
Test or outer covering composed of sand grains bound together by organic cement (e.g., of testate amebas).

Areola (pl. areolae)
Small area between or about structures (e.g., around a vesicle), esp. a colored ring; the regularly repeated perforation through the siliceous layer of a diatom frustule; connection scar (hilum) at point of contact between spores (e.g., in acrasids). See *Hilum*. Illustration: Velum.

Argentophilic (adj.)
Referring to part of cell or tissue that stains black with silver stains.

Argillaceous (adj.)
Claylike in texture or structure.

Armored plates
Latitudinal series of articulated plates that make up the dinomastigote theca.

Articulate (adj.)
Segmented or jointed in appearance; bearing joined segments.

Asexual reproduction
Increase in number of individuals in the absence of conjugation, fertilization, or any other sexual process.

Assemblage
Group of relatively homogeneous organisms; a group of fossils that occurs at the same stratigraphic level.

Assimilatory hairs
Filaments or rows of cells capable of assimilating nutrient materials from hosts or from the environment (e.g., in the phaeophyte order Cutleriales).

Asters
Stellate, polar, paired structures of animal eggs and other mitotically dividing cells; conspicuous but ephemeral star-shaped microtubule-organizing centers usu. at distal ends of the mitotic spindle. Illustration: Mitosis.

Astral microtubules
Microtubules that arise from a microtubule-organizing center (aster) at the spindle poles.

Astropyle
Main opening of the central capsule of phaeodarian actinopods; usu. accompanied by two or more secondary openings, the parapylae.

Athecate (adj.)
Lacking a theca or covering. See *Theca*.

ATP
Adenosine triphosphate. Molecule that is the primary energy carrier for cell metabolism and motility.

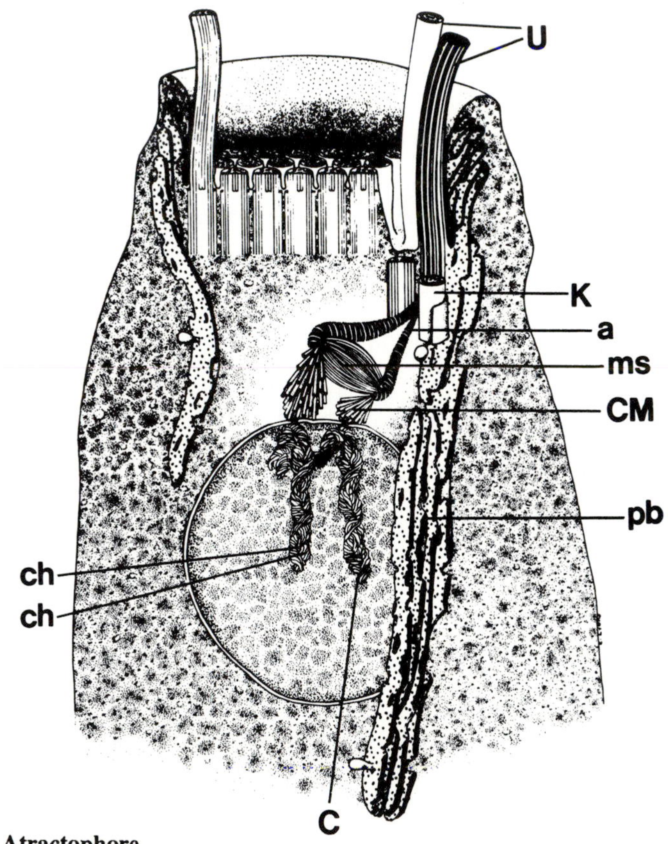

Atractophore
Anterior region of the parabasalian *Lophomonas striata*. a= atractophore; C=chromosome; ch=chromatid; CM=centromeric microtubules; K=kinetosome; ms=mitotic spindle; pb=parabasal body (golgi apparatus); U=undulipodia.

Atractophore
Fibrillar rodlike structure arising from the kinetosome that is or serves the role of centriole or centrocone in the formation of the mitotic spindle (e.g., in some trichomonad mastigotes, foraminiferans, and radiolarian actinopods).

Attenuate (v.)
Become noticeably reduced (e.g., in diameter, narrowing to a point; in size, quantity, strength, force, or severity); esp. of light intensity.

Aufwuchs community
Periphyton. Interacting microorganisms on rocks, plants, and other surfaces on the bottoms of streams and lakes; communities of organisms surrounding submerged vegetation or roots of vegetation in shallow fresh-water environments. See *Microbenthos*. Illustration: Habitat.

Autecology (adj. autecological)
Ecological studies dealing with a single species and its relationship to the biological and physicochemical aspects of its environment.

Autocolonies
Offspring colonies, miniatures of parent colonies, formed by reproduction (multiple fission) of parental colonies (usu. in algae, e.g., the chlorophyte *Pediastrum).*

Autogamy (adj. autogamous)
Pedogamy. Self-fertilization; a type of karyogamy characterized by the union of two nuclei both derived from a single parent nucleus.

Autolysin
Substance that enzymatically degrades glycoprotein-type cell walls.

Autopoiesis (adj. autopoietic)
Self maintenance; set of principles defining life and pertaining to membrane-bounded, self-limited, internally organized systems that dynamically maintain their identity in the face of external fluctuations and limitations. Autopoietic entities have the capability continually to replace and repair their constituent parts ultimately at the expense of solar energy. See *Metabolism.*

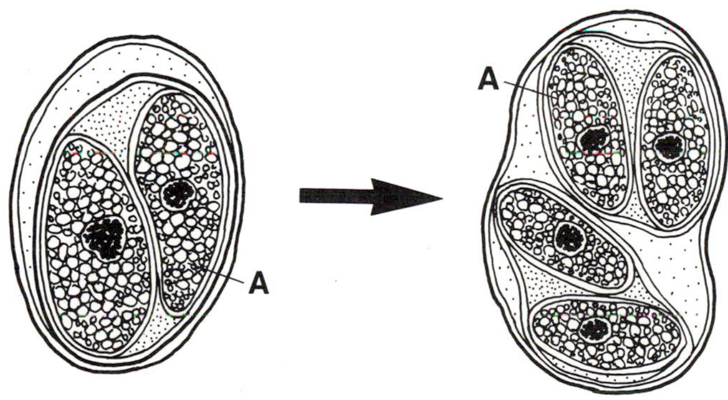

Autospore
The glaucocystid *Glaucocystis nostochinearum*. A=autospores.

Autospore
Offspring cell produced within parental cell wall that resembles parent cell at time of release except that it is smaller (typical of chlorophytes of the genus *Chlorella).*

Autotrophy (n. autotroph; adj. autotrophic)
Mode of nutrition; pertaining to organisms that synthesize organic compounds using an inorganic source of carbon (e.g., carbon dioxide). Strict autotrophs also derive energy and electrons from sources other than organic compounds, i.e., from sunlight or from the oxidation of hydrogen, ammonia, or other inorganic compounds (see Table 1). See *Biotrophy, Heterotrophy, Saprotrophy.*

Auxiliary cell
Cell in rhodophytes to which the diploid zygote is transferred, and from which the sporophyte is generated. Illustration: Florideophycidae.

Auxiliary zoospore
Zoospore within the zoosporangium bearing apically or subapically inserted undulipodia that are retracted upon encystment; characteristic of some members of the Saprolegniaceae (phylum Oomycota). See *Principal zoospore.*

Auxospore
Diatom cell released from its rigid siliceous test; often the zygotic product of fertilization.

Auxotrophic mutant
Microorganism capable of growth only when minimal medium is supplemented with a specific substance (e.g., vitamin, amino acid) not required for growth of wild-type strains.

Axenic (adj.)
Pure, lacking strangers; esp. a culture containing only a single identified strain or species of organism.

Axil
The angle between the upper surface of a leaf (or thallus of an alga) and the stem (or main branch) that bears it.

Axonemal dense substance
Axosome; interaxonemal substance. Electron-opaque material in which microtubules are embedded at the base of the axonemes of undulipodia or in the outer sheet of the centrosphere in actinopods. Illustrations: Interaxonemal substance, Kinetid.

Axoneme
Microtubular axis or shaft, exclusive of the covering membrane, extending the length of an undulipodium (cilium, flagellum, or axopod) composed of the [9(2) + 2] arrangement of microtubules. Each of the nine doublets is comprised of a complete A-tubule and an incomplete B-tubule. See *Microtubular rod.* Illustrations: Actinopoda, Anisokont, Bodonidae, Dinomastigota, Diplomonadida, Interaxonemal substance, Kinetid, Kinetosomes, Mitotic apparatus, Paraxial rod.

Axoplast
Central granule. In actinopods, microtubule-organizing center from which axonemes of axopods arise; devoid of inner differentiation. See *Centroplast.* Illustrations: Actinopod, Actinopoda, Interaxonemal substance.

Axopod
Axopodium. Cell process stiffened by a microtubular shaft or

axoneme; characteristic of actinopods; used primarily in feeding but also in "walking" by the heliozoan *Sticholonche zanclea.* Illustrations: Acantharia, Actinopoda, Centroplast.

Axopodium (pl. axopodia)
See *Axopod.*

Axosome
Electron-opaque fuzzy structure at the base of the central tubules of an undulipodium. See *Axonemal dense substance.* Illustrations: Kinetid, Kinetosomes.

Axostylar cap
Proteinaceous material covering anterior end of axostyle. Illustration: Parabasalia.

Axostyle
Axial motile organelle of metamonads (pyrsonymphids) and parabasalians composed of a patterned array of microtubules and their cross-bridges that runs from the apical end to (and sometimes through) the posterior pole of the organism. Illustrations: Parabasalia, Pyrsonymphida.

Azygospore
Parthenogenetically produced zygospore; characteristic of endomycorrhizal symbionts.

B

B form
In foraminifera with alternation of generations, the asexual organism, the agamont.

B-tubule
Tubule (subfiber) of doublet of axoneme of undulipodium; the outermost or incomplete microtubule of axoneme doublets or the central tubule of centriolar triplets is called the B-tubule. Illustrations: Kinetid, Kinetosomes.

Backing membrane
Part of the endoplasmic reticulum in blastocladialean chytridiomycote zoospores that extends part way around the side body complex.

Bacterium (pl. bacteria)
Microorganism with prokaryotic cell organization. Illustrations: Acanthopodium, Endocytosis, Genophore.

Bacterized medium (pl. bacterized media)
Nutritional fluid or agar containing bacteria (living or dead) as a food source.

Bactivory (n. bactivore; adj. bactivorous)
Mode of nutrition; referring to organisms that feed on bacteria (see Table 2).

Ballistospore
Propagule (spore) that is violently discharged for long distances (up to several meters) from its point of origin (e.g., in protostelid plasmodial slime molds).

Banded root fiber system
See *Kinetodesma.*

Banded roots
Kinetodesmal fibers. Any of several types of rootlet fibers associated with kinetosomes which, ultrastructurally, have a striated appearance. Parts of kinetids. See *Kinetodesma.* Illustrations: Organelle, Undulipodial rootlet.

Basal apparatus
See *Kinetid.*

Basal bodies
See *Kinetosomes.*

Basal disc
Any plate-shaped structure at base of cell process. See *Adhesive disc.* Illustration: Sporangium.

Basal plate
Electron-dense, platelike kinetid component positioned at the proximal end of and perpendicular to the kinetosome.

Basal swelling
Enlargement of volume at base of structure, often applied to peduncles, undulipodia, or other vertically extended structures.

Base plate scale
Base of spined scale on organic portion of a coccolith.

Basipetal (adj.)
Proceeding from the apex toward the base.

Basipetal development
Process in which sporangia are made in basipetal sequence from an undifferentiated hypha terminated by a sporangium (e.g., oomycotes). See *Determinate sporangium, Percurrent development.*

Basiphyte
Plant on which an epibiont or an epiphyte lives.

Bathyal (adj.)
Bathyl. Refers to upper part of an aphotic benthic zone, generally the continental slope, at depths from 1,000 to 3,000 meters, in which algae and plants are excluded because solar radiation cannot penetrate. Illustration: Habitat.

Bathyl (adj.)
See *Bathyal.*

Benthos (adj. benthic, benthonic)
Community of organisms near the bottom or attached to the bottom of an ocean, sea, lake, or other aquatic environment. Illustration: Habitat.

Beta spores
Small, colorless spermatia in sexual species of the rhodophyte *Porphyra.*

Biflagellate
See *Bimastigote.*

Bifurcate (adj.; n. bifurcation)
Having two branches or peaks; forked.

Biliproteins
See *Phycobiliproteins.*

Biloculine (adj.)
Referring to foraminiferan test in which each chamber is added to the previous chamber so that only two final chambers are externally visible.

Bimastigote
Biflagellate. Cell possessing two undulipodia, one of which may be nonemergent; adjective referring to such a cell. See *Nonemergent undulipodium.*

Binary fission
Mode of reproduction; division of parent prokaryotic or eukaryotic cell into two roughly equal-size offspring cells. Illustration: Foraminifera.

Binucleate (adj.)
Containing two nuclei.

Bioassay
Determination of an unknown concentration of a substance, such as a drug, by comparing its effect on a test organism with the effect of a known standardized concentration.

Biogenic (adj.)
Pertaining to a structure (e.g., stromatolite), substance (e.g., amino acid), or pattern (e.g., laminated sediment) produced by organisms.

Bioluminescence
Luminescence. Emission of light by living organisms (e.g., some marine dinomastigotes).

Biomass
Total weight of all organisms at a given time in a particular area, volume, or habitat, generally expressed in such units as grams/meter^2, pounds/acre, or kilograms/hectare.

Biomineralization
Formation of minerals by living organisms. Two kinds are known: biologically controlled or matrix-mediated biomineralization, i.e., intracellular precipitation of a given mineral type under genetic control of the cell (magnetite in magnetotactic bacteria, calcite by coccolithophorids), and biologically induced (e.g., production of acid), which changes local pH that in turn causes potentially mineralizable material to precipitate (e.g., extracellular precipitation of iron and manganese oxides by bacteria; precipitation of amorphous calcium carbonates in lakes due to algal activity). See *Mineral.*

Biostratigraphy
Study of the geological arrangement of sedimentary layers (strata), or the origin, composition, distribution, and succession of strata that contain fossils or remnants of fossils. Biostratigraphy, which especially employs fossil foraminifera and coccolithophorids, is exceedingly important in relative dating, reconstruction of environments of deposition, and thus in petroleum exploration.

Biota
Sum of animals (fauna), plants (flora), and microbiota on Earth. Term microbiota is preferable to microfauna (e.g., in reference to intestinal symbionts, ciliates, motile bacteria) or microflora (e.g., for bacteria).

Biotope
Habitat. Environment surrounding a community of organisms.

Biotrophy (n. biotroph; adj. biotrophic)
Mode of nutrition; pertaining to organisms that derive carbon and energy from living food sources (see Table 2); e.g., many symbionts and pathogens are biotrophs. See *Autotrophy, Saprotrophy, Symbiotrophy.*

Biozone
Biostratigraphic unit; biochron; range zone (e.g., sedimentary rock deposits formed during the life span of a certain fossil form); rocks identified by the occurrence in them of a specific kind of fossil; valuable for establishing intercontinental geological correlations.

Biphasic medium (pl. biphasic media)
Culture medium that has two phases (e.g., agar overlain by a liquid medium). See *Culture medium.*

Bipolar bodies
Diplosomes. Xenosomes or organelles found in the cytoplasm of kinetoplastid mastigotes; in bodonids, they appear as encapsulated Gram-negative bacteria; in trypanosomatids, they appear to be derived from Gram-negative bacteria that have lost their characteristic cell walls.

Biraphid (adj.)
Having a raphe running along the apical axis on both the epivalve and hypovalve; descriptive of diatoms.

Birefringence
Splitting of a light beam into two components which travel at different velocities. The principle of birefringence is employed in differential interference, polarizing, and phase contrast microscopy.

Biserial (adj.)
Biseriate; organized in two rows or series (e.g., foraminiferan test with this organization).

Biseriate (adj.)
See *Biserial.*

Bisporangium (pl. bisporangia)
Sporangium, the contents of which divide to form two spores.

Bispore (adj. bisporal)
Structure or organism making two kinds of spores.

Bisporic generation
Two-spored generation; a generation marked by the production of two types of spores.

Bladder
Saclike or vesicular structure.

Blade
Flat part of algal thallus (e.g., of kelp or other foliaceous algae).

Blastular embryo
Diploid product of fertilization that forms a hollow ball (blastula); defining characteristic of members of the animal kingdom.

Bleached mutant
Altered photosynthetic organism (e.g., *Euglena gracilis)* that has permanently lost its chloroplasts and accompanying plastid deoxyribonucleic acid.

Blepharoplast
Kinetosome or other conspicuous microtubule-organizing center involved in cell division as determined by light microscopic observations of live cells (e.g., in *Stephanopogon,* other mastigotes, cycads, and ferns).

Bloom
Dense growth of a population in aqueous media, in aquaria or nature; characteristic of certain species of planktonic algae, dinomastigotes, ciliates; often detected by discoloration of water; usu. self-limiting and of short duration.

Blue-green algae (sing. blue-green alga)
Cyanobacteria; Cyanophyceae. The terms Cyanophyceae, cyanophytes, and blue-green algae have been replaced by the term cyanobacteria, which recognizes the fundamental bacterial (prokaryotic) nature of these organisms.

Boreal (adj.)
Northern; pertaining to the forest areas and tundras of the northern temperate zone and arctic region. Illustration: Habitat.

Bothrosome
See *Sagenogen.*

Brackish (adj.)
Referring to water with a salinity intermediate between that of seawater (3.4%) and of standard fresh water.

Bradyzoite
Zoite in latent phase; slowly developing merozoite of apicomplexans. See *Merozoite, Zoite.*

Brevetoxin complex
Fish toxin produced by the dinomastigote *Ptychodiscus (Gymnodinium) brevis.* Illustration top of page.

Brine
Seawater that, because of evaporation or freezing, contains dissolved salts in concentrations higher than 3.4%.

Brittleworts
Calcareous charophytes, diatoms; obsolete name.

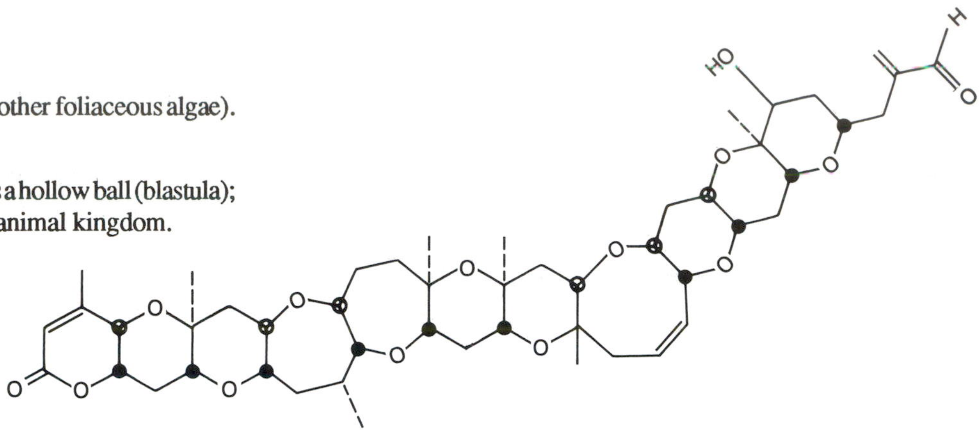

Brevetoxin complex
Brevetoxin B, a polyether alcohol from the dinomastigote *Ptychodiscus brevis.*

Buccal cavity
Ingestion apparatus; mouth; oral apparatus; peristome. Pouch or depression toward the apical end of the cell and/or on the ventral side containing compound ciliary organelles that lead to the cytopharyngeal/cytostomal area (e.g., ciliates). Illustration: Ciliophora.

Budding
Mode of reproduction by outgrowth of a protrusion (one or more buds) smaller than the parental cell or body that only slowly reaches parental size. See *Exogenesis.* Illustration: Foraminifera.

Bulbil
Asexual reproductive organ that forms on the rhizoids of some species of charophytes (phylum Chlorophyta), appearing as a white star or sphere.

Bulla (pl. bullae)

Blisterlike structure or large vesicle. In foraminifera, it may partially or completely cover the primary or secondary aperture(s).

C

C-axis
Longest axis of the hexagonal pattern, perpendicular to the surface of some foraminiferan tests. See *A-axis.*

C-tubule
One of three tubules forming the kinetosome; the incomplete microtubule comprising the outermost kinetosomal (or centriolar) triplet. See *A-tubule, B-tubule.* Illustration: Kinetosomes.

Caducous (adj.)
Becoming detached; falling off prematurely, used originally of floral organs. See *Deciduous*.

Calcareous (adj.)
Containing calcium, usu. in the form of $CaCO_3$.

Calcite
Mineral made of calcium carbonate ($CaCO_3$), crystallized in hexagonal form; the major component of common limestone, chalk, and marble; the material from which foraminiferan tests and coccoliths are composed. See *Aragonite*.

Calyptolith
Callote (skullcap)-shaped coccolith; holococcolith having the form of an open cap or basket (e.g., in *Sphaerocalypta* and *Calyptosphaera*). Illustration: Coccolith.

Canal
Channel-shaped or tubular structure; in euglenids, the tubelike feature connecting the reservoir or anterior invagination to the outside, open only at its anterior end. See *Furrow*. Illustrations: Euglenida, Myxozoa.

Canal raphe
Structure of diatoms; raphe type consisting of cylindrical canal usu. located on a crest or keel. Inner wall possesses row of rounded apertures that connect to the internal protoplasm.

Canaliculate (adj.)
Channeled or grooved longitudinally.

Cancellate (adj.)
Chambered; reticulate.

Caneolith
Elliptical discoid heterococcolith with petal-shaped upper and lower rims and a central area filled with slatlike elements (e.g., *Syracosphaera*). Illustration: Coccolith.

Cap
Reproductive structure in life cycle stage of acetabularians that becomes filled with nuclei. Illustration: Acetabularian life history.

Cap ray
Chamber in acetabularian reproductive structure. Illustration: Acetabularian life history.

Capillitium (pl. capillitia)
Anterior part of a myxomycote sporophore that consists of nonprotoplasmic threadlike structures.

Capitulum (pl. capitula)
Cell in the antheridium of charophytes from which the antheridial filaments arise; amorphous material capping proximal ends of nematodesmata in some hypostome ciliates.

Capsalean (adj.)
Referring to a loose grouping of spherical cells of colonial coccoid cyanobacteria or algae.

Capsular wall
Walls of spherical or nearly spherical structures. In acantharian actinopods, a perforated, fibrillar cover that limits the endoplasm and through which ectoplasm is emitted; in myxozoan spores, wall of the polar capsule consisting of two layers; the inner is electron-lucent and alkaline-hydrolysis-resistant, whereas the outer layer is electron-dense and proteinaceous. Illustrations: Acantharia, Actinopod, Sporoblasts.

Capsule
See *Polar capsule*.

Capsulogenesis
Process of formation of capsulogenic cell that gives rise to multicellular capsule in myxozoans. See *Polar capsule*.

Capsulogenic cell
In myxozoan sporoblasts, the cell that produces the polar capsule in its cytoplasm.

Carbon fixation
Uptake and conversion of carbon dioxide (CO_2) into organic compounds.

Carina (pl. carinas or carinae; adj. carinal)
Keel-shaped structure or process (e.g., foraminiferan test).

Carinal band
Carinate. Foraminiferan shell having a keel or flange at the margin.

Carinate
See *Carinal band*.

Carnivory (n. carnivore; adj. carnivorous)
Mode of nutrition; referring to organisms that are heterotrophic and often predatory (see Table 2) (e.g., ciliates and other protoctists that feed on zoomastigotes or metazoans); generally refers to a holozoic and predatory, rather than parasitic or histophagous, mode of nutrition. See *Osmotrophy, Phagotrophy*.

Capillitium
The myxomycote *Hemitrichia calyculata*.

Carotenoids

Generally yellow, orange, or red isoprenoid (C_{40}) pigments (e.g., carotene, fucoxanthin) found in the plastids (and often the cytoplasm) of virtually all phototrophic organisms as part of their photosynthetic apparatus and in many heterotrophs.

Carpogonium (pl. carpogonia; adj. carpogonial)

Female gametangium in rhodophytes; the flask-shaped egg-bearing portion of the female reproductive branch; a carpospore-containing oogonium, usu. with a trichogyne. Illustrations: Florideophycidae, Rhodophyta.

Carposporangium (pl. carposporangia)

Sporangium derived directly or indirectly from the zygote nucleus produced in the carposporophyte generation in rhodophytes. Can release diploid carpospores (products of mitosis) or haploid carpotetraspores (products of meiosis). Illustration: Florideophycidae.

Carpospore

Alpha spore. Spore of rhodophytes, typically diploid, released from a carposporangium. Illustrations: Florideophycidae, Rhodophyta.

Carposporophyte

Diploid red algal organism produced after fertilization, a phase characterized by the presence of carposporangia (i.e., composed of gonimoblast filaments bearing carpospores in florideophycidean rhodophytes). Illustration: Florideophycidae.

Carpotetraspores

Meiotic products formed in carposporangia; carpotetraspores germinate to give rise to gametophyte thalli in some rhodophytes.

Carrageenan

Sulfated polymer of α-1,3- and β-1,4-linked D-galactopyranose units; type of phycocolloid produced by some rhodophytes and marketed commercially for the production of ice cream and other products.

Cartwheel structure

Portion of kinetosome; refers to the appearance in ultrastructural cross section of the microtubular wheel, radial spokes, axle, and dynein arms. Illustration: Kinetid.

Catenate (adj.)

Referring to cells or other structures arranged end-to-end like beads in a chain.

Caudal appendage

Tail-end structure (e.g., caudal cilium, caudal undulipodium); in ciliates, distinctly longer somatic cilium (occasionally more than one) at or near the posterior or antapical pole, sometimes used in temporary attachment to the substratum. Illustrations: Diplomonadida, Oral region.

Caudate (adj.)

Having a "tail" or a caudal appendage.

Caudo-frontal association

See *Syzygy.*

Caulerpicin

Toxin produced by *Caulerpa.*

cDNA

Complementary DNA; DNA sequence manufactured from a messenger RNA using the viral enzyme reverse transcriptase. Such a copy lacks the introns (intervening sequences) of the natural gene, since the mRNA sequences corresponding to the introns have been removed by splicing following transcription.

Celestite

Usu. white mineral made of strontium sulfate ($SrSO_4$) comprising the spines of some acantharian actinopods.

Cell cycle

Repeating sequence of growth and division of a cell consisting of interphase, G1 (growth phase 1); S (DNA synthesis); G2 (growth phase 2); and M (mitosis); characteristic of plants, animals, fungi, and some protoctists. Extreme variation in cell cycle theme occurs in protoctists.

Cell division

Division of cell to produce two or more offspring cells. See *Cytokinesis, Karyokinesis.*

Cell envelope

See *Plasma membrane.*

Cell junction

Any of a number of connections between cells in multicellular organisms (e.g., desmosomes, septa, plasmodesmata, pit connections, and others). Especially developed in animals (e.g., gap junctions and septate junctions).

Cell membrane

See *Plasma membrane.*

Cell plate

Phragmoplast. Collection of vesicles that forms between telophase nuclei, oriented by microtubules, in the development of a new cell wall; the phragmoplast is characteristic of some taxa of chlorophytes and of all plants. Illustration: Phragmoplast.

Cellular slime molds

Members of the phyla Acrasea or Dictyostelida; heterotrophic protoctists which during the course of their life cycle move from independently feeding and dividing amebas into a slimy mass and eventually transform into a stalked structure that produces cysts capable of germinating into amebas.

Cellulose plates

Surface covering on dinomastigotes. See *Armored plates.* Illustration: Epicone.

Central capsule

Double or single membranous structure that delimits the ectoplasm from the endoplasm in actinopods. Illustration: Actinopoda.

Central granule

See *Axoplast.* Illustration: Centrosphere.

Centric (adj.)
Descriptive of diatoms with radially symmetrical valves.

Centrifugal cleavage
The progressive development of cleavage furrows from the central region of the body toward the periphery; usu. refers to algal thalli. See *Centripetal cleavage*.

Centriolar plaques
Flattened microtubule-organizing centers at the spindle poles to which the spindle microtubules attach, associated with the nuclear membrane; on the ultrastructural level they are observed to reproduce by extension and duplication (e.g., in yeast).

Centriole
Barrel-shaped cell organelle 0.25 µm (diameter) x 4 µm (length). Kinetosome lacking an axoneme; a [9(3) + 0] microtubular structure that forms at each pole of the mitotic spindle during division in most animal cells. Observed to reproduce by developmental cycle (e.g., in which new centriole appears at right angles to parental one). Illustrations: Hyphochytriomycota, Merogony, Mitosis, Mitotic apparatus, Nuclear cap, Organelle.

Centripetal cleavage
The progressive development of cleavage furrows from the peripheral regions of the body toward the center; usu. refers to algal thalli. See *Centrifugal cleavage*.

Centrocone
Division center; cone-shaped extranuclear microtubular bundle, at the apex of which is a [9(1) + 1] centriole (nine singlet microtubules surrounding a single axial tubule); formed during mitosis in apicomplexans and probably arising from a microtubule-organizing center.

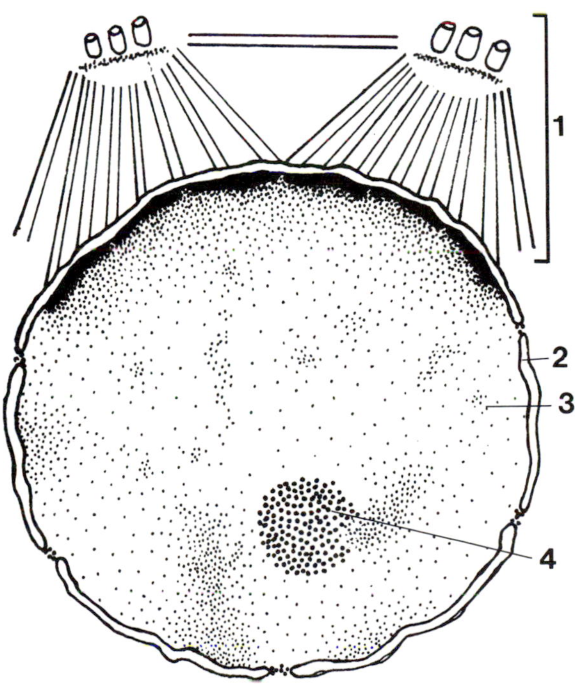

Centrocone
Two centrocones of the apicomplexan *Grebnickiella gracilis*. 1=centrocone; 2=nuclear membrane; 3=nucleus; 4=nucleolus.

Centromere
Structure attaching chromosomes to microtubules of mitotic spindle. Microtubule-capturing center located on chromosomes. Centromeric connections to the spindle are required for chromatid segregation. The centromere, as a region of the chromosome deduced from genetic behavior, is sometimes distinguished from kinetochore as a structure observable in the electron microscope. Some authors consider centromere synonomous with kinetochore. In some parabasalians, centromeres are embedded in the nuclear membrane. See *Kinetochore*. Illustrations: Chromosome, Mitotic apparatus.

Centroplast
Single, central microtubule-organizing center from which axonemes of axopods arise in certain actinopods; it is a tripartite disc consisting of an electron-lucent exclusion zone and interaxonemal substance sandwiched between two caps of electron-dense material (e.g., centroaxoplasthelid heliozoans). See *Axoplast*. See also illustration: Interaxonemal substance.

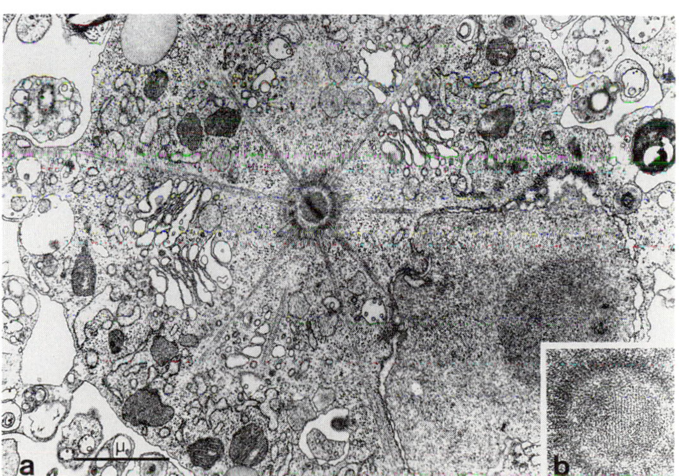

Centroplast
Microtubule-organizing centroplast from which the axopods diverge; eccentric nucleus, and golgi bodies of the heliozoan actinopod *Heterophrys marina*. a. TEM, x 20,000. b. Detail of the centroplast. TEM, x 50,000.

Centrosphere
In actinopods, translucent, spherical area in which a centroplast resides; the centrosphere is divided into two sheets: 1) a clear exclusion zone and 2) an interaxonemal zone containing material (axonemal dense substance or interaxomenal substance) in which the axonemes are rooted. Illustration next page.

CER
See *Chloroplast endoplasmic reticulum*.

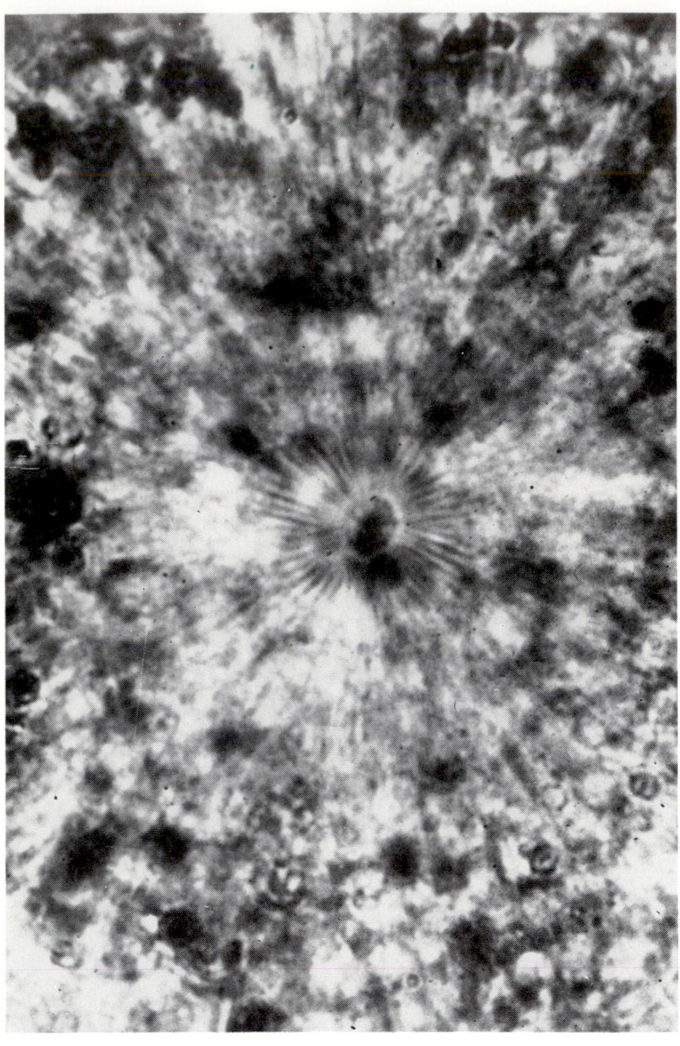

Centrosphere
Microtubule-organizing center of the heliozoan actinopod *Cienkowskya* displaying the central granule and the striated centrosphere. PC.

Chagas' disease
South American human trypanosomiasis; disease found in Central and South America caused by infection with *Trypanosoma cruzi* carried by "kissing bugs" *(Triatomine hemipterano).*

Chalk
Limestone (which is mostly $CaCO_3$) consisting largely of microscopic coccolith blades and spines.

Chamber
Portion or subdivision (e.g., of a test of foraminifera). Illustration: Granuloreticulosa.

Chasmolith (adj. chasmolithic)
Ecological term referring to microorganisms living in rock crevices produced by erosion or by endolithic organisms. See *Endolith, Epilith, Lithophile.* Illustration: Habitat.

Chemosynthate
Any metabolic product of chemoautotrophy (chemosynthesis); total chemosynthate contains sugars, amino acids, and other products of metabolism.

Chemotaxis (adj. chemotactic)
Movement either toward or away from a chemical stimulus (chemotactic agent).

Chemotaxonomy
Grouping into higher taxa of organisms based on their chemical characteristics.

Chert
Siliceous rock (including flint) of microcrystalline quartz; the embedding matrix for many well-preserved microfossils. Material of which radiolarite is composed.

Chiasma (pl. chiasmata)
Region of contact between homologous chromatids when crossing over has occurred during meiosis; these regions resemble the letter *chi* ("X").

Chitin
Hard organic polysaccharide composed of ß-1,4 linked acetylglucosamine units. Chitin is found in cell walls of some rhodophytes, some chlorophytes, some chytridiomycotes, and in threads secreted by diatoms and other protoctists.

Chitinozoa
An extinct group, probably protoctists, that left organic microfossil remains in rocks of Proterozoic and early Paleozoic age.

Chlamydospore
Asexual spherical structure of fungi or funguslike protoctists originating by differentiation of a hyphal segment (or segments) used primarily for perennation, not dissemination (e.g., monoblepharidalean chytridiomycotes).

Chlorophylls
Green lipid-soluble pigments required for photosynthesis; all are composed of closed tetrapyrroles (porphyrins or chlorins) chelated around a central magnesium atom; comprise part of thylakoid membrane in all photosynthetic plastids.

Chloroplast
Green plastid; membrane-bounded cell organelle containing lamellae (thylakoid membranes), chlorophylls *a* and *b*, usu. carotenoids and other pigments, proteins, and nucleic acids in a nucleoid and ribosomes. Illustrations: Chlorophyta, Chlorarachnida, Periplastidial compartment, Prymnesiophyta, Reservoir.

Chloroplast endoplasmic reticulum
CER. Plastid endoplasmic reticulum of some algae; an "extra" layer of ribosome-studded membrane surrounding the plastid. Illustration: Prymnesiophyta.

Chloroplast lamellae (sing. chloroplast lamella)
Thylakoid membranes in chloroplasts, some of which stack to form grana (in some algae and most plants).

Chloroxybacteria (sing. chloroxybacterium)
Prochlorophyta. Chlorophyll *a*-, chlorophyll *b*-containing oxygenic phototrophic prokaryotes that lack phycobiliproteins (e.g., *Prochloron, Prochlorothrix,* and an open-ocean dwelling marine coccoid).

Choanomastigotes
Choanoflagellates; class of marine heterotrophic mastigotes or sessile colonial organisms in the phylum Zoomastigina. Cells enclosed by an organic (theca) or siliceous (lorica) structure with collars of tentacles. Term for a stage in the development of trypanosomatid mastigotes (class Kinetoplastida, phylum Zoomastigina) in which the kinetoplast lies anterior to the nucleus and the associated undulipodium emerges at the anterior extremity through an expanded undulipodial pocket. Illustrations: Choanomastigota, Kinetoplastida.

Chondriome
Mitochondriome. Complete set of mitochondria or mitochondrial genetic complement of a cell.

Chromatic adaptation
Alteration in the relative quantities of photosynthetic pigments in response to changes in light quality and intensity (leading to reduction or increase in light absorption) usu. observed as color changes in algae and cyanobacteria.

Chromatic granules
Colored bodies. See *Hydrogenosomes.*

Chromatid
Half chromosome. Chromatids segregate from each other in late metaphase/early anaphase of mitosis, whereas in meiosis they move jointly to the same pole as entire chromosomes segregate from each other. Illustration: Atractophore.

Chromatin
Eukaryotic DNA complexed with histone (and/or other basic proteins) to form the nucleosome-studded DNA strands that usually "condense" (coil and become deeply stainable) to form chromosomes during mitotic cell division. Illustrations: Ciliophora, Karyoblastea, Microgamete, Prymnesiophyta.

Chromatoid bodies
Ribonucleoprotein structures in *Entamoeba* cysts that form from ribosomes in the trophozoite cytoplasm.

Chromatophore
Pigment-containing structure or organelle; the colored portion of a cell or organism.

Chromophilic (adj.)
Referring to the tendency of a structure or tissue to become colored by taking up stain in a cytological or histological preparation. Chromophilic bodies are cell structures with affinity for stain.

Chromophore
Colored portion of molecule; molecule (purified substance) that is colored.

Chromophyte algae (sing. chromophyte alga)
Algae containing plastids with chlorophylls *a* and *c* (lacking chlorophyll *b*) and certain carotenoids as accessory pigments. Chrysoplast-containing algae (e.g., chrysophytes, diatoms, xanthophytes, and phaeophytes).

Chromosome
Intranuclear organelle made of chromatin (DNA, histone and nonhistone protein) and containing most of the cell's genetic material; usu. visible only during mitotic nuclear division. In dinomastigotes, the chromosomes, which have a peculiar composition, tend to be visible throughout the life cycle of the cells. Illustration next page. See also illustrations: Mitosis, Organelle.

Chrysolaminarin
Leucosin. Storage polymer, colorless and usu. found in membranous vacuoles, composed of ß-1,3 or ß-1,6 linked glucopyranoside units; found in diatoms, chrysophytes, and phaeophytes. Illustrations: Chrysophyta, Eyespot.

Chrysophytes
Informal name of members of the phylum Chrysophyta. Illustrations: Chrysophyta, Mastigoneme.

Chrysoplast
Golden-yellow plastid of chromophyte algae (e.g., diatoms or chrysophytes) that contains chlorophylls *a* and *c*. Illustrations: Chrysophyta, Eyespot.

Chrysoplast endoplasmic reticulum
Membranes studded with ribosomes, surrounding the plastid of chrysophytes (the chrysoplast). Illustration: Eyespot.

Chute
Canal-like, membranous structure associated with the nematocyst/taeniocyst complex of the dinomastigote *Polykrikos.* Illustration: Taeniocyst.

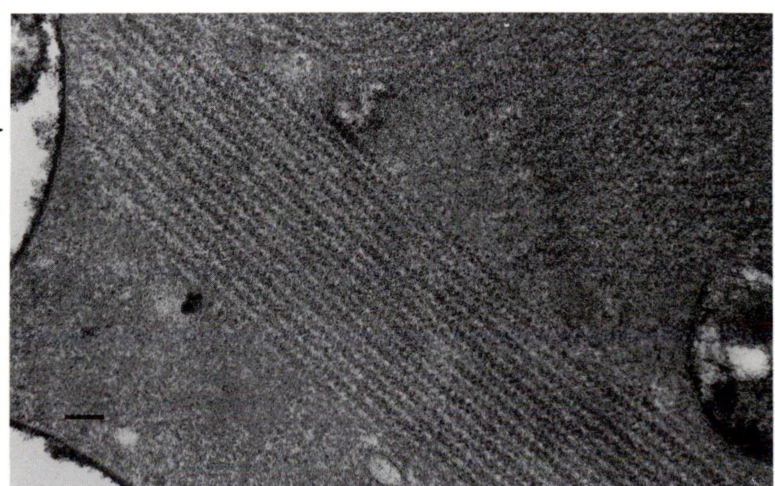

Chromatoid bodies
Developing chromatoid body from parasitic rhizopod *Entamoeba histolytica.* EM. Bar=0.1 μm.

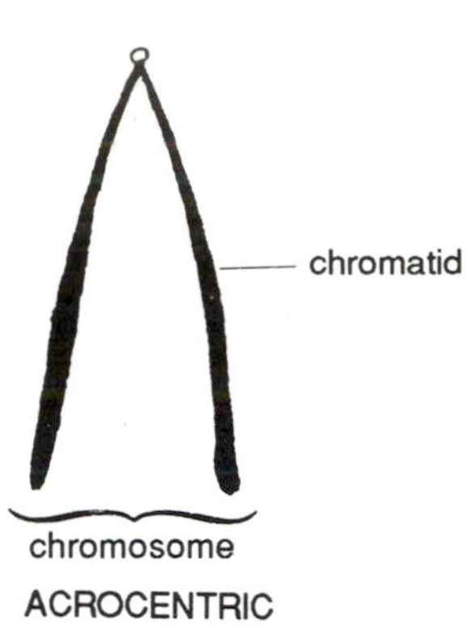

Chromosome
Chromosomal nomenclature.

Chytrid
Common name limited to organisms in the order Chytridiales of the phylum Chytridiomycota (and not the entire phylum, referred to as chytridiomycotes). Illustration: Mastigoneme.

Ciguatera poisoning
Illness resulting from human ingestion of marine fish taken from coral reefs or areas of red tide. The toxin is produced by dinomastigotes comprising red tide (e.g., *Peridinium*, *Gymnodinium*).

Ciliary axoneme
Shaft of cilium. Illustrations: Ciliophora, Undulipodium.

Ciliary necklace
Structure of membrane particles seen with the electron microscope at the base of the axonemal membrane. Arranged in single rings, double rings, and other conformations; these "necklaces" may be of taxonomic significance. See *Undulipodial bracelet*. Illustration: Kinetid.

Ciliature
General term referring to the position or arrangement of undulipodia of ciliates.

Cilium (pl. cilia)
Undulipodium. Organelle of motility that protrudes from the cell, comprised of an axoneme covered by the plasma membrane. The term is used to refer to undulipodia of ciliates and of animal tissue cells. Composed of the [9(2) + 2] microtubular configuration. Illustrations: Ciliophora, Cortex, Mitotic apparatus, Undulipodium.

Cingulum (pl. cingula; adj. cingular)
Girdle region of the dinomastigote cell, the constriction running transversely; the girdle region of the frustule connecting the two distal valves in diatoms. See *Girdle*. Illustrations: Dinomastigota, Epicone, Thecal plate.

Circadian rhythm
The occurrence of a phenomenon in live cells (e.g., cell division, maximum photosynthetic rate, bioluminescence, or enzyme production) with a periodicity of approximately 24 hours.

Cirrus (pl. cirri)
Polykinetid. Tuft-shaped organelle formed from bundles of undulipodial axonemes covered by a common membrane in ciliates; functions primarily in locomotion, but also in feeding.

Cisterna (pl. cisternae)
Flattened membranous vesicle, such as those comprising the golgi apparatus or endoplasmic reticulum. Illustrations: Coated vesicles, Funis.

Cisternal membrane
Membrane surrounding the cisternae of the endoplasmic reticulum or the golgi apparatus.

Clade
Branch on a phylogenetic tree consisting of a taxon (or set of directly related taxa) and its descendants. Also refers to the peripheral bifurcations in ebridian skeletons. Illustration: Ebridians.

Cladistic analysis
Cladistics; a subfield of the biological science of systematics; the formal taxonomic examination of clades or branches on evolutionary trees; a method of arranging taxa by the analysis of primitive and derived characteristics to reflect phylogenetic relationships between extant organisms.

Cladogram →
Phylogenetic tree that is derived from cladistic analysis.

Class
Taxon more inclusive than order and less inclusive than phylum in the systematic hierarchy. For list of protoctist classes, see Tables 4, 5 and 7.

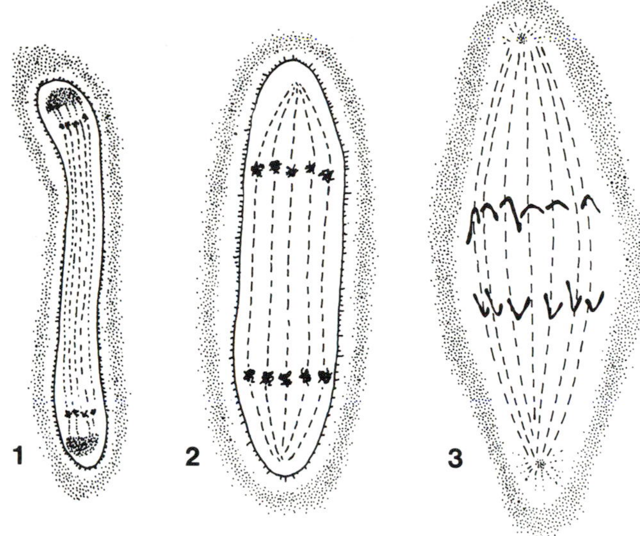

Closed mitosis
Three types of mitosis during anaphase. 1=promitosis (closed); 2=mesomitosis (closed acentric); 3=metamitosis (open).

Clast (adj. clastic)
Small piece of rock; small product of crustal erosion. See *Sand*, *Silt*.

Cline
Front. Gradient; gradation of morphological differences in a species or population of organisms over a geographic area. See *Lysocline*, *Nutricline*, *Thermocline*.

Clonal culture
Culture of genetically identical offspring organisms produced by cell division of a single parent cell.

Clone
Offspring produced from a single parental individual in the absence of sexual processes.

Closed mitosis
Cryptomitosis. Any mitosis (karyokinesis) during which the nuclear envelope is preserved intact throughout the division process. See also illustration: Mitosis.

Cnidocyst
Nematocyst. Complex extrusome produced by dinomastigotes such as *Nematodinium*.

Coated vesicles
Vesicular structures surrounded by a layer of the protein clathrin, arising from endocytosis or by the budding of portions of intracellular membranes (e.g., from golgi apparatus); function in transport of substances into, out of, and between cells. Illustration next page.

Coccalean (adj.)
Spherical in form; usu. refers to algae.

Coccidian life history
Stages in the development of an apicomplexan, member of a large, economically important group of parasites of animals. Illustration next page.

Siphonocladales
Dasycladales
Caulerpales
Ulvales
Ulotrichales
Pyramimonas
7
10
7,9,11
9
3
7
2-8
1
Halosphaera

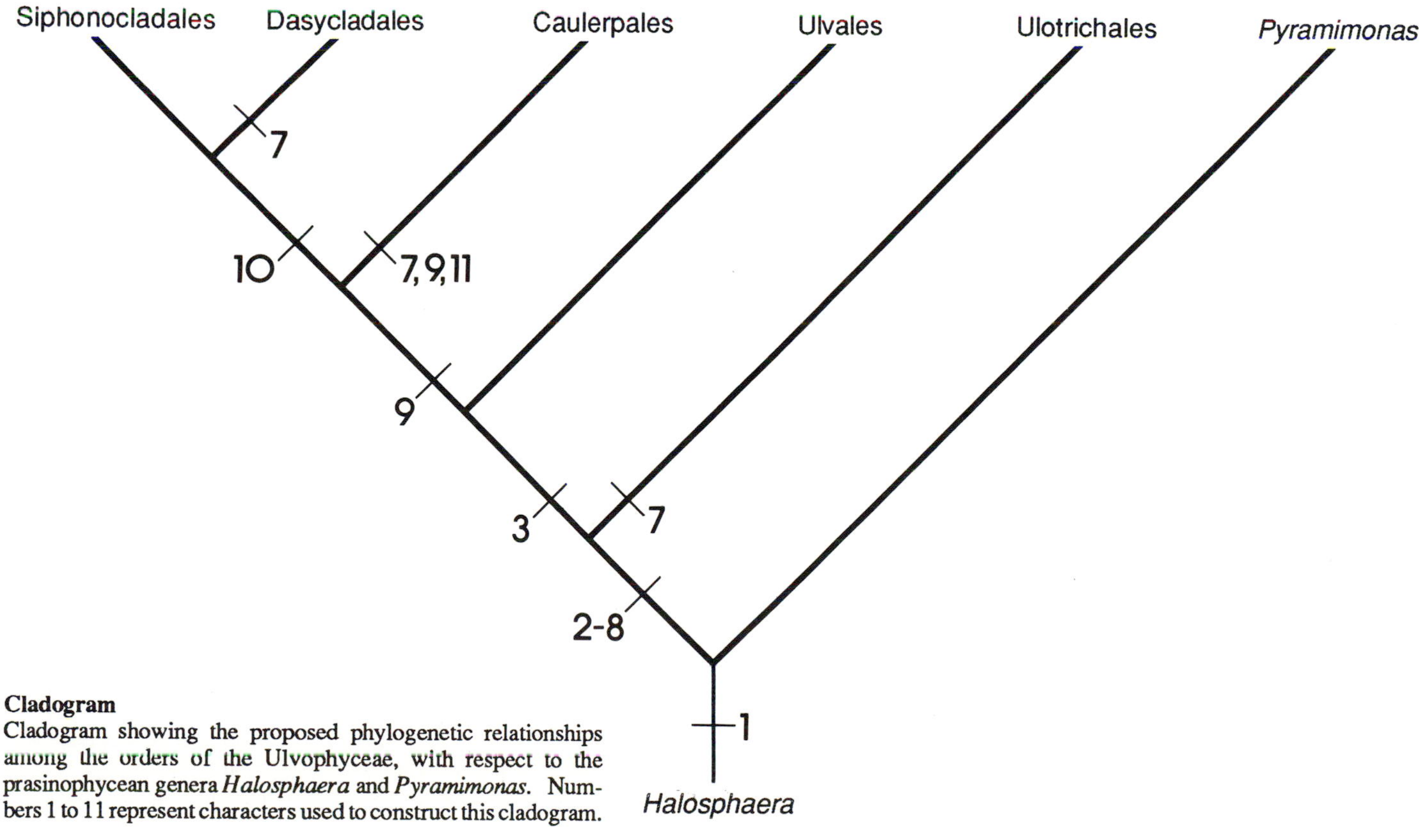

Cladogram
Cladogram showing the proposed phylogenetic relationships among the orders of the Ulvophyceae, with respect to the prasinophycean genera *Halosphaera* and *Pyramimonas*. Numbers 1 to 11 represent characters used to construct this cladogram.

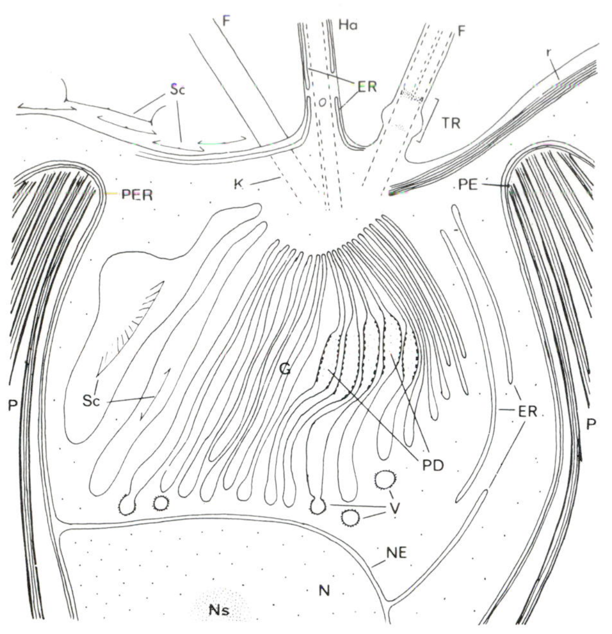

Coated vesicles
Anterior portion of the prymnesiophyte *Chrysochromulina*. ER=endoplasmic reticulum; F=undulipodia; G=golgi body (dictyosome); Ha=haptonema; K=kinetid; N=nucleus; NE=nuclear envelope; Ns=nucleolus; P=plastid; PD= dilations of the golgi cisternae; PE=plastid envelope; PER=plastid endoplasmic reticulum; r=kinetid root; Sc=scales; TR=transition region of the undulipodium; V=coated vesicles.

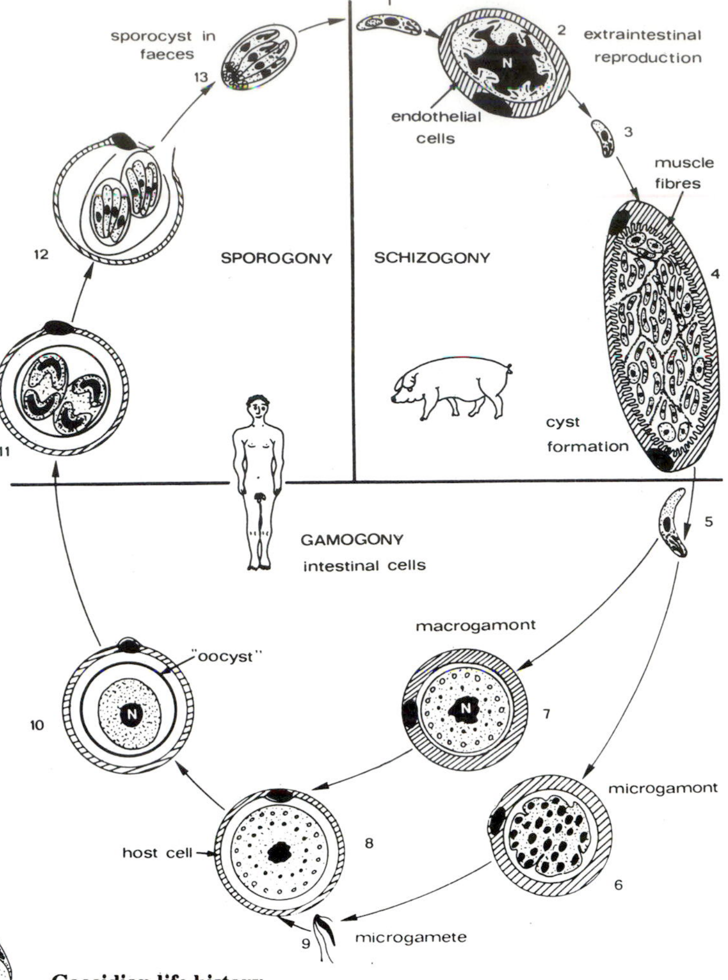

Coccidian life history
The life cycle of the apicomplexan *Sarcocystis suihominis*. 1. sporozoite. 2. meront with a large nucleus. 3. merozoite. 4. intramuscular encysted meront. 5. intracystic merozoite. 6. microgamont. 7. macrogamont. 8. macrogamete. 9. microgamete. 10. zygote. 11. differentiation of two sporocysts in the oocyst. 12. rupture of the host cell and of the oocyst wall releasing the two sporocysts, each containing four sporozoites. 13. released sporocyst.

Coccoid (adj.)
Spherical or approximately spherical in form.

Coccolith
Scale; calcified structure, essentially platelike, but often elaborated, found externally on some prymnesiophyte algae (coccolithophorids); made of $CaCO_3$, usu. deposited as calcite on an organic substructure or matrix; often abundant as fossil remains

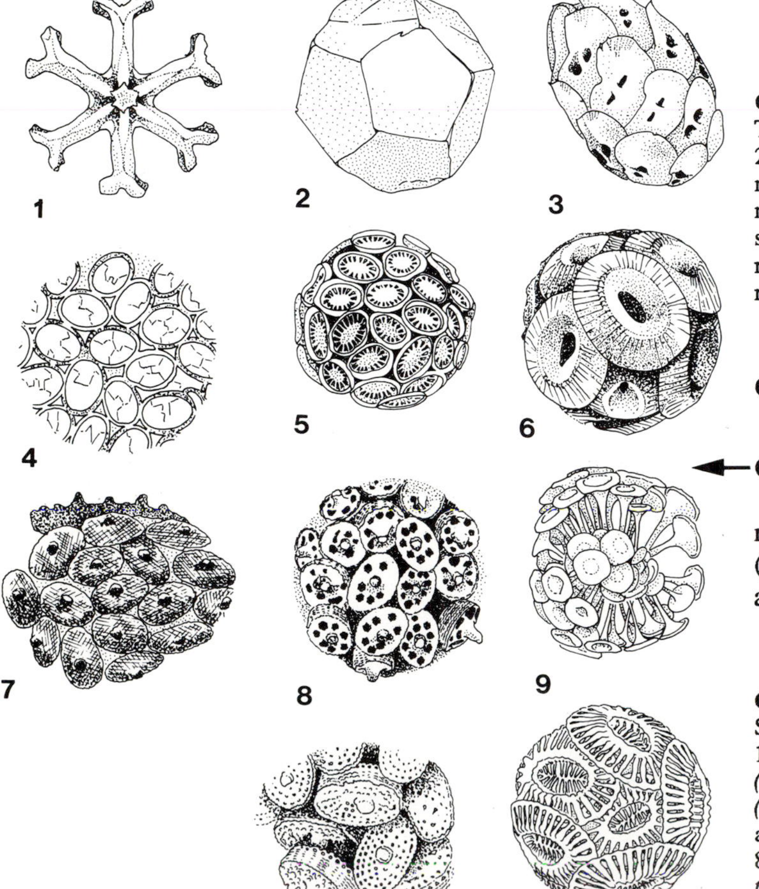

Coccolith
Scale types and representative species of prymnesiophyte algae. 1=discoaster (*Rhabdosphaera* aff. *claviger*); 2=pentalith (*Braarudosphaera bigelowii*); 3=helicoid placolith (lopodolith) (*Helicosphaera carteri*); 4=cricolith; 5=caneolith (*Syracosphaera* aff. *nodosa*); 6=placolith (*Coccolithus pelagicus*); 7=calyptolith; 8=zygolith (*Corisphaera multipora*); 9=rhabdolith (*Discosphaera tubifer*); 10=lapidolith (*Laminolithus hellenicus*); 11=*Emiliana huxleyi*.

of coccolithophorids in chalk. See *Calyptolith, Caneolith, Cricolith, Crystallolith, Discoaster, Helicoid placolith, Heterococcolith, Holococcolith, Lopodolith, Pentalith, Placolith, Rhabdolith,* and *Zygolith.* See also illustration: Prymnesiophyta.

Coccolith vesicle
Modified endoplasmic reticulum in which the coccoliths of coccolithophorids form.

Coccolith vesicle-reticular body system
See *Cv-rb system.*

Coccolithogenesis
Intracellular process of the formation of coccoliths.

Coccolithophorid
Prymnesiophyte alga bearing coccoliths.

Coccolithosome
Granular particle 25 nm in diameter located in the golgi apparatus of coccolithophorids; precursor of the coating that surrounds the coccoliths.

Coccosphere
Total coccolith covering of a coccolithophorid; a cell covering of coccoliths in which the coccoliths hold together to form an intact shell of scales.

Coccus (pl. cocci; adj. coccal)
Spherical structure; spherical bacterium.

Coelopodium (pl. coelopodia)
Structure involved in prey capture in polycystine actinopods consisting of thickened envelopes of cytoplasm; serves to enclose the appendages of such larger prey as copepods.

Coelozoic parasite
Parasite of the coelom or body cavity of metazoans. See *Histozoic parasite.*

Coenobium (pl. coenobia; adj. coenobial, coenobic)
Colony containing a fixed number of cells prior to its release from the parent colony (e.g., *Volvox*).

Coenocyst
Multinucleate thick-walled algal cyst; propagule resistant to desiccation.

Coenocyte (adj. coenocytic)
Plasmodium; syncytium. Multinucleate structure (thallus) lacking septa or walls; thallus with siphonous, syncytial, or plasmodial organization.

Coevolution
The simultaneous development of morphological or physiological features in two or more populations or species that, by their close interaction, exert selective pressures on each other.

Coiled fiber
See *Transitional helix.*

Collar
Inverted cone-shaped structure at cell apex; may be protoplasmic (e.g., in choanomastigotes) or mineralized. Illustrations: Choanomastigota, Stomatocyst.

Colony (adj. colonial)
Group of cells or organisms of the same species, derived from the same parent(s) and living in close association as a unit, each member capable of further reproduction.

Colony inversion
The turning inside out, from undulipodia facing inward, to undulipodia facing outward, of a colony; characteristic of the coenobia of volvocalean chlorophytes.

Columella (pl. columellae)
Structure arising from the stalk of a myxomycote sporangium and extending into the spore.

Commensalism
Ecological term referring to facultative associations between members of different species in which one associate obtains nutrients or other benefits from the other without damaging or benefiting it.

Communities
Interacting populations of organisms of different species, found in the same place at the same time (e.g., termite hindgut protist communities).

Competitive exclusion principle
Gause's Law. The principle that the degree of niche overlap of two species will influence the domination of one species by the other.

Complexity
See *Genomic complexity.*

Compound rootlet
Rootlet made of microtubules that extends laterally from the rhizostyle in cryptomonads. Complex proximal structure of kinetids.

Compound zoospore
See *Synzoospore.*

Concentric fibrils
Small, solid, long, thin structures arranged as one circle inside another. Illustration: Chytridiomycota.

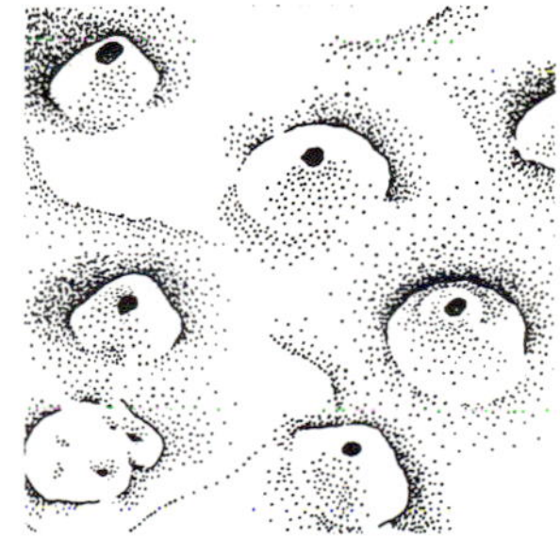

Conceptacle
Conceptacle-containing receptacles on the surface of the rhodophyte alga *Heteroderma* sp. (Corallinales).

Conceptacle
External cavity visible with the naked eye as a receptacle. Contains reproductive cells, usu. on the surface of algal thalli;

found in phaeophytes such as the Fucales and Ascoseirales and rhodophytes such as the Corallinales. Receptacle contains conceptacles.

Conchocelis phase
Microscopic, branched, filamentous, endolithic, sporophytic phase of conchospores in the life history of the rhodophyte *Porphyra* and other Bangiales. Illustration: Rhodophyta.

Conchosporangium (pl. conchosporangia)
Type of enlarged sporangium, usu. produced in series, by the conchocelis phase of some rhodophytes (e.g., Bangiales).

Conchospores
Spores produced during the conchocelis phase in bangiophycidean rhodophytes; spores produced and released singly by a conchosporangium. Illustration: Rhodophyta.

Confluent (adj.)
Growing, running, or flowing together, as in the intermingling of the mucilagenous sheaths of certain algae or the growth of cells on nutrient agar plates.

Congeneric (adj.; n. congener)
Referring to members of the same genus.

Conglomerate
Coarse-grained sedimentary rock, composed of rock fragments larger than 2 millimeters embedded in a fine-grained sand or silt matrix.

Conidiogenesis
Process by which individual conidia form.

Conidiophore
Spore-bearing structure, usu. of fungi; the subtending hypha or stalk to a conidium or group of conidia.

Conidium (pl. conidia)
Exogenously produced spore, usu. deciduous; in oomycotes equivalent to a caducous sporangium.

Conjugation
Copulation; mating. In prokaryotes, cell contact during the transmission of genetic material from donor to recipient; in eukaryotes, the fusion of nonundulipodiated gametes or gamete nuclei or the fusion of structures leading to fusion of gametes or gamete nuclei. See *Lateral conjugation, Scalariform conjugation.* Illustrations: Genophore, Life cycle.

Conjugation tube
Joined outgrowths in conjugating green algae from adjacent cells in which gametes fuse or through which they move prior to fusion.

Connecting fiber
Fibrillar or amorphous structure linking triplets of different kinetosomes with each other; any filament (long solid structure) connecting other entities in cells or between cells.

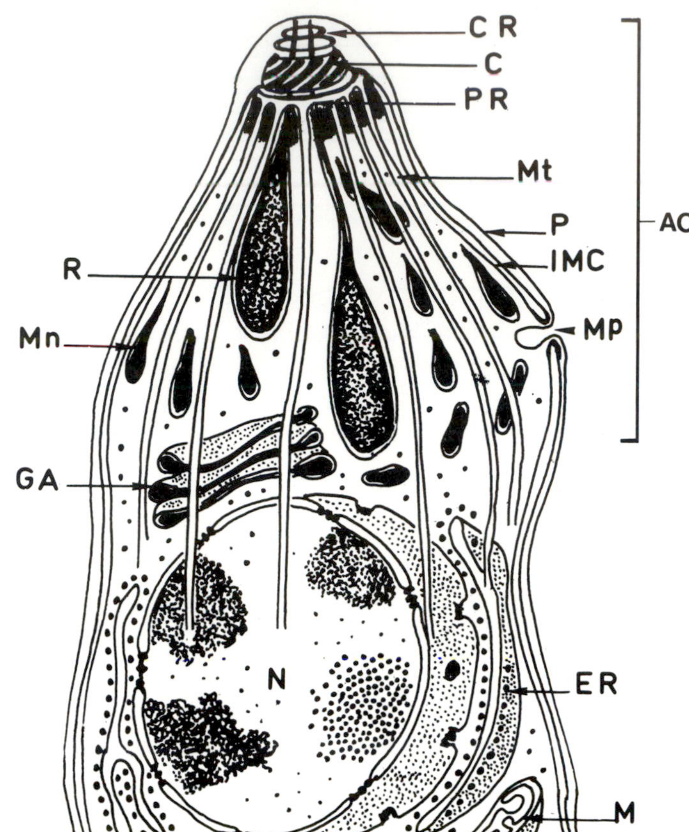

Conoid
The anterior region of an apicomplexan zoite. AC=apical complex; C=conoid; CR=conoidal rings; ER=endoplasmic reticulum; GA=golgi apparatus; IMC=inner membrane complex of the pellicle; M=mitochondrion; Mn=microneme; MP=micropore; Mt=microtubules; N=nucleus; P=plasmalemma; PR=polar rings; R=rhoptry.

Conoid
Apical cone-shaped structure made up of several spirally arranged microtubules; part of the apical complex in apicomplexans. See also illustrations: Apicomplexa, Merogony.

Conspecific (adj.)
Referring to members of the same species.

Contamination
Presence in growth medium of organisms other than those desired.

Continental shelf
That part of the edge or margin of a continent between the shoreline and the continental slope; characterized by a very gentle slope of 0.1°. Illustration: Habitat.

Continental slope
That part of the edge or margin of a continent between the continental shelf and the continental rise (or oceanic trench); characterized by greater angle to the horizontal than the continental shelf. Illustration: Habitat.

Continuous culture
Cultivation of organisms or cells in which the growth rate is

maintained constant through continuous addition of fresh medium and continuous removal of cell or organism-containing, spent medium.

Contophora
A large group crossing taxonomic boundaries encompassing all algae in which the thylakoids are assembled in groups (grana); i.e., all algae except rhodophytes.

Contractile vacuole
Expulsion vesicle. Vacuole in the cortex or ectoplasm of protoctists that functions in osmoregulation of the cytoplasm by alternately dilating and contracting to excrete water from the cell against an osmotic gradient. See also illustrations: Epiplasm, Eyespot, Interphase, Oral region.

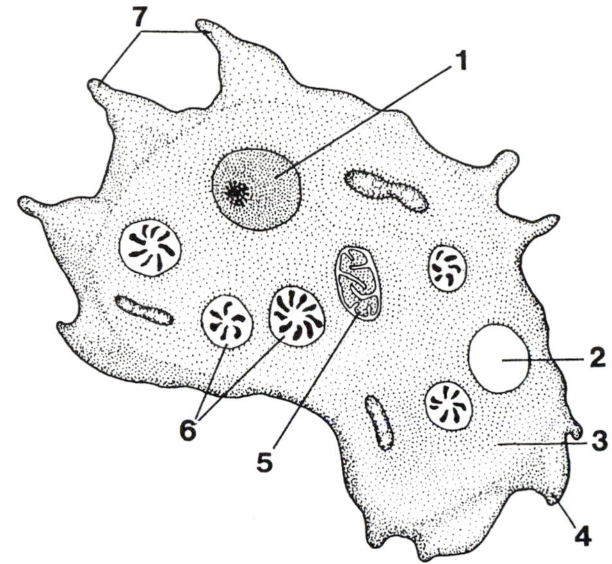

Contractile vacuole
The rhizopod *Amoeba proteus*. 1=nucleus; 2=contractile vacuole; 3=endoplasm; 4=ectoplasm; 5=mitochondrion; 6=food vacuoles with partially digested prey; 7=pseudopodia.

Convergent evolution
See *Parallel evolution*.

Coprolite
Fossil of lithified feces (animal excrement).

Coprophile (adj. coprophilic, coprophilous)
Ecological term referring to organisms that live on or attached to dung or fecal pellets.

Coprozoic (adj.)
Referring to organisms living in feces.

Copula (pl. copulae)
See *Girdle band*.

Copulation (v. copulate)
Mating, the fusion of gamonts or gametes. See *Conjugation*. Illustration: Life cycle.

Core
Core sample; generally refers to a cylindrical section of rock or sediment collected with a coring device.

Corona (pl. coronas or coronae)
Crown or crown-shaped structure.

Coronula (pl. coronulae)
Little crown-shaped structure (e.g., charophytes).

Cortex
Morphological descriptive term referring to the outer layer of a cell, organism, or organ; usu. made of proteinaceous or polysaccharide complexes; in ciliates, highly structured fibrillar outer covering, one to several micrometers thick, in which the undulipodia are embedded; in algae, tissue underlying the epidermis. See also illustrations: Actinopoda, Oral region.

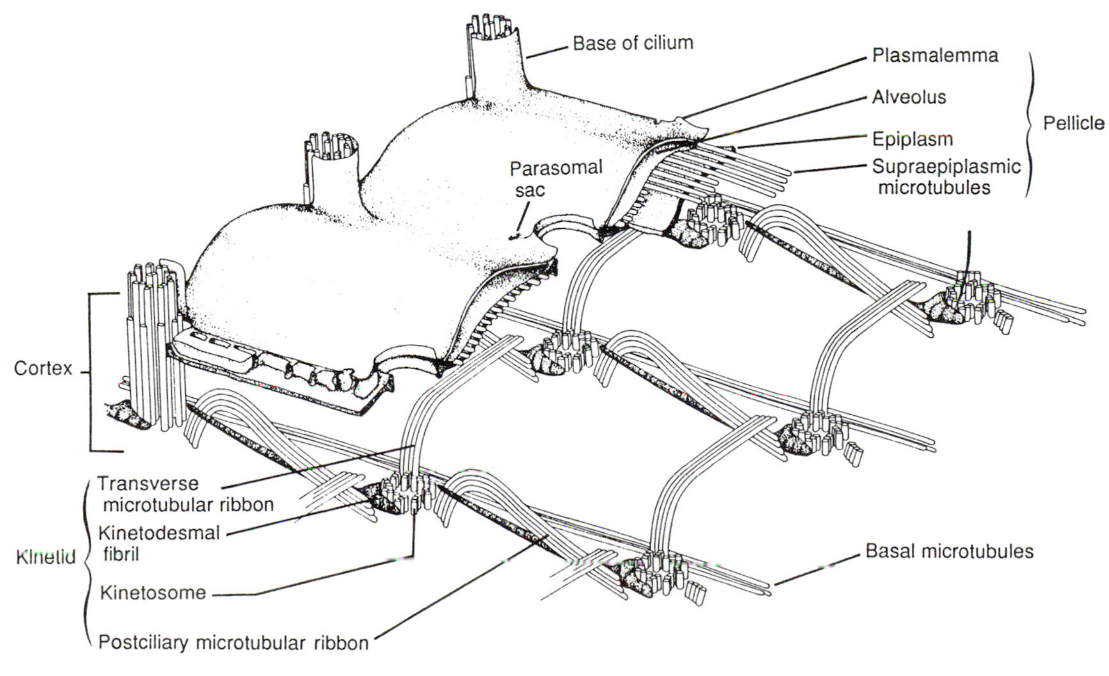

Cortex
Generalized somatic cortex of a ciliate.

Corticating (adj.; n. cortication)
Cortex-forming. Cortication refers to the secondarily formed outer cellular covering of algal thalli (e.g., charophytes, phaeophytes, rhodophytes).

Corticolous (adj.)
Referring to organisms living on the bark of trees.

Cosmopolitan (adj.)
Referring to the growth or occurrence of organisms in all or most parts of the world; widely distributed.

Costa (pl. costae; adj. costate)
Highly motile nonmicrotubular intracellular rod in zoomastigotes (e.g., parabasalians); elongated solid thickening (fibula) of the valve in a diatom frustule; attachment band, connected at both ends to coiled filaments that confer elasticity to the cortex in acantharian actinopods. Rib or ridge (e.g., foraminifera). See *Costal strip, Subraphe costae*. Illustrations: Choanomastigota, Pedicel.

Costal strips
Siliceous strips which join to form costae, which in turn make up a basketlike lorica in some choanomastigotes.

Crampon
Branched stalk base in dictyostelids.

Craticulum (pl. craticula)
Irregular siliceous plate forming an internal shell in certain pennate diatom frustules.

Crenulate (adj.)
Wavy, ruffled; describing a surface with notches or small waves.

Cresta (pl. crestae)
Fibrillar, noncontractile structure, found below the basal portion of the trailing undulipodium in devescovinid mastigotes (Parabasalia).

Cribrate (adj.)
Sievelike, profusely perforated; having a cribrum (e.g., a closing plate (velum) of the pores (areolae) of a diatom wall with regularly arranged perforations in the silica); aperture composed of many rounded holes grouped together over a defined area.

Cricolith
Elliptical heterococcolith with the elements arranged peripherally on a base-plate scale. Coccolith with $CaCO_3$ elements stacked to form a simple tube (e.g., *Hymenomas carterae*). Illustration: Coccolith.

Cristae (sing. crista)
Inwardly-directed tubular or pouchlike folds of the inner membrane of a mitochondrion; the site of ATP production during aerobic metabolism; rich in respiratory enzymes, cristae may be vesicular (discoid), tubular, or vermiform (platelike or flattened). See *Tubular cristae, Vermiform cristae, Vesicular cristae.*

Cross-banded roots
Basal part of kinetid structure; undulipodial rootlets with a striated appearance.

Crown cells
Cells that make up the coronula (corona) (e.g., in charophytes).

Cruciate (adj.)
Cross-shaped; as in the microtubules of the kinetid structure of some chlorophytes or the contents of a tetrasporangium that are oriented at right angles to each other.

Cruciform mitosis
Cross-shaped appearance of the nucleus in metaphase; characterized by an elongated nucleolus arranged perpendicularly to the chromosomes at the equatorial plate; cruciform nuclear division (e.g., in plasmodiophorids). Illustration: Plasmodiophoromycota.

Crude culture
See *Agnotobiotic culture.*

Crustose (adj.)
Crustlike.

Crustose thallus
Growing hyphae or trichomes, usu. of alga or lichen body, that together form a crust.

Cryophile (adj. cryophilic)
See *Psychrophile.*

Cryoplankton
Plankton of polar or other cold regions.

Cryopreservation
Viable preservation of organisms, tissues, or cells by suspension in appropriate solutions and storage at extremely cold temperatures.

Cryptobiosis
Suspended or deathlike condition generally brought on by starvation, desiccation, or freezing, reversible by anabiosis.

Cryptobiotic cysts
Cysts capable of resuscitation; "suspended life" in which respi-

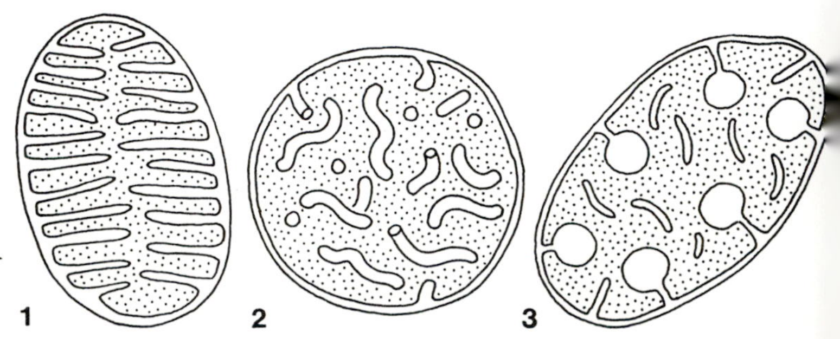

Cristae
Cross sections of mitochondria. 1=vermiform (platelike or flattened); 2=tubular; 3=vesicular (discoid).

ration and other metabolic activities are scarcely discernible but reversible by anabiosis. See *Anabiosis, Cryptobiosis.*

Cryptomitosis
Closed mitosis. Mitosis in which the nuclear membrane remains intact.

Cryptomonad
Informal name of cryptophytes, or members of the phylum Cryptophyta. Illustrations: Cryptophyta, Mastigoneme.

Cryptopleuromitosis
Mitosis in which a bilaterally symmetric mitotic spindle is located entirely outside the nucleus and the nuclear membrane remains intact. Characteristic of some parabasalians, diatoms, etc.

Cryptostomata (sing. cryptostoma)
Small cavities on the surface of the thallus containing rows of sterile hairs in phaeophytes (e.g., *Adenocystis, Scytothamnus, Splachnidium).*

Crystallolith
Coccolith type made of disc-shaped rhombohedrons (e.g., *Crystallolithus);* holococcolith with the crystals deposited on the distal surface of an organic scale.

Culture
Laboratory-maintained population of organisms that survives on culture medium and that is transferred by inoculation.

Culture medium (pl. culture media)
Medium. Liquid or solid material providing nutrients for the survival in laboratory culture of protoctists or other organisms.

Cumatophyte
Alga, usu. brown or red, living exposed to surf (e.g., the phaeophyte *Postelsia).*

Cuneate (adj.)
Narrowly triangular with the acute angle toward the base; wedge-shaped (cuneiform).

Curved vane assembly
Cytoskeletal support element for the ingestion apparatus of phagotrophic euglenids; four long equidistantly spaced sheets, "j"-shaped in cross section, radiating out from four microtubules immediately adjacent to the cytopharynx.

Cuticle
Waxy or fatty layer on the outer wall of epidermal cells. In protists sometimes synonymous with cortex.

Cv-rb system
Coccolith vesicle-reticular body system. Membrane system associated with coccolith formation; includes golgi and vacuoles. Illustration: Prymnesiophyta.

Cyanelles
Endocyanome. Intracellular structures considered by some to be cyanobacterial symbionts and by others to be organelles derived from symbiotic cyanobacteria, active in oxygenic photosynthesis (e.g., in glaucocystophytes). Cyanelles are distinguished from rhodoplasts by having at least remnants of cell wall material. Illustrations: Glaucocystophyta, Thylakoid.

Cyanobacteria (sing. cyanobacterium)
Chlorophyll *a*-, phycobiliprotein-containing, oxygenic photosynthetic bacteria; formerly called blue-green algae; phototrophic prokaryotes that use water (some may use sulfide) as an electron donor in the reduction of CO_2, produce oxygen in the light, have paired thylakoids, and are unicellular, or form filaments or thalli. Some filamentous cyanobacteria differentiate specialized cells (heterocysts) for nitrogen fixation; some have gliding motility. The most widespread phylum of phototrophic aerobic prokaryotes, cyanobacteria, initiated the rise of gaseous oxygen in Earth's atmosphere some two billion years ago.

Cyanophyceae
Class in the botanical division Cyanophyta of the Plant kingdom; obsolete term for cyanobacteria.

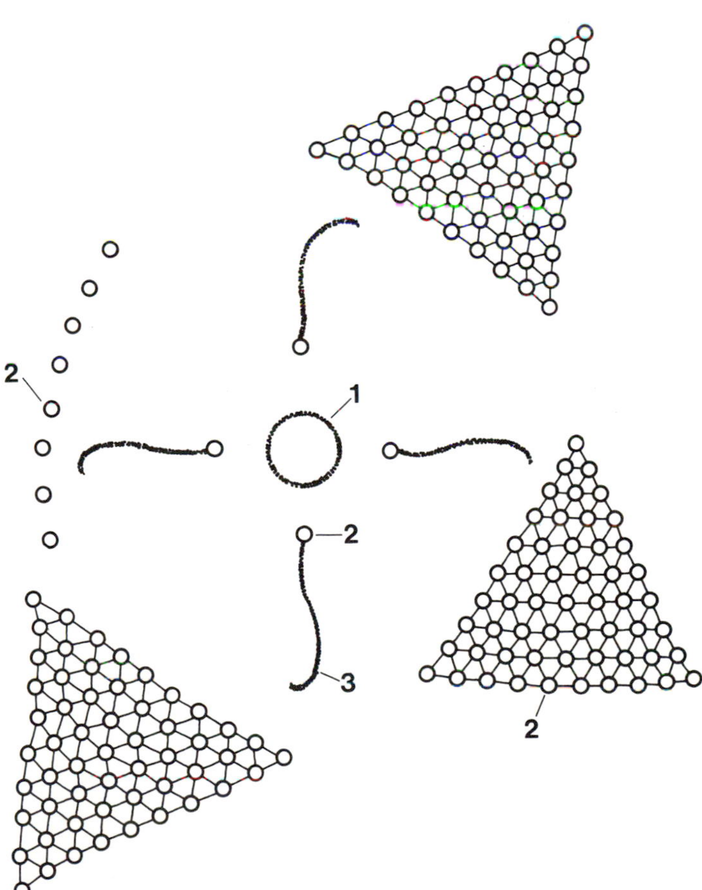

Curved vane assembly
Section through the cytopharyngeal region of a euglenid. 1=cytopharynx; 2=microtubules; 3=curved vane.

Cyclical transmission
Cycle of development of a heteroxenous parasite in which the parasite undergoes a cycle of development in one host before it infects and develops in the alternate host (e.g., *Trypanosoma brucei* undergoes cycle of development in tsetse flies before infecting mammalian host).

Cyclosis
Cytoplasmic streaming; protoplasmic streaming. Circulation of cell cytoplasm, characteristic of eukaryotes; internal cell motility based on nonmuscle actinomyosin fibrous protein complexes.

Cymose renewal
Lateral renewal. Hyphae or sporangial hyphae produced in an arrangement in which each main axis is terminated by a single sporangium; secondary and tertiary axes may also end in sporangia (e.g., oomycotes). See *Sequential zoosporangium formation.*

Cyrtos
Microtubular apparatus surrounding the cytopharynx (e.g., hypostome ciliates).

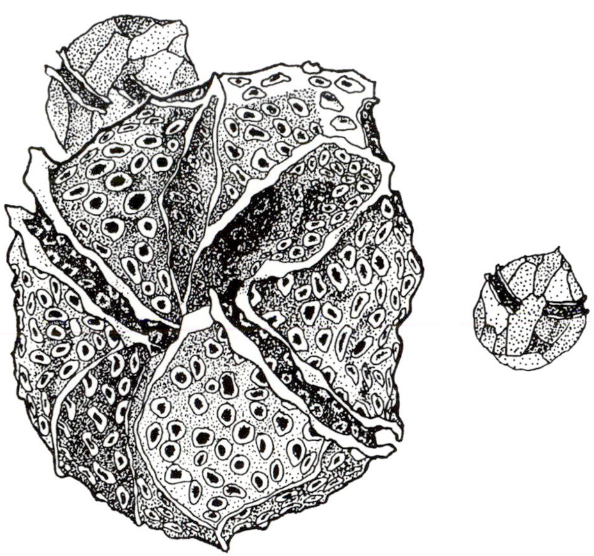

Cyst
Hypnocyst (resting cyst) of a dinomastigote.

Cyst
Kind of propagule; morphological manifestation of "resting state" in protoctist life cycles; formation of structures may or may not be associated with sexual phenomena. Resistant, sporelike, frequently thick-walled structure independently evolved in many protoctists. Nonmotile, dehydrated, usually resistant to environmental change and inactive. In the life cycle of many protoctists the cyst is generally considered to serve an important role in either protection or dispersal. Cysts are often formed in response to environmental conditions, esp. starvation and desiccation. The organism typically rounds up and becomes surrounded by one or more layers of secreted cystic envelopes or walls, which may be sculptured on the outside and with or without an emergence pore. See *Cryptobiosis, Ectocyst, Endocyst, Gametocyst, Gamontocyst, Macrocyst, Mesocyst, Microcyst, Multiplicative cyst, Oocyst, Pansporoblast, Propagule, Resistant cyst, Sclerotium, Sorocyst, Sporocyst, Stomatocyst, Temporary cyst,* and *Trophocyst.* See also illustrations: Acetabularian life history, Coccidian life history, Cystosorus, Plasmodial Slime Molds, Polykinetoplastic, Residual body, Stachel.

Cystocarp
Carposporophyte and surrounding tissue or cells provided by the gametophyte in rhodophytes; reproductive structure on the spore-forming female gametophyte.

Cystogenesis
Process by which cysts are formed.

Cystogenous plasmodium (pl. cystogenous plasmodia)
Plasmodium that forms cysts (e.g., in plasmodiophorids). Illustration: Synaptonemal complex.

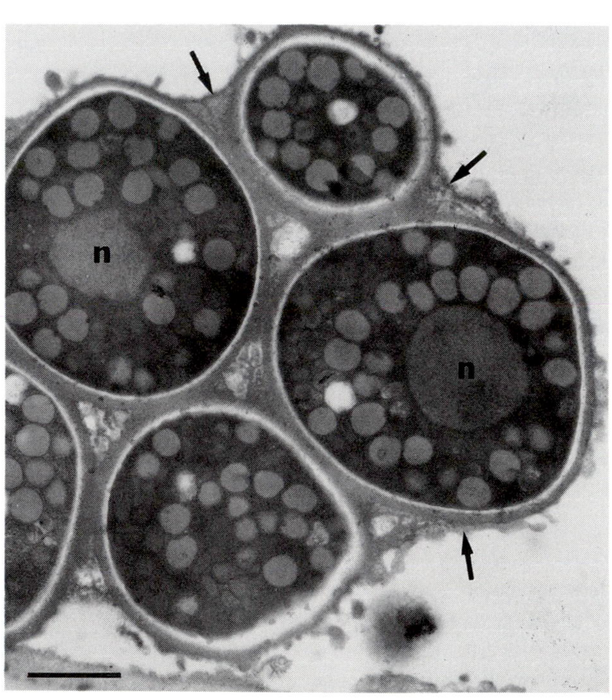

Cystosorus
Portion of mature cystosorus of the plasmodiophorid *Ligniera verrucosa* in the root hair cells of *Veronica persica.* The walls of contiguous cysts are fused and finely verrucose (arrows). n=nuclei. Bar=1.0 µm.

Cystosorus (pl. cystosori)
Structures into which cysts may be united, the presence and morphology of which are of taxonomic significance (e.g., in plasmodiophorids). See also illustration: Plasmodiophoromycota.

Cytobionts
Cellular symbionts. See *Endocytobiont, Xenosomes.*

Cytochromes
Low-molecular-weight proteins conjugated to iron-chelated tetrapyrroles (e.g., iron porphyrins), chromophores often yellow in color; cytochromes act as electron carriers in respiration and photosynthesis.

Cytokinesis
Cytoplasmic division, exclusive of nuclear division (karyokinesis); also used as synonym of cell division. Illustrations: Foraminifera, Phragmoplast.

Cytolysis
Rupturing of cells (e.g., toxicysts induce cytolysis).

Cytopharynx
Cell "throat"; region through which particulate food travels after passing through cytostome (e.g., in ciliates). Illustrations: Bodonidae, Ciliophora, Endocytosis, Retortamonadida.

Cytoplasm
Fluid portion of cell containing enzymes and metabolites in solution. Illustrations: Raphidophyta, Stachel, Thylakoid.

Cytoplasmic inheritance
Non-Mendelian (non-nuclear, nonchromosomal) inheritance of distinctive genetic traits. Often associated with the inheritance of plastids or mitochondria, or correlated with the presence of viral, bacterial, or other endocytobionts.

Cytoplasmic membrane
See *Plasma membrane.*

Cytoplasmic streaming
See *Cyclosis.*

Cytoproct
Cytopyge. Cell "anus"; anal pore; generally permanent (when present) slitlike opening (though actually usu. closed) near the posterior end of the cell, through which egesta may be discharged. In some ciliate species, located in or just to the left of the posterior portion of kinety number one, the cytoproct is a portion of the cortex with taxonomic significance. Its edges, resembling a kind of pellicular ridge and reinforced with microtubules, are argentophilic. Illustration: Oral region.

Cytopyge
See *Cytoproct.*

Cytoskeleton
Asymmetric scaffolding, often associated with cell motility inside eukaryotic cells. Microfilaments and microtubules and their associated proteins provide a dynamic framework which influences the shape of protoctists. Secreted organic or inorganic materials in, on, or below the surface of a protoctist may also contribute to the cytoskeleton.

Cytosome
Ingestive apparatus of euglenids.

Cytostomal groove
Depression or opening of cell through which food particles pass. Illustration: Funis.

Cytostome
Cell "mouth." A two-dimensional, usu. permanently open aperture (e.g., in *Noctiluca).* In ciliates, the cytostome may open directly to the exterior or be sunken in a cavity such as an atrium, vestibulum, buccal or peristomial cavity; the end of the ribbed wall in the ciliate cortex, i.e., the level in the ciliate cortex at which pellicular alveolar sacs are no longer present. Illustrations: Bodonidae, Ciliophora, Oral region, Pyrsonymphida.

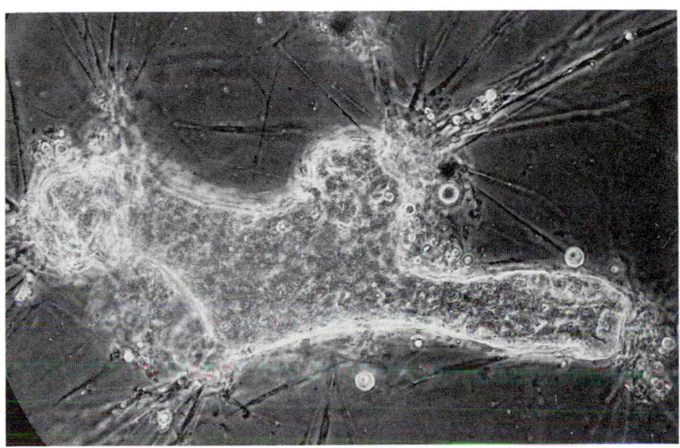

Cytotomy
The granuloreticulosan *Allogromia laticollaris* undergoing cytotomy. PC.

Cytotomy
"Cell cutting," multiple fission; cytokinesis delayed with respect to karyokinesis resulting in the formation of several offspring simultaneously. A subcategory of plasmotomy. In some monothalamic, multinucleate foraminifera with organic tests, the whole cell divides unequally by binary fission to form multiple buds. See also illustration: Foraminifera.

D

Dactylopodium (pl. dactylopodia)
Digitiform (finger-shaped) determinate pseudopods, typical of some *Mayorella* spp. (phylum Rhizopoda).

DAP pathway
See *Diaminopimelic acid pathway.*

DBV
See *Dense body vesicle.*

Deciduous (adj.)
Becoming detached when fully developed. See *Caducous.*

Decomposer
Osmotrophic organism that converts polymeric organic material into monomers by secretion of extracellular digestive enzymes.

Defined medium (pl. defined media)
Culture medium in which the precise chemical nature of the ingredients and their starting concentrations has been identified.

Definitive host
Host in which a symbiont attains sexual maturity (e.g., the coccidian *Aggregata eberthi* in cuttlefish). See *Intermediate host*.

Dehiscence
Opening of a structure by drying or programmed death of certain structures or cells (e.g., to allow the escape of reproductive bodies contained within).

Dendrogram
Branching graphic representation of taxonomic arrangement; "family tree" based on numerical relationships (i.e., derived from quantification of the similarities and differences among organisms).

Dendroid (adj.)
Dendritic. Shaped like a tree; treelike.

Dense body vesicle
DBV. Membrane-bounded vesicle, associated with phosphoglucan metabolism, found in heterokont protoctists (e.g., oomycotes). Its appearance in thin section using the transmission electron microscope changes with metabolic activity. Sometimes the DBV is electron-lucent with one or more central or eccentrically placed electron-opaque zone(s); sometimes it has close-packed lamellar formations between electron-opaque and electron-lucent zones. At oospore formation DBVs coalesce to form a single, large, membrane-bounded inclusion known as the ooplast.

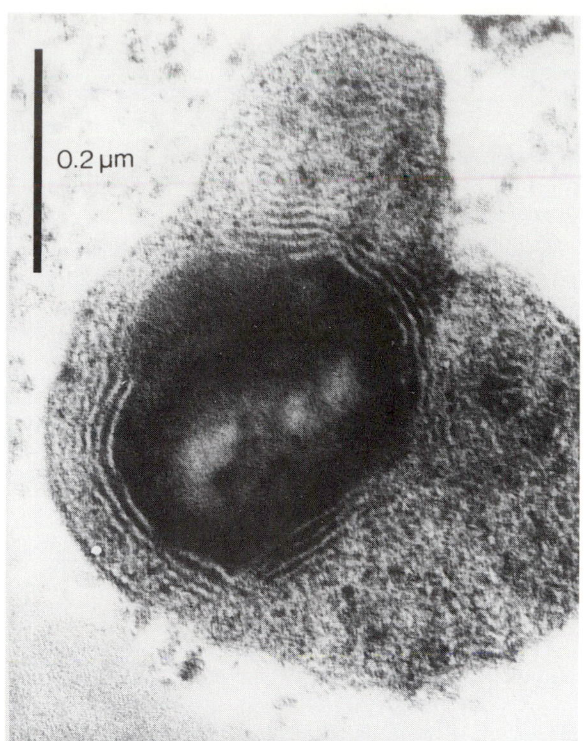

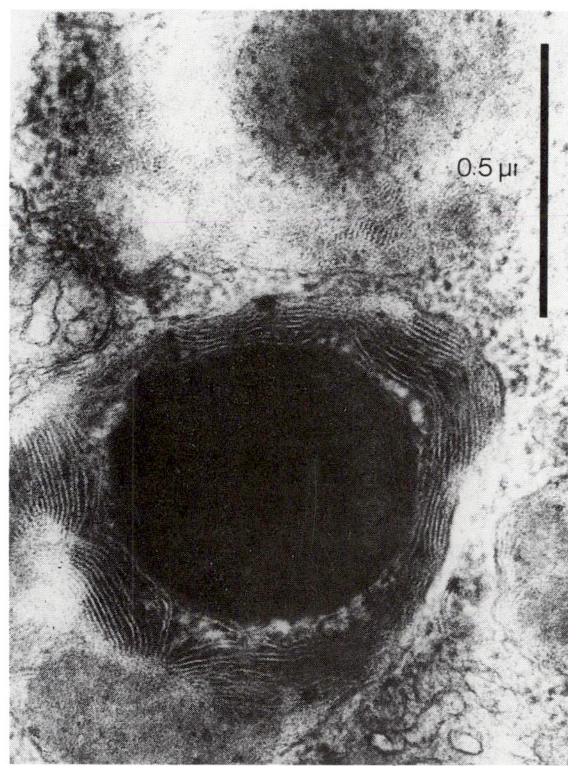

Dense body vesicle
Dense body vesicles in the oogonium of the oomycote *Achlya radiosa*.

Desert
An area of low moisture due to low rainfall (i.e., fewer than 25 cm annually), high evaporation, or extreme cold, and which supports only specialized vegetation; wind often produces distinctive erosional features (e.g., dunes). Illustration: Habitat.

Desmid
Unicellular or filamentous conjugating green alga of the families Mesotaeniaceae or Desmidiaceae in which amastigote ameboid gametes conjugate.

Desmodexy, Law of
The invariant position of the kinetodesma to the right (not the left) of its kinety in ciliates.

Desmokont
Member of a subgroup (Desmophyceae) of the dinomastigotes characterized by two apically inserted undulipodia.

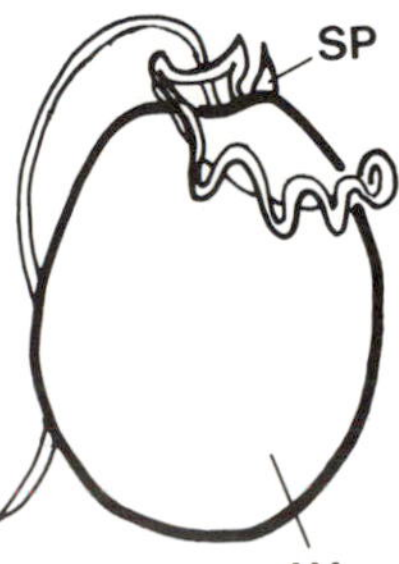

Desmokont
Apical (anterior) position of undulipodia of a desmokont dinomastigote (e.g., *Prorocentrum*). LV=left valve; SP=spine.

Desmoschisis
Cell division in which the parental wall forms part of the wall of the progeny (e.g., thecate dinomastigotes, chlorosarcinalean chlorophytes). See *Eleutheroschisis.*

Desmose (desmos)
Part of kinetid structure. Connecting fiber; composite fibrillar connection of unknown nature or function between two adjacent kinetosomes or among several kinetosomes that form a localized group (e.g., in the blepharoplast complex of many mastigotes); absent in ciliates.

Desmosome
Type of cell junction in animal tissues; morphologically and compositionally distinct area of cell membrane at which tissue cells of animals firmly adhere. Desmosome-like structures bind the undulipodium to the cell body of trypanosomatids. Illustration: Paraxial rod.

Determinate growth
Pertaining to a growth style, like that of a chytrid thallus, a heterotrichous ciliate, or volvocalean chlorophyte, in which growth stops after reaching a determined size. See *Indeterminate growth.*

Determinate sporangium (pl. determinate sporangia)
Sporangium that terminates the axis (e.g., oomycotes). See *Basipetal development, Percurrent development.*

Deuteromerite
Posterior portion of the trophozoite in some gregarine apicomplexans that is separated by a transverse septum from the nucleus-containing protomerite (the anterior cell).

Diadinoxanthin
Carotenoid found in the plastids of several types of algae (e.g., euglenids, xanthophytes, and eustigmatophytes).

Diagenesis (adj. diagenetic)
Geological term for physical and chemical alterations in sediments after their deposition and prior to their lithification.

Diakinesis
Last stage of meiotic prophase I, in which bivalents and chiasma disappear as homologues begin to segregate. See *Meiosis.*

Diaminopimelic acid pathway
DAP pathway. Biosynthetic metabolic pathway forming the amino acid lysine; pathway characteristic of bacteria, some protoctists, and plants. See *Alpha aminoadipic acid pathway.*

Diapause
Temporary suspension in growth and development in insects and other animals.

Diatomite
Sedimentary rock formed from diatom frustules; when poorly lithified, it is equivalent to diatomaceous earth.

Diatoms
Members of the phylum Bacillariophyta; unicellular and colonial aquatic algal protoctists renowned for their siliceous tests (frustules). Illustration: Bacillariophyta.

Diatoxanthin
Carotenoid found in the plastids of several protoctists (e.g., euglenids, xanthophytes, eustigmatophytes, and diatoms).

Dichotomous (adj.)
Referring to the branching into two equal or nearly equal parts.

Dichotypical (adj.)
Referring to the condition in desmids (conjugating green algae) in which one semicell resembles members of one species and the other resembles members of a different species.

Diclinous (adj.)
Referring to antheridia and oogonia on separate hyphae (e.g., the oomycote *Pythium lutarium).* See *Monoclinous.*

Dictyosome
Golgi apparatus; golgi body. Botanical term for this elaboration of the endomembrane system. Illustrations: Coated vesicles, Organelle.

Diel movement
Locomotion that follows a 24-hour cycle. See *Circadian rhythm.*

Diffuse growth
Generalized, indeterminate growth, characteristic of protoctists such as the plasmodial stage of myxomycotes, labyrinthulids, some chrysophytes, etc.

Digenetic (adj.)
See *Heteroxenous.*

Dikaryon (pl. dikarya; adj. dikaryotic)
Cell or organism with cells containing a pair of nuclei (fungi); typically each is derived from a different parent. Illustration: Life cycle.

Dikinetid
Kinetid composed of two kinetosomes and associated structures. Illustration: Oral region.

Dimorphism (adj. dimorphic)
Two forms; referring to an organism that, during the course of its life cycle, develops two different types of normal morphologies. Also refers to two genetic types of individuals in a population (e.g., sexual dimorphism or seasonal dimorphism). See *Polymorphism.*

Dinokaryon (pl. dinokarya; adj. dinokaryotic)
Dinonucleus. Unique nucleus of dinomastigotes characterized by densely packed chromosomes that persist during interphase. The atypical chromosomes contain DNA with small (25Å) unit fibrils and lack conventional histone protein that makes up nucleosomes, distinguishing dinokarya from nuclei of other protoctists.

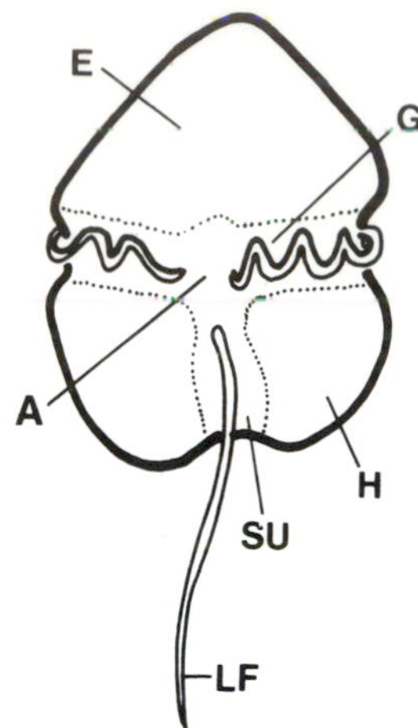

Dinokont
Ventral view of a dinokont dinomastigote. A=acrobase; E=epicone (episome); G=girdle (cingulum); H=hypocone (hyposome); LF=longitudinal undulipodium; SU=sulcus.

Dinokont
Organism with one undulipodium located in a transversely aligned groove, the other undulipodium beating in a longitudinally aligned groove. Characteristic of a subgroup of the dinomastigotes.

Dinomastigote
Dinoflagellate. Member of the phylum Dinomastigota. Illustrations: Desmokont, Dinokont, Dinomastigota, Mastigoneme.

Dinomastigote life history
Stages in development of dinomastigotes correlating environment and morphology.

Dinomitosis
Closed extranuclear pleuromitosis; the characteristic mitosis of dinomastigotes. See *Dinokaryon.*

Dinonucleus (pl. dinonuclei)
See *Dinokaryon.*

Dinospore
Dinomastigote propagule; spore issued from successive multiple fissions, esp. in parasitic dinomastigotes.

Dioecious (adj.)
Referring to organisms that have male and female structures on different individual members of the same species. See *Diclinous, Monoecious.*

Diphasic life cycle
Life cycle with two distinct parts; in symbionts, can refer to two distinct hosts or tissues of attachment.

Diplobiontic (adj.)
Having two free-living phases in the life history of an organism. See *Haplobiontic.*

Diplohaplontic (adj.)
Referring, in algae, to an organism with separate haploid and diploid stages in its life history which may or may not be morphologically distinguishable.

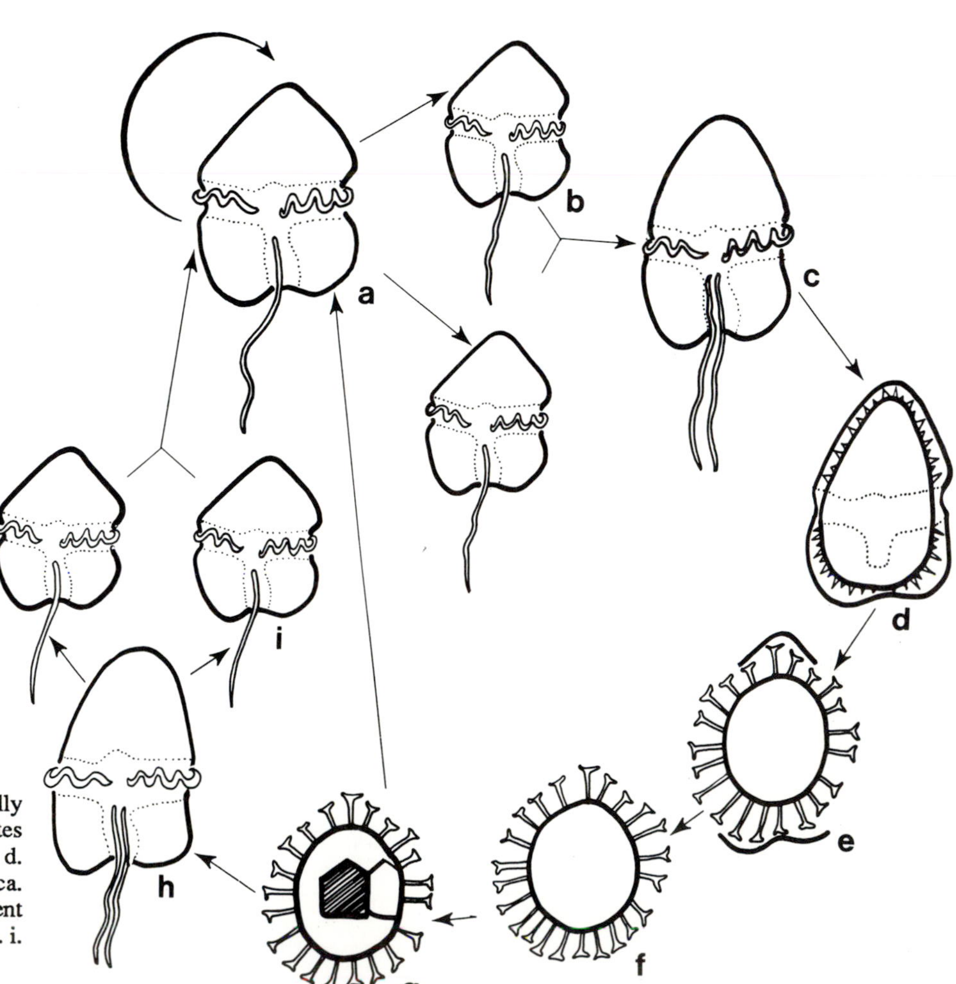

Dinomastigote life history
A common dinomastigote life cycle. a. asexually reproducing motile cell (mastigote). b. gametes (either iso- or anisogametes). c. planozygote. d. hypnocyst (resting cyst) formation within the theca. e. theca discarded. f. dormancy. g. excystment through the archeopyle. h. meiocytic planozygote. i. meiotic division.

Diploid (adj.)
Referring to eukaryotic cells in which the nucleus contains two complete sets of chromosomes, abbreviated 2N. See *Euploid, Haploid, Polyploid.* Illustration: Life cycle.

Diplokaryon (pl. diplokarya; adj. diplokaryotic)
Two diploid nuclei inside single cells characteristic of some microsporans *(Thelohania, Pleistophora, Tuzetia)*; a microsporan with such a nuclear arrangement.

Diplontic (adj.)
Referring to life cycles of organisms in which individual cells are diploid throughout their life history. Refers to organisms that undergo gametic meiosis such that haploidy is limited to the gamete stage. See *Haplontic.*

Diplophase
That part of the life cycle in which organisms are diploid, their cells each containing two complete sets of chromosomes. Illustration: Life cycle.

Diplosomes
See *Bipolar bodies.*

Diplotene
Diplonema; stage in meiosis just prior to diakinesis in which doubled bivalents become clearly visible. See *Meiosis.*

Diplozoic (adj.)
Having a double body form (e.g., as a result of incomplete cell division) as in diplomonad zoomastigotes. See *Monozoic.*

Discharge vesicle
Membrane, usu. continuous with the inner zoosporangium wall and papilla in chytridiomycotes, that is laid down during or after sporangial discharge; zoospore delimitation is completed within it.

Discoaster
Star-shaped coccolith. Illustration: Coccolith.

Discobolocyst
Ejectile organelle originating in the golgi that on discharge forms a firm ring with a gelatinous head; function unknown; restricted to mastigotes, especially chrysomonads.

Disporous (adj.)
Referring to the presence, in microsporans and myxozoans, of cells in which two spores have been produced within a single pansporoblast.

Distromatic (adj.)
Referring to a thallus only two cell layers thick.

Division
Botanical term for taxonomic group equivalent to phylum. See *Fission.*

Division center
Microtubule-organizing center of mitosis; centriole, centriolar plaque, centrocone, or any one of a number of structures found at the poles of mitotically dividing cells. Illustration: Mitosis.

Dixenous (adj.)
See *Heteroxenous.*

DNA complexity
See *Genomic complexity.*

DNA hybridization
Hybridization. *In vitro* analytical tool involving pairing of complementary DNA and RNA strands to produce a DNA-RNA hybrid or the partial pairing of complementary DNA strands from different genetic sources. Can be used to determine genetic relatedness between organisms and for purification of messenger RNA.

Dormancy
Resting stage; stage in propagule development of lowered metabolism and resistance to environmental extremes of temperature, desiccation, etc. Illustration: Dinomastigote life history.

Dorsiventral (adj.)
Referring to structures or tendencies (e.g., flattening) that extend from the dorsal toward the ventral side; also, having distinct dorsal and ventral surfaces.

Dourine
Venereally transmitted disease of horses caused by *Trypanosoma equiperdum.*

Dual nuclear apparatus
Dimorphic nuclei of heterokaryotic cells (e.g., ciliates or foraminifera).

Dune
A low mound, ridge, bank, or hill of loose, windblown granular material (generally sand, sometimes volcanic ash), either bare or with vegetation, capable of movement but always retaining its characteristic shape. Illustration: Habitat.

Dysaerobic (adj.)
Geological or ecological term referring to aquatic environments with low oxygen, or transition zones between oxic and totally anoxic sediments.

Dyskinetoplastic (adj.)
Pertaining to members of the kinetoplastids grown in culture in which the kinetoplast has become unstainable and invisible (either because its contents have become dispersed throughout the mitochondrion or because the structure has been lost as a result of faulty kinetoplast reproduction).

Dystrophic (adj.)
Ecological term (meaning ill-nourished) referring to lakes with very low lime content and containing very high quantities of humus (organic matter). Also refers to bay lakes with colored water and limited inorganic nutrient composition.

E

Ecad
See *Ecotype*.

Ecdysis
The act of shedding an outer cuticular layer; in dinomastigotes, shedding of theca prior to division.

Echinate (adj.)
Spiny.

Ecosystem
Communities of plants, animals, and microorganisms together with their immediate environment, capable of the complete cycling of the biological elements (C, N, O, P, S) (e.g., forests, deserts, or ponds). The metabolism and community interactions in an ecosystem are such that cycling within is more rapid than cycling between ecosystems.

Ecotype
Ecad. Genetic race, strain, or variety of organisms that has developed an identifiable morphological response to its environment.

Ectocarpin
Chemotactic pheromone produced by the female gametes of the phaeophyte *Ectocarpus* causing accumulation of sperm at the source of the pheromone.

Ectocyst
Outermost of the three layers surrounding a cyst. See *Endocyst, Mesocyst*.

Ectoparasite (adj. ectoparasitic)
Ecological term referring to the topology of parasites and hosts: a parasite that lives upon the surface of its host. See *Endoparasite, Epibiont*.

Ectoplasm
Outermost, relatively rigid and transparent, granule-free layer of the cytoplasm of many cells (e.g., amebas). See *Hyaloplasm, Stereoplasm*. Illustrations: Acantharia, Actinopoda, Contractile vacuole.

Ectoplasmic network
Extracellular matrix; branching and anastomosing, hyaline, membrane-bounded network of ectoplasmic filaments devoid of cytoplasmic organelles that functions as an attachment and absorbing structure and is produced by specialized organelles, called sagenogens, on the cell surface of labyrinthulomycotes. In labyrinthulids, the ectoplasmic network completely surrounds the cells and joins them in a common network through which the cells move by a gliding locomotion. In thraustochytrids, the ectoplasmic network arises from one side of each cell and does not surround it. Illustration: Sagenogen.

Edaphic (adj.)
Referring to soil.

Egg
Female gamete, nonmotile and usually larger than the male gamete. See *Oosphere*. Illustration: Life cycle.

Ejectile body
See *Ejectosome, Extrusome*.

Ejectosome (ejectisome)
Ejectile body. Any organelle forcibly ejected from a cell (e.g., trichocysts); in cryptomonads, the ribbonlike extrusome that is coiled and contained in a vesicle. Illustrations: Cryptophyta, Extrusome, Undulipodial rootlet.

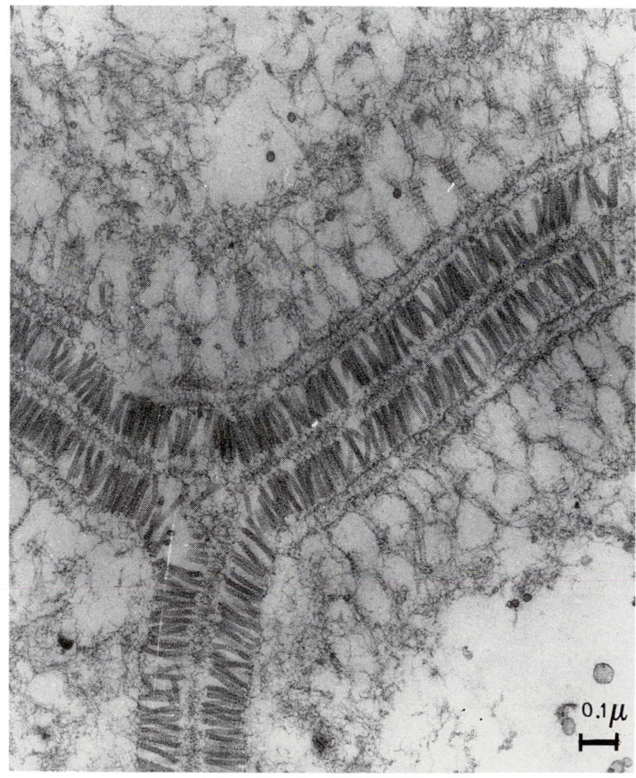

Elastic junctions
The periplasmic cortex of the chaunacanthid actinopod *Gigartacon mulleri*. TEM.

Elastic junctions
Structures in periplasmic cortex of chaunacanthid acantharians (phylum Actinopoda) consisting of microfibrils interconnected in very precise patterns which confer elasticity and compensate for the contractions and relaxations of the myonemes. See also illustrations: Acantharia, Perispicular cone.

Electrolyte
Salt. A substance that dissociates into its constituent ions in aqueous solution.

Electrophoresis
Gel electrophoresis. Technique used to separate macromolecules on the basis of electrical charge or size. The movement of charged molecules in solution in an electrical field. Often the

solution is held in a porous support medium such as a gel made of agarose or polyacrylamide.

Eleutheroschisis
Cell division in which the walls of offspring cells are entirely new and free from parental walls (e.g., thecate dinomastigotes). See *Desmoschisis.*

Embryo
Early developmental stage of a plant or animal individual that develops by cell and nuclear divisions from a zygote (fertilized egg). See *Blastular embryo.*

Embryophyte
Plant. Embryo enclosed in maternal tissue, usu. developing into an adult. Phototrophic organism; sporophyte growing from an embryo which is dependent for its nutrition on parental tissue during development. Includes all bryophytes and tracheophytes.

Emergent flagellum (pl. emergent flagella)
See *Emergent undulipodium.*

Emergent undulipodium (pl. emergent undulipodia)
Undulipodium in a bimastigote that protrudes (e.g., the undulipodium that extends beyond the canal in euglenids). See *Nonemergent undulipodium.* Illustration: Keel.

Encyst (v.; n. encystment, encystation)
To form or become enclosed in a cyst. Illustrations: Amebomastigota, Opalinata, Stachel, Trophozoite.

Endemic (adj.)
Referring to populations of organisms or viruses, including disease agents, constantly present (often in low numbers) only in a limited geographical area.

Endobiont (adj. endobiotic)
Endosymbiont. Ecological term describing the topology of partners in an association in which one partner lives within the other partner (the host); may be intra- or extracellular. See *Epibiont.* Illustration: Habitat.

Endochite
Innermost layer of the fucalean oogonium in phaeophytes (e.g., *Fucus*).

Endocyanome
Cyanelles. All the connected cyanelles of a glaucocystophyte.

Endocyst
Innermost of the layers surrounding a cyst (e.g., heliozoan oocyst); composed of a layer of fibers and golgi membranes. See *Ectocyst, Mesocyst.*

Endocytic (adj.)
Inside a cell; intracellular; pertaining to topological relations between associates in which one organism lives inside another cell. Endocytobiotic.

Endocytobiology
The study of intracellular symbionts and cell organelles. *Endocytobiology and Cell Research* journal, organ of the International Society of Endocytobiology, published by Tübingen University Press, Germany.

Endocytobiont
Intracellular symbiont.

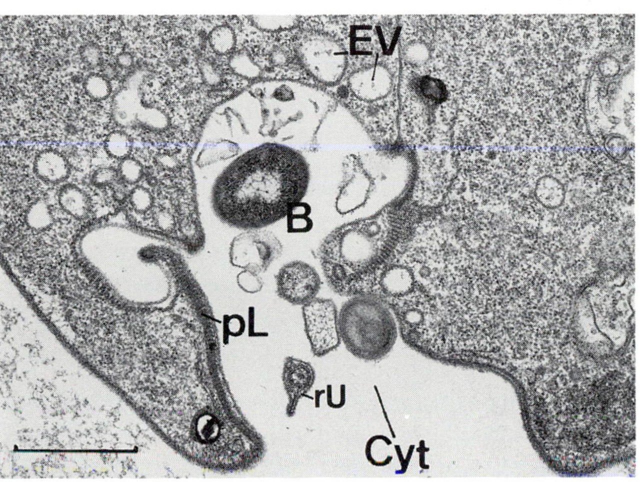

Endocytosis
Transverse section of the cytopharyngeal region in the retortamonad *Chilomastix caulleryi* where endocytosis occurs. B=bacterium; Cyt=cytostomal ventral pouch; EV=endocytotic vesicle; pL=striped lamina; rU=recurrent undulipodium. TEM. Bar=1 μm.

Endocytosis
Intake of extracellular material through invagination and pinching off of the plasma membrane; includes intake of fluid (pinocytosis), particulate matter (phagocytosis), and neighboring cell material in tissues (endocytosis *sensu stricto).*

Endocytotic vesicles
Cell membranes involved in particle uptake. Illustrations: Endocytosis, Retortamonadida.

Endodyogeny
Endogenesis resulting in the production of two offspring cells within the parent cell (e.g., in coccidian apicomplexans). See *Endopolygeny.* Illustration: Life cycle.

Endogenesis
Endogenous budding; endogenous cleavage; endogenous multiplication; endogeny. Process by which offspring cells are formed inside the parent cell. See *Endodyogeny, Endopolygeny.*

Endogenous budding
See *Endogenesis.*

Endogenous cleavage
See *Endogenesis.*

Endogenous multiplication
See *Endogenesis.*

Endogeny (adj. endogenous)
See *Endogenesis.*

Endolith (adj. endolithic)
Ecological term describing microorganisms living in tiny openings in rocks or rock crevices that have been produced by the metabolic activities of the endolithic organisms themselves. See *Chasmolith, Epilith, Lithophile.* Illustration: Habitat.

Endomitosis
Endoreplication. Duplication of chromosomes in the absence of karyokinesis and not followed by chromatid segregation; the process thus leads to polytene chromosomes rather than polyploidy.

Endomycorrhizal fungi (sing. endomycorrhizal fungus)
Fungal symbionts (usually zygomycotes) of plants that penetrate tissues of the roots and form a specialized swollen type of root tissue that augments nutrient uptake from the soil. VA (vesicular-arbuscular) mycorrhizae.

Endoparasite (adj. endoparasitic)
Ecological term describing the topology of parasites and hosts in which the parasite lives within its host, either extra- or intracellularly; endobiotic parasite. See *Ectoparasite, Endobiont.*

Endophyte (adj. endophytic)
Ecological term referring to the topology of symbiotic associations with plants. Refers to fungi, protoctists, or bacteria living within the tissue of plants or other photosynthetic organisms. Since "-phyte" may refer to fungi, protoctists, and bacteria, which are not plants, the term should be replaced with endobiont, endosymbiotic bacteria, or other specific name. Illustration: Habitat.

Endoplasm
Inner central portion of the cytoplasm of cells (such as amebas), more fluid than the ectoplasm. See *Rheoplasm.* Illustrations: Acantharia, Actinopoda, Contractile vacuole.

Endoplasmic reticulum (pl. endoplasmic reticula)
ER. Extensive endomembrane system found in most protoctist, plant, and animal cells in places continuous with the nuclear membrane, the golgi apparatus, the outer membranes of other organelles, and the plasma membrane; called rough (RER) if coated with ribosomes, and smooth (SER) if not. See *Chloroplast endoplasmic reticulum.* See also illustration: Organelle.

Endopolygeny
Endogenesis characterized by the production of several offspring cells within the parent cell (e.g., in *Toxoplasma, Chlorella).* See *Endodyogeny.* Illustration: Life cycle.

Endoreplication
Endomitosis. In ciliates, may refer to the reproduction of nuclei inside other nuclei.

Endosome
Karyosome; nucleolus. Body, or bodies, into which the nucleolar material is organized and which contains ribosomal precursors. Also, a vesicle resulting from endocytosis. Illustration: Reservoir.

Endospore
Spore formed by successive cell divisions within a parent wall. In Actinosporea (phylum Myxozoa), an envelope of one or two modified cells housing the sporoplasm within the sporal cavity between the episporal cells; in dinomastigotes, the thick inner wall of the three-layered cell wall of a hypnozygote; in conjugating green algae, the inner layer of the zygospore wall. In oospores of oomycotes, thick inner wall that is in part a carbohydrate-reserve wall precursor. See *Exospore, Mesospore.* Illustration: Oospore.

Endospore cell
Parent cell inside which spore(s) form; referring primarily to bacteria. Cells that make up the covering in which sporoplasms of actinosporean myxozoans originate; may persist in mature spores in some genera.

Endosymbiont (adj. endosymbiotic)
Endobiont. Ecological term referring to the topology of association of partners, a member of one species living inside a member of a different species. May be intracellular or extracellular. See *Xenosomes.* Illustrations: Karyoblastea, Kinetoplastida, Pyrsonymphida.

Endosymbiosis
Ecological term referring to the topology of an association of partners; the condition of one organism living inside another. Includes intracellular symbiosis (endocytobiosis) and extracellular symbiosis.

Endothelial cells
Cells of the tissue that lines the blood vessels of mammals. Illustration: Coccidian life history.

Endozoic (adj.)
Entozoic. Ecological term referring to any organism that lives inside an animal. Illustration: Habitat.

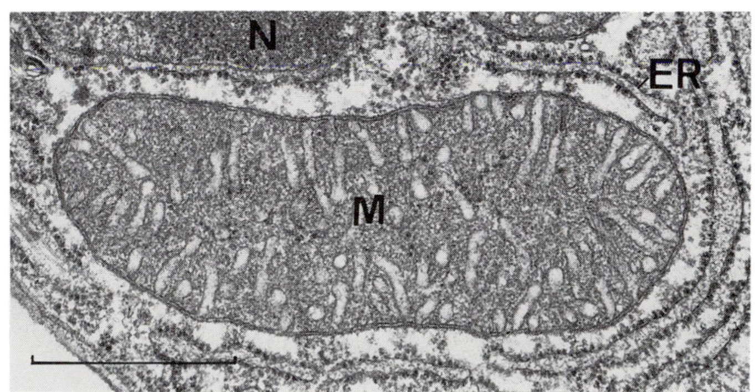

Endoplasmic reticulum
Mitochondrion with tubular cristae of the proteromonad zoomastiginid *Proteromonas lacertae-viridis*, surrounded by endoplasmic reticulum (ER). M=mitochondrion; N=nucleus. TEM. Bar=0.5 μm.

Endozoite
Zoite. Usu. in apicomplexans; trophic, motile individual formed by endogenesis.

Entosolenian tube
In some foraminifera, an internal tubelike extension from the aperture.

Entozoic (adj.)
Endozoic. Ecological term referring to any organism that lives in an animal (may also refer specifically to the gut).

Enucleation
Anucleation; removal of the nucleus from a cell.

Envelope
See *Lorica, Plasma membrane*. Illustration: Organelle.

Epibiont (adj. epibiotic)
Ecological term describing the topology of association of partners in which one organism lives on the surface of another organism. See *Endobiont*. Illustration: Habitat.

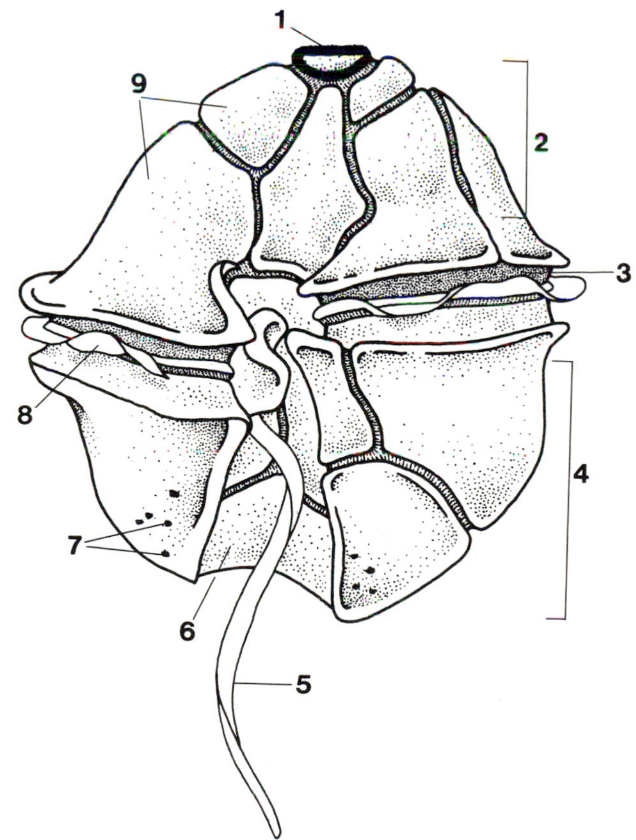

Epicone
The dinomastigote *Gonyaulax tamarensis*. 1=apical pore; 2=epicone; 3=girdle (cingulum); 4=hypocone; 5=longitudinal undulipodium; 6=sulcal groove; 7=trichocyst plates; 8=transverse undulipodium; 9=cellulose plates.

Epicone
Episome. Upper body; upper surface or hemisphere of a dinomastigote cell, anterior to the cingulum. Illustration bottom of page.

Epicontinental (adj.)
Pertaining to extensive marine environments, i.e., inland seas, formed on the surface of continental masses. Characteristic of the lower Paleozoic Era and later.

Epilimnion
Between the surface and the thermocline, the uppermost layer of water in a thermally stratified lake, characterized by uniform temperature and generally warmer than elsewhere in the lake; oxygen-rich layer of water that overlies the metalimnion in a thermally stratified lake. See *Hypolimnion, Metalimnion*. Illustration: Habitat.

Epilith (adj. epilithic)
Ecological term referring to the biota living on the surface of rocks and stones. See *Chasmolith, Endolith, Lithophile, Saxicolous*. Illustration: Habitat.

Epimastigote
Stage in development of kinetoplastids in which the kinetoplast lies anterior to the nucleus and the associated undulipodium emerges laterally to form an undulating membrane along the anterior part of the body, usu. becoming free at its anterior end. Illustrations: Kinetoplastida, Trypomastigote.

Epimerite
Anchoring organelle in the anterior region of septate gregarine apicomplexans, set off from the rest of the body by a septum. See *Mucron*. Illustration: Sporozoite.

Epipelon (adj. epipelic)
Ecological term referring to biota living attached to the surface of marine or freshwater mud or sand. Illustration: Habitat.

Epiphyte (n. epiphyton; adj. epiphytic)
Ecological term referring to the topology of association of partners, one of which is a plant (or traditionally an alga). The second partner grows on the plant using it for support but not nutrition; epiphyton also refers to communities of microbes growing on algae in aquatic environments. Term appropriate only if host is member of Plant kingdom. Illustration: Habitat.

Epiplasm
Fibrous or filamentous layer of cytoplasm closely applied to the innermost plasma membrane; in ciliates, a layer under the pellicle, comprising a part of the cortex. Illustration next page. See also illustration: Cortex.

Epipsammon (adj. epipsammic)
Ecological term referring to biota living on or in fine interstices of sand grains (from *psammon*, meaning sand).

Episeme
Change in a seme; evolutionary alteration in a trait. See *Seme*.

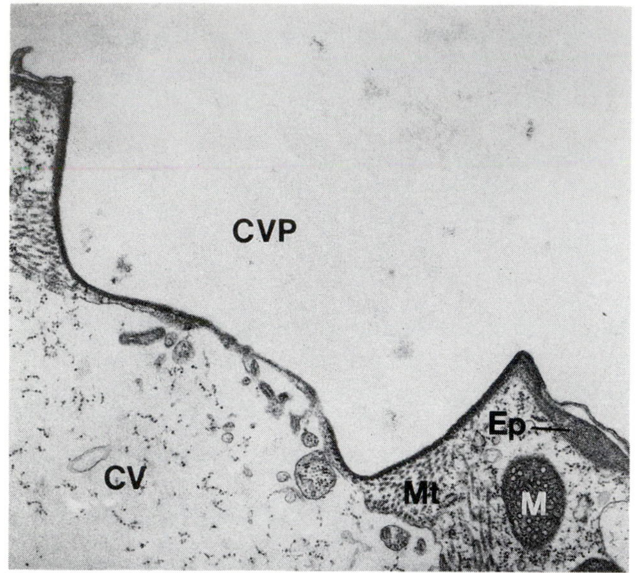

Epiplasm
Contractile vacuole pore (CVP) and adjacent epiplasm of the ciliate *Colpoda*. CV=contractile vacuole; Ep=epiplasm; M=mitochondrion; Mt=microtubules. TEM, x 20,000.

Episome
Small genome; genetic element (stretch of DNA sequence capable of coding for a product), usu. of bacteria; may be integrated or attached to genophore or replicate independently of the genophore (even at faster rates than the genophore). In dinomastigotes, ill-advised synonym for epicone. See *Epicone.*

Episporal cell
Modified valve cell, three of which make up the epispore that houses the sporoplasm in actinosporean myxozoans (e.g., *Tetractinomyxon*).

Epispore
Spore wall in actinosporean myxozoans, consisting of three valve or episporal cells often bearing long posterior processes; enucleate half of diploid sporoblast that encloses the sporoplasm in haplosporidians. In oomycote oospores, outer wall formed from zygote wall that overlies the endospore and underlies the exospore when present. Illustration: Oospore.

Epithallium (pl. epithallia)
See *Epithallus.*

Epithallus (pl. epithalli)
Epithallium. Part of the growing thallus in which the cells or filaments are developed outwardly from an intercalary meristem (e.g., coralline rhodophytes).

Epitheca (pl. epithecae)
Epivalve and adjacent portion of girdle in dinomastigotes; anterior portion of a thecate (armored) dinomastigote; a covering for the epicone. Illustrations: Bacillariophyta, Dinomastigota.

Epithelium (pl. epithelia)
Type of animal tissue that lines the surface of kidneys, or other organs. Epidermis of plants. Illustrations: Sporozoite, Trophozoite.

Epitope
Antigenic determinant. Part of a molecule recognizable by an antibody (e.g., part of an amino acid residue or a few amino acid residues in a protein); the portion of a macromolecule to which an antibody binds.

Epivalve
Upper test or shell, found opposite to and usu. larger than the hypovalve in diatom frustules. Illustrations: Bacillariophyta, Frustule, Hypovalve.

Epizoic (adj.; n. epizoon)
Ecological term referring to the topology of association of partners in which an organism lives on the surface of an animal. See *Epibiont, Epiphyte.*

Epizootic (adj.)
Pertaining to a widespread occurrence of an infectious disease of animals other than people.

Equatorial groove
See *Girdle.*

Equatorial plate
See *Metaphase plate.*

Equipotential genomes
Genomes (total genetic material of cell) resulting from cell division of a parent cell in which both offspring cells are capable of the same extensive further development.

ER
See *Endoplasmic reticulum.*

Estuary
The seaward end or the widened funnel-shaped tidal mouth of a river valley where fresh water mixes with and measurably dilutes seawater and where tidal effects are evident. Illustration: Habitat.

Etiolation
Bleached condition of photosynthetic eukaryotic organisms growing in the dark characterized by poorly developed plastids and their lack of chlorophyll. Stem elongation and poor leaf development accompanies etiolation in plants.

Etiological agent
Causative agent (e.g., of a disease).

Eucarpic (adj.)
Referring to development in certain protoctists (e.g., oomycotes) and fungi that form reproductive structures on limited portions of the thallus, such that the residual nucleate protoplasm remains capable of further mitotic growth and regeneration. See *Holocarpic.*

Euglenid
Any member of the phylum Euglenida. Euglenoid refers to *Euglena*-like features. Illustration: Mastigoneme.

Euglenoid (adj.)
See *Euglenid.*

Euglenoid motion
Metaboly. Peculiar flowing, contracting, expanding ("crawling") movement on surfaces displayed by euglenids capable of changing shape, i.e., those not restricted by too rigid a pellicle. Illustration: Pellicle.

Euhaline (adj.)
Ecological term referring to salinity of water in the normal oceanic range, i.e., between 3.3 and 3.8 percent salt as sodium chloride. See *Hyperhaline, Oligohaline.*

Eukaryote (eucaryote)
Organism comprised of cell(s) with membrane-bounded nuclei. Most contain microtubules, membrane-bounded organelles (i.e., mitochondria and plastids), and chromatin organized into more than a single chromosome.

Eukinetoplastic (adj.)
Pertaining to kinetoplastid mastigotes in which the DNA of the kinetoplast (kDNA) forms a single stainable mass located close to the kinetosome(s).

Eulittoral (adj.)
See *Intertidal, Littoral.*

Euphotic zone
Photic zone. Ecological term referring to the illuminated portion of a water column, soil profile, microbial mat, etc.; the layer in which, because of the penetration of light, photosynthesis can occur. Illustration: Habitat.

Euplankton
Ecological term referring to aquatic organisms that spend their entire lives suspended in a water column.

Euploid (adj.)
Possessing a chromosome set that is either the haploid complement or an exact multiple of the haploid complement (e.g., diploid, triploid, etc.); not aneuploid.

Euryhaline (adj.)
Ecological term referring to organisms that tolerate and grow under wide ranges of salinity. See *Stenohaline, Euhaline.*

Eurythermal (adj.)
Eurythermic. Ecological term referring to an organism that tolerates and grows under a wide range of temperatures. See *Stenothermal.*

Eurythermic (adj.)
See *Eurythermal.*

Eutrophy (adj. eutrophic)
Ecological term referring to waters rich in dissolved nutrients (e.g., nitrate, phosphate) for phototrophs. See *Oligotrophy.*

Evaporite flat
Open area covered with nonclastic sedimentary rocks composed primarily of minerals produced from saline solution that became concentrated by evaporation of the solvent.

Evolute test
Foraminiferan test in which each succeeding whorl does not embrace earlier whorls, so that all chambers are visible.

Exclusion zone
Layer of the centroplast of heliozoan actinopods. Illustration: Interaxonemal substance.

Excyst (v.; n. excystment, excystation)
Process of leaving the cyst stage; cyst germination. Illustrations: Amebomastigota, Dinomastigote life history, Sporocyst.

Exocytosis
Cell secretion; process of eukaryotic cells involving intracellular motility in which substances are eliminated to the exterior by emptying them from a vesicle that fuses with the plasma membrane, forming a cuplike depression. See *Endocytosis.*

Exogenesis
Production of smaller cells at the periphery of the parent cell; a type of budding (e.g., in suctorian ciliates).

Exogenous (adj.)
Referring to origin or development on or from the outside.

Exon
Segment of DNA that is both transcribed to RNA and translated into protein. Gene or part of gene. See *Intron.*

Exospore
Externally borne reproductive cell; not necessarily heat- or desiccation-resistant; in dinomastigotes, the thick outer layer of the triple-layered cell wall of the hypnozygote; in conjugating green algae, the outermost layer of the zygospore wall. In oospores of some oomycotes, outermost wall formed from condensed periplasm. See *Endospore, Mesospore.* Illustration: Oospore.

Exotoxin
Soluble poisonous substance passing into the host or the environment during growth of an organism (e.g., red tide dinomastigotes).

Expression-linked copy
ELC. Duplicate copy of a trypanosome variant surface glycoprotein DNA sequence (gene) expressed when transposed to a telomeric site on the chromosome.

Expulsion vacuole
See *Contractile vacuole.* Illustration: Plasmodiophoromycota.

Expulsion vesicle
See *Contractile vacuole.*

Extant (adj.)
Living; still in existence.

Extinct (adj.)
No longer existing.

Extrinsic encystment
Formation of cysts during the exponential phase of population growth; sexual resting cysts are produced (e.g., in the chrysophyte *Dinobryon cylindricum*). See *Intrinsic encystment.*

Extrusive organelle
See *Extrusome.*

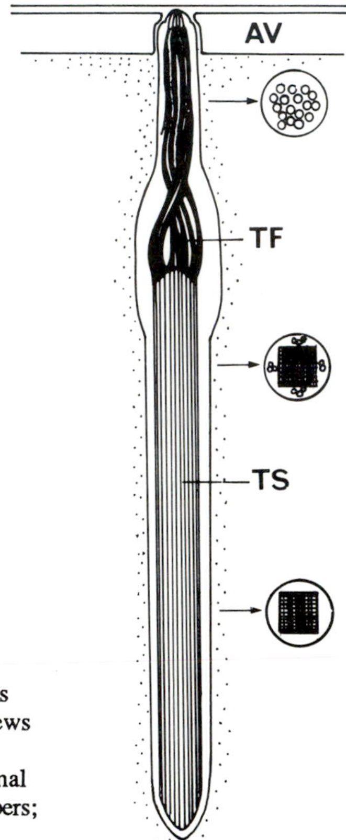

Extrusome
Longitudinal section of a trichocyst from a dinomastigote. The circles contain cross-sectional views of different regions of the trichocyst. AV=amphiesmal vesicle; TF=trichocyst fibers; TS=trichocyst shaft.

Extrusome
Extrusive organelle. Membrane-bounded structure, the contents of which are extruded by protoctists in response to a variety of stimuli (e.g., predators, prey, changes in acidity). Extrusomes are derived from vesicles of the golgi system and are anchored to the cell membrane by proteinaceous particles; generalized term referring to various, probably nonhomologous, structures. See *Cnidocyst, Discobolocyst, Ejectosome, Kinetocyst, Mucocyst, Nematocyst, Polar capsule, Toxicyst, Trichocyst.* See also illustration: Taeniocyst.

Eyespot
Stigma. Small, pigmented, and probably light-shielding structure in certain undulipodiated protists (e.g., euglenids, eustigmatophytes, labyrinthulid zoospores). See also illustrations: Chlorophyta, Chrysophyta, Eustigmatophyta, Phaeophyta, Xanthophyta.

F

Facies
Part of a sedimentary rock unit characterized by lithological and biological features and segregated from other parts of the unit, usu. seen in the field as a coherent rock layer.

Facultative (adj.)
Optional; e.g., a facultative autotroph is an organism that, depending on conditions, can grow either by autotrophy or by heterotrophy. See *Obligate.*

Falx (pl. falces)
Sickle-shaped structure of opalinids. Specialized area of the cortex along the front edge of the body; a region of kinetidal proliferation that results in the increased length of the kineties; the falx is usually bisected during the symmetrogenic fission of the organisms.

Fascicles
Bundles (e.g., oomycote mastigonemes or suctorian ciliate tentacles).

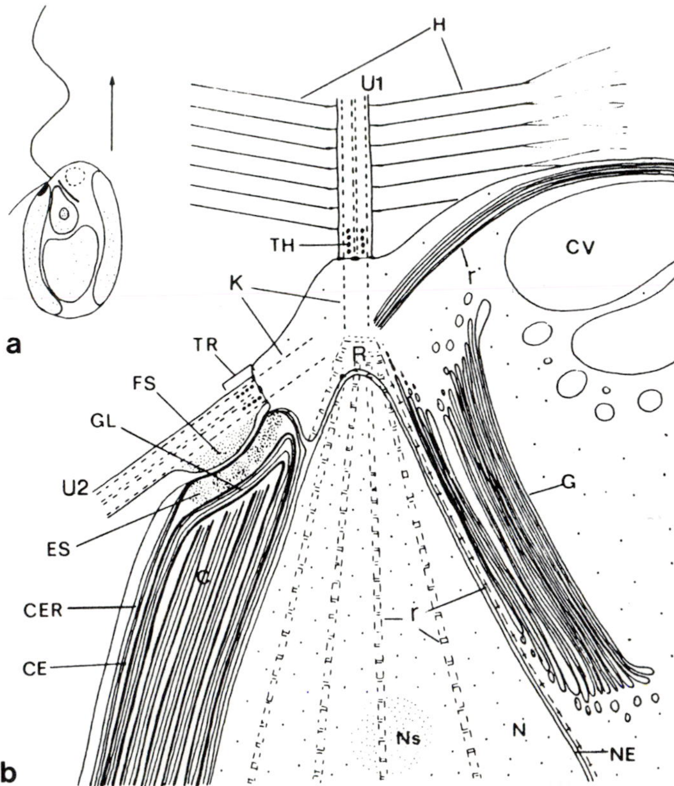

Eyespot
Basic organization of a chrysophycean cell. a. Undulipodia and other important organelles as seen with the light microscope (chrysoplasts, eyespot, nucleus, golgi body, chrysolaminaran vacuole). b. Anterior region of the cell. C=chrysoplast; CE=chrysoplast envelope; CER=chrysoplast endoplasmic reticulum; CV=contractile vacuole; ES=eyespot; FS=undulipodial swelling; G=golgi body; GL=girdle lamella; H=undulipodial hairs; K=kinetosomes; N=nucleus; NE=nuclear envelope; Ns=nucleolus; R=rhizoplast; r=microtubular undulipodial roots of rhizoplast; TH=transitional helix; TR=transitional region; U1=anteriorly directed mastigonemate undulipodium; U2=laterally directed smooth undulipodium.

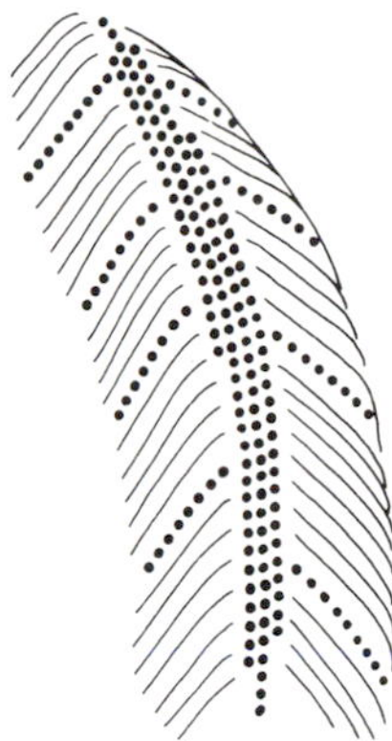

Falx
Kinetid pattern on portion of *Opalina* cortex consisting of falx and pellicular folds, and (origin) of rows of undulipodia extending down both sides or surfaces of the body.

Fathom
Unit of water depth (1 fathom = 2 meters).

Fauna
Animal life. Inappropriate term for protoctists and bacteria.

Feeding veil
Pallium. Cytoplasmic sheet extended from the sulcus of some nonphotosynthetic dinomastigotes during extracellular digestion when feeding on diatoms or other dinomastigotes.

Female
Gamont. Gender of individual that produces ovaries, eggs, or other sexual organs and receives the male sperm.

Fenestra (pl. fenestrae)
Foramen. Opening in a surface; small "window" (e.g., lesion in nuclear membrane in essentially closed mitosis). See *Polar fenestrae*.

Fermentation
Nutritional mode: enzyme-mediated pathway of catabolism of organic compounds in which other organic compounds serve as terminal electron acceptors (process that yields energy and organic end products in the absence of oxygen).

Ferruginous (adj.)
Made of or containing iron; having the reddish brown color of iron rust.

Fertile sheet
Cell layers lining the inside of the conceptacle from which the reproductive structures, antheridia and oogonia, are produced in fucalean phaeophytes.

Fertilization
Syngamy or karyogamy. Fusion of two haploid cells, gametes, or gamete nuclei to form a diploid nucleus, diploid cell, or zygote. Illustrations: Life cycle, Rhodophyta, Sporocyst, Trophozoite.

Fertilization cone
Cytoplasmic cone originating at the posterior end of the female gamete of hypermastigotes (e.g., the parabasalian *Trichonympha*).

Fertilization tube
Structure facilitating fertilization, e.g., structure forming in laterally fused, mating dinomastigote gametes beneath the kinetosomes into which nuclei migrate and fuse during fertilization. Illustration: Oomycota.

Feulgen stain
Red stain requiring hydrolysis of deoxyribose and formation of Schiff base that is quantitatively specific for chromatin DNA. Named for R. Feulgen, a German cytologist at the beginning of the 20th century.

Fibril
See *Filament*.

Fibrillar kinetosome props
See *Kinetosome props*.

Fibrillar rhizoplast
See *Rhizoplast*.

Fibrous lamina (pl. fibrous laminae)
Thick microfibrillar network coating the inner surface of the nuclear membrane in the acantharian actinopod *Haliommatidium*.

Fibula (pl. fibulae)
Clasp of buckle-shaped, elongated structure (e.g., bar running beneath the displaced raphe on the side of valve in such diatoms as *Nitzschia* and *Hantzschia*). See *Subraphe costae*.

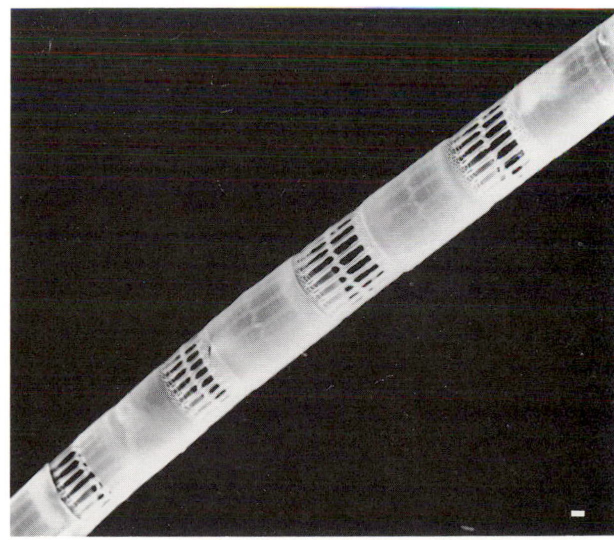

Filament
Frustule of the diatom *Sceletonema*. SEM. Bar=10 μm.

Filament
Fibril. General descriptive term for thread-shaped solid structure including morphology of growing organisms consisting of a single row of cells; e.g, two- to three-nanometer filaments lacking actin seen in motile systems of various protoctists including acantharians and heliozoan actinopods, dinomastigotes, and ciliates; in vorticellid ciliates, a 20,000-dalton protein called spasmin

has been identified which corresponds to the three nanometer microfilaments of their contractile stalk. See also illustration: Funis.

Filopodium (pl. filopodia)
Very thin pseudopods which may be stiffened by one or very few microtubules (e.g., those of desmothoracid heliozoans that pass through openings in the central capsule). Illustrations: Actinopoda, Kinetocyst.

Filose (adj.)
Terminating in a threadlike process.

Filose pseudopod
Cell protrusion or retractile process ending in a filamentous wisp; esp. the motile organelles of amebas.

Fimbriate (adj.)
Bordered by or decorated with tiny fibers or fibrils. Illustration: Ligula.

Fine structure
See *Ultrastructure.*

Fission
Division of any cell or organism; reproduction by division of cells or organisms into two or more parts of equal or nearly equal size. Longitudinal fission: division through long axis (e.g., most mastigotes); transverse fission: division through small equatorial plane of ovoid organisms (e.g., all ciliates). See *Binary fission, Homothetogenic fission, Interkinetal fission, Mitosis, Multiple fission, Perkinetal fission, Polytomic fission, Symmetrogenic fission.*

Fjord
A long, narrow, winding, V-shaped and steep-walled, generally deep inlet or arm of the sea between high cliffs or slopes along a mountainous coast, typically with a shallow sill or threshold of solid rock or earth material submerged near its mouth and becoming deeper farther inland. Illustration: Habitat.

Flabelliform (adj.)
Fan-shaped.

Flagellar apparatus
See *Kinetid.*

Flagellar bracelet
See *Undulipodial bracelet.*

Flagellar groove
See *Undulipodial groove.*

Flagellar hairs
See *Undulipodial hairs.*

Flagellar pocket
See *Undulipodial pocket.*

Flagellar root
See *Undulipodial root.*

Flagellar rootlet
See *Undulipodial rootlet.*

Flagellar transition zone
See *Transition zone.*

Flagellate
See *Mastigote.*

Flagellum (pl. flagella)
Bacterial flagellum; prokaryotic extracellular structure composed of homogeneous protein polymers, members of a class of proteins called flagellins; moves by rotation at the base; relatively rigid rod driven by a rotary motor embedded in the cell membrane that is intrinsically nonmotile and sometimes sheathed. Undulipodium, an intrinsically motile intracellular structure used for locomotion and feeding in eukaryotes; composed of a standard arrangement of nine doublet microtubules and two central microtubules composed of tubulin, dynein, and approximately 200 other proteins, none of them flagellin; no flagellum (but every undulipodium) is underlain by a kinetosome. See Introduction (p. xviii) for an explanation of the restriction of the term flagellum. Illustrations: Organelle, Undulipodium.

Flange
Projecting rim which provides strength or support to a structure. Illustration: Diplomonadida.

Flimmer
See *Mastigoneme.*

Flora
Plant life. Inappropriate for protoctists, fungi, and bacteria.

Flotation chamber
Gas-filled portion of a cell lending buoyancy (e.g., the final chamber of the foraminiferan *Rosalina bulloides* which adds buoyancy such that it floats among the plankton).

Foliose (adj.)
Leafy; pertaining to leaflike growth. Growing hyphae or trichomes, usually of lichens or algae, that together form a leafy or leaflike structure.

Fomite
Inanimate object that transmits infective stages of parasites or pathogens.

Foramen (pl. foramina)
Fenestra. Small opening, orifice, or perforation; in foraminifera, an opening in a septum separating two chambers.

Foraminiferan test →
Shell or covering of members of the phylum Granuloreticulosa.

Fragmentation
Means of asexual reproduction in which the breakup of a parental thallus or filament gives rise to a new individual (e.g., some conjugating green algae, large foraminifera).

Fresh water
Water containing only small quantities of dissolved salts or

other minerals, such as water of streams and inland lakes. Illustration: Habitat.

Frond

Leaflike structure; any divided thallus (or leaf).

Front

See *Cline*.

Frontal syzygy

See *Syzygy*.

Fructification

See *Fruiting body*.

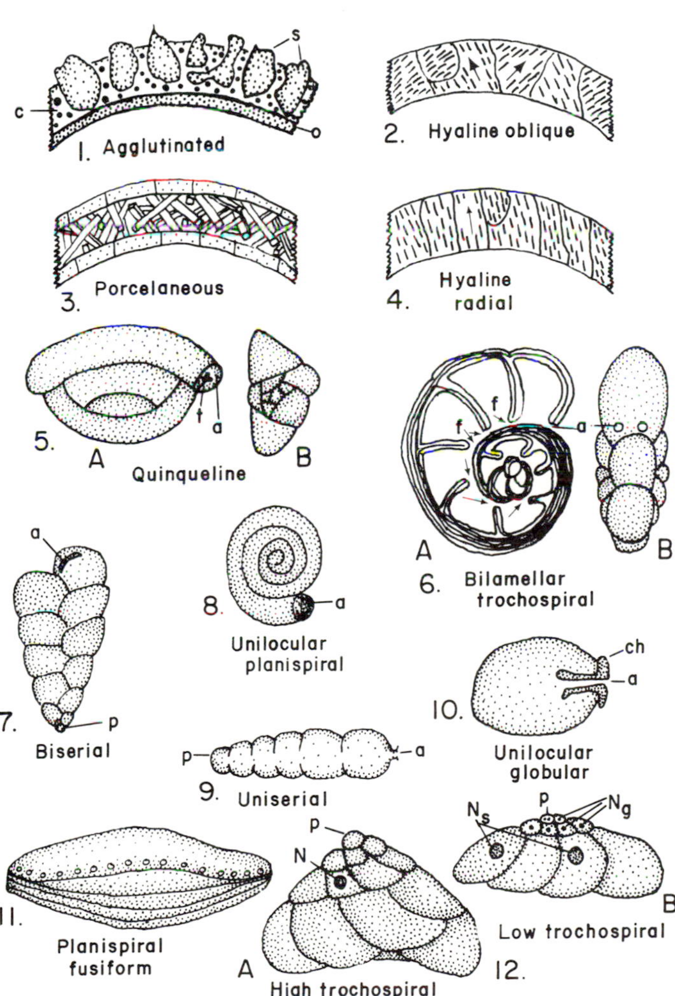

Foraminiferan test

1-4. Types of calcareous granuloreticulosan walls. Arrows indicate crystal orientation as determined by polarized light. 5-12. Test morphologies. (5a. Lateral view. 5b. Cross-sectional view. 6a. Diagrammatic cross section. 6b. Apertural view; arrows show foramina. 10. Example: *Allogromia laticollaris*. 12. Example: *Glabratella sulcata:* a. gamont; b. agamont.) a=aperture; c=cement; ch=collar; f=foramina; N=nucleus; N_g=generative nuclei; N_s=somatic nuclei; o=organic layer; p=proloculum; s=sand grain; t=tooth.

Fruit

Botanical term describing structures of angiosperm plants, i.e., matured ovary or ovaries of one or more flowers and their associated structures.

Fruiting body

Fructification; sorocarp; sporocarp. Structure that contains spores, cysts, or other propagules. This term, derived from botany and ambiguously applied, should be replaced with appropriate protoctistological alternatives.

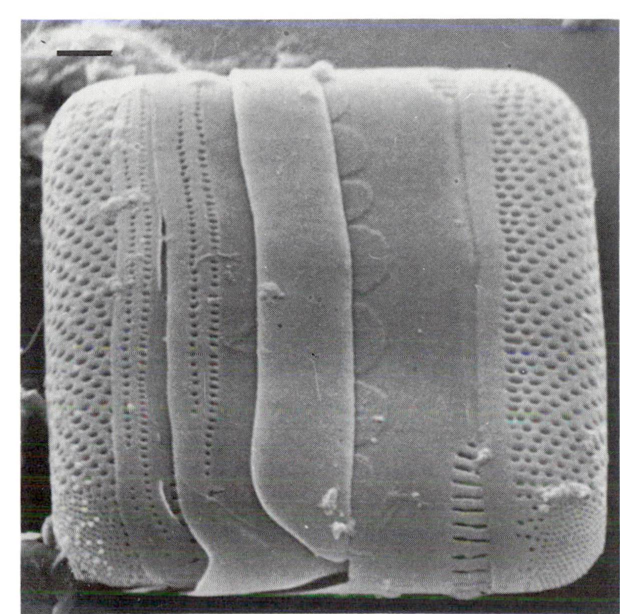

Frustule

The diatom *Grammatophora*. The epivalve to the right has 4 girdle bands, which partly obscure the hypovalve to the left. The 2 girdle bands of the (incomplete) hypocingulum underlie the epivalve. a. SEM. Bar=10 μm; b. drawing.

Frustule

Siliceous cell wall or test of a diatom, composed of two valves.

Fucan
See *Fucoidan.*

Fucoidan
Fucan. Sulfated polysaccharides found in phaeophytes, containing L-fucose.

Fucosan vesicle
See *Physode.*

Fucoserraten
Pheromone produced by eggs of *Fucus* that attracts sperm.

Fucoxanthin
Carotenoid, usu. in chrysoplasts such as those of diatoms, phaeophytes, and some dinomastigotes.

Fultoportula (pl. fultoportulae)
Organelle of some centric diatoms surrounded by basal pores and buttresses; may bear an external tube that continues internally.

Fungi (sing. fungus; adj. fungal)
Members of the Kingdom Fungi; osmotrophic, chitinous-walled eukaryotic organisms that develop from spores; they lack both embryos and undulipodia at all stages of their life cycle. See *Higher fungi, Lower fungi.*

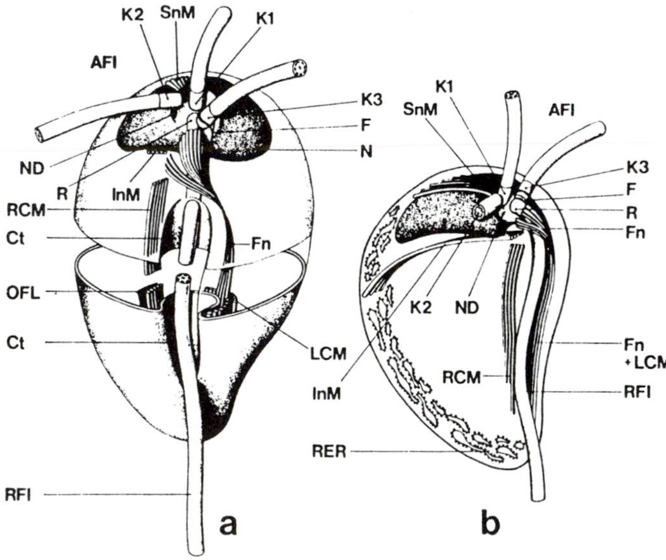

Funis
The diplomonad *Enteromonas*: a. ventral view; b. lateral view. AFl=anterior undulipodium; Ct=cytostomal groove; F=filaments; Fn=funis; InM=infranuclear microtubule bands; K1, K2, K3, R=kinetosomes; LCM, RCM=accessory microtubule bands; N=nucleus; ND=depression in nucleus; OFL=osmiophilic filamentous layer; RER=cisternae of rough endoplasmic reticulum; RFl=recurrent undulipodium; SnM=supranuclear microtubule band.

Funis
Ribbon of microtubules paralleling the recurrent undulipodium (or its intracellular axoneme) to the posterior end of the cell in diplomonads. See also illustration: Diplomonadida.

Furcellaran
Sulfated polysaccharide phycocolloid produced by the rhodophyte *Furcellaria;* wall component with mucilaginous properties, similar to carrageenan.

Furrow
Long narrow structure that differs from a canal in that it is open along its length. Illustration: Undulipodial rootlet.

Fusiform (adj.)
Spindle-shaped; tapering at each end. Illustration: Foraminiferan test.

Fusion cell
Cell produced by the union of the protoplasts of two or more cells.

Fusion competence
State of a gamete that is capable of undergoing sexual fusion; exposure of one or both gametes to pheromones may be required.

Fusule
Complex structure perforating the skeleton and through which pass axopodial axonemes; strand of cytoplasm that connects the region inside the capsule to that outside in polycystine actinopods. Illustration: Actinopoda.

G

G1
Growth phase 1 or gap 1; stage of interphase of mitotic cell cycle preceding DNA synthesis ("S" phase), during which growth occurs.

G2
Growth phase 2 or gap 2; stage in cell cycle, following DNA synthesis but before mitosis, during which growth occurs. During this stage, protein synthesis and increase in organelle number are observed, chromatin condenses, and microtubules are polymerized from tubulin prior to spindle formation.

Gall
Hypertrophy, often spherical or irregular-shaped; growth on plants caused by penetration of plant tissues by xenogenous organisms (e.g., insects, fungi, protoctists, or bacteria).

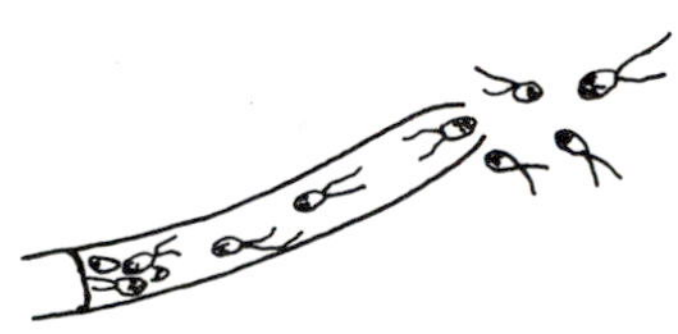

Gametangium
Release of gametes from a gametangium of the chlorophyte *Cladophora kuetzingianum* Grunow.

Gametangium (pl. gametangia)
Any structure in which gametes or gametic nuclei are generated and from which they are released. See *Sporangium.*

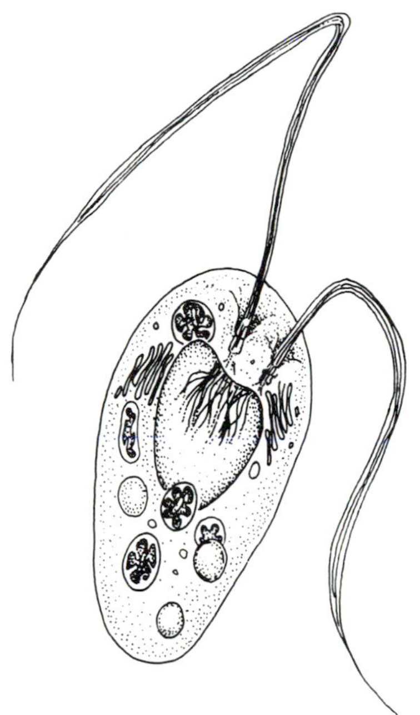

Gamete
Reproductive cell of the type produced by the granuloreticulosans *Hastigerina pelagica* and *Allogromia laticollaris*.

Gamete

Mature haploid reproductive cell or nucleus capable of fusion with another gamete, usu. of a different mating type, to form a diploid zygote nucleus. See *Anisogametes, Anisoplanogametes, Hologamete, Isogametes, Isoplanogametes, Macrogamete, Microgamete*. See also illustrations: Acetabularian life history, Dinomastigote life history, Foraminifera, Gametangium, Life cycle, Opalinata.

Gametic meiosis

Life cycle in which meiosis immediately precedes gamete formation (e.g., most animals and protoctists such as diatoms). See *Zygotic meiosis*.

Gametocyst

Cyst of gamete(s) or cyst forming immediately after syngamy (e.g., in the hypermastigote *Trichonympha*). Illustrations: Acetabularian life history, Life cycle.

Gametocyte

Gamont composed of a single cell. Illustration: Life cycle.

Gametocytotomont

Cell whose multiple division product is a gametocyte. Illustration: Foraminifera.

Gametogamy (adj. gametogamous)

Fusion of gametes. See *Syngamy*.

Gametogenesis

Production of gametes by cell differentiation.

Gametogony

Formation of gametes by multiple fission; in apicomplexans, often as a result of schizogony.

Gametophyte generation

Haploid gamonts. Since the concept originates from a botanical tradition, it refers to the individual plant or alga composed of haploid cells which produce gametes. Characteristic of all plants, many rhodophytes, and phaeophytes with life cycles having alternation of generations. The gametophyte generation usu. begins with the germination of spores that were produced by meiosis; it terminates with fertilization and diploid zygote formation. See *Sporophyte generation*. Illustrations: Florideophycidae, Life cycle.

Gamogony

Gamogonic process; sexual phase in which gametes are eventually produced; series of karyokineses and/or cytokineses leading to gamonts, individuals that produce gamete nuclei or gametes capable of fertilization. Illustrations: Coccidian life history, Foraminifera.

Gamont

Reproducing organism or cell at a stage in its life cycle during which it produces gametes or other sexual structures. See *Agamont, Gametocyte, Gametophyte*. Illustrations: Foraminifera, Life cycle, Opalinata, Trophozoite.

Gamontocyst

Cyst formed around gamonts; when gamonts are single-celled, gamontocysts are gametocysts. In gregarine apicomplexans, a cyst forms around two conjugating gamonts (engaging in syzygy), and fertilization of two ameboid gametes (products of the gamonts) takes place within the gamontocyst.

Gamontogamy (adj. gamontogamous)

Mating of gamonts. Copulation, sexual intercourse, conjugation. Fusion of two or more gamonts is followed by gametogamy (e.g., foraminifera, some gregarines, and some zoomastiginids). Illustration: Life cycle.

Gause's law

See *Competitive exclusion principle*.

Gel electrophoresis

See *Electrophoresis*.

Generative cell

Cell capable of further growth (e.g., free, uninucleate cell within a large myxozoan trophozoite (plasmodium) that gives rise to a pansporoblast); cell capable of further division or of fertilization followed by further division.

Generative nuclei (sing. generative nucleus)

Nuclei capable of further growth and karyokinesis (e.g., small compact nuclei in heterokaryotic foraminifera which are the antecedents of nuclei of the next generation).

Genetic locus (pl. genetic loci)

Position on a linkage group that can be determined by recombination analysis of inherited traits displaying distinguishable genetic alternatives (alleles).

Genetic marker

Gene determining a distinguishable phenotype that can be used to identify a cell or individual that carries it; may also be used to identify a nucleus, chromosome, or locus.

Geniculum (pl. genicula; adj. geniculate)
Uncalcified portion of a thallus between segments of articulated coralline rhodophytes. See *Intergeniculum.*

Genome
Sum of all genes of an organism or organelle. Illustration: Genophore.

Genomic complexity
Complexity; DNA complexity. Measure of the amount of DNA that is present in single copy (unique sequences rather than repeat DNA). This is determined by the kinetics of reassociation of denatured double-stranded DNA and represented as the combined length in nucleotide pairs of all unique DNA fragments.

Geosyncline (adj. geosynclinal)
Very large (hundreds of kilometers long) troughlike depression in the Earth's surface filled with layered sedimentary rocks and produced by orogeny.

Geotaxis
Directed locomotion toward the gravitational center of Earth. Gravitational response dependent on gravitational sensor (e.g., barium sulfate crystals in *Chara).*

Geotropism
Growth directed toward the gravitational center of Earth. Gravitational response dependent on gravitational receptor (e.g., modified plastids in grasses).

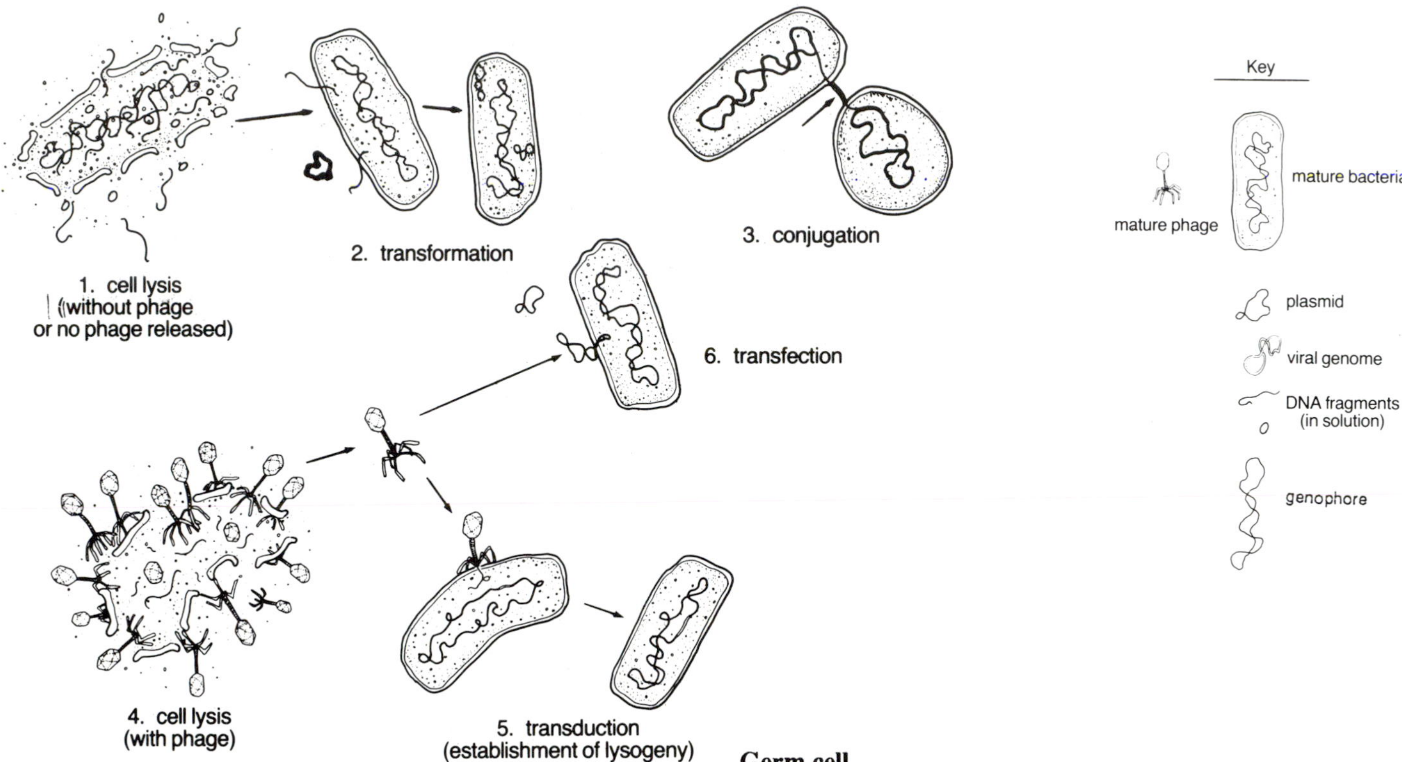

Genophore
Types of bacterial and viral gene exchange.

Genophore
Gene-bearing structure of prokaryotes and certain organelles ("bacterial chromosome," a term to be avoided) (e.g., DNA-containing nucleoid of bacteria, mitochondria, or plastids). Nucleoids are structures visible by microscopy whereas genophores are their equivalents that are inferred from genetic investigation. See *Nucleoid.* See also illustration: Organelle.

Genotype (adj. genotypic)
Genetic make-up of an organism with respect to specific traits, in contrast to the physical appearance of those traits (phenotype).

Germ cell
Sex cell. Cell requiring fertilization before it can grow or reproduce (e.g., ovum (egg), spermatozoan (sperm), gamete). See *Somatic cell.*

Germ tube
Tube-shaped structure capable of further growth (e.g., in chytridiomycotes, hyphochytrids, and oomycotes, a short, hyphalike process that develops upon germination of a spore; usu. gives rise to more hyphae).

Germinate (v.)
Begin to grow (e.g., from a spore or cyst).

Germination chamber
Structure in which growth begins (e.g., in the chrysophyte *Dinobryon,* a chamber formed from the porus of a germinating stomatocyst, into which go the four offspring protoplasts and from which they eventually emerge).

Germling
Bud or newly attached developing propagule capable of growth into an adult of some stage. Illustration: Rhodophyta.

Giant cell
Usu. large cell among those of normal size (e.g., in dictyostelids, zygote that engulfs and digests other cells to attain a large size; in dasycladalean chlorophytes, large cells produced prior to the division of the primary nucleus). Illustration: Dictyostelida.

Girdle
Cingulum; equatorial groove. Midlatitude feature in a spherical organism or structure (e.g., portion of a diatom frustule between the valves, space between the hypocone and epicone of a dinomastigote test). Illustrations: Epicone, Hypovalve, Thecal plate.

Girdle band
Copula. Band-shaped midlatitude structure in spherical structures. Silica band in diatoms, an overlapping series of which intervene between the epivalve and hypovalve; sometimes used synonomously with girdle (cingulum) of dinomastigote tests. See *Girdle lamella*. Illustrations: Frustule, Ligula.

Girdle groove
Surface groove (girdle) in which the transverse undulipodium lies in many dinomastigotes. See *Cingulum*.

Girdle lamella (pl. girdle lamellae)
Band of thylakoids just inside the plastid membrane and arranged peripherally in some classes of algae (e.g., phaeophytes; girdle band of xanthophytes). Illustration: Eyespot.

Glabrous zone
Hairless zone (e.g., nonciliated region used for ingestion in some karyorelictid ciliates).

Gliding motility
Motility of cell or organism always in contact with a solid surface (e.g., glass, rocks, conspecifics) in the absence of external appendages; occurs in both prokaryotes and eukaryotes (e.g., diatoms, labyrinthulids) but differs in mechanisms. Mechanisms of bacterial gliding (e.g., myxobacteria, filamentous cyanobacteria) are unknown. In the labyrinthulids, motility is thought to be related to the presence of a calcium-dependent contractile system of actinlike proteins; in diatoms, actin microfibrils lying in the cytoplasm beneath the raphe slits have been implicated.

Glycocalyx (pl. glycocalyxes or glycocalyces)
Covering; sheath; coat or wall (e.g., surface coating secreted by many of the "naked" rhizopod amebas that covers the plasma membrane; polysaccharide components found outside the bacterial inner lipoprotein membrane). Illustrations: Karyoblastea, Organelle.

Glycogen body
Structure composed of the carbohydrate glycogen. Illustrations: Bicosoecids, Karyoblastea.

Glycolipids
Class of organic compounds composed of a mixture of small carbohydrate and lipid molecules; lipids with sugar esters.

Glycosome
Organelle; peroxisomelike microbody peculiar to kinetoplastid mastigotes that lacks peroxidase but contains enzymes of the glycolytic metabolic pathway. Illustrations: Paraxial rod, Trypomastigote, Zoomastigina.

Glycostyle
Flexible surface projection arising from the cell membrane (e.g., glycocalyx of some amebas, measuring 110 to 120 nm in length, may facilitate ingestion of food particles, including bacteria, because of its stickiness).

Glyoxylate shunt
Biochemical pathway of photorespiration in which organic carbon is converted to amino acids via glyoxylate.

Glyoxysomes
Organelles; membrane-bounded microbodies harboring the enzymes of glyoxylate metabolism. Illustration: Organelle.

Gnotobiotic (adj.)
Term denoting that the biological composition of a preparation or medium is known; germ-free. See *Agnotobiotic culture*.

Golden yellow algae
Chrysophytes (algae classified as Chrysophyta or Chrysophyceae) or haptophytes (prymnesiophytes).

Golgi apparatus
Dictyosome; golgi body. Portion of the endomembrane system of nearly all eukaryotic cells visible with the electron microscope as membranous structure of flattened saccules, vesicles, or cisternae, often stacked in parallel arrays; involved in elaboration, storage, and secretion of products of cell synthesis; prominent in many protoctists (e.g., parabasalians) and less prominent in others (e.g., ciliates). Cis golgi refers to the face of the membrane where vesicles coalesce to form the cisterna; trans golgi refers to the secreting (maturing face) of the golgi apparatus. Cis and trans cisternae contain different enzymes. Illustration: Organelle.

Golgi body
See *Golgi apparatus*.

Gonimoblast
See *Gonimoblast filament*.

Gonimoblast filament
Gonimoblast. In rhodophytes, a filament bearing one or more carpospores or the collection of these filaments that make up the carposporophyte. Illustration: Florideophycidae.

Gonocyte
Dividing cell yielding, by multiple fission, offspring cells capable of propagation (e.g., parasitic colonial dinomastigote cell that gives rise to dinospores during palisporogenesis; in apostome ciliates, offspring cells produced by palintomy).

Gonomere
The ellobiopsid *Thalassomyces marsupii*. An adjacent trophomere is seen through a gonomere (top center). DIC. Bar=100 μm.

Gonomere
Terminal, globular reproductive segment borne on branches called trophomeres in ellobiopsids. See also illustration: Ellobiopsida.

Gonospore
See *Polyspore*.

Gonyautoxins
See *Saxitoxin complex*.

Granellae (sing. granella)
Crystals of barium sulfate (barite) found in large numbers in the cytoplasm of xenophyophorans.

Granellare
The plasma body ("protoplasm") of a xenophyophoran together with its surrounding tubes which are yellowish and branched in varying degrees.

Granule
Small spherical structure, often unidentified. Illustrations: Euglenida, Reservoir, Thylakoid.

Granuloplasm
Granular endoplasm.

Granuloreticulopodium (pl. granuloreticulopodia)
Anastomosing pseudopod (e.g., in phylum Granuloreticulosa).

Granum (pl. grana)
Stack of thylakoids inside plastids, formed by fusion of membranes of adjacent thylakoids.

Grex
See *Slug*.

Gross culture
Agnotobiotic culture; crude culture; culture containing other organisms in addition to the one of interest.

Growth
Increase in size and volume of a cell or of an organism by a number of processes alone or in combination (e.g., increase in size, uptake of water, increase in number of cells, etc.). Apical growth refers to growth at tip; basal growth to growth at base; intercalary growth to growth localized at points between base and tip or between two other nongrowing points.

Gullet
Oral cavity (e.g., canal and reservoir of euglenids or cryptomonads). Illustration: Cryptophyta.

Guluronic acid
Carbohydrate component of alginate.

Gyrogonite
Whorled, ovate carbonate fossils interpreted to be remains of the female gametangia of charophytes (phylum Chlorophyta). Most are of Devonian period, approximately 420 million years old.

H

H-pieces
Cell wall units in some filamentous xanthophytes (e.g., *Tribonema)* composed of the joined half-walls of adjacent cells and forming structures that are H-shaped in optical section.

Habit
Descriptive term referring to morphology of growth (e.g., bushy, capsalean, crustose, filamentous, foliose, filamentous, single-celled, viny, etc.).

Habitat
Immediate surroundings of a population or community. Habitats for protoctists are shown here. See *Biotope*.

Hadal (adj.)
See *Abyssal*.

Haematochrome
See *Hematochrome*.

Halophile (halophil; adj. halophilic)
Ecological term referring to organisms requiring high salt concentrations for growth, including those flourishing in saline environments (e.g., salt-requiring bacteria and protists like *Dunaliella* and *Tetramitus)*. Illustration: Habitat.

Haplobiontic (adj.)
Haplomitotic. Referring to the life cycle of organisms that possess only one morphologically distinct stage; life cycle in which there is only one growing phase; cells in this phase usually have haploid nuclei (e.g., oomycotes). See *Diplobiontic*.

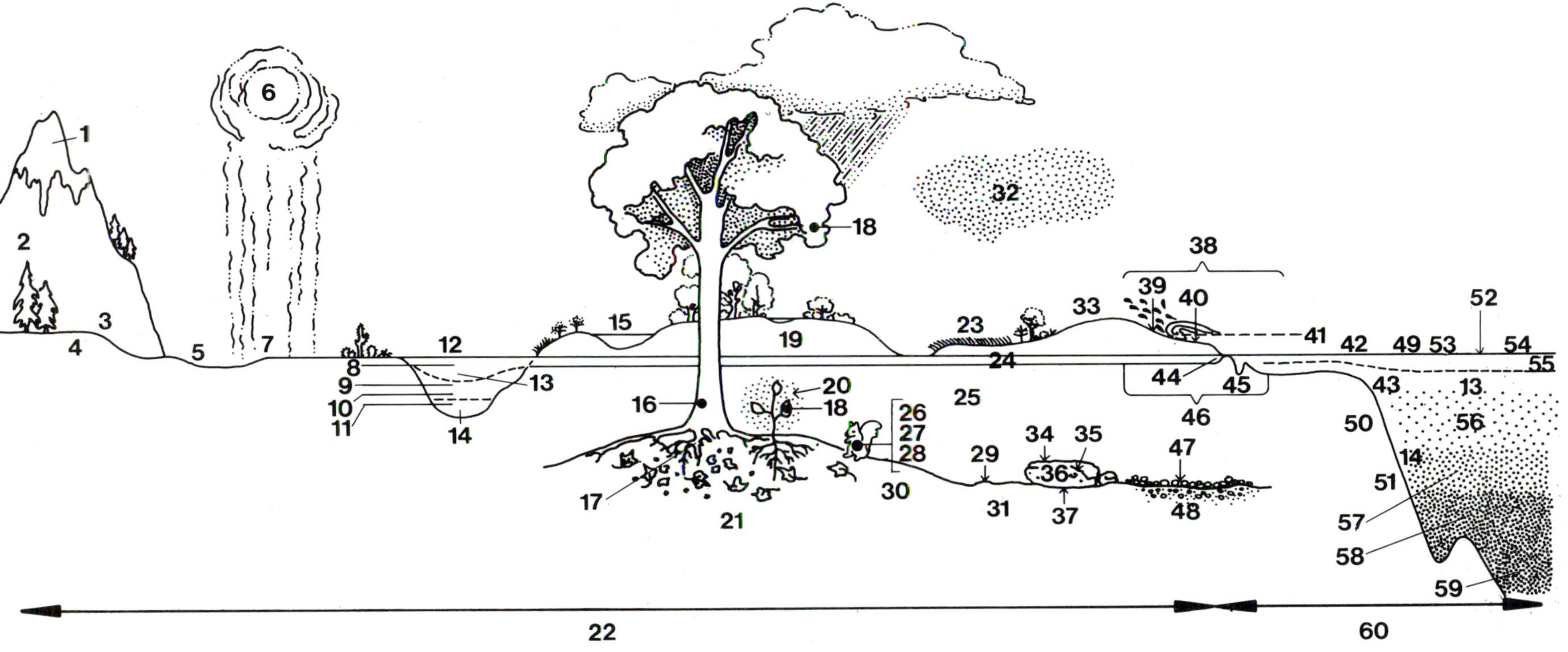

Habitat

1=snow; 2=alpine; 3=tundra; 4=boreal; 5=playa; 6=sun; 7=desert; 8=epilimnion; 9=thermocline; 10=hypolimnion; 11=anoxic layer; 12=lake; 13=planktonic; 14=benthic; 15=pond; 16=lignophilic; 17=endophytic; 18=epiphytic; 19=lentic; 20=metaphyton; 21=soil and litter; 22=fresh water; 23=salt marsh; 24=river; 25=Aufwuchs (periphyton); 26=endobiotic; 27=endozoic; 28=intrazoic; 29=epipelic; 30=infaunal; 31=mud; 32=airborne dust; 33=dune; 34= epilithic; 35=endolithic/chasmolithic; 36=rock; 37=hypolithic; 38=littoral; 39=supralittoral; 40=midlittoral; 41=high tide; 42=low tide; 43=sublittoral; 44=estuary; 45=fjord; 46=intertidal; 47=psammophilic; 48=sand; 49=neritic; 50=continental shelf; 51=continental slope; 52=neuston; 53=oceanic; 54=pelagic; 55=euphotic zone; 56=mesophotic zone; 57=oligophotic zone; 58=aphotic zone; 59=abyssal (bathyal); 60=marine.

Haploid (adj.)
Monoploid. Referring to eukaryotic cells or organisms composed of cells in which the nucleus contains one single complete set of chromosomes, abbreviated 1N. See *Diploid.* Illustration: Life cycle.

Haplomitotic (adj.)
See *Haplobiontic.*

Haplontic (adj.)
Referring to the life cycle of organisms in which individual cells are haploid throughout their history. Diploidy is limited to stage immediately preceding meiosis. See *Diplontic.*

Haplophase
That part of the life cycle in which organisms are haploid, their cells each containing a single complete set of chromosomes. Illustration: Life cycle.

Haplosporosome
Electron-dense membrane-bounded organelle with unknown function; generally spherical but sometimes having profiles that are oblate, spheroidal, vermiform, pyriform, or cuneiform. Another unit membrane (which distinguishes the organelle from other membrane-bounded, electron-dense inclusions in other eukaryotic cells) is found internally in various configurations free of the delimiting membrane (e.g., of haplosporidians and possibly myxozoans and paramyxeans). Illustration: Haplosporidia.

Haplostichous (adj.)
Referring to thallus composed of free or consolidated filaments lacking a true parenchymatous organization (e.g., typical of some phaeophyte orders).

Hapteron
Kelp holdfast; multicellular attaching organ.

Haptomonad
Protist cell attached to any substratum by modified undulipodium (e.g., many prymnesiophytes; the attached, nonswimming stage of trypanosomatids). Illustration: Kinetoplastida.

Haptonema (pl. haptonemata) →
Microtubular appendage, cell organelle, usu. coiled, often used as a holdfast; associated with the undulipodia in prymnesiophytes. Ultrastructure reveals outer sheath of three concentric membranes and an axoneme of [6(1) + 0]: an inner circle of six or seven microtubules surrounded by a cylinder of endoplasmic reticulum; haptonemata may be long and coiling or reduced in length and substructure. See also illustration: Coated vesicles.

Haptonematal root
Kinetid of haptonema of prymnesiophytes; root fiber system of haptonema.

Haptonematal scales
Scales of the axonemal membrane of haptonema.

Haptonemid
See *Haptophyte.*

Haptophyte
Haptonemid. Prymnesiophytes that form haptonemata: coccolithophorids and their relatives; prymnesiophytes with or without coccoliths. Those golden yellow algae that produce coccoliths which may fossilize as marine calcium carbonate sediments. Illustration: Mastigoneme.

Haustorium (pl. haustoria)
Absorbing organelle of osmotrophic protoctists and fungi formed from a projection of a hypha; haustoria penetrate plant cell walls invaginating, but not puncturing, the cell membrane.

Helicoid placolith
Lopodolith. Coccolith subtype; placolith with helical shape; heterococcolith composed of two plates or shields interconnected by a tube (e.g., *Helicosphaera).* See *Placolith.* Illustration: Coccolith.

Helicoid spiral test
See *Trochospiral test.*

Hematochrome
Haematochrome. Astaxanthin or 3,3'-dihydroxy-4,4'-diketo-ß-carotene; red-to-orange-colored pigment found in some euglenids and some chlorophytes.

Hematozoic (adj.)
Referring to mode of heterotrophic nutrition; blood-eating (e.g., vertebrate bloodstream parasites).

Hemiautospore
See *Aplanospore.*

Herbivory (n. herbivore; adj. herbivorous)
Mode of nutrition referring to organisms feeding on plants (see Table 2).

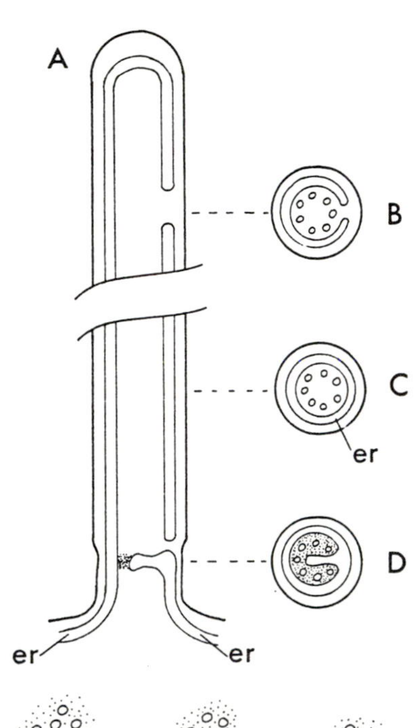

Haptonema
a. Longitudinal section of a haptoneme showing endoplasmic reticulum. The microtubules are not shown in this view. Transverse sections through the shaft (b-d) and at various levels within the cell (e-g) show the arrangement of microtubules. er=endoplasmic reticulum.

Hermaphroditic (adj.; n. hermaphrodite)
Monoecious. Referring to organisms that have both male and female structures on the same individual.

Heterococcolith
Coccolith composed of distinguishable sub-elements; heterococcoliths include placoliths and cricoliths and are made up of morphologically diverse calcite structures; coccoliths in which the crystals show a variety of form and modification. See *Holococcolith.*

Heterodynamic flagella
See *Heterodynamic undulipodia.*

Heterodynamic undulipodia
Undulipodia on the same cell but with different patterns of beating. See *Homodynamic undulipodia.* Illustrations: Euglenida, Reservoir.

Heterogamy
State in which morphologically distinguishable gametes are produced by members of a single species (e.g., anisogamy and oogamy).

Heterogeneric (adj.)
Referring to organisms from different genera.

Heterogenomic (adj.)
From different genomes (e.g., microsporidians are heterogenomic relative to their host tissue).

Heterokaryon (pl. heterokarya)
Cell or organism with cells containing a pair of nuclei in which it can be shown that each is derived from a genetically distinct parent. See *Dikaryon.*

Heterokaryotic (adj.; n. heterokaryosis)
Exhibiting nuclear dimorphism or two or more genetically different nuclei in common cytoplasm (e.g., ciliates and foraminifera with their large and small nuclei and basidiomycete fungi). See *Homokaryotic.*

Heterokont (adj.)
See *Anisokont.*

Heterokontimycotina
Proposed subdivision to include fungi-like protoctists possessing or thought to be derived from ancestors which once possessed heterokont undulipodia at some stage in their life cycle, i.e., oomycotes, labyrinthulids, and hyphochytrids.

Heteromorphic life cycle
Life cycle in which the different phases are morphologically distinct; e.g., alternation of generations in which the diplophase and haplophase are morphologically distinguishable. See *Isomorphic life cycle.*

Heteroplastidy
Simultaneous occurrence in one cell or organism of two kinds of plastids (e.g., chloroplasts and starch-storing leucoplasts).

Heteroside
Chemical compound: type of complex carbohydrate composed of a monosaccharide (hexose) and a nonsugar compound (organic acid, polyol).

Heterospecific (adj.)
Referring to organisms from different species.

Heterothallism (adj. heterothallic)
Protoctist life cycle in which two different clones are required for sexual fusion. Single propagules give rise to individuals of a single mating type; the condition of species in which the sexes (mating types) are segregated in separate clones or thalli; two different clones or thalli of compatible mating types are required for fertilization. See *Homothallism.* Illustration bottom of page.

Heterotrich (adj. heterotrichous; n. heterotrichy)
Filamentous, hairy, or undulipodiated cell or structure with hairs, filaments, or undulipodia of more than a single type (e.g., heterotrichous ciliates; filamentous algal morphology composed of both an erect and prostrate portion).

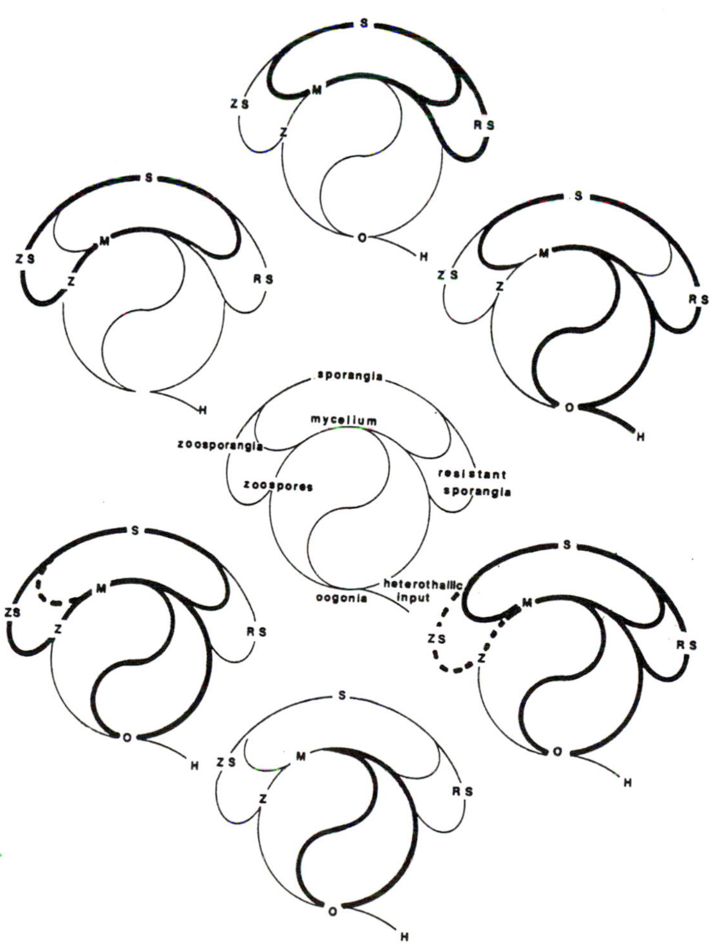

Heterothallism
Various life histories of the Oomycota. H=heterothallic input; M=mycelium; O=oogonia; RS=resistant sporangia; S=sporangia; Z=zoospores; ZS=zoosporangia.

Heterotrophy (n. heterotroph; adj. heterotrophic)
Mode of nutrition in which organisms obtain carbon, electrons, and energy from preformed organic compounds (see Tables 1 and 2). Examples of heterotrophs include algivores, biotrophs, carnivores, necrotrophs, osmotrophs, parasites, phagotrophs, and saprobes.

Heterotype
Minor variable antigen type (VAT) (i.e., trypanosomatid parasites). See *Homotype*.

Heteroxenous (adj.)
Digenetic; dixenous. Descriptive of symbiotrophs with development in their life history in two different types of host. See *Homoxenous parasite, Polyxenous parasite*.

Higher fungi (sing. higher fungus)
Ascomycetes, basidiomycetes, and deuteromycetes (term defined by omission: all fungi except protoctistan "lower fungi"). It is inadvisable to use the terms "higher" and "lower" because of their ambivalence and anthropocentrism.

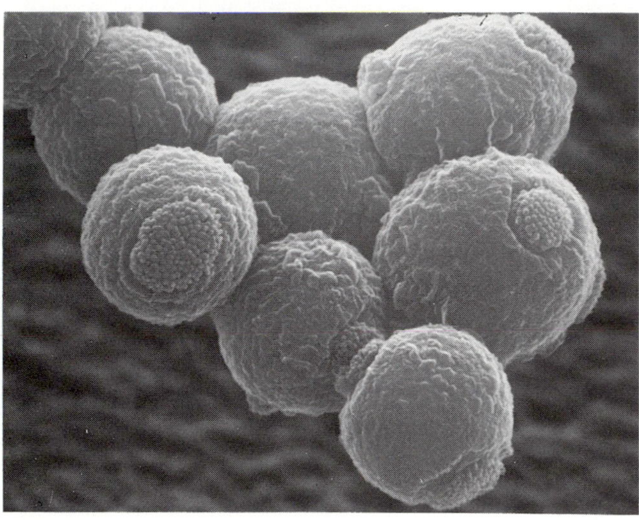

Hilum
Spores of the acrasid *Acrasis rosea*. SEM, x 1,400.

Hilum (pl. hila; adj. hilar)
Areola. Spore-connection scar that appears to be composed of numerous globular particles (e.g., protostelids and acrasids). Scar on a seed marking point of attachment to ovule in plants.

Histones
Lysine-, arginine-rich proteins that complex with nuclear DNA in eukaryotes to form the nucleosome substructure and therefore components of chromatin. Quantity and quality of this class of basic nucleoproteins varies greatly in protoctists whereas histones of plants, animals, and fungi are very similar to each other.

Histophagy (adj. histophagous)
Mode of nutrition; heterotrophy of microorganisms that ingest tissues of animals (see Table 2). See *Carnivory, Holozoic, Osmotrophy, Phagotrophy.*

Histozoic parasite
Ecological term referring to symbiotrophs that live in animal tissues (e.g., myxozoan parasites). See *Coelozoic parasite*.

Holdfast
Peduncle. Attachment structure that may be organ or organelle.

Holocarpic (adj.)
Referring to mode of development in which the thallus is entirely converted into one or more reproductive structures; entire cell is used for production of spores, normally simultaneously but occasionally sequentially (e.g., oomycotes). See *Eucarpic*.

Holococcolith
Coccolith type not composed of sub-elements; holococcoliths include calyptroliths, crystalloliths, and zygoliths and are composed of homogeneous microcrystals; coccoliths in which the $CaCO_3$ is deposited as uniform rhombohedral or hexagonal crystals showing little modification. See *Heterococcolith*.

Hologamete
Gamete of the same size and structural features as growing cells of the same species.

Hologenous sperm formation
Spermatogenesis in which the entire protoplasm of a microspore converts to form a sperm (e.g., in centric diatoms).

Holotrophic (adj.)
See *Holozoic*.

Holozoic (adj.)
Holotrophic; phagotrophic. Referring to mode of heterotrophic nutrition involving motile pseudopods in which food is obtained by ingestion of relatively large, solid organic particles (e.g., live bacteria or protists). See *Carnivory, Histophagy, Osmotrophy, Phagotrophy.*

Homodynamic flagella
See *Homodynamic undulipodia*.

Homodynamic undulipodia
Undulipodia on the same cell with the same pattern of beating. See *Heterodynamic undulipodia*.

Homogenomic (adj.)
From the same genome (e.g., heterokarya of ciliates).

Homokaryotic (adj.; n. homokaryosis)
Referring to cells that have only a single kind of nucleus as determined by genetics and morphology (the number of nuclei may be greater than one per cell). See *Heterokaryotic*.

Homology (adj. homologous)
Structure or physiology of common evolutionary origin but not necessarily identical in present structure and/or function (e.g., haptonemata and sperm tails are homologous structures); term in molecular biology referring to the degree of sequence similarity of DNA from different sources.

Homothallism (adj. homothallic)
Descriptive term of protoctist life cycle in which members of a single clone are adequate to ensure fertilization. Homothallism describes the sexual system in protoctists and fungi in which single propagules give rise to individuals of compatible mating types such that the sexual process of gamontogamy and/or gametogamy can occur by mating of members of a clone or cells from a single thallus. Homothallic clones or thalli are therefore self-compatible. See *Heterothallism.*

Homothetogenic fission
Type of transverse binary fission such that a point-to-point correspondence (homothety) is maintained between structures in both progeny (e.g., most ciliates). See *Perkinetal fission, Symmetrogenic fisson.*

Homotype
Major variable antigen type (trypanosomatid mastigotes); as homotype is destroyed by host vertebrate antibody response, heterotype multiplies to become dominant, forming the new homotype. See *Heterotype.*

Homoxenous parasite
Ecological term describing a parasite which completes its life history in a single host. See *Heteroxenous.*

Hormogonium (pl. hormogonia)
Type of propagule; short filaments that break off parent organism, disperse, and are capable of further growth (e.g., some cyanobacteria and algae).

Host
Organism that provides nutrition or lodging for symbionts or parasites. The larger member of a symbiotrophic association. Illustrations: Coccidian life history, Cystosorus, Macrogamont, Microgamete, Sporocyst, Stachel.

Humic acids
Alkaline- or water-soluble compounds extractable from humus.

Humus
Layer of loose organic debris on land composed of sufficiently decayed organic materials such that their origins are obscured; organic rich soil.

Hyaline (adj.)
Glassy, translucent, or transparent; descriptive of ameba cytoplasm or tests of foraminifera. Hyaline tests may be subdivided into subtypes (radial, oblique, compound) depending upon the orientation of the crystal laths. Illustration: Foraminiferan test.

Hyaloplasm
Clear, organelle-free cytoplasm (e.g., ectoplasm of ameba pseudopods).

Hyalosome
Lens portion of a dinomastigote ocellus (e.g., some members of the Warnowiaceae).

Hybridization
See *DNA hybridization.*

Hydrogenosomes
Chromatic granules; paracostal granules; paraxostylar granules. Type of microbody; membrane-bounded organelles of zoomastiginids, *Entamoeba*, and anaerobic ciliates, 0.5-2.0 μm; especially in trichomonads and other mastigotes, which, under anaerobic conditions, generate H_2. Hydrogenosomes have been called anaerobic mitochondria.

Hyperhaline (adj.)
Hypersaline. Referring to water of higher than typical marine salinity; i.e., greater than about 3.5% sodium chloride and other salts. See *Euhaline, Euryhaline, Oligohaline.*

Hyperparasitism
Ecological term describing the topology of symbiotrophs in which a symbiont itself maintains a second symbiont; threeway symbiosis (e.g., a microsporan parasite of a myxozoan parasite of a fish).

Hypersaline (adj.)
See *Hyperhaline.*

Hyperseme
Increase in number or size of a trait of evolutionary significance (e.g., number of undulipodia in parabasalians). See *Seme.*

Hypertrophy
Abnormal enlargement of a body part or structure (e.g., a gall).

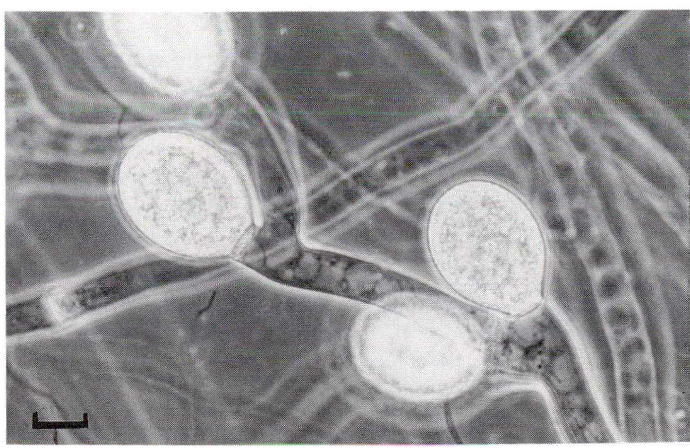

Hypha
Resting spores and filamentous hyphae of the chytridiomycote *Allomyces* sp. DIC. Bar=1.0 μm.

Hypha (pl. hyphae; adj. hyphal)
Long slender threadlike cells, walled syncytia, or parts of cells comprising the body of most fungi and many protoctists (e.g., oomycotes or elongate cells of the medulla of kelps and fucoids). See also illustration: Oomycota.

Hyphochytrid
Informal name of members of the phylum Hyphochytriomycota. Illustrations: Hyphochytriomycota, Mastigoneme.

Hypnocyst
Resting cyst of a dinomastigote. Illustration: Dinomastigote life history.

Hypnospore
Thick-walled aplanospore (e.g., dinomastigotes).

Hypnozygote
Thick-walled zygote; fossilized form of dinomastigote encountered in core samples; interpreted to be nonmotile zygote of dinomastigotes with a three-layered outer wall.

Hypocingulum (pl. hypocingula)
Lower portion of the girdle (cingulum) adjacent to the hypotheca (e.g., dinomastigotes). In diatoms, cingulum of the hypovalve. Illustration: Frustule.

Hypocone
Hyposome. Lower surface or hemisphere posterior to the girdle (cingulum) (e.g., dinomastigotes). Illustration: Epicone.

Hypogynous (adj.)
Botanical term meaning below the ovary or female reproductive structure; as applied to protoctists, refers to antheridia of oomycote *Apodachlya;* in rhodophytes, hypogynous cell subtends a carpogonium in a carpogonial filament. Term should be avoided when referring to protoctists.

Hypolimnion
The lowermost layer of water in a lake, characterized by uniform temperature, generally colder than elsewhere in the lake; often relatively stagnant, oxygen-poor water. See *Epilimnion.* Illustration: Habitat.

Hypolith (adj. hypolithic)
Ecological term referring to organisms dwelling on the underside of rocks. See *Epilith.* Illustration: Habitat.

Hyposeme
Decrease in number or size of a trait of evolutionary significance (e.g., reduction of oral apparatus of *Mesodinium rubrum* relative to other mesodinia). See *Seme.*

Hyposome
See *Hypocone.*

Hypothallus (pl., hypothalli; adj. hypothallial)
Thin, often transparent deposit at the base of sporocarps (e.g., myxomycotes); lowermost tissue in crust on which one or more layers of filaments are oriented parallel to substrate in rhodophytes; medulla.

Hypotheca (pl. hypothecae)
Hypovalve and hypocingulum portion of diatom frustule; posterior portion of a thecate dinomastigote cell. Illustration: Dinomastigota.

Hypovalve
Lower shell, opposite to and usu. smaller than the epivalve (e.g., diatom frustules). See also illustrations: Bacillariophyta, Frustule.

Hystrichosphere
Fossil dinomastigote cyst; thick-walled, nearly spherical structure bearing characteristic projections and markings including an apparent excystment aperture (archeopyle). Fossil objects (Proterozoic to Recent) lacking enough detail to be identified as hystrichospheres are classified as acritarchs.

I

Ichnofossil
Trace fossil. Mark left by extinct organisms resulting from their life activities (e.g., *Scolithus*, a vertical tube, burrow, or trail). Geologists often give these genus and species names.

Ichthyotoxin
Any substance toxic to fish.

Imbrication pattern
Overlapping pattern.

Immunofluorescence
Visual detection by fluorescence microscopy of the presence and distribution of specific antigens on or in cells using antibodies bound to fluorescent molecules.

Imperfect fungi (sing. imperfect fungus)
Fungi in which sexual stages have not been observed; form phylum Deuteromycota. Often closely related to identifiable asco- or basidiomycota.

Imperforate (adj.)
Lacking opening or aperture.

Inbreeding
Mating and production of offspring by organisms known to be derived from recorded common ancestors.

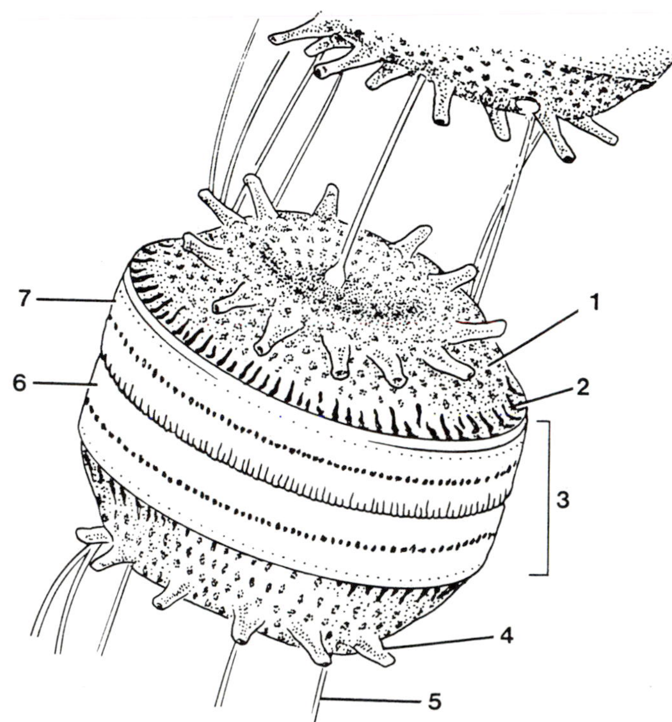

Hypovalve
The diatom *Thalassiosira nordenskjøldii.* 1=mantle; 2=striae; 3=girdle or pleura; 4=spine; 5=seta; 6=hypovalve; 7=epivalve.

Indeterminate growth
Growth that continues indefinitely under optimum conditions (e.g., acellular slime molds, labyrinthulids, mycelial fungi). See *Determinate growth.*

Indeterminate sporangial renewal
Continued cell division such that growth and development are not necessarily terminated by the appearance of a sporangium (e.g., oomycotes). See *Cymose renewal, Internal renewal.*

Inducing factors
Pheromones (e.g., substances that stimulate the sexual maturation of immature individuals).

Infaunal (adj.)
Ecological term referring to organisms covered by sand, mud, or other sediment. Illustration: Habitat.

Infection
Initiation of symbiotrophic (including necrotrophic) relationship among organisms of different species. Illustration: Plasmodiophoromycota.

Infectious germ
Ecological term describing the stage in which a symbiotroph is capable of continuing growth; propagule stage in which parasite is infectious (e.g., apicomplexans, motile zoite; myxozoans, ameboid sporoplasms within the spore).

Infraciliature
Kinetids taken together; that layer of cortex containing undulipodial substructure. Assemblage of all kinetosomes and associated subpellicular microfibrillar and microtubular structures (i.e., ciliates, opalinids).

Ingestatory apparatus
See *Ingestion apparatus.*

Ingestion apparatus
Oral apparatus; ingestatory apparatus. Entire complex of structures and organelles involved in or directly related to the mouth *sensu lato* and its ingestatory function (e.g., multiple in suctorians and absent in astomatous ciliates and opalinids; cryptomonad gullet). Illustration: Euglenida.

Inner membrane complex
Cellular structure: flattened vesicles forming a double membrane lining the plasma membrane.

Inoculum (pl. inocula)
Starter; a subpopulation, usu. of microorganisms, used to transfer a culture for continued growth on fresh culture medium.

Insolation
Incoming solar radiation; solar radiation received at the Earth's surface; the flux of direct solar radiation incident upon a horizontal surface.

Interaxonemal substance
See *Axonemal dense substance.*

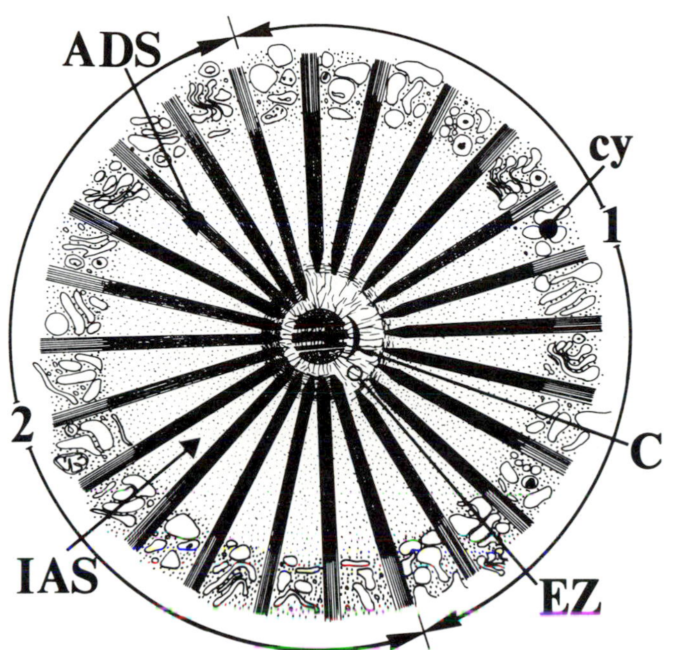

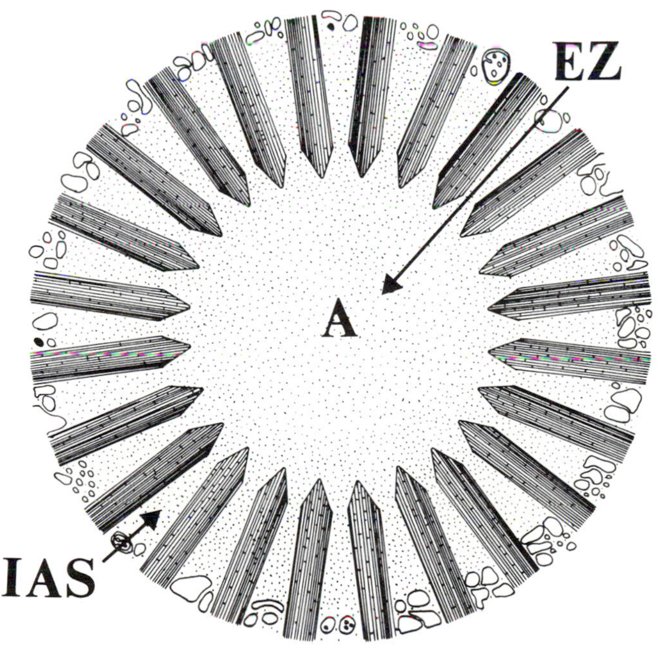

Interaxonemal substance
Internal structure of Centroaxoplasthelida (top) and Axoplasthelida (bottom) heliozoans. 1=Acanthocystidae ultrastructure; 2=Heterophryidae ultrastructure; A=axoplast; ADS=axonemal dense substance in which axonemes are embedded; C=centroplast consisting of tripartite disk sandwiched between two electron-dense caps; cy=cytoplasm; EZ=clear exclusion zone; IAS=interaxonemal substance.

Intercalary (adj.)
Between nodes. See *Growth.*

Intercalary band
Zone, often striated, around margins of dinomastigote thecal plates where cell growth occurs by addition of wall material. Illustration: Thecal plate.

Intercameral (adj.)
Between chambers (e.g., tests of foraminifera).

Intergeniculum (pl. intergenicula)
Between genicula; calcified section between uncalcified joints; inflexible region of the axis of segmented thalli (e.g., coralline rhodophytes or calcified chlorophytes). See *Geniculum.*

Interiomarginal aperture
Opening in a final chamber bounded in part by the wall of an earlier chamber in tests of foraminifera.

Interkinetal fission
Fission between the kineties; descriptive of longitudinal cell division (e.g., mastigotes, opalinids). See *Perkinetal fission.*

Interkinetal space
Nonciliated region of a ciliate cortex that lies between kineties. Illustration: Oral region.

Intermediate host
Animal host in which only the asexual or immature stages of the symbiont occur (e.g., coccidian *Aggregata eberthi* in crab). See *Definitive host.*

Internal renewal
Proliferation through the sporangial septum of the sporangiophore producing a new hypha, a sporangium of undetermined size, or a combination of these (e.g., oomycotes). See *Indeterminate sporangial renewal, Percurrent development.*

Internal toothplate
Projection into the aperture of foraminiferan test. The internal portion of the toothplate usu. extends as far as the previous foramen did through the chamber.

Internode
Portion of stem or thallus lying between the nodes or joints (e.g., in the chlorophyte *Chara* or the rhodophyte *Ceramium*).

Interphase
Growth stage in the cell cycle of eukaryotes between successive mitoses in which the processes of mRNA transcription and protein synthesis are most active. Chromatin is uncondensed and invisible or difficult to see and stain. Illustration shows interphase nucleus.

Interseptar (adj.)
Between septae or partitions, referring to the position of a structure.

Interstitial (adj.)
Existing in small or narrow spaces between things or parts; ecological term referring to organisms or material between sand grains or mud particles. See *Psammophile.*

Intertidal (adj.)
Eulittoral; littoral. Ecological term referring to the areas situated between the tides and therefore covered with seawater at high tide and exposed at low tide. Synonym of littoral in one of its senses: the benthic ocean environment or depth zone between high water and low water.

Intervening sequence
See *Intron.*

Interzonal spindle
Array of microtubules extending from one end of a cell to the other, i.e., those microtubules extending in telophase between the two offspring nuclei; distinguished from kinetochoric or chromosomal microtubules that only extend from the kinetochores to the spindle poles.

Intracristal filament
Filament occurring within the inner membrane (cristae) of mitochondria.

Intraerythrocytic (adj.)
Ecological term referring to the topology of symbiotrophs in red blood cells (erythrocytes) of vertebrates (e.g., the intraerythrocytic stage of apicomplexans involves multiple fissions which lead to destruction of the red blood cells and poisoning by the breakdown products of erythrocytes).

Intratissular (adj.)
Topological term referring to endosymbionts located within a tissue. See *Pseudointratissular.*

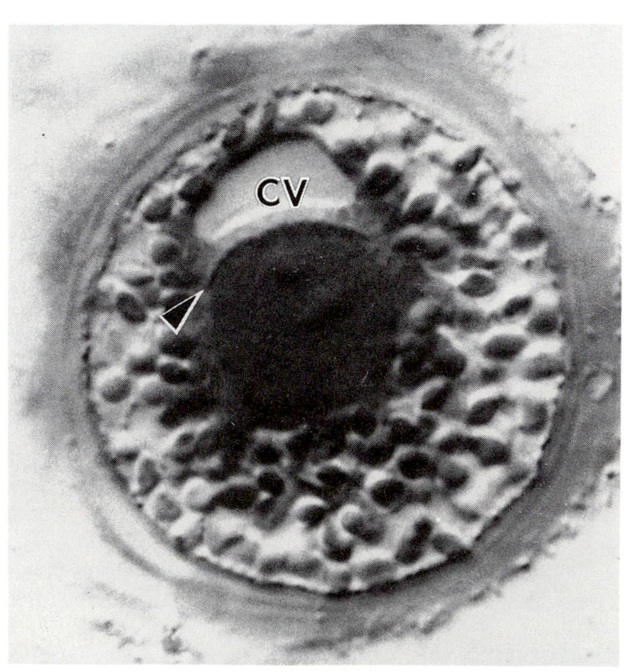

Interphase
Section through palmelloid cell of the raphidophyte *Vacuolaria virescens.* Layers of mucilage surround the cell. Large numbers of disk-shaped plastids are present between the nucleus and the cell membrane. Arrow=golgi; CV=contractile vacuole. DIC, x 1,320.

Intrazoic (adj.)
Topological term referring to organisms located inside animals. See *Endozoic, Entozoic, Holozoic.* Illustration: Habitat.

Intrinsic encystment
Formation of cysts induced by nutrient depletion (e.g., in the chrysophyte *Dinobryon cylindricum).*

Intron
Intervening sequence. Untranslated sequence of DNA that forms part of a gene and is removed by splicing of the corresponding mRNA after transcription. See *Exon.*

Intussusception
Invagination; the assimilation into a structure of new material and its dispersal among pre-existing material.

Involucre
Nongrowing (sterile) group of cells or filaments that form envelopes around growing (fertile, reproductive) structures.

Involute (adj.)
Referring to tests that are curled spirally; having the whorls closely coiled; curled inward; having the edges rolled over the upper surface toward the midrib. Each whorl may completely embrace and cover earlier whorls so that only the final whorl is externally visible (e.g., tests of foraminifera).

Iodine test
Test for the presence of starch: treatment of cells or tissues with weak aqueous solution of iodine-potassium iodide (Lugol's solution) in which colorless or white starch grains turn blue to black.

Ion exchange chromatography
Technique for separating and identifying the components from mixtures of molecules using a resin that has a higher affinity for some charged organic ions than it has for others.

Ionophores
Class of bacterially derived compounds including antibiotics that facilitate the movement of mono- and divalent cations across biological membranes.

Isoaplanogametes
Nonmotile gametes of equal size (e.g., hyphochytrid *Anisolpidium ectocarpii).*

Isoenzymes
Isozymes. Variants of a given enzyme occurring within a single organism, having the same affinity for substrate but differing sufficiently in molecular structure so that their separation is possible (usu. by electrophoresis).

Isofilar (adj.)
Referring to filamentous structure composed of stretches of equal or nearly equal width along its length (e.g., microsporan polar tube in the everted state or myxozoan polar filament). See *Anisofilar.*

Isogametes (adj. isogametous)
Gametes similar in size and morphology to the corresponding gametes of the opposite mating type. See *Anisogametes.* Illustration: Acetabularian life history.

Isogamonts (adj. isogamontous)
Gamonts of a given species that are the same in size or form. See *Anisogamonts.*

Isogamy (adj. isogamous)
Pairing of gametes alike in morphology and size to opposite mating types (isogametes). See *Anisogamy.*

Isokont (adj.)
Referring to an undulipodiated cell bearing undulipodia of equal length (e.g., *Chlamydomonas).* See *Anisokont.*

Isolate
Population or strain of organisms under investigation in the laboratory.

Isomorphic life cycle
Life cycle having alternation of generations in which individuals (e.g., gametophyte, sporophyte) are morphologically similar. See *Heteromorphic life cycle.*

Isoplanogametes
Undulipodiated gametes (swarmers) of equal size destined for sexual reproduction. See *Anisoplanogametes, Isoaplanogametes.*

Isoprenoids
Class of organic compounds synthesized from multiples of a ubiquitous five-carbon compound precursor (isopentenyl pyrophosphate). Includes carotenoids, phytol, terpenes, steroids, and many other important biochemicals.

Isotherm
A line drawn on a map or chart linking all points with the same mean temperature for a given period or the same temperature at a given time.

Isotope fractionation
Any process leading to the selective incorporation of certain isotopes of an element (e.g., photosynthetic organisms fractionate carbon: they preferentially incorporate a ratio of ^{12}C to ^{13}C greater than that present in the atmosphere).

Isozymes
See *Isoenzymes.*

Isthmus (pl. isthmuses or isthmi)
Equatorial region connecting two half cells (e.g., desmids).

K

K
Saturation density. In population biology the density of organisms at which population growth no longer occurs, i.e., the reproduction rate equals the mortality rate.

Kappa particles
Cytoplasmic particles correlated with the capacity of *Paramecium* to kill conspecifics. Bacteria *(Caedibacter)* of *Paramecium.* See *Xenosomes.*

Karyogamy
Fusion of nuclei; usu. follows syngamy and leads to zygote production. Illustration: Life cycle.

Karyokinesis
Division of the nucleus to form two offspring nuclei. Illustration: Foraminifera.

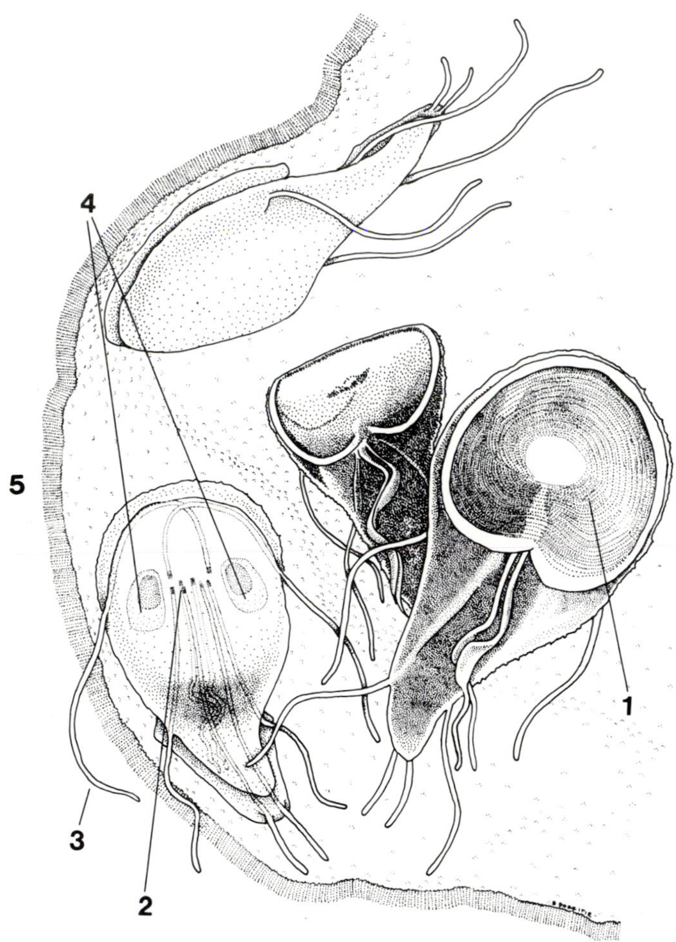

Karyomastigont system
The diplomonad zoomastiginid *Giardia.* 1="sucker" (adhesive disc) that holds *Giardia* onto intestinal lining where it feeds by absorption; 2=kinetosome; 3=undulipodium; 4=nuclei; 5=host intestinal lining.

Karyomastigont system
Mastigont system with its associated nucleus (or nuclei) (e.g., diplomonads). See *Mastigont system.*

Karyonide
Clonal population of organisms, maintained in laboratories, in which no nuclear reorganization has been allowed to occur.

Karyosome
See *Endosome, Nucleolus.*

Karyotype
Total chromosome complement of an animal, plant, fungus, or protoctist as seen in fixed and stained preparations of condensed chromosomes using the light microscope; karyotyping is a fixation and staining procedure used to determine characteristic morphology and number of chromosomes for a species.

kDNA
Kinetoplast DNA. Circular DNA molecules of trypanosomatids of two size classes, maxicircles and minicircles; usu. catenated together in a network. See *Kinetoplast.*

Keel
The euglenid *Petalomonas,* showing single emergent undulipodium, subapical opening of canal, and prominent keels or ridges on typically ovoid, flattened cell. x 500.

Keel
Longitudinal plate or timber extending along the center of the bottom of a ship and projecting outward into the ocean; in protoctists, any projection resembling a keel (e.g., peripheral thickening of a foraminiferan test or a ridge on the valves of some pennate diatoms).

Kelp
Large phaeophytes that are members of the Laminariales and *Durvillaea* spp.

Kinete
Motile form of the zygote of hematozoan apicomplexans; the kinete has zoite-like features such as pellicle, subpellicular microtubules, rhoptries, and micronemes.

Kinetid
Basal apparatus; flagellar (undulipodial) apparatus. Kinetosomes and their associated tubules and fibers present in all undulipodiated cells; unit of organization of the ciliate cortex. The functional organellar complex, including undulipodia, is usu. responsible for locomotion. Synonyms include: basal apparatus; flagellar apparatus; flagellar root system; proboscis root; root fiber system; undulipodial apparatus; kinetosomal territory; ciliary corpuscle. Kinetids always consist of at least one kinetosome, but may have pairs or occasionally more than two kinetosomes (e.g., they may be dikinetids or polykinetids). Structures associated with the kinetosomes of ciliates usually include cilia, unit membranes, alveoli, kinetodesmata, and various ribbons, bands, or bundles of microtubules (e.g., postciliary microtubules and some nematodesmata). Root microtubules of kinetids may

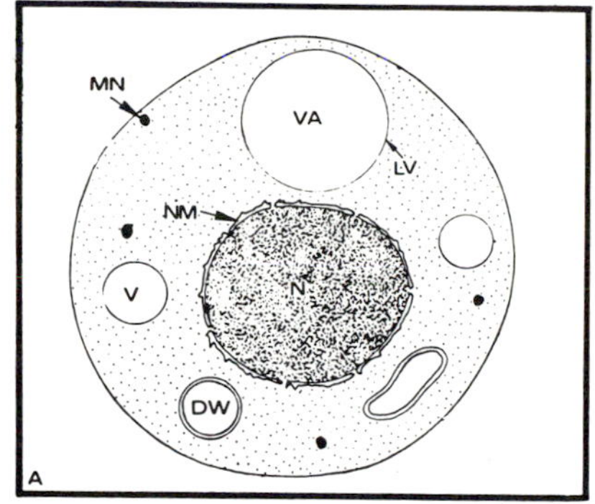

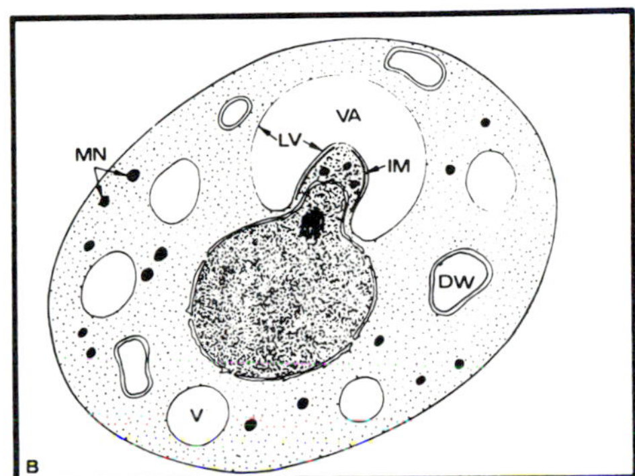

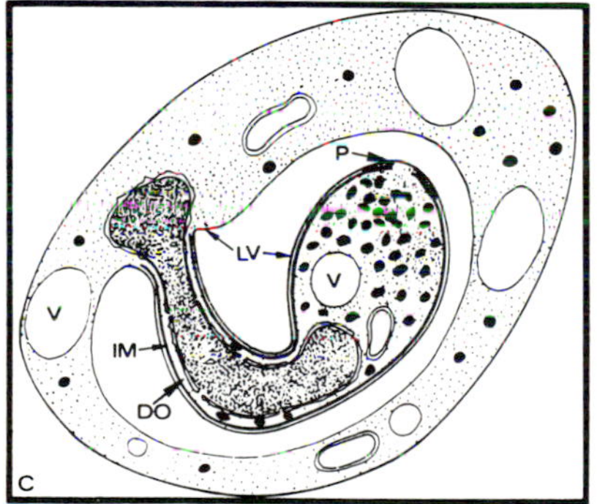

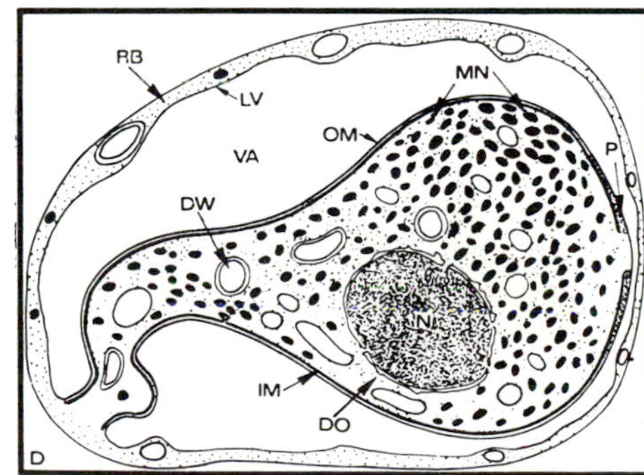

Kinete

Transformation of the zygote of the hematozoan apicomplexan *Theileria* sp. (a) into a motile kinete (d). The kinete protrudes (b) in a vacuole previously differentiated in the zygote cytoplasm (c). DO=kinete; DW=mitochondrion; IM, OM=inner and outer membranes of the pellicle; LV=membrane of the vacuole that gives rise to the outer membrane of the pellicle; MN=micronemes; N=nucleus; NM=nuclear envelope; P=polar ring; RB=residual body; V, VA=vacuole.

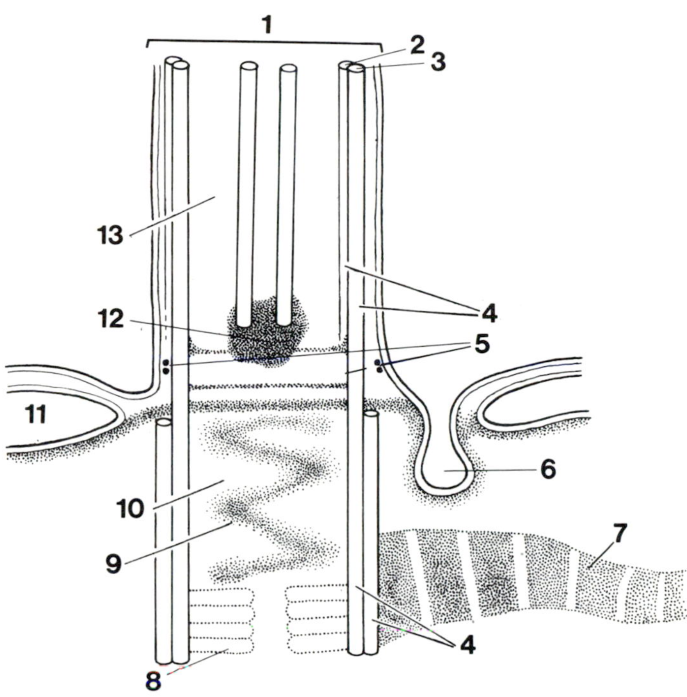

Kinetid

Longitudinal section of a ciliate kinetid. 1=undulipodium; 2=B-tubule; 3=A-tubule; 4=microtubules; 5=ciliary necklace; 6=parasomal sac; 7=kinetodesmal fiber; 8=cartwheel; 9=NP; 10=[9(3)+0] kinetosome; 11=alveolus; 12=axosome (axonemal dense substance); 13=[9(2)+2] axoneme.

be laterally associated microtubules that originate at kinetosomes in definite numbers and follow a defined path within the cell (e.g., ciliates). Some kinetids are also comprised of microfibrils, myonemes, parasomal sacs, mucocysts, or trichocysts. Details of the kinetid are essential for taxonomic and evolutionary studies of motile protoctists. See *Kinety, Monokinetid, Oral kinetid, Polykinetid, Somatic kinetid.* See also illustrations: Cortex, Rhizoplast, System I fiber, Transverse ribbon.

Kinetid root

See *Banded roots.* Illustration: Coated vesicles.

Kinetochore

Centromere. Microtubule-organizing center usu. located at a constricted region of a chromosome that holds chromatids together. Kinetochores, morphologically visible manifestations of centromeres, are the site of attachment of microtubules forming the spindle fibers during nuclear division (mitosis and meiosis). In general, kinetochores reproduce in synchrony with the chromosomes and divide in two at metaphase, one new centromere segregating with each chromatid to the poles of mitotic spindle. "Centromere" is a synonym, or if distinguished, centromeres are deduced from genetic behavior whereas kinetochores are directly visible by electron microscopy. Illustrations: Chromosome, Mitotic apparatus.

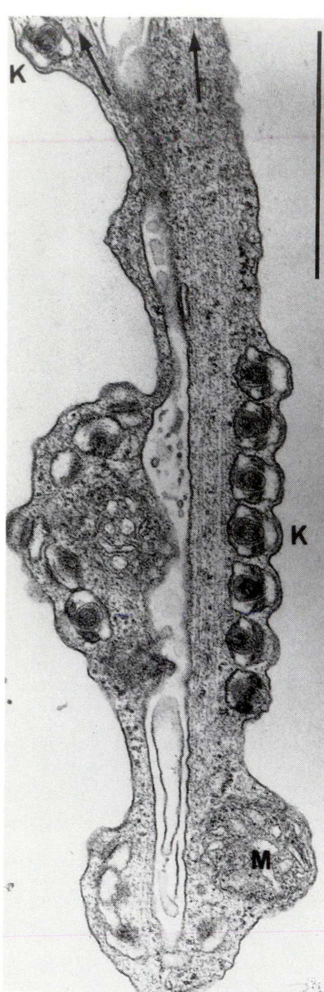

Kinetocyst
Longitudinal section through filopods of the heliozoan actinopod *Clathrulina elegans*. K=kinetocysts; M=mitochondria. TEM. Bar=1 µm.

Kinetocyst
Extrusome with a complex substructure of missile-like differentiations embedded in a fuzzy material (e.g., heliozoan and polycystine actinopods).

Kinetodesma (pl. kinetodesmata)
Banded root; banded root fiber system; kinetodesmal fiber; kinetodesmos; striated fiber. Periodically striated, longitudinally orientated, subpellicular fiber (or component fibrils) arising close to the base of a somatic kinetosome (posterior one, if paired), near its microtubular triplets 5-8 and extending anteriorly toward or parallel to the organism's pellicular surface and always on the right side of its kinety ("law of desmodexy"). Structure is diagnostic for ciliates. Striated fibers showing exception to these characteristics and orientation are not true kinetodesmata. Kinetodesmata of a length greater than the interkinetosomal distance along the kinety overlap, shingle-fashion, producing a bundle of fibers. These are well developed in apostome, hymenostome, and scuticociliate ciliates. They are present as large and heavy bundles in certain astomous ciliates. Illustration: Cortex.

Kinetodesmal (striated) fiber
See *Kinetodesma*. Illustrations: Kinetid, Kinetosomes.

Kinetodesmal fibril
Portion of kinetids of ciliates; striated rootlet fibril originating near triplets 5, 6, 7, and 8 of the ciliate kinetosome and extending into the cortex. Illustrations: Cortex, Transverse ribbon.

Kinetodesmos (kinetodesmose)
See *Kinetodesma.*

Kinetoplast
Modified mitochondrion; intracellular DNA-containing structure, often near a kinetosome, characteristic of the kinetoplastids (class Kinetoplastida in phylum Zoomastigina); the mitochondrial DNA that characterizes the kinetoplast is usu. associated with the mitochondrial envelope apposed to the kinetosome(s). Kinetoplasts reproduce prior to the nuclei in cell division. Illustrations: Bodonidae, Kinetoplastida, Polykinetoplastic, Trypomastigote, Zoomastigina.

Kinetoplast DNA
See *kDNA*.

Kinetosome props
Fibrillar kinetosome props. Fibrillar, often coarse structures of kinetids of chytridiomycote zoospore that connect at an angle of about 45° to the nine C-tubules of the kinetosome triplet tubules and extend to the plasma membrane. See *Transition fibers*. Illustration: Chytridiomycota.

Kinetosomes →
Basal bodies. Intracellular organelles not membrane-bounded, characteristic of mastigotes and all other undulipodiated cells. Microtubule structures, cylinders about 0.25 µm in diameter and up to 4 µm long. Their microtubules are organized in the [9 (3) + 0] array; all undulipodia are underlain by kinetosomes. These basal organelles are necessary for the formation of all undulipodia; kinetosomes differ from centrioles (which share cross section characteristic of a circle of nine triplets of microtubules) in that from them extend [9(2) + 2] axonemes. The term kinetosome, because of its precision, is preferable to basal body. See also illustrations: Anisokont, Atractophore, Cortex, Eyespot, Funis, Karyomastigont system, Kinetid, Mitotic apparatus, Multilayered structure, Organelle, System I fiber, Undulipodial rootlet, Undulipodium.

Kinety (pl. kineties)
Structure of the ciliate cortex; a row of kinetids. Kineties, typically oriented longitudinally, are composed of kinetids (single, paired, or occasionally several kinetosomes, their axonemes, and other associated cortical structures). These rows are bipolar (though some may be interrupted, fragmented, intercalated, partial, shortened, etc.), with an asymmetry allowing recognition of anterior and posterior poles of the organism; kinety sometimes also refers to linearly aligned buccal infraciliar structures. Illustrations: Ciliophora, Oral region.

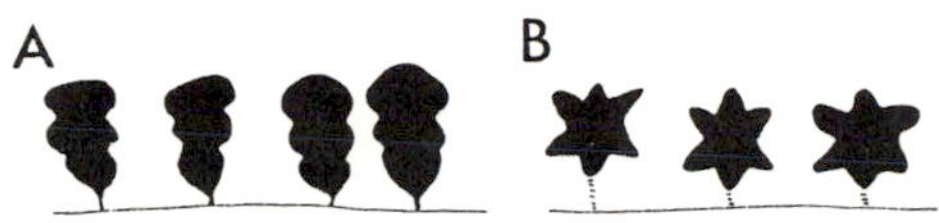

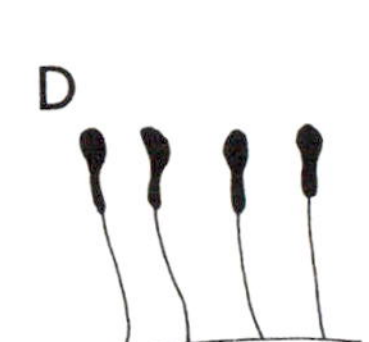

Knob scales
Knob scales of prymnesiophytes in the order Pavlovales. a. Undulipodial scales of *Pavlova helicata*. b. Undulipodial scales of *P. calceolata*. c. Body scales of *P. gyrans*. d. Body scales of *P. salina*.

Knob scales

Scales; type of undulipodial surface structure in a subgroup of the prymnesiophytes.

Koch's postulates

Criteria for proving that a specific type of microorganism causes a specific disease; formulated by Robert Koch. The postulates state that the microbe should be found in diseased animals, but not healthy ones; that the organism must be grown in pure culture away from animal; that this culture should cause the disease when injected into a healthy animal; and that the organism must be reisolatable from the experimental animals, reculturable, and visibly identifiable as the original organism.

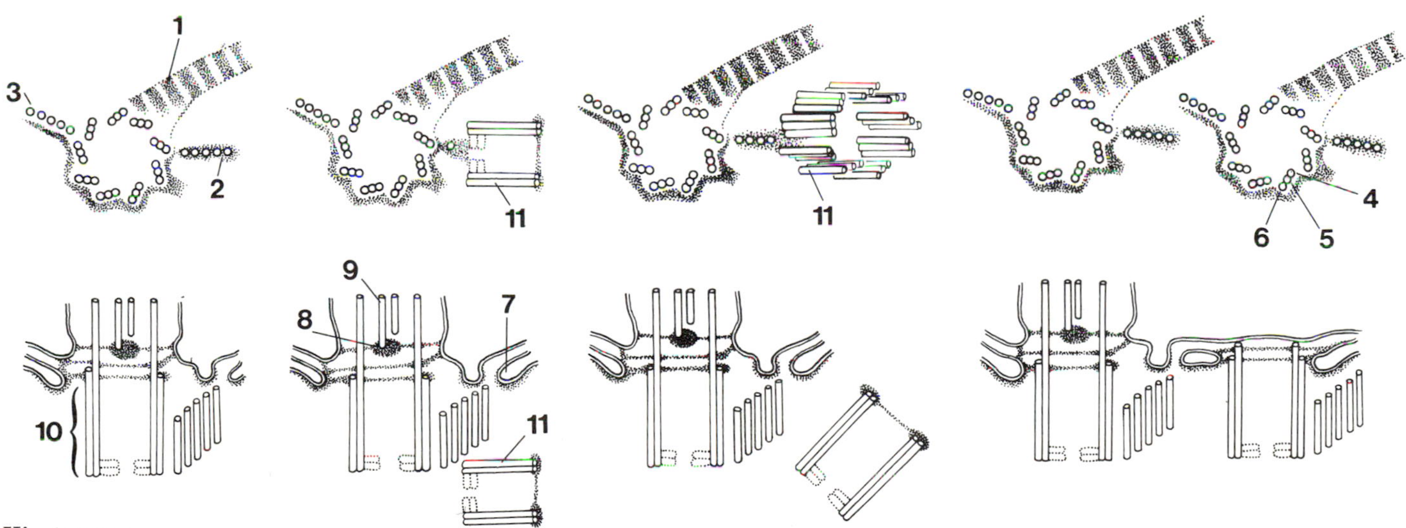

Kinetosomes
Reproduction of ciliate kinetids. 1=kinetodesmal fiber; 2=radial transverse microtubules; 3=postciliary microtubules; 4=A-tubule; 5=B-tubule; 6=C-tubule; 7=alveolar sac; 8=axosome; 9=axoneme; 10=kinetosome; 11=prokinetosome (procentriole).

Kofoid system

System of thecal plate designation in thecate dinomastigotes, devised by Charles Kofoid.

Kombu

Edible seaweed, commonly prepared from kelp.

L

L zone

Acantharian actinopod myoneme structure; L zones, as seen by electron microscopy, consist of repeated clear areas separated by thin dark transverse lines called T bands. L zone length varies with the degree of contraction of the myoneme. See *T bands*. Illustration: Myonemes.

Labiate process

Lip-shaped structure. Siliceous tube or opening which projects inward, or even outward, from the valve surface of diatoms. The labiate process terminates in a longitudinal slit surrounded by two liplike structures.

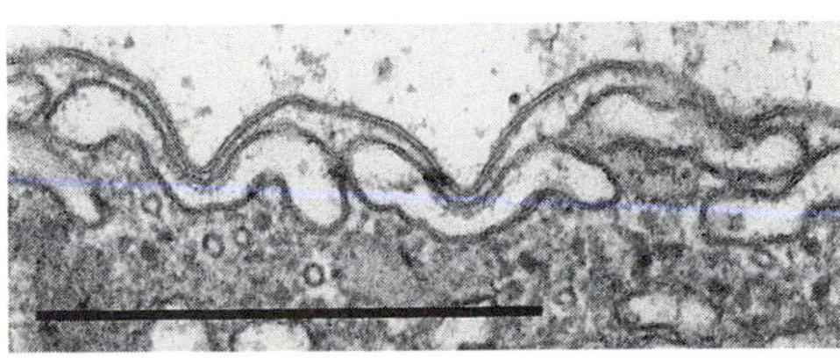

Lacunae
Pellicle of the glaucocystid *Glaucocystis nostochinearum*, strain IABH 2344. A layer of flat vesicles (lacunae) is associated with microtubules beneath the plasma membrane. TEM. Bar=0.5 µm.

Lacunae (sing. lacuna)

Cell structures; layers of flat vesicles underneath the cell membrane (plasmalemma) forming the lacunar system.

Lacustrine (adj.)

Ecological term pertaining to lakes.

Lacustrine ooze

Loose sediment, usu. rich in small clasts, microbes in suspension, and organic matter. Slimy or muddy material at the bottom of lakes.

Lag phase

The period just after inoculation of a microbial population prior to the detection of exponential growth rate.

Lageniform (adj.)
Morphological descriptive term meaning flask-shaped.

Lake
An inland area of open, relatively deep water (fresh or saline) whose surface dimensions are sufficiently large to sustain waves. Illustration: Habitat.

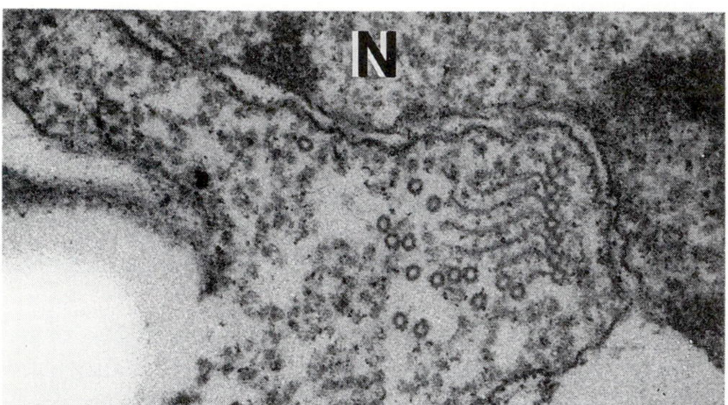

Lamella
Section through the rhizostyle of the cryptomonad *Chilomonas paramecium* showing winglike lamella associated with microtubules. N=nucleus. TEM, x 60,000.

Lamella (pl. lamellae; adj. lamellate)
Morphological descriptive term referring to a flat thin scale or flattened saclike structure. See *Thylakoid.* See also illustration: Dense body vesicle.

Lamellopodium (pl. lamellopodia)
Ameboid cell process: broad, flat pseudopodium.

Lamina (pl. laminae)
Morphological descriptive term referring to a thin plate or scale; layer. Illustrations: Endocytosis, Retortamonadida.

Laminarase
Enzyme that degrades laminarin.

Laminarin
Carbohydrate stored as food in phaeophytes; polymer of glucose and mannitol with ß-1,3 (and some ß-1,6) glycoside linkages.

Lanceolate (adj.)
Morphological descriptive term meaning shaped like a lance head, i.e., tapering to a point at the apex and sometimes also at the base.

Lapidolith
Coccolith type in which the layers of elements are parallel to the coccolith base; holococcolith (e.g., *Laminolithus hellenicus*). Illustration: Coccolith.

Latent forms
Stages in the life cycle of an organism that are more or less dormant; cysts or resting spores (e.g., the sporocyst or other stages of apicomplexans that develop slowly and persist for some time without growth).

Lateral conjugation
Gamontogamy in which conjugation tubes link gametes from adjacent cells in the same filament (e.g., the conjugating green alga *Spirogyra* spp.).

Lateral crest
Ridge that supports the adhesive disc (e.g., diplomonads).

Lateral renewal
See *Cymose renewal.*

Laver
Dried edible preparations of algae such as *Ulva* (chlorophyte) and *Porphyra* (rhodophyte).

Lectin
Protein capable of agglutinating certain cells by binding to specific carbohydrate receptors on the surface of these cells.

Leishmaniasis
Infection by the genus *Leishmania* (trypanosomatids) that inhabit macrophages of vertebrate blood.

Lens
Translucent structure capable of light refraction. Illustration: Ocellus.

Lentic (adj.)
Ecological term referring to organisms inhabiting standing water. Illustration: Habitat.

Lenticular (adj.)
Morphological descriptive term referring to the shape of a double-convex lens.

Leptotene
Leptonema; first stage of meiotic prophase I, in which chromosomes begin to condense and form threads. See *Meiosis.*

Leucoplast
Cell organelle of algae; colorless or white, often starch-storing plastid.

Leucosin
See *Chrysolaminarin.*

Life
See *Autopoiesis.*

Life cycle
Events throughout the development of an individual organism correlating environment and morphology with genetic and cytological observations (e.g., ploidy of the nuclei, fertilization, meiosis, karyokinesis, cytokinesis, etc.). Illustration next page.

Life history
Events throughout the development of an individual organism correlating environment with changes in external morphology, formation of propagules, and other observable aspects.

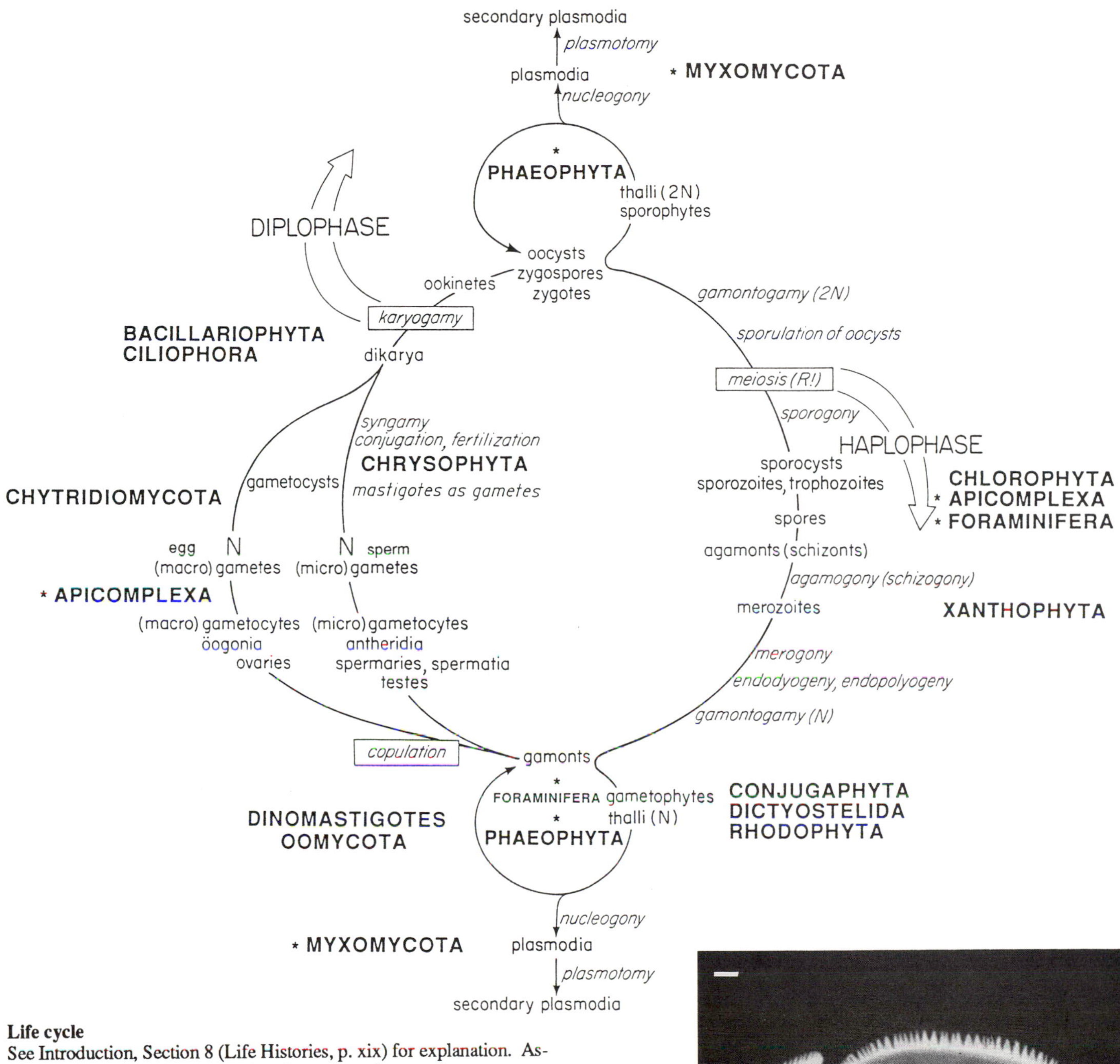

Life cycle
See Introduction, Section 8 (Life Histories, p. xix) for explanation. Asterisks denote phyla demonstrating alternation of generations.

Ligand
Molecule that binds to a complementary site.

Lignophile (adj. lignophilic)
Ecological term referring to organisms living in or on the surface of wood of shrubs or trees. Illustration: Habitat.

Ligula (pl. ligulae)
Tonguelike extension of the girdle band which fits beneath the split in the adjacent band in diatoms.

Limacine movement
Monopodial locomotion; limax movement. Sluglike pattern of locomotion produced by single eruptive anterior ectoplasmic pseudopod of an ameba.

Ligula
Fimbriate girdle band of the diatom *Pleurosira*. The tongue or ligula (arrow) inserts between the open ends of the adjacent girdle band. SEM. Bar=10 µm.

Limax (adj.)
Morphological descriptive term meaning "shaped like a slug."

Limax ameba
Elongate, usu. monopodial morphotype typical of the amebas of amebomastigotes and acrasids.

Limestone
Sedimentary rock consisting primarily of calcite (calcium carbonate, $CaCO_3$) with or without magnesium carbonate. Limestone is the most widely distributed of the carbonate rocks.

Linellae (sing. linella)
Long, thin threads composed of a cementlike matter found outside the granellare of xenophyophorans; regarded as an organic part of the test.

Linkage
Linkage group; genetic term describing the condition in which traits are inherited together and thus the genes for these traits are inferred to be physically close together on the same chromosome (linked). In genetically well-mapped eukaryotic organisms the number of linkage groups corresponds to the number of chromosomes, since linked genes tend to segregate together. In viruses and prokaryotes the single linkage group corresponds to the genophore.

Lipids
Class of organic compounds soluble in organic but not aqueous solvents; includes fats, waxes, steroids, phospholipids, carotenoids and xanthophylls.

List
Cellulosic extension of the cell wall in some armored dinomastigotes usu. extending out from the cingulum and/or sulcus. Illustration: Thecal plate.

Lithology
The study of rocks on the basis of such characteristics as color, grain size, and mineralogical composition.

Litholophus (pl. litholophi)
Modification of the cell shape of chaunacanthid actinopods. Initially spherical with radial spicules, the acantharian progressively takes the shape of a closed umbrella with all the spicules lying parallel preparatory to cyst formation.

Lithophile (adj. lithophilic)
Ecological term referring to organisms dwelling on stones and rocks. See *Chasmolith, Endolith, Epilith, Saxicolous.*

Lithosome
Organelle; vesicular, membrane-bounded cytoplasmic inclusion, comprised of inorganic material laid down in concentric layers (e.g, the prostomate ciliates; acantharian actinopods, in which it is composed of $SrSO_4$).

Litter
Layer of loose organic debris on land, composed of freshly fallen leaves or only slightly decayed materials in which the remains of organisms are detectable. Illustration: Habitat.

Littoral (adj.)
Intertidal. Ecological term referring to that portion of sandy, muddy, and rocky coasts that lies between high- and low-water marks. The underwater zone between the shore and a depth of about 200 m. See *Sublittoral, Supralittoral.* Illustration: Habitat.

Lobocyte
Free cell with a supposedly phagocytic, scavenger function inside large myxosporean trophozoite (plasmodium) of the genus *Sphaeromyxa.*

Lobopodium (pl. lobopodia)
Lobular, more-or-less rounded or cylindroid pseudopod; used in both locomotion and feeding.

Lobose (adj.)
Having many or large lobes; esp. in reference to the pseudopods of amebas in the order Lobosa (phylum Rhizopoda). Illustration: Rhizopoda.

Locule (adj. loculate)
Chamber having a constricted opening on one side and a velum on the other side (e.g., diatom frustule; compartment of a reproductive organ in algae).

Longitudinal fission
Cell division along the longitudinal axis of an asymmetric cell. See *Interkinetal fission, Symmetrogenic fission.*

Longitudinal flagellum (pl. longitudinal flagella)
See *Longitudinal undulipodium.*

Longitudinal undulipodium (pl. longitudinal undulipodia)
Longitudinally aligned undulipodium of dinomastigotes that originates and lies partially within the sulcus.

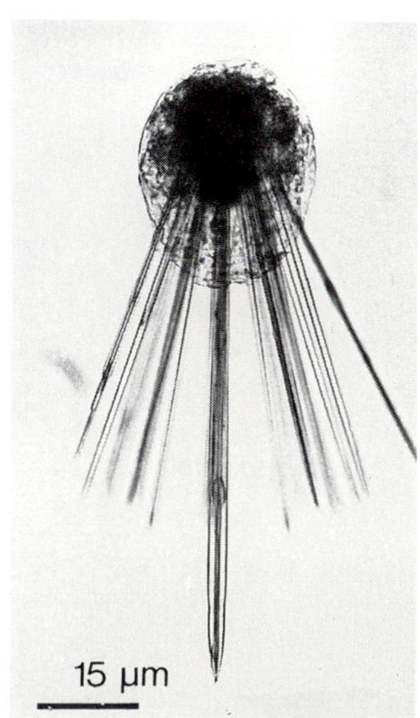

Litholophus
Litholophus of the acantharian actinopod *Heteracon biformis* with all the spicules oriented to the posterior region. PC.

Lopodolith
Helicoid placolith. See *Coccolith.* Illustration: Coccolith.

Lorica (pl. loricae; adj. loricate)
General term for structure external to the cell membrane in many protoctists: envelope; test; shell; valve; sheath; protective covering secreted and/or assembled which may be calcareous, proteinaceous, chitinous, pseudochitinous, siliceous, or tectinous in nature or made up of foreign matter such as siliceous sand grains, diatom frustules, coccoliths, or debris. Illustrations: Bicosoecids, Choanomastigota, Pedicel.

Lower fungi (sing. lower fungus)
Term used to group three mastigote protoctistan groups (chytridiomycotes, hyphochytriomycotes, and oomycotes) with the zygomycote amastigote fungi to distinguish them from the "higher" fungi (asco- and basidomycotes). It is inadvisable to use such nontaxonomic general terms as "higher" and "lower" because of their ambivalence and anthropocentrism.

Luciferases
Enzymes that catalyze the oxidation of luciferins in reactions that generate visible light (bioluminescence). Many types of enzymes and substrates exist. Eukaryotic luciferases always require ATP and oxygen; prokaryotic luciferases may not be ATP-dependent.

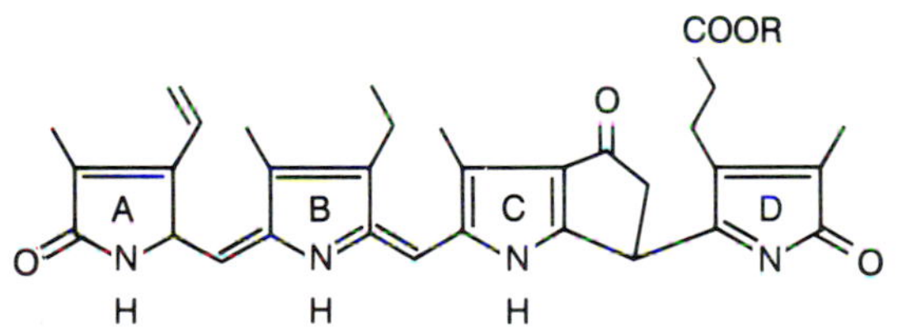

Luciferin
Luciferin from the dinomastigote *Pyrocystis noctiluca.*

Luciferin
Any of a number of organic compounds of luminescent organisms that are substrates in luciferase reactions.

Lugol's solution
See *Iodine test.*

Luminescence
See *Bioluminescence.*

Lyophilization
Freeze-drying; method for preservation of resistant protoctists.

Lysis
Cell disintegration following rupture of the cell membrane. Illustration: Genophore.

Lysocline
Depth in the ocean below which calcium carbonate ($CaCO_3$) skeletons dissolve because of hydrostatic pressure. Neither living nor dead foraminifera or other protoctists with calcareous tests are found beneath this depth.

Lysogenic conversion
Change in phenotype of a bacterium that accompanies lysogeny, i.e., the process in which the genetic material of a virus is incorporated into the genetic material of its host bacterium (e.g., in some bacteria, toxin production occurs only when the appropriate virus is incorporated into the genophore).

Lysosome
Membrane-bounded organelle containing releasable hydrolytic enzymes. Illustration: Organelle.

Lysozyme
Enzyme hydrolyzing the peptidoglycans of cell walls and hence used to break open bacteria (e.g., in egg white, tears).

M

Macroalgae (sing. macroalga)
Algae visible to the naked eye; large algae in contrast to microscopic algae. See *Macrophyte.*

Macrocyst
Large cyst (e.g., multicellular, irregularly circular or ellipsoidal resting structure about 25-50 µm in size with three distinct walls formed during the sexual cycle in some dictyostelids; protoplasmic, walled, usu. multinucleate portions of a myxomycote sclerotium). Illustrations: Dictyostelida, Plasmodial Slime Molds.

Macrogametangium (pl. macrogametangia)
Algal sexual structure (e.g., gametangium containing relatively large locules and thus producing the larger (macro-) gametes).

Macrogamete
Large, usu. female gamete (e.g., apicomplexans). See *Microgamete.* Illustrations: Coccidian life history, Life cycle, Macrogamont, Sporocyst.

Macrogametocyte
See *Macrogamont.*

Macrogamont
Macrogametocyte. Descriptive term for the larger of the two gamonts in anisogamontous gamontogamy. Female gamont that transforms into a single macrogamete (e.g., in coccidian apicomplexans). See *Microgamont.* Illustration next page. See also illustrations: Coccidian life history, Life cycle.

Macronucleus (pl. macronuclei)
Somatic nucleus. Larger of the two kinds of nuclei in ciliate cells; site of messenger RNA synthesis; containing more than two (and often hundreds of) copies of genes, it is required for growth and division. See *Micronucleus.* Illustration: Ciliophora.

Macrophagy
Mode of heterotrophic nutrition in which organisms feed on food particles large with respect to their own size.

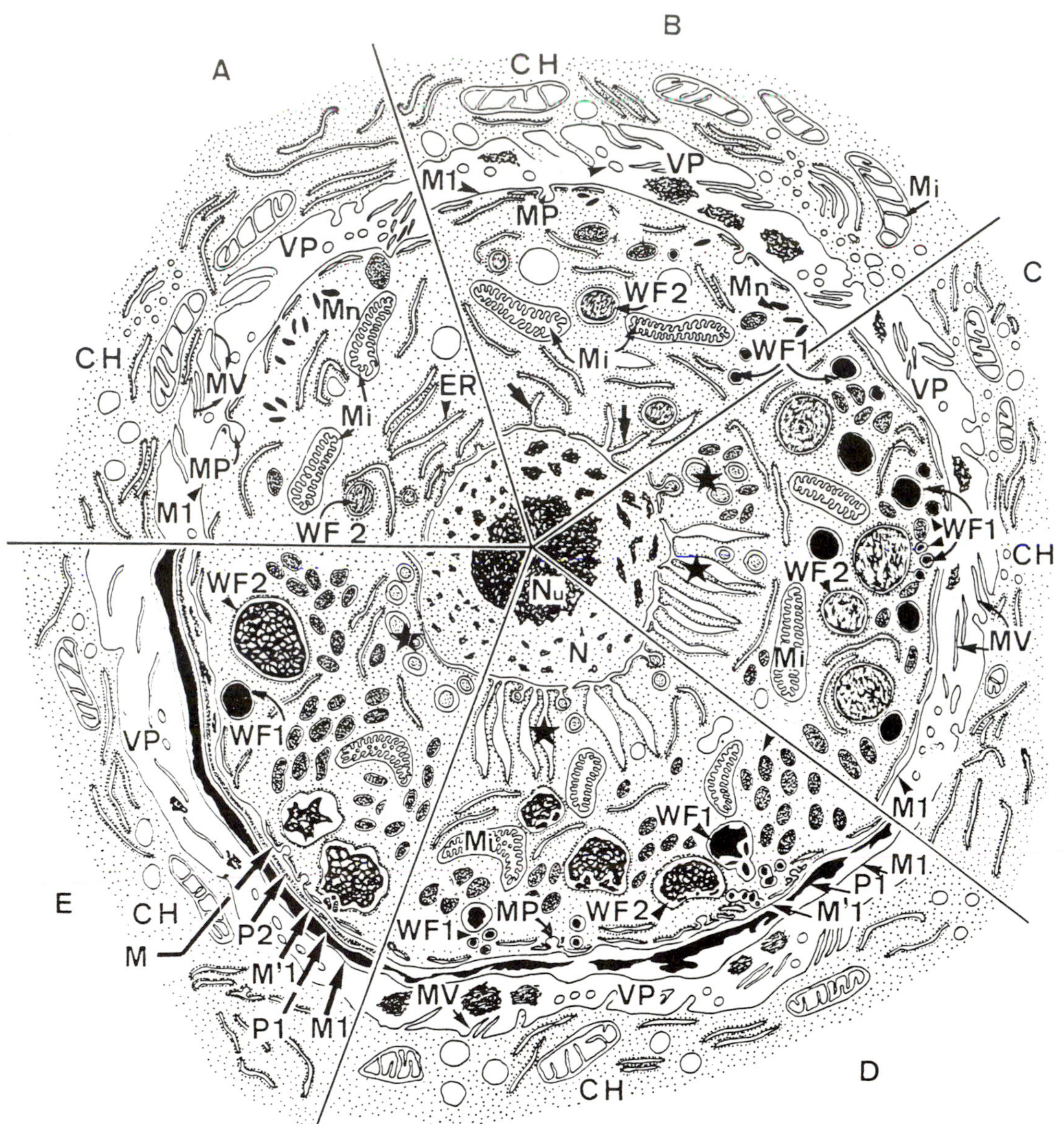

Macrogamont
Transformation of the macrogamont of the coccidian apicomplexan *Eimeria acervulina* into an oocyst. a, b. formation of the wall-forming bodies (WF1, WF2) in maturing macrogamont; c. macrogamete characterized by evaginations of the nuclear envelope (stars); d. the fertilized macrogamete begins to secrete the oocyst wall; e. the oocyst. CH=host cell; ER=endoplasmic reticulum; M, M1, M'1=membranes of the cyst wall; Mi=mitochondrion; Mn=micronemes; MP=micropore; MV=microvilli; N=nucleus; Nu=nucleolus; P1, P2=cyst walls; VP=parasitophorous vacuole.

Macrophyte (adj. macrophytic)
Literally "large plant" but inadvisably used for large algae. Usu. refers to phaeophytes, rhodophytes, and large chlorophytes.

Macroschizozoite
Life cycle stage of apicomplexans; zoite produced by large schizont.

Macrosclerotia (sing. macrosclerotium)
Large sclerotia formed by phaneroplasmodia or aphanoplasmodia in myxomycotes.

Macrosporangium (pl. macrosporangia)
Megasporangium containing large spores (macrospores) as opposed to sporangium of similar dimensions containing small spores.

Macrospore
Large spore, usu. in contrast to microspores made by the same species.

Macrostome
Inducible morph of certain *Tetrahymena* ciliates in which cell develops a large oral apparatus correlated with carnivorous feeding. See *Microstome*.

Macrothallus (pl. macrothalli)
Large conspicuous flat-structured morph of relatively large thallus (e.g., rhodophytes, chlorophytes, or phaeophytes). See *Microthallus*.

Macrozoospore
Zoospore of large size relative to others produced by the same organism (e.g., in the prymnesiophyte *Phaeocystis*). See *Microzoospore*.

Maerl
See *Marl.*

Magnetotaxis (adj. magnetotactic)
Directed locomotion in a magnetic field toward a magnetic pole (e.g., south- or north-seeking; as in magnetite-containing bacteria or *Chlamydomonas*).

Maintenance culture
Collection of mixed microorganisms (e.g., a polyxenic culture of live protoctists maintained in the laboratory by periodic addition of water or new medium). See *Culture.*

Maintenance culture medium (pl. maintenance culture media)
Liquid or solid material providing nutrients and osmotic conditions for the maintenance (but not necessarily the growth) of microorganisms in the laboratory.

Male
Gamont. Gender of individual that produces sperm or other usu. motile gamete and donates it to the female.

Mannan
Polysaccharide component of walls of some rhodophytes and chlorophytes which yields mannose upon hydrolysis; often a ß-1,4 mannopyranoside.

Mannitol
A polyol, i.e., a 6-carbon sugar alcohol storage product of phaeophytes.

Mannuronic acid
Algal metabolite; acid derivative of the sugar mannose.

Mantle
Outermost portion of valve that is bent at approximately 90° and connected to the girdle (e.g., diatoms). Inner shell lining of bivalve mollusks. Illustration: Hypovalve.

Manubrium (pl. manubria)
Columnar cell connecting the pedicel to the shield cell in the antheridia of charophytes.

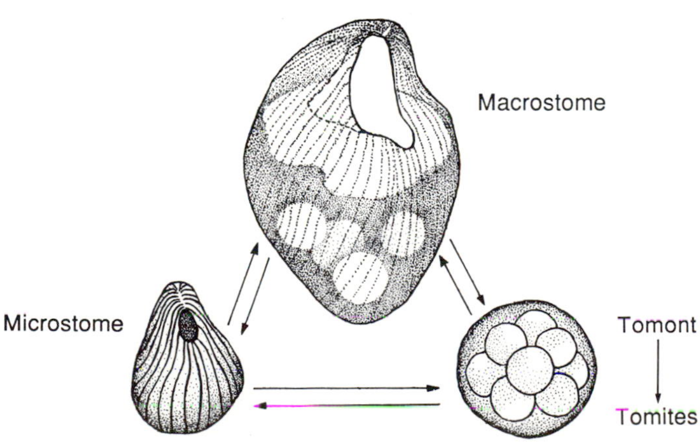

Macrostome
Life history of the macrostome-microstome ciliate *Tetrahymena vorax.*

Marine
Of the sea or ocean; inhabiting, found in, or formed in the sea. Illustration: Habitat.

Marine snow
Irregularly shaped particulate matter, up to several millimeters in diameter, that precipitates in the ocean, falling toward the bottom. Includes living colonies of protoctists and organic remains of protoctists.

Marker species
Fossil species found in sedimentary facies and used to correlate (and date) that facies over relatively long distances.

Marl
Maerl. Unconsolidated poorly lithified clastic sediment, a mixture of clay and calcium carbonate, usu. including shells and sometimes including living fragments of deep water crustose coralline rhodophytes; formed under marine and esp. freshwater conditions.

Mastigoneme (adj. mastigonemate)
Flimmer; tinsel. Fine hairlike projection that extends laterally from undulipodia; mastigonemes differ in detail in various protoctist groups; they are probably formed from proteins synthesized on the ribosomes of the outer nuclear membrane. Illustration next page. See also illustrations: Anisokont, Hyphochytriomycota.

Mastigont system
Intracellular organellar complex found in many mastigotes (e.g., parabasalians, diplomonads, retortamonads). Organelles associated with undulipodia, the mastigont system may include the kinetids with their undulipodia, undulating membrane, costa, parabasal bodies, and axostyle. See *Karyomastigont system.*

Mastigote
Flagellate (see Introduction). Eukaryotic microorganism motile via undulipodia. Illustrations: Dinomastigote life history, Mastigoneme, Trypomastigote.

Mastigote division
Karyokinesis and cytokinesis of undulipodiated protists (e.g., cryptomonads, euglenids, proteromonads). Illustration next page.

Mating
Syngamy. See *Conjugation.*

Mating type
Strain of organisms incapable of sexual fusion with each other but capable of sexual reproduction with members of another strain of the same organism.

Maxicircle
Large circular DNA molecule (20-38 kb) in the kinetoplast of trypanosomatids, corresponding to mitochondrial DNA of other eukaryotes; held together in a network by minicircle DNA.

MC
See *Microtubule-organizing center.*

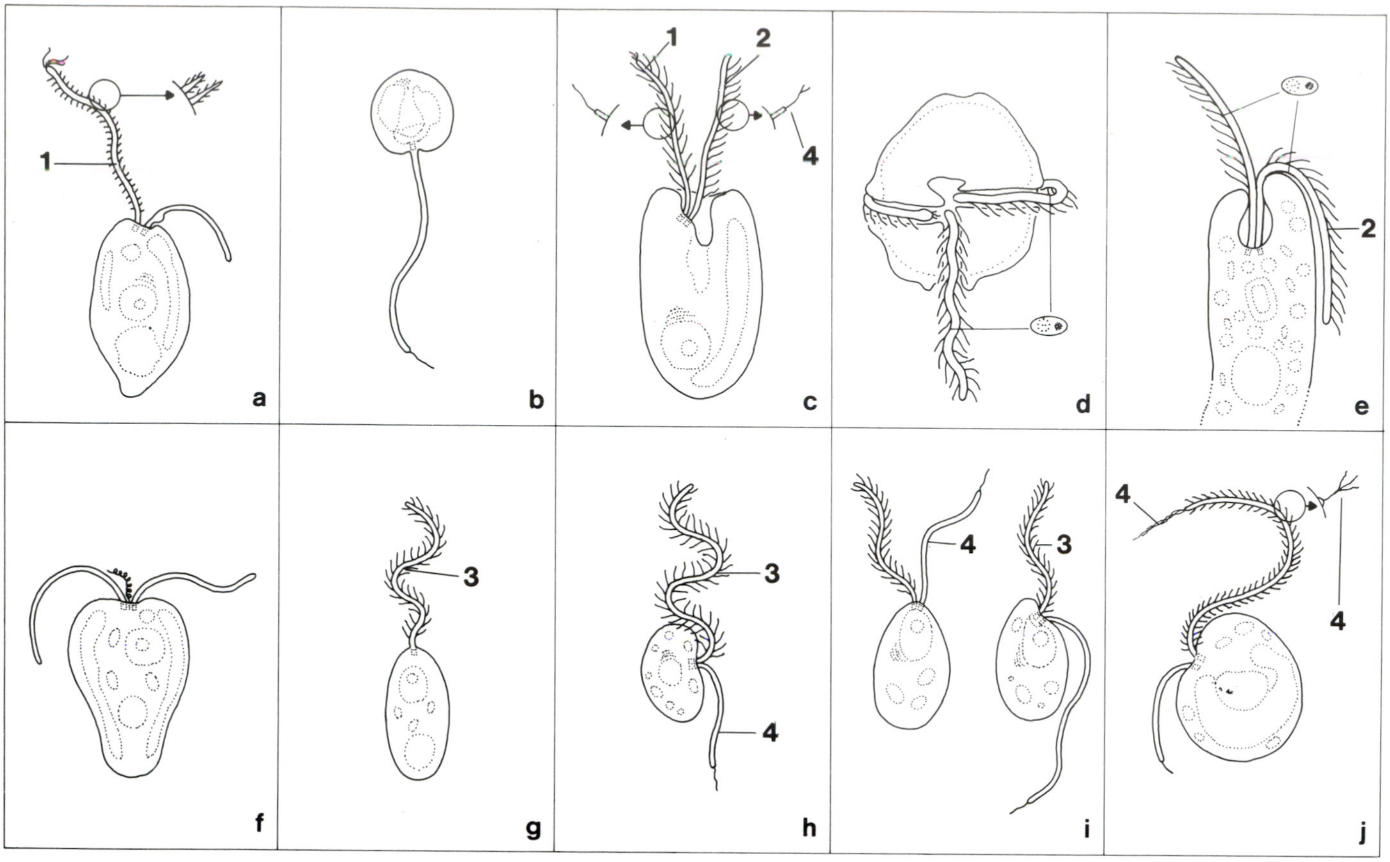

Mastigoneme

Mastigotes. a. Chrysophyte. b. Chytridiomycote. c. Cryptomonad. d. Dinomastigote. e. Euglenid. f. Prymnesiophyte (haptophyte). g. Hyphochytrid. h. Labyrinthulomycote. i. Oomycote. j. Phaeophyte. 1=pantacronematic; 2=stichonematic; 3=pantonematic; 4=acronematic.

Mechanical transmission

Transmission from one host to another of a symbiont that does not undergo a cycle of development in the vector but retains its morphological and physiological state (e.g., *Trypanosoma evansi* that moves between vampire bats and ungulates).

Median bodies

Structures composed of cytoskeletal proteins; incipient adhesive disc of an offspring mastigote formed prior to mitosis (e.g., diplomonads: distinguishes species of *Giardia*).

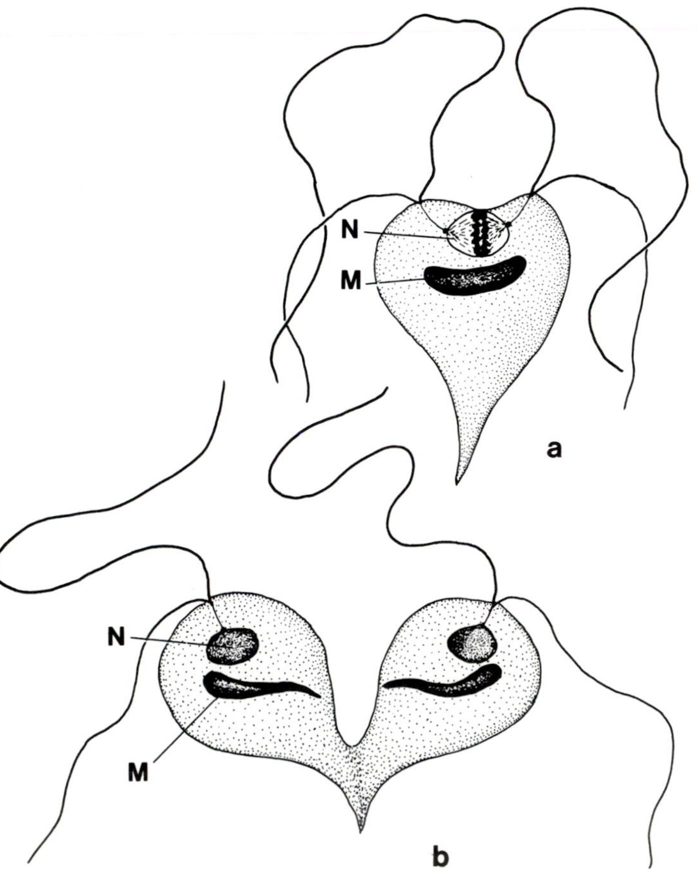

Mastigote division

Two division stages of the proteromonad zoomastiginid *Proteromonas*. a. An intranuclear spindle is formed, the chromosomes are arranged in a metaphasic plate, and each pole of the spindle seems to be attached to a rhizoplast (spindle pole body) anchored to each set of undulipodia. b. As the nucleus and the very elongated mitochondrion divide, the two offspring cells are attached only by their posterior end. M=mitochondria; N=nucleus.

Mediocentric (adj.)
Referring to chromosomes with centrally, or nearly centrally, located kinetochores (centromeres). Illustration: Chromosome.

Medium (pl. media)
See *Culture medium.*

Medulla (pl. medullas or medullae)
Morphological term referring to the central region of an organ (e.g., adrenal) or organism (e.g., thallus of lichen or alga). See *Cortex.*

Megacytic zone
Region of expansion between adjacent plates where new thecal material is added to allow enlargement of the cell in armored dinomastigotes.

Megalospheric test
Gamont generation test with a large initial chamber in foraminifera.

Megasporangium (pl. megasporangia)
See *Macrosporangium.*

Meiocyte (adj. meiocytic)
Cell destined to undergo meiosis. Illustration: Dinomastigote life history.

Meiosis (adj. meiotic)
One or two successive divisions of a diploid nucleus that result in the production of haploid nuclei. In organisms with gametic meiosis, meiotic divisions precede the formation of gametes (e.g., diatoms). In organisms with zygotic meiosis the zygote undergoes meiosis immediately after it forms (e.g., volvocalean algae). Prophase is much longer than in mitosis and can be divided into five consecutive stages: leptonema, zygonema, pachynema, diplonema, and diakinesis. During most meioses homologous chromosomes pair, forming the synaptonemal complex. Meiosis is found in all plants and animals, most if not all sexual fungi, and in many, but by no means all, protoctists. Protoctists display extremely varied patterns of life cycles that involve meiosis and fertilization. See also illustrations: Acetabularian life history, Dinomastigote life history, Life cycle, Plasmodiophoromycota, Trophozoite.

Meiosporangium (pl. meiosporangia)
Sporangium in which meiosis occurs.

Meiospore
Spore produced by meiosis.

Melanosome
Black or brown pigmented body containing the tyrosine-derived polymer melanin. Light-sensitive portion of the ocellus onto which the hyalosome focuses (e.g., pigmented portion of certain dinomastigote cells); also, any of the melanogenic granules of pigment-producing cells from their earliest recognizable (unpigmented) stage to their completely or partially electron-dense and definitively patterned stage. Illustration: Ocellus.

Meristem
Undifferentiated tissue composed of rapidly growing cells; region on protoctist thallus at which new cells arise.

Meristoderm
Superficial layer of rapidly dividing and growing cells covering the thalli of laminarialean phaeophytes.

Merogenous sperm formation
Spermatogenesis in which a portion of the protoplasm of a microspore converts to form a sperm and the remainder is discarded (e.g., centric diatoms).

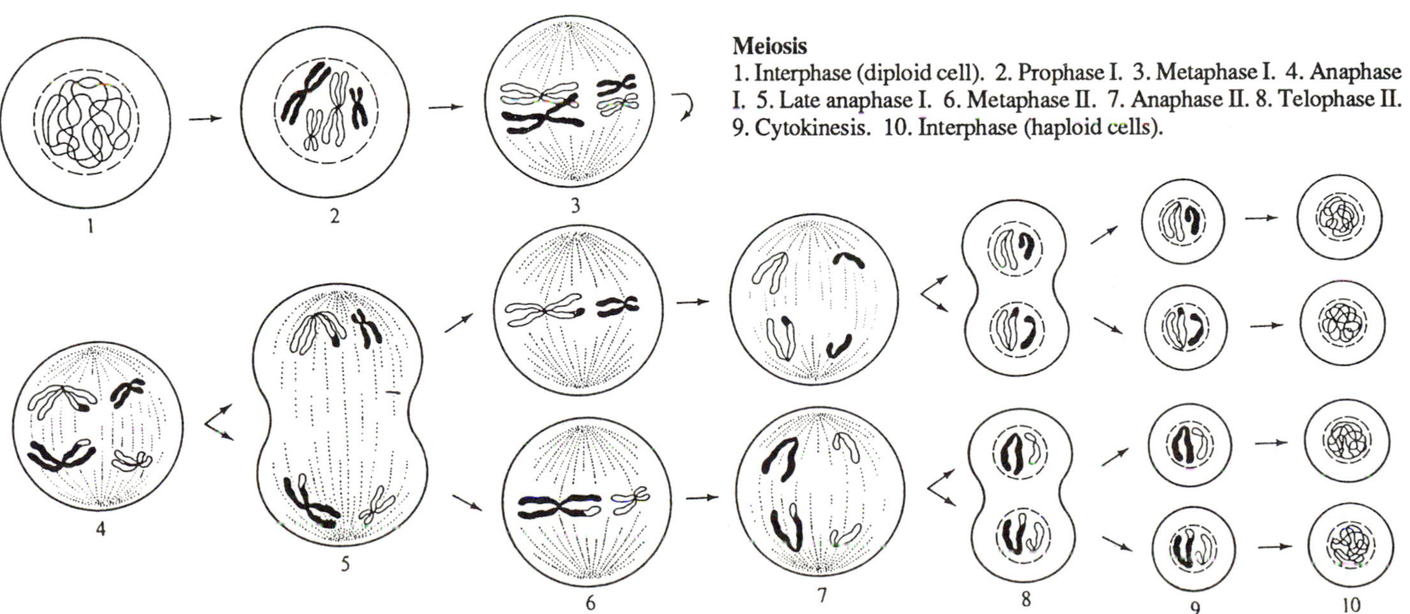

Meiosis
1. Interphase (diploid cell). 2. Prophase I. 3. Metaphase I. 4. Anaphase I. 5. Late anaphase I. 6. Metaphase II. 7. Anaphase II. 8. Telophase II. 9. Cytokinesis. 10. Interphase (haploid cells).

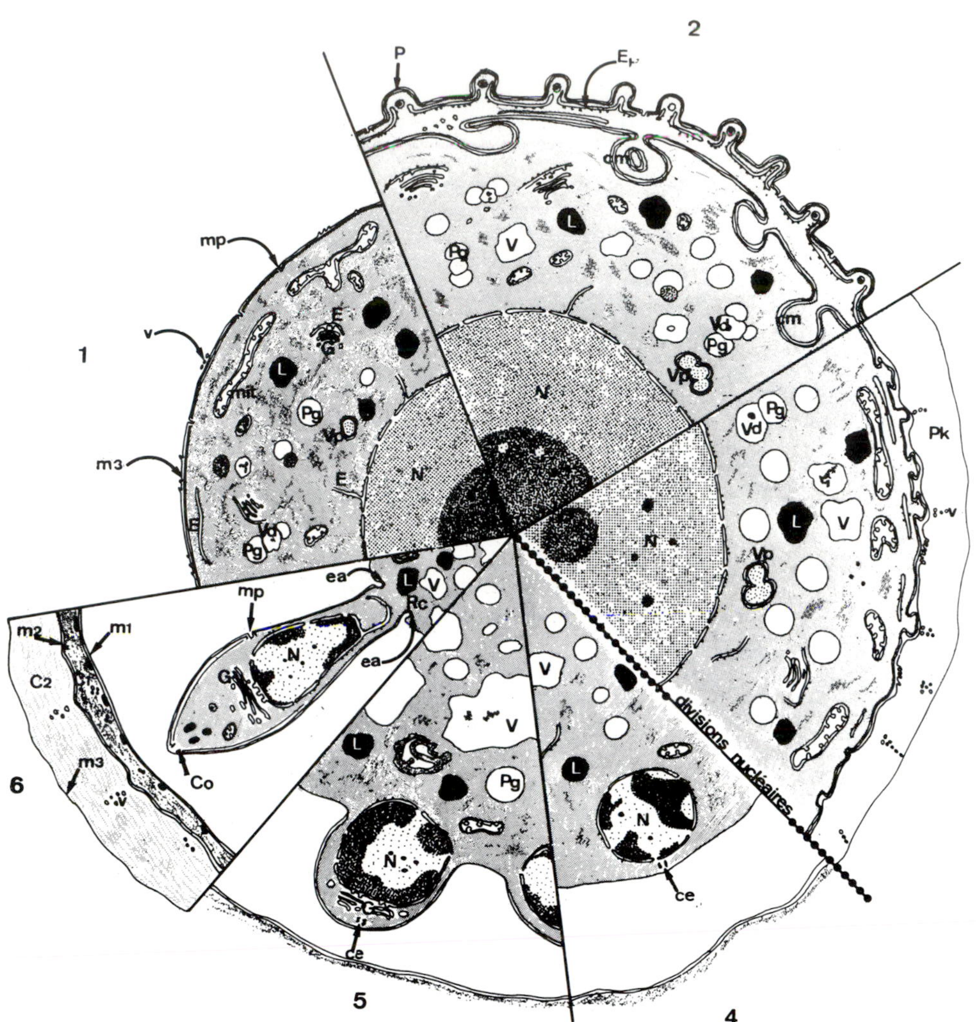

Merogony
Merogony in the coccidian *Aggregata eberthi*. 1. Young meront with a smooth wall. 2. Older meront with outer protrusions. 3. Secretion of the cyst wall. 4, 5, 6. Formation of the merozoite. C1, C2= mucous layers of the cyst wall; ce= centriole; cm=cell membrane; Co=conoid; E=endoplasmic reticulum; ea=annular thickening of the stalk connecting the developing merozoite with the residual cytoplasm (Rc); G=golgi apparatus; L=lipids; mit=mitochondrion; mp=micropore; m1, m2, m3=membranes of the cyst wall; N=nucleus; P=protrusion of the meront wall; Pg=amylopectin granule; Pk=cyst wall; V=vacuole.

Merogony
Multiple fission of apicomplexans; schizogony resulting in merozoites which themselves undergo multiple cell divisions. See also illustration: Life cycle.

Meront
Life history stage. Active cell produced by merogony from a trophozoite. Meronts later, by multiple fission, develop into merozoites (e.g., apicomplexans and microsporans). Illustrations: Coccidian life history, Merogony.

Meroplankton (adj. meroplanktonic)
Ecological term referring to neritic organisms that spend part of their life history as plankton and part in benthic communities.

Merotomy
Division of cells ("cutting up") into portions with or without nuclei.

Merozoite
Schizozoite. Life history stage. Mitotic product of meront. Merozoites may differentiate into gamonts (e.g., *Plasmodium*). Illustrations: Coccidian life history, Life cycle, Merogony, Sporocyst.

Mesocyst
Middle of three layers surrounding a cyst (e.g., heliozoan oocyst where the mesocyst is surrounded by siliceous scales). See *Ectocyst, Endocyst*.

Mesokaryotic (adj.; n. mesokaryote)
Dinokaryotic. Referring to dinomastigote with nuclei that lack conventional histones and have permanently condensed chromosomes. Literally "between prokaryotic and eukaryotic." See *Noctikaryotic*. Illustration: Dinomastigota.

Mesomitotis
Dinomastigote karyokinesis; mitosis of the mesokaryotic nucleus in which chromosomes remain condensed and attached to nuclear membrane, breakdown of the nuclear envelope and nucleolus is delayed, and centrioles are lacking. See *Metamitosis, Promitosis*. Illustration: Closed mitosis.

Mesophotic zone
Ecological term referring to the region in the water column between the compensation depth (at which the rates of respiration and photosynthesis in phytoplankton are equal over a 24-hour period) and the depth to which no surface light reaches. Dimly lit

lower portion of the photic zone, in which light is limited but photosynthesis still leads to net productivity. Illustration: Habitat.

Mesosaprobic (adj.)
Mesotrophic. Ecological term referring to an aquatic environment containing a moderate amount of dissolved organic matter (moderately polluted). See *Eutrophy, Oligotrophy.*

Mesospore
Thin middle wall of the three-layered cell wall of hypnozygote (dinomastigote); the middle layer of the zygospore wall (conjugating green algae). See *Endospore, Exospore.*

Mesotrophic (adj.)
See *Mesosaprobic.*

Metabolism
The sum of enzyme-mediated biochemical reactions that occur continually in cells and organisms and provide the material basis of autopoiesis.

Metaboly
See *Euglenoid motion.*

Metacentric (adj.)
Morphological term pertaining to a mitotic spindle that radiates from centrioles that lie in the same plane as the metaphase chromosomes (e.g., in some chlorophytes); also, referring to chromosomes with centromeres that lie exactly or nearly halfway between the chromosome ends or telomeres. Illustration: Chromosome.

Metachronal waves
Synchronous waves of movement of adjacent undulipodia resulting from a tight viscous-mechanical coupling (hydrodynamic linkage). Symplectic metachronal waves refer to those which pass over the field of undulipodia in the same direction as the effective stroke of the beat. Antiplectic metachronal waves are those which pass in a direction more or less opposite to the effective stroke.

Metacyclic stage
The stage in the development of a parasitic protoctist in its invertebrate host (vector) just before transfer to the vertebrate; this stage is normally infective to the vertebrate host.

Metacyclogenesis
Formation of metacyclic stage.

Metagenesis (adj. metagenic)
Alternation of generations, esp. of a sexual and an asexual generation, in heterotrophic organisms.

Metalimnion
Water at thermocline in lake or other thermally stratified body of water. See *Epilimnion, Hypolimnion.*

Metamitosis
Conventional metazoan mitosis in which there are centrioles or other conspicuous microtubule-organizing centers at the poles and loss of the nucleolus and nuclear envelope. See *Mesomitosis, Promitosis.* Illustration: Closed mitosis.

Metaphase plate
Equatorial plate; nuclear plate. Transient structure observable in many but not all dividing protoctist cells; plane in the equatorial region of the mitotic spindle at which the chromosomes align by way of their movement during the metaphase stage of mitosis or meiosis. Illustrations: Mastigote division, Mitosis.

Metaphyton
Ecological term referring to the biota, esp. the microbiota, surrounding plants (metaphytes). Illustration: Habitat.

Metazoa
Members of the Kingdom Animalia. All organisms developing from a blastular embryo, itself derived from an egg usu. fertilized by a sperm. Metazoan bodies are made of cells differentiated into tissues and organs and usu. have a digestive cavity with specialized cells. Excludes all "protozoa" - a term that includes many different protoctist phyla.

Microalgae (sing. microalga)
Microscopic algae (as opposed to large algae).

Microbe
Microorganism. Any live being not visible to the naked eye and thus requiring visualization by microscopy.

Microbenthos
Bottom-dwelling microbes, small animals, or microbial communities in fresh or marine waters. See *Aufwuchs, Periphyton.*

Microbial mat
Laminated organosedimentary structure composed of stratified communities of microorganisms, usu. dominated by phototrophic bacteria, esp. cyanobacteria. Types range from soft, brightly colored layered sandy sediment to lithified carbonates; living precursors of stromatolites.

Microbiota
Sum of microorganisms in a given habitat (e.g., termite intestinal microbiota); term preferable to microflora, which implies plants, or microfauna, which implies animals.

Microbody
Small intracellular structure; any of a number of organelles of eukaryotic cells bounded by a single membrane and containing a variety of enzymes. Microbodies, usu. associated with one or two cisternae of the endoplasmic reticulum, include glycosomes, glyoxysomes, hydrogenosomes, and peroxisomes. Illustration next page. See also illustrations: Acantharia, Bodonidae, Chytridiomycota, Nuclear cap.

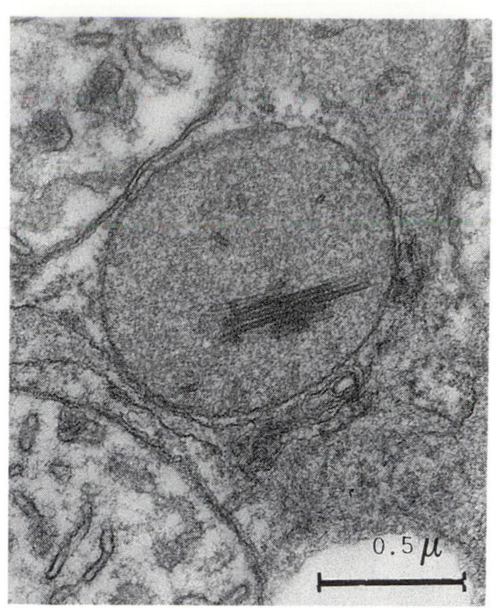

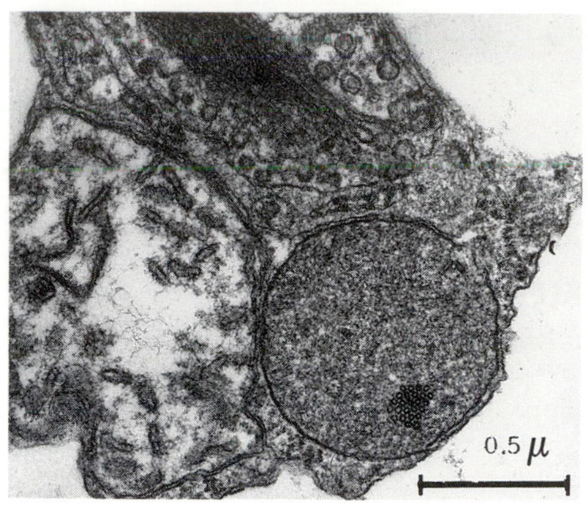

Microbody
Microbodies in longitudinal and transverse section of acantharian actinopods; mitochondria with tubular cristae are also visible. TEM.

Microbody-lipid globule complex

MLC. Conspicuous structure, of unknown function, near the kinetosome of zoospores of chytridiomycotes. Illustration: Chytridiomycota.

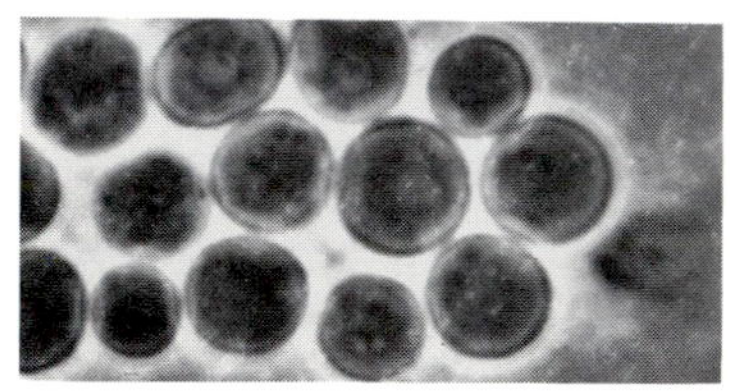

Microcyst
The acrasid *Copromyxella filamentosa*. PC, x 1,500.

Microcyst

Type of cyst or spore. Microcysts are encysted myxamebas (e.g., myxomycotes and acrasids). A two-layered fibrillar wall composed mostly of cellulose comprises the microcyst of dictyostelids, particularly *Polysphondylium* and most of the smaller species of *Dictyostelium* (but not *D. discoideum* and other large species). The wall is secreted in response to adverse environmental conditions, esp. the presence of ammonia. See also illustrations: Dictyostelida, Plasmodial Slime Molds.

Microfauna

See *Microbiota*.

Microfilament

Very small filament or microfibril; general term describing any solid, thin, fibrous, proteinaceous structure, generally those in the cytoplasm of eukaryotic cells, some of which are composed of actin and participate in motility.

Microflora

See *Microbiota*.

Microgamete

Small gamete (e.g., in apicomplexans the male gamete). See *Macrogamete*. See also illustrations: Coccidian life history, Life cycle, Sporocyst.

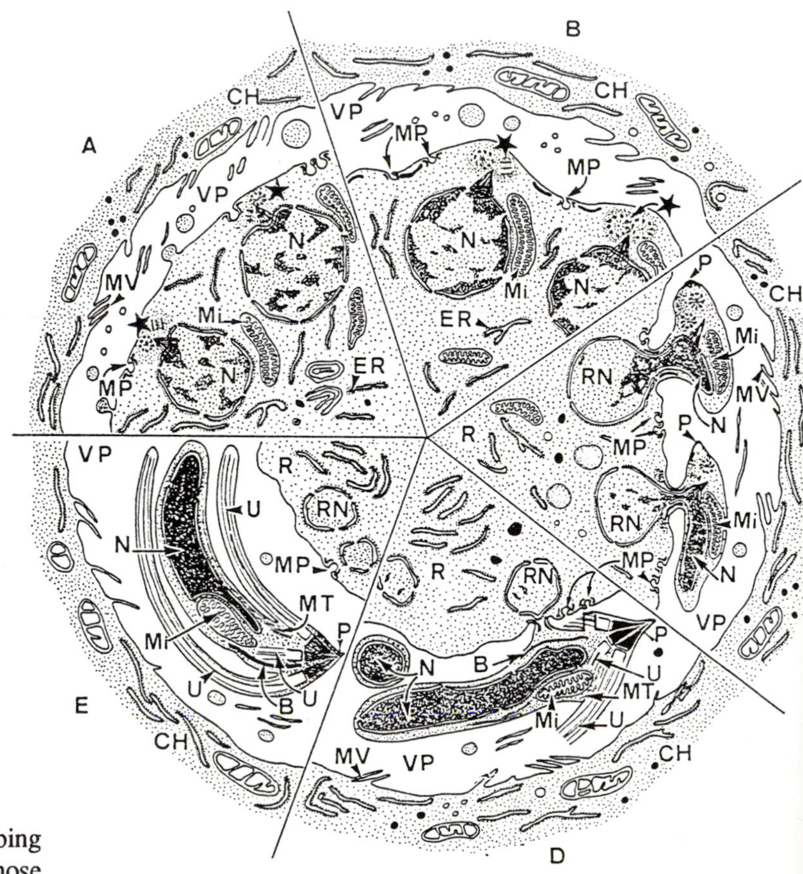

Microgamete
The differentiation of the microgametes in the coccidian apicomplexan *Eimeria acervulina:* a. microgamont with gametic nuclei; b. condensation of the chromatin; c. incorporation of the denser part of the nucleus in the budding microgamete; d. the nearly mature microgamete still connected with the residual body; e. the mature microgamete. B=dense rod; CH=host cell; ER=endoplasmic reticulum; Mi=mitochondrion; MP=micropore; MT=microtubules; MV=microvilli; N=nucleus; P=perforatorium; R=residuum; RN=nuclear residuum; U=undulipodium; VP=parasitophorous vacuole; star=centrocone.

Microgametocyte
See *Microgamont.*

Microgamont
Microgametocyte. The smaller of the two gamonts in anisogamontous gamontogamy (e.g., in apicomplexans, the gamont which produces microgametes). See *Macrogamont.* Illustrations: Coccidian life history, Life cycle, Microgamete.

Microgranular test
Test of foraminifera composed of equidimensional subspherical granules of calcite packed closely together without detectable cement. In many forms there are two layers: an outer layer of irregularly packed granules and an inner, highly ordered, packed layer.

Microheterotroph
Term used by oceanographers and ecologists that groups prokaryotes with nonphotosynthetic or mixotrophic protists to form an arbitrary grouping of diverse facultatively heterotrophic microorganisms fewer than 8 μm in diameter.

Micrometer
Micron. A millionth of a meter; linear unit of measurement (see Table 3).

Micron
See *Micrometer.*

Micronemes
Dense bodies in apicomplexan zoites, most abundant in apical complex area and probably corresponding to secretions of the golgi apparatus. Illustrations: Conoid, Kinete, Macrogamont.

Micronucleus (pl. micronuclei)
Small nucleus; the smaller of the two types of nuclei in ciliates. The ciliate micronucleus does not synthesize messenger RNA; it is usu. diploid and may undergo meiosis prior to syngamy and autogamy; required for all sexual processes, it is not always necessary for growth or cell division. See *Macronucleus.* Illustration: Ciliophora.

Micronutrient
Trace element. Mineral or element required only in minute quantities for microbial growth (e.g., iron, magnesium, cobalt, or zinc).

Microorganism
See *Microbe.*

Micropaleontology
Subdiscipline of geology: study of fossil microbes and the microscopic parts of fossil organisms (e.g., pollen and spores).

Micropore
Small opening. Illustrations: Conoid, Macrogamont, Merogony, Microgamete.

Microschizozoite
Life cycle stage of apicomplexans; zoite produced by small schizont.

Microsource
Brightly fluorescent spherical body, about 0.5 μm in diameter, from which light flashes emanate; microsources are distributed primarily in the cortical cytoplasmic region of bioluminescent dinomastigote cells. See *Scintillons.*

Microspheric test
Foraminiferan test with a small initial chamber. The overall size of the test is generally larger than a test of megalospheric generation. Commonly part of the agamont generation.

Microspine
Small pointed structure such as that found decorating protist tests.

Microsporangium (pl. microsporangia)
Structure that harbors microspores.

Microspore
Small spore; haploid spore that develops from the microspore parent cell and develops into a male gametophyte (e.g., rhodophytes, plants); product of division of a cell that undergoes meiosis giving rise to four sperm cells (e.g., diatoms); reproductive structure formed in sporophores (e.g., slime molds).

Microstome
Inducible morph of *Tetrahymena* ciliates in which the cell develops a small oral apparatus correlated with bactivorous feeding. See *Macrostome.* Illustration: Macrostome.

Microthallus (pl. microthalli)
Small, inconspicuous phase in the life history of some rhodophytes, phaeophytes, or chlorophytes that alternates with the macrothallus. See *Macrothallus.*

Microtrophy (adj. microtrophic)
Nutritional mode referring to heterotrophic organisms (i.e., animals, protists) that feed on microbes.

Microtubular bundles
Fibrous structures (on inspection by electron microscopy) composed of longitudinally aligned 24Å microtubules. Illustrations: Actinopod, Retortamonadida.

Microtubular fiber
Thin structure (as seen by light microscopy). Electron microscopy reveals it to be a microtubule bundle. Fiber associated with microtubule. Illustrations: Actinopod, Retortamonadida.

Microtubular rod
Axoneme. Bundle of parallel microtubules that stiffens axopods of actinopods or tentacles of suctorian ciliates.

Microtubule (adj. microtubular)
Slender, hollow structure made primarily of tubulin proteins (α-tubulin and ß-tubulin) each with molecular weight of about 50 kilodaltons, arranged in a heterodimer. Microtubules are of varying lengths but usu. invariant in diameter at 24-25 nm; substruc-

ture of axopods, mitotic spindles, kinetosomes, undulipodia, haptonemata, nerve cell processes, and many other intracellular structures; their formation is often inhibitable by colchicine, vinblastine, podophyllotoxin, and other microtubule polymerization-inhibiting drugs. Illustrations: Atractophore, Cortex, Epiplasm, Kinetid, Kinetosomes, Nuclear cap, Paraxial rod, Somatonemes, Spindle pole body, Transverse ribbon, Undulipodium.

Microtubule
Isolated microtubules, 24nm in diameter and variable in length. SEM.

Microtubule-organizing center

MC, MOC, or MTOC. Granulofibrosal material visible with the electron microscope from which microtubules arise and grow. May be associated with centrioles, centriolar plaques, kinetosomes, or other intracellular organelles. Illustrations: Centroplast, Centrosphere.

Microvillus (pl. microvilli)

Cytoplasmic projection from epithelial cells; may contain microfibrils or microtubules. Illustrations: Macrogamont, Microgamete.

Microzooplankton

Ecological term grouping small motile heterotrophs passively moved by currents in aquatic environments. Term should be restricted to microscopic organisms in the Kingdom Animalia. Unidentified suspended protoctists are more accurately referred to by size categories such as microplankton or nanoplankton.

Microzoospore

Zoospore of small size relative to others produced by same organism. See *Macrozoospore.*

Midlittoral (adj.)

Ecological term referring to the central region of the intertidal zone. Illustration: Habitat.

Mineral

Naturally formed chemical element or compound having a definite chemical composition (e.g., calcite, strontium sulfate, silica). See *Biomineralization.*

Minicircles

One type of organization of DNA of the kinetoplastids; small circular kinetoplast DNA molecules (0.46-2.5 kb) of unknown function composing the bulk of kDNA and catenated with maxicircles to form a network.

Mitochondria (sing. mitochondrion)

Membrane-bounded intracellular organelles containing enzymes and electron transport chains for oxidative respiration of organic acids and the concomitant production of ATP. Mitochondria have DNA, messenger RNA, and small ribosomes and are thus capable of protein synthesis; they are nearly universally distributed in protoctists but notably absent in *Pelomyxa*, some rhizopods, parabasalians, and certain other protist taxa. Illustrations: Cristae, Organelle.

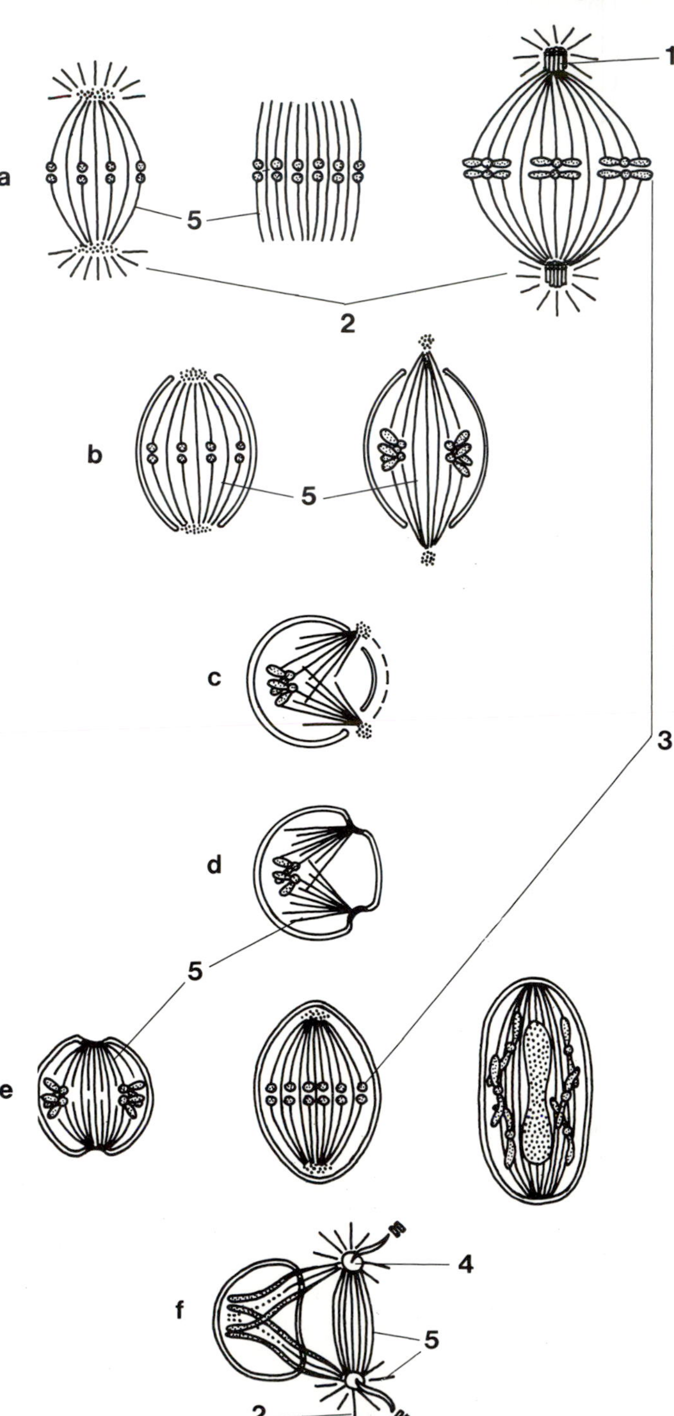

Mitosis
Chromosomes in metaphase. a. orthomitosis, open. b. orthomitosis, semi-open. c. pleuromitosis, semi-open. d. pleuromitosis, closed intranuclear. e. orthomitosis, closed intranuclear. f. pleuromitosis, closed extranuclear. 1=centriole; 2=asters; 3=chromosomes on metaphase plate; 4=division center; 5=microtubules.

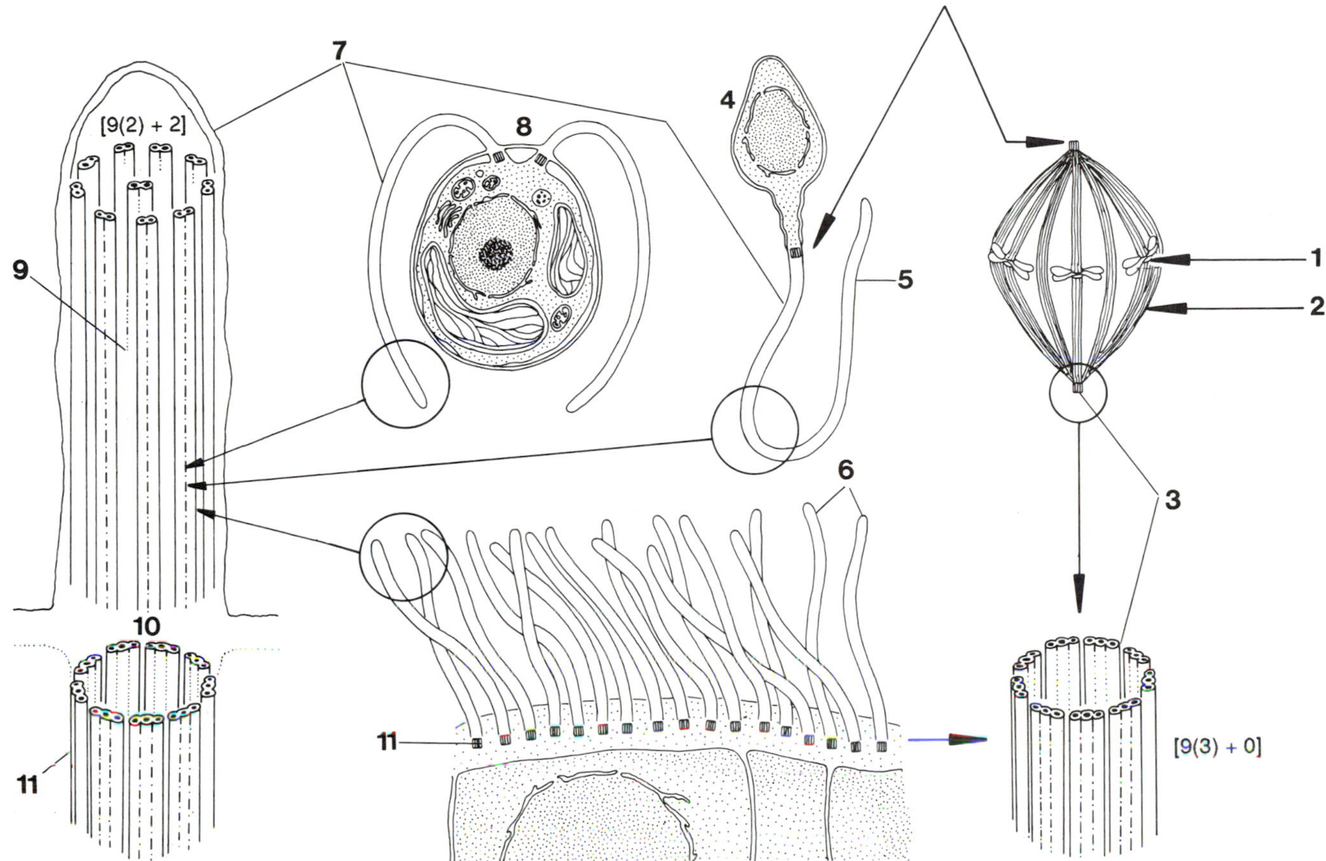

Mitotic apparatus
Microtubular structures: undulipodia, cilia, and mitotic spindles. [9(2)+2] axoneme microtubule arrangement and [9(3)+0] centriole-kinetosome microtubule arrangement are indicated. 1=kinetochore (centromere); 2=mitotic spindle; 3=centriole; 4=sperm; 5=sperm tail; 6=cilia; 7=undulipodia; 8=alga; 9=axoneme; 10=transition zone; 11=kinetosome.

Mitochondriome
See *Chondriome.*

Mitosis (adj. mitotic)
Nuclear division; karyokinesis. Although protists vary widely in details of the mitotic process, generally four stages are recognizable: prophase, in which the centriole divides and the attached pairs of duplicate chromosomes condense; metaphase, in which the chromosomes move and align at the equatorial plane of the nucleus; anaphase, in which the chromatids separate at their kinetochores and move to opposite poles; and telophase, in which the chromosomes return to their extended state. The result is two separate, identical groups; since the chromosomes replicate once before mitosis, and only one division occurs, the ploidy of the nucleus is unaltered by mitosis. See *Cryptomitosis, Cryptopleuromitosis, Cytokinesis, Mesomitosis, Metamitosis, Pleuromitosis, Promitosis.* See also illustrations: Amebomastigota, Closed mitosis, Opalinata, Plasmodiophoromycota.

Mitosporangium (pl. mitosporangia)
Sporangium in which spores are produced by mitotic cell divisions.

Mitotic apparatus
Microtubules, kinetochores, centrioles, centrosomes, and any other transient proteinaceous structures associated with mitotic cell division. Illustration above.

Mitotic oscillator
Hypothetical regulator of synchronous nuclear division in plasmodia of plasmodial slime molds.

Mitotic spindle
Transient microtubular structure that forms between the poles of nucleated cells and is responsible for chromosome movement. Illustrations: Atractophore, Mitotic apparatus.

Mixis (adj. mictic)
Two-parent sex; syngamy or karyogamy leading to fertilization to form an individual with two different parents. See *Apomixis.*

Mixotrophy (adj. mixotrophic)
Nutritional mode: facultative chemoheterotrophy in a photoautotrophic organism (see Table 2).

MLC
See *Microbody-lipid globule complex.*

MLS
See *Multilayered structure.*

MOC
See *Microtubule-organizing center.*

Monad (adj. monadoid)
Zoid. Single cell; free-living, unicellular, usu. undulipodiated organism or stage of an organism; mastigote.

Moniliform (adj.)
Descriptive morphological term referring to components arranged in a linear order like beads on a string.

Monocentric (adj.)
Referring to development of thallus in funguslike protoctists (e.g., oomycotes, chytridiomycotes) and algae. Thalli with a single central structure into which nutrients flow and from which reproductive structures are initiated; a monocentric thallus may be holocarpic or eucarpic. See *Polycentric.*

Monoclinous (adj.)
Referring to antheridia and oogonia originating from the same hypha (e.g., the oomycote *Aphanomyces stellatus).* See *Diclinous, Monoecious.*

Monoclonal antibody
Antibody derived from a single clone of vertebrate plasma cells that has highly specific antigen-binding properties.

Monodisperse (adj.)
Referring to polymers that are homogeneous in molecular weight.

Monoecious (adj.)
Hermaphroditic; monoclinous. Referring to organisms that have male and female structures on the same individual. See *Dioecious.*

Monogenetic (adj.)
See *Monoxenous.*

Monokinetid
Kinetid containing one kinetosome and one each of associated structures. See *Dikinetid.* Illustration: Oral region.

Monolamellar (adj.)
Referring to form: structure with one lamella or layer.

Mononucleate (adj.)
Referring to a cell containing a single nucleus.

Monophyletic (adj.)
Evolutionary term referring to a trait or group of organisms that evolved directly from a common ancestor. See *Polyphyletic.*

Monoploid (adj.)
See *Haploid.*

Monopodial (adj.)
Increasing in length by apical growth; in algae, a type of growth in which the primary axis is maintained as the main line of growth and secondary laterals (offshoots) are produced from the primary axis. See *Limacine movement, Limax ameba.*

Monoraphid (adj.)
Referring to a diatom frustule with a single raphe.

Monospecific (adj.)
Belonging to a single species (e.g., a monospecific bloom consists of a single species).

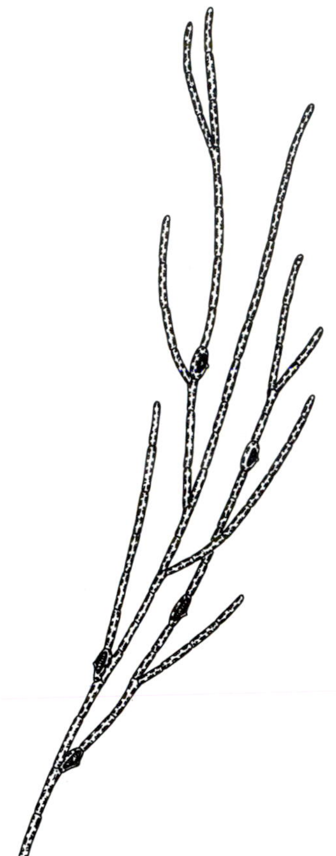

Monosporangium
Branched uniseriate thallus with monosporangia of the rhodophyte *Rhodochaete.* x 80.

Monosporangium (pl. monosporangia)
Sporangium that produces a single spore, i.e., a monospore.

Monospores
Nonmotile spores produced by mitosis and cytokinesis one at a time from a sporangium; asexual, naked spores of bangiophycidean rhodophytes. Illustration: Rhodophyta.

Monostromatic (adj.)
Morphological term describing a structure composed of a single layer of cells.

Monothalamic (adj.)
Monothalamous. Bearing only a single chamber; refers to foraminifera that form a test with only one chamber.

Monothalamous (adj.)
See *Monothalamic.*

Monoxenic (adj.)
Referring to laboratory growth of two species of organisms, one of which is usu. studied from a biochemical or ecological viewpoint. A monoxenic culture may involve, for example, a

ciliate plus one "stranger" bacterium, alga, yeast, or other ciliate species; the second organism may be unwanted (a contaminant) or may be present in the medium to serve as food for the first. See *Axenic*.

Monoxenous (adj.)
Monogenetic. Referring to the life cycles of symbiotrophic protoctists that occur in only one kind of host. See *Heteroxenous*.

Monozoic (adj.)
Referring to single body form (e.g., resulting from complete cell division). See *Diplozoic*.

Morph
Form. Organism or structure with distinguishable size and shape. Environmentally induced form of organism.

Morphology
Study of form or results from a study of form.

Morphometrics
Subfield of morphology; quantitative study of form, size, and shape variation within a species or strain of organism.

Morphotype
Typical morph. Term also used when taxonomic identification is temporarily uncertain.

Mouth
See *Buccal cavity, Oral region*.

MTOC
See *Microtubule-organizing center*.

Muciferous bodies
Mucocyst. Intracellular organelles (less complex than trichocysts) producing or filled with mucilage or mucus. Seen in euglenids, dinomastigotes, chrysophytes, and prymnesiophytes. May be extrusomes.

Mucilage
Mucous material, generally composed of polysaccharides. Illustrations: Interphase, Palmelloid.

Mucocyst
Mucus body; muciferous body; mucus trichocyst. Mucilage-containing extrusome; subpellicularly located, saccular or rod-shaped organelle or paracrystalline structure, dischargeable through an opening in the pellicle as an amorphous, mucuslike mass (e.g., dinomastigotes, pseudociliates, raphidophytes); in ciliates probably involved in cyst formation (in some species) among other possible functions; not a trichocyst although sometimes used synonomously. Illustration: Raphidophyta.

Mucron
Anteriorly located attachment organelle not separated from the rest of the body by a septum (e.g., in gregarine apicomplexan families Ganymedidae and Lecudinidae); analogous structures are found in some symbiotic ciliates and mastigotes. See *Epimerite*.

Mucus body
See *Mucocyst, Muciferous body*.

Mucus trichocyst
See *Mucocyst*.

Mud
A slimy and sticky or slippery mixture of water, slime, and finely-divided particles (silt size or smaller) of aluminosilicate clays or other minerals.

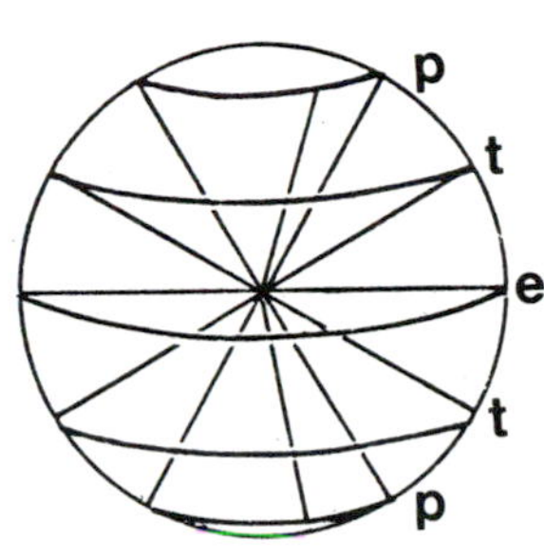

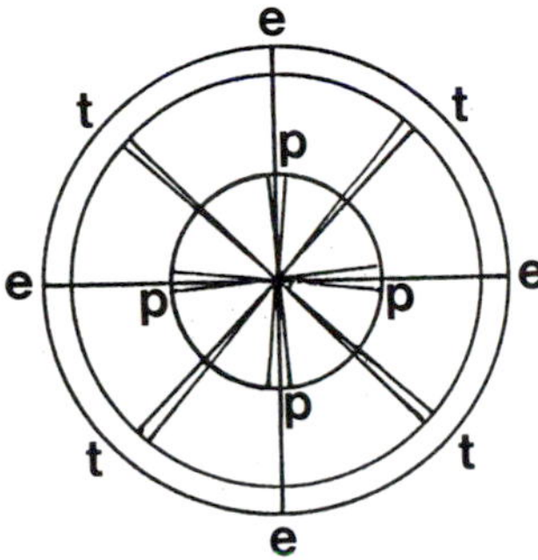

Müller's Law
Spicular arrangement in acantharian actinopods according to Müller's Law. e=equatorial spicules; p=polar spicules; t=tropical spicules.

Müller's Law
Law describing the unique radially symmetrically ordered skeleton of acantharians (phylum Actinopoda) in which the cell may be conceived as a globe from whose center spicules radiate and pierce the surface at fixed latitudes and longitudes. If there are twenty spicules, then there are five quartets, one equatorial, two polar, and two tropical, piercing the globe at latitudes 0°, 30°N, 30°S, 60°N, and 60°S. Longitudes of the piercing points are 0°, 90°W, 90°E, 180°, 45°W, 45°E, 135°W, and 135°E for their respective quartets. Variations in shape of cell, thickness, length, or number of spicules are still grouped by some elaboration of Müller's Law.

Multilayered structure
MLS. Type of kinetid characteristic of chlorophytes (e.g., charophyte motile cells) containing a band of microtubules that overlies several layers of parallel plates; proximal portion of the kinetid consists of several layers of which one layer has regularly spaced lamellae oriented perpendicularly to the overlying rootlet microtubules. Illustration next page.

Multilocular (adj.)
Multiloculate. Referring to foraminifera characterized by many cells or chambers.

Multiloculate (adj.)
See *Multilocular*.

Multinucleate (adj.)
Referring to cells or tissue containing more than a single nucleus in a membrane-bounded space. See *Coenocyte, Plasmodium, Syncytium*.

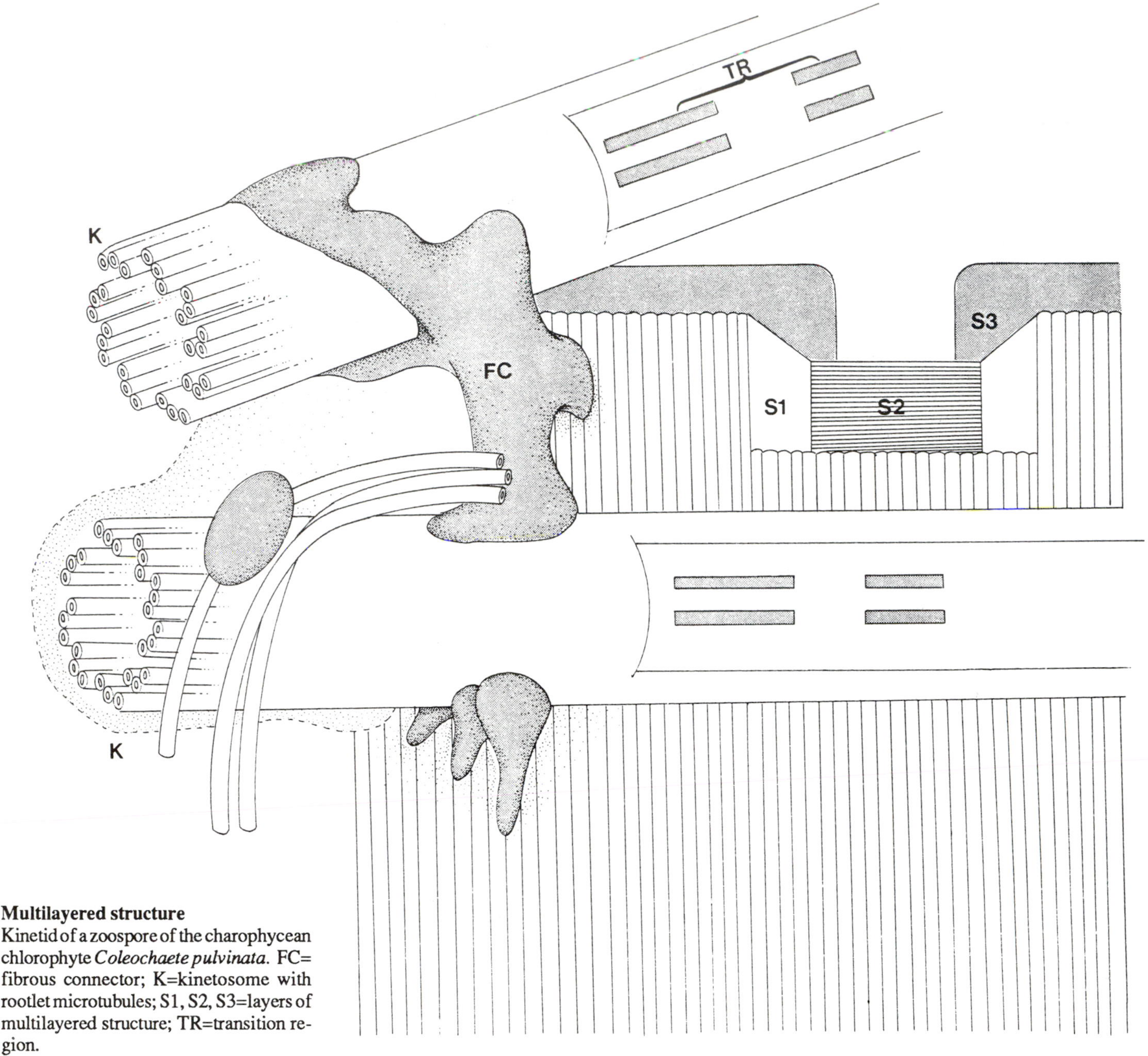

Multilayered structure
Kinetid of a zoospore of the charophycean chlorophyte *Coleochaete pulvinata*. FC= fibrous connector; K=kinetosome with rootlet microtubules; S1, S2, S3=layers of multilayered structure; TR=transition region.

Multiple fission

Karyokinesis followed by a delay in cytokinesis such that when cytokinesis occurs, 2^n offspring are produced at once, where n represents the number of generations cytokinesis was delayed. See *Progressive cleavage*. Illustration: Kinetoplastida.

Multiplicative cyst

Cyst in which multiple fissions (mitotic cell divisions) occur.

Multipolar nucleus (pl. multipolar nuclei)

Dividing nucleus containing spindle microtubules oriented toward more than two poles.

Multiseriate (adj.)

Pluriseriate. Morphological term referring to structures (e.g., trichomes, filaments, algal "hairs") composed of more than a single row of cells.

Mural pores

Minute openings in tests of many foraminifera.

Mutant

Organism bearing an altered gene expressed in its phenotype; organism demonstrating a heritable, detectable, structural or chemical change.

Mutualism (adj. mutualistic)

Ecological term referring to associations between organisms that are members of different species such that the associated partners leave more offspring per unit time when together than when they are growing separately.

Mycelium (pl. mycelia; adj. mycelial)
Threadlike material (hyphae) that together forms a matted tissuelike structure that makes up the body of most fungi and some protoctists (e.g., chytridiomycotes, oomycotes). Illustration: Heterothallism.

Mycology
Study of fungi. Subfield of biology that traditionally included study of funguslike protoctists (e.g., chytridiomycotes, plasmodiophorids).

Mycophagy
Mode of nutrition in which organisms feed on fungi.

Mycosis
Disease caused by a fungus.

Mycovirus
Virus of a fungus.

Myonemes
"Muscle threads." Contractile ribbonlike or cylindrical organelles found in acantharian actinopods and some dinomastigotes and ciliates. Consist of densely packed 2-3 nm microfibrils, exhibiting long clear L zones cross-striated by thin dark T bands (e.g., ciliates). Myonemes may play a part in buoyancy regulation (e.g., acantharians) and are responsible for cell contraction (e.g., in dinomastigotes and some ciliates). Term applied to cell structures that are probably unrelated. See *L zone, T bands*. See also illustrations: Acantharia, Perispicular cone.

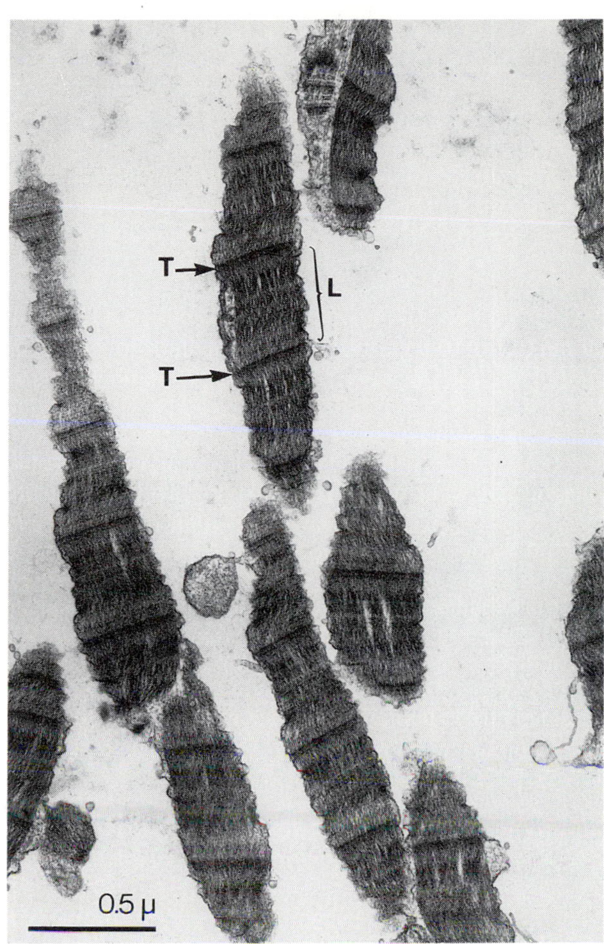

Myonemes
The microfibrillar structure of myonemes of the symphiacanthid actinopod *Acantholithium stellatum*. L=L zones; T=T bands. TEM.

Myxameba
Ameboid stage of plasmodial (myxomycote) slime molds in which cells lack cell walls and feed by phagocytosis; this stage gives way to the formation of a plasmodium and later a stalked sporocarp. Illustration: Plasmodial Slime Molds.

N

Naked (adj.)
Wall-less; lacking a cell wall, scales, or decorations; also, ciliates denuded of cilia.

Nanofossils
Microfossils of the smallest kind, usu. 1-20 µm in size.

Nanoplankton (adj. nanoplanktonic)
Planktonic protists in the 1-20 µm size range; plankton with dimensions of fewer than 70 to 75 µm that tend to pass through plankton nets.

NAO
See *Nucleus-associated organelle.*

Nebenkörper (German, meaning neighboring body)
Paranucleus; parasome. Feulgen-positive body found alongside the nucleus in *Paramoeba,* possibly xenosome of symbiotic origin from bacteria.

Necrosis (adj. necrotic)
Death of cells, a piece of tissue, or an organ in an otherwise living organism.

Necrotrophy (n. necrotroph; adj. necrotrophic)
Nutritional mode in which a symbiotroph damages or kills its host; parasitism or pathogenesis (see Table 2).

Nectomonad
Free-swimming (as opposed to attached) stage in the life cycle of trypanosomatids.

Negative staining
Technique in electron microscopy in which a sample is mixed with a stain (e.g., phosphotungstic acid) and sprayed onto a grid; because the stain surrounds the contours of the sample, organismal structure appears light against a dark background.

Nemathecium (pl. nemathecia)
Raised or wartlike area on the surface of the thallus of some florideophycidean rhodophytes; contains reproductive organs.

Nematocyst
Cnidocyst. Modified cell with a capsule containing a threadlike stinger used for defense, anchoring, or capturing prey; some con-

tain poisonous or paralyzing substances (e.g., in all coelenterates and ctenophores; analogous organelles found in some dinomastigotes and some karyorelictid and suctorian ciliates). Illustration: Taeniocyst.

Nematodesma (pl. nematodesmata)
Part of kinetids of certain ciliates. Bundle of microtubules, usu. hexagonally packed, that originates in association with the kinetosome and forms part of the wall of the cytopharyngeal apparatus.

Nematogene
Organelle giving rise to the nematocyst in dinomastigotes.

Neontology
The study of extant species, in contrast with paleontology.

Neosaxitoxin
See *Saxitoxin complex.*

Neoseme
Appearance of a new trait of evolutionary importance (seme). See *Seme.*

Neritic (adj.)
Ecological term referring to the region of shallow ocean water along a sea coast; between low-tide level and 200 meters or between low-tide level and approximately the edge of the continental shelf; also referring to the organisms living in that environment. See *Pelagic.* Illustration: Habitat.

Neuston (adj. neustonic)
Ecological term referring to the surface biota of aquatic environments; those dwelling at the interface between atmosphere and water. Illustration: Habitat.

Niche
Role performed by members of a species in a biological community.

Noctikaryotic (adj.)
Referring to dinomastigotes (e.g., *Noctiluca)* in which the nucleus changes from the usual mesokaryotic condition (dinokaryotic) to a conventional eukaryotic appearance during the life cycle.

Nodes
Protruberances found on the umbilical surfaces of certain foraminifera (e.g., Glabratellidae); sites on an algal axis from which new growth arises (e.g., charophytes).

Nonclastic (adj.)
Sediment that is chemically precipitated in place (e.g., halite). See *Clast.*

Nonemergent flagellum (pl. nonemergent flagella)
See *Nonemergent undulipodium.*

Nonemergent undulipodium (pl. nonemergent undulipodia)
Undulipodium lacking an emergent axoneme, reduced to a kinetosome (centriole) only in extreme cases; short undulipodium (e.g., in euglenids). See *Emergent undulipodium.*

Nongeniculate (adj.)
Referring to a structure not formed by joints; nonarticulated; lacking segmentation; also, not bent abruptly at an angle.

Nori
Edible, dried preparation of the rhodophyte *Porphyra.*

NP
Abbreviation of nucleoprotein. Illustration: Kinetid.

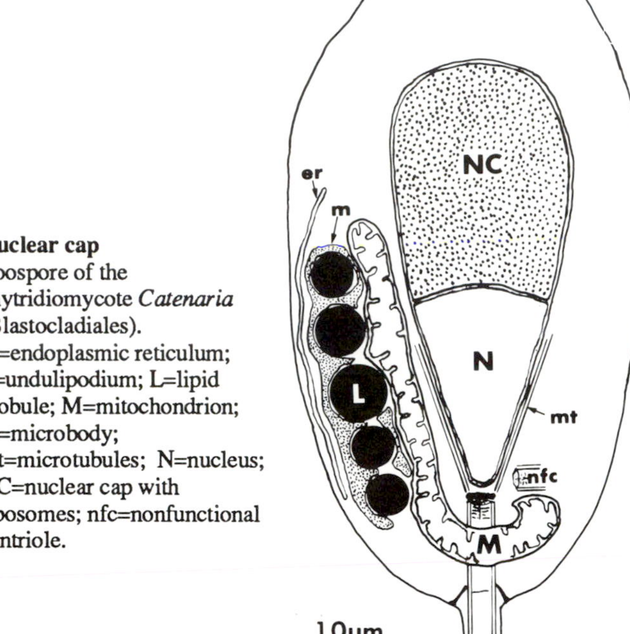

Nuclear cap
Zoospore of the chytridiomycote *Catenaria* (Blastocladiales). er=endoplasmic reticulum; F=undulipodium; L=lipid globule; M=mitochondrion; m=microbody; mt=microtubules; N=nucleus; NC=nuclear cap with ribosomes; nfc=nonfunctional centriole.

Nuclear cap
Crescent-shaped, membrane-bounded sac surrounding a third or more of the zoospore nucleus of the blastocladialean chytridiomycotes; the nuclear cap apparently contains all the cell's ribosomes.

Nuclear cyclosis
Intranuclear movement by means of filaments which apparently use the proteins associated with intracellular motility: actin, myosin, and tubulin; the enlargement and slow swiveling movement and rotation of the nucleus is associated with the first meiotic division (e.g., some suctorian ciliates and dinomastigotes).

Nuclear dualism
Heterokaryosis. Possession of two functionally different nuclei in the same cell; characteristic of ciliate cells and a few foraminifera.

Nuclear envelope
Nuclear membrane. Double membrane structure, often containing many pores, surrounding the nucleoplasm. Structural criterion defining eukaryotes. Illustration: Organelle.

Nuclear membrane
See *Nuclear envelope.*

Nuclear plate
See *Metaphase plate.*

Nucleogony
Multiple karyokineses to produce many small nuclei at once. Illustration: Life cycle.

Nucleoid
DNA-containing structure of prokaryotes, not bounded by a membrane. See *Genophore*. Illustration: Organelle.

Nucleolar-organizing center
See *Nucleolar-organizing chromosome.*

Nucleolar-organizing chromosome
Nucleolar-organizing center. Chromosome or chromatin with long secondary constrictions (nucleolar-organizing regions); site of formation of new nucleoli that are precursors to RNA subunits of ribosomes.

Nucleolar substance
Stainable material present during or after mitosis and derived from the nucleolus.

Nucleolus (pl. nucleoli)
Endosome; karyosome. Structure in the cell nucleus composed of RNA and protein, precursor material to the ribosomes.

Nucleomorph
Organelle surrounded by a double membrane resembling a small nucleus, lying between the plastid ER and plastid membrane in cryptomonads; a membrane-bounded, nucleic acid-containing organelle in the periplastidial compartment, thought to be remnant nucleus of a eukaryotic photosynthetic endosymbiont. Illustration: Cryptophyta.

Nucleonema (pl. nucleonemata)
Network of strands consisting of granular material, located at nucleolar surface (e.g., *Pelomyxa palustris).*

Nucleoplasma
Fluid contents of the nucleus of any eukaryote.

Nucleus (pl. nuclei)
Membrane-bounded, spherical, DNA-containing organelle, universal in protoctists. Chromatin (DNA, protein) organized into chromosomes; site of DNA synthesis and RNA transcription. Nuclear membranes bear pores. Definitional for eukaryotes. Illustration: Organelle.

Nucleus-associated organelle
NAO; spindle pole body. Microtubule-organizing center just outside, on, or associated with the nuclear membrane of some protoctists (e.g., rhodophytes) and fungi. One is found at each of the poles during mitotic division.

Nudiform replication
Term applied to loricate choanomastigotes (Acanthoecidae) indicating the absence of bundles of component costal strips in one of the two offspring cells resulting from a cell division. See *Tectiform replication.*

Nutricline
Ecological term referring to gradients of nutrient concentration in aquatic environments.

O

Obligate (adj.)
Compulsory or mandatory as opposed to optional or facultative; e.g., an obligate anaerobe can survive and grow only in the absence of gaseous oxygen. See *Facultative.*

Oceanic (adj.)
Pertaining to those areas of the ocean deeper than the littoral and neritic; open ocean depths. Illustration: Habitat.

Ocelloid
See *Ocellus.*

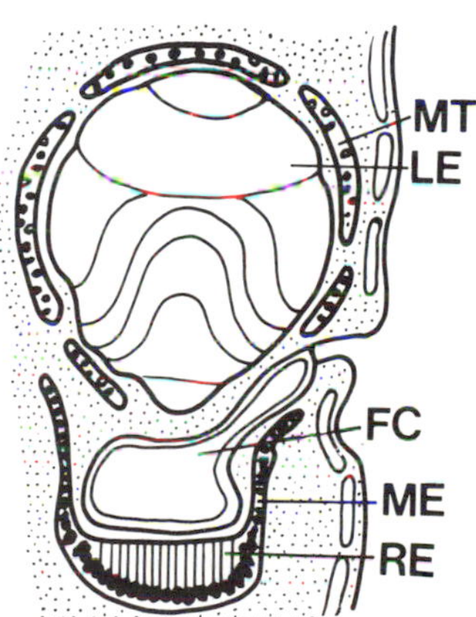

Ocellus
Portion of the dinomastigote *Nematodinium*. FC=fluid cavity (ocular chamber); LE=lens; ME=melanosome; MT=mitochondrion; RE=retinoid.

Ocellus (pl. ocelli; adj. ocelloid)
Ocelloid. Complex light-perceiving organelle in a few dinomastigote genera, consisting of a large refractive lens (hyalosome) and a pigment-containing cup (melanosome) *(e.g., Warnowia, Erythropsidinium);* slightly raised area of a valve that is externally rimmed and encloses an area of fine pores (porelli) in diatoms. See also illustration: Porelli.

Ocular chamber
Component of dinomastigote ocellus; chamber with a canal extending to the sulcus. Illustration: Ocellus.

Offspring
Progeny. Filial products. "Daughter cells" should be referred to as offspring cells. "Daughter nuclei" should be referred to as offspring nuclei. (Term "daughter" should be avoided in cases where the female gender of the offspring has not been established.)

Oligohaline (adj.)
Oligosaline. Ecological term referring to marine environments with low salinities, i.e., less than about 3.3% salt. See *Euhaline, Euryhaline, Hyperhaline.*

Oligophotic zone
The region in aquatic environments below the mesophotic zone, in which the organisms are limited by insufficient sunlight for optimal growth but in which sufficient incident radiation penetrates so that some photosynthesis is possible. Illustration: Habitat.

Oligosaline (adj.)
See *Oligohaline*.

Oligotrophy (adj. oligotrophic)
Ecological term referring to clear water, i.e., an aquatic environment deficient in inorganic and organic nutrients, and usu. containing high concentrations of dissolved oxygen. See *Eutrophy, Polysaprobic*.

Omnivory (n. omnivore; adj. omnivorous)
Heterotrophic mode of nutrition; ingestion of plant, fungal, and/or animal food.

Ontogeny
Development of an individual organism (e.g., animal from fertilized egg to death).

Oocyst
Encysted zygote (e.g., coccidian apicomplexans and heliozoan actinopods). Illustrations: Coccidian life history, Life cycle, Macrogamont, Sporocyst.

Oogamy (adj. oogamous)
Fusion of a nonmotile large egg (female gamete) with a small motile sperm (male gamete); extreme form of anisogamy (e.g., some protoctists, most animals).

Oogenesis
Development of ova (egg cells) (e.g., animal eggs prior to fertilization).

Oogonial cavity
Space enclosed by the oogonial wall; may be completely filled by an oospore (plerotic) or only partially filled by one or more oospores (aplerotic) (e.g., oomycotes).

Oogonioplasm
Cytoplasm of the oogonium.

Oogonium (pl. oogonia)
Uninucleate or coenocytic cell (or the cell wall) that generates female gamete(s) (e.g., oomycotes). Illustrations: Antheridium, Dense body vesicle, Heterothallism, Life cycle, Oomycota.

Ookinete
Motile zygote. Illustration: Life cycle.

Oomycote
Informal name for members of the phylum Oomycota. Illustrations: Mastigoneme, Oomycota.

Ooplast
Organelle formed in the oospore as a result of coalescence of the dense body vesicles (e.g., oomycotes). Illustration: Oospore.

Oosphere
Egg. Unfertilized and unpenetrated female gamete containing a single, haploid nucleus (e.g., oomycotes).

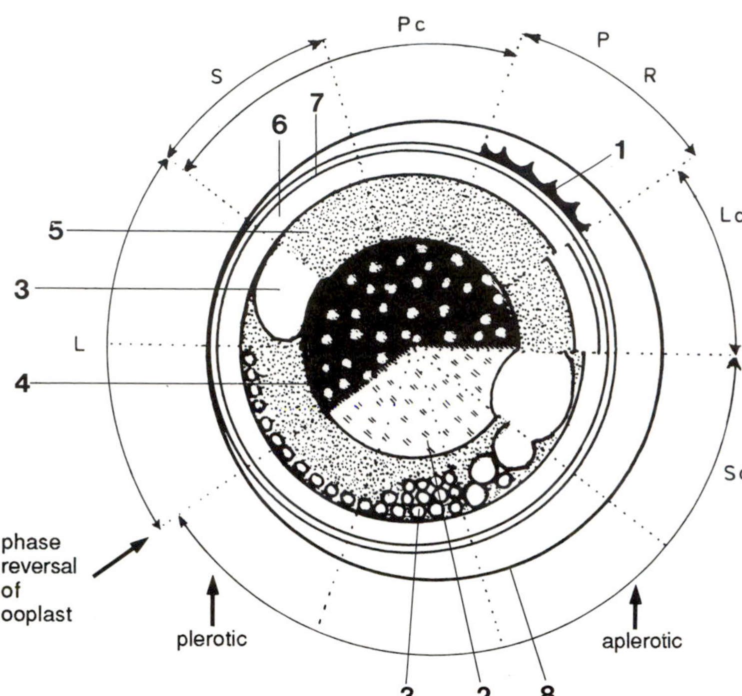

Oospore
Variations in oospore wall and protoplasmic structure of seven orders of oomycotes as indicated by double-ended arrows. L=Leptomitales; Lc=Leptolegniellaceae; P=Peronosporales; Pc=Pythiales; R=Rhipidiales; S=Sclerosporales; Sa=Saprolegniales. 1=exospore wall; 2=fluid granular ooplast; 3=oil droplet; 4=solid ooplast; 5=dispersed oil droplets; 6=endospore wall; 7=epispore wall; 8=border of oogonial cavity.

Oospore
Thick-walled spherical structure developing from an oosphere after fertilization in oomycotes. See *Aplerotic, Plerotic*. See also illustration: Oomycota.

Oosporogenesis
Oospore formation.

Operculum (pl. opercula)
Lid, covering, or flap of an aperture.

Opisthe
Posterior offspring of transverse binary fission of the parental organism (e.g., ciliates). See *Proter*.

Opisthokont (adj.)
Referring to the morphology of a posteriorly undulipodiated mastigote (e.g., chytrid zoospores, some dinomastigotes); pertaining to the insertion of the undulipodium at the posterior pole of the cell (in relation to movement).

Opisthomastigote

Stage in development of a trypanosomatid in which the kinetoplast lies behind the nucleus and the associated undulipodium emerges at the anterior extremity from a long, narrow, undulipodial pocket. Illustration: Kinetoplastida.

Oral apparatus

See *Buccal cavity, Ingestion apparatus, Oral region.*

Oral cavity

Mouth opening.

Oral kinetid

Kinetid within the oral region of the ciliate cortex.

Oral opening

Mouth cavity. Illustration: Ciliophora.

Oral region

Oral area; mouth; ingestion apparatus; oral apparatus. General term for that part of a protist cell bearing the ingestion apparatus or cytostome; sometimes used in a nonspecific way. See *Somatic region.*

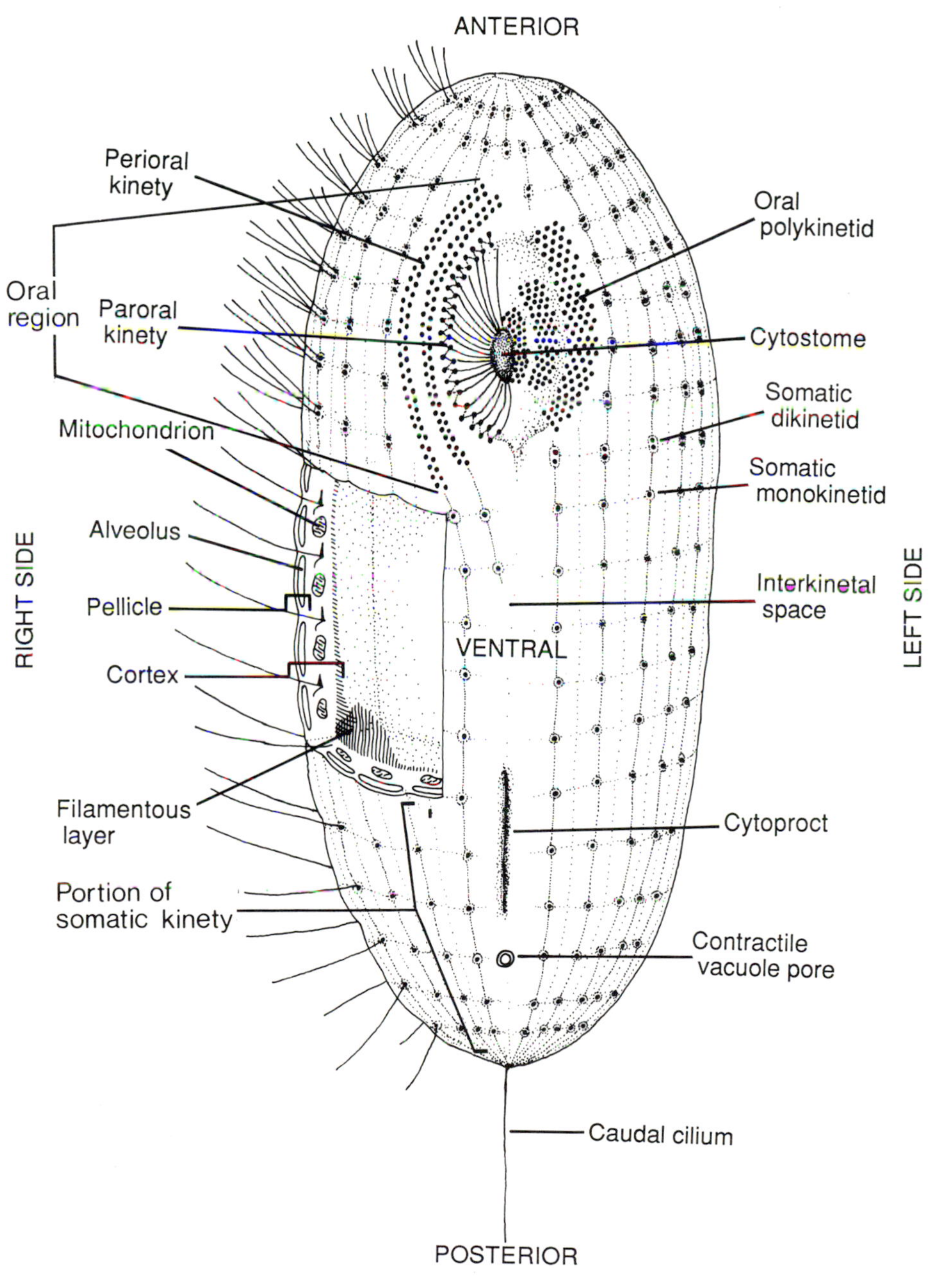

Oral region
Somatic kinetids and oral region (perioral kinety, paroral kinety, three oral polykinetids, and cytostome) of a generalized cyrtophoran ciliate. The somatic kinetids are aligned in files called kineties. Kinetosomes are represented as dots.

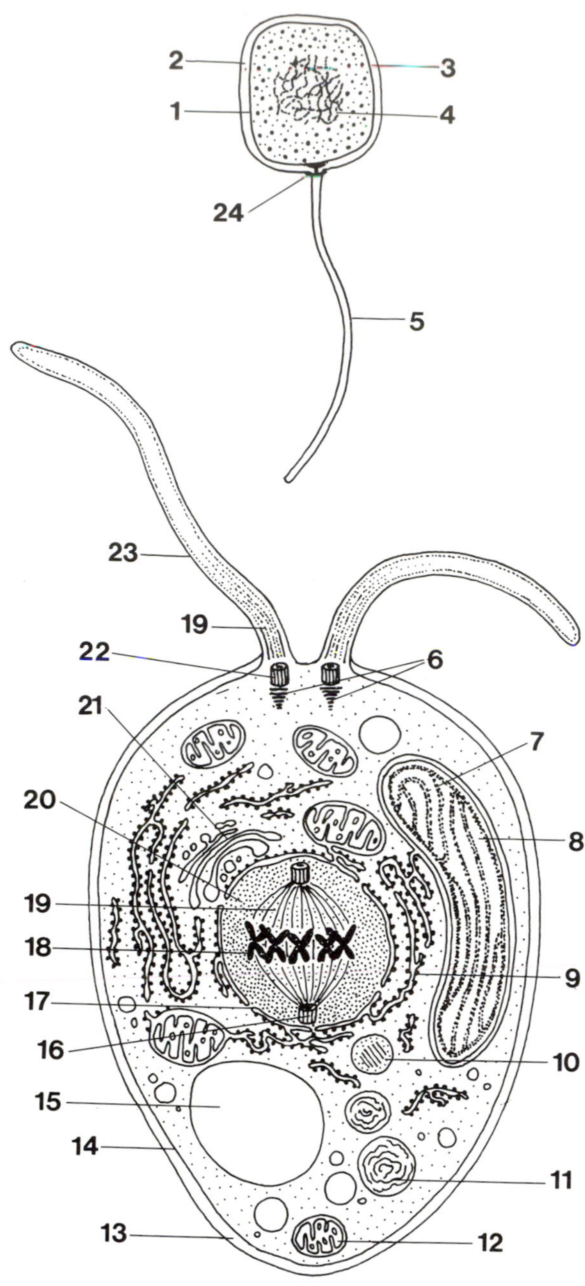

Organelle
Generalized prokaryotic (top) and eukaryotic (bottom) cell structure. 1=plasma membrane; 2=cell wall; 3=small ribosomes; 4=nucleoid (genophore); 5=flagellum; 6=banded roots; 7=thylakoids; 8=plastid; 9=endoplasmic reticulum with ribosomes; 10=peroxisome (glyoxysome); 11=lysosome; 12=mitochondrion; 13=cell wall (glycocalyx); 14=plasma membrane (envelope); 15=vacuole; 16=centriole; 17=nuclear pore; 18=chromosome; 19=microtubules; 20=nucleus; 21=golgi apparatus (dictyosome); 22=kinetosome; 23=undulipodium; 24=rotary motor.

Organelle
Distinctive structure detected by microscopy inside a cell. Some, like mitochondria, nuclei, and plastids, are double membrane-bounded and capable of division. Others, like carboxysomes, ribosomes and liposomes, are visualizable as locally high concentrations of certain enzymes and other macromolecules.

Organic test
Covering or shell of an organism composed of organic materials (e.g., chitin, cellulose, or the complex of protein and mucopolysaccharide of foraminifera known as tectin).

Orogeny
Mountain-building processes.

Orthomitosis
Karyokinesis; mitotic cell division in which spindle tubules are parallel to each other. Illustration: Mitosis.

Osmiophilic (adj.)
Osmium-loving. Tendency to stain black with osmium tetroxide, esp. characteristic of electron microscopy preparations. Illustrations: Funis, Hyphochytriomycota.

Osmoregulation
Maintenance of constant internal salt and water concentrations in an organism, requiring the input of energy.

Osmotrophy (n. osmotroph; adj. osmotrophic)
Mode of heterotrophic nutrition in which organisms take in soluble organic compounds; osmotrophic organisms absorb food in the dissolved state from the surrounding medium directly by osmosis, active transport, or pinocytosis (e.g., fungi, some protoctists) (see Table 2). See *Carnivory, Histophagy, Holozoic, Macrophagy, Phagotrophy.*

Out-group
Cladistic concept; species or higher monophyletic taxon that is examined in a phylogenetic study to determine which of two homologous characters may be inferred to be symplesiomorphic (apomorphic). One or several out-groups may be examined for each decision. The most critical out-group comparisons involve the sister group of the taxon studied.

Outbreeding
Mating and production of offspring by organisms not known to have a traceable common ancestor (genetically unrelated organisms).

Ovary (pl. ovaries)
Multicellular sex organ of female animals and plants. In flowering plants, an enlarged basal portion of a carpel or of a gynoecium composed of fused carpels that becomes the fruit. Term not appropriate to protoctists. Illustration: Life cycle.

Oyster spat
Juvenile oyster.

P

Pachynema
Stage in prophase of meiotic cell divisions in which chromosomes are tightly packed. Illustrations: Plasmodiophoromycota, Synaptonemal complex.

Pachytene
Pachynema; stage in meiotic prophase I in which pairs of homologous chromosomes shorten and thicken. See *Meiosis*.

Paedogamy
See *Pedogamy*.

Paleoecology
Subfield of paleontology: attempt to reconstruct past communities of organisms and their environments by study of their fossil remains.

Paleontology
Study of past life on Earth primarily by investigation of fossil remains; subfield of geology, essential to evolutionary biology.

Palintomy
Rapid sequence of binary fissions, typically within a cyst and with little or no intervening growth, resulting in production of numerous, small offspring cells. Common in various parasitic protists (e.g., ciliates and dinomastigotes); produces tomites in apostome ciliates. Illustrations: Opalinata, Tomont.

Palisporogenesis
Specialized type of division in some blastodinian dinomastigotes in which the first division results in an individual that continues to feed on the host (trophocyte) and one that is responsible for the subsequent division (gonocyte).

Pallium (pl. pallia)
See *Feeding veil*.

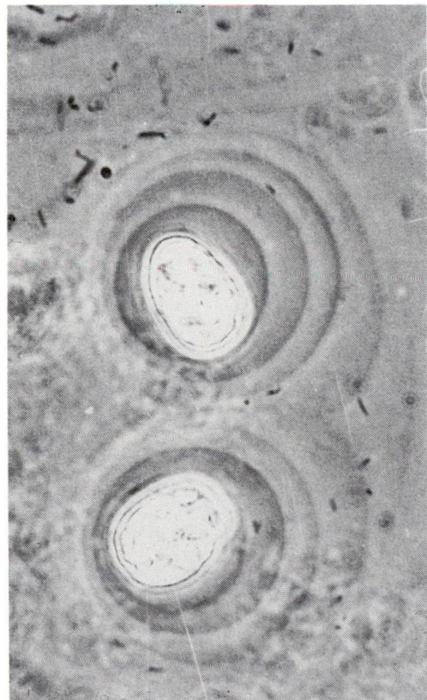

Palmelloid
The cryptomonad *Cryptomonas erosa*. Palmelloid colony with multiple layers of mucilage surrounding the cells. PC, x 900.

Palmelloid (adj.)
Referring to colony morphology characteristic of many algae in which nonmotile cells are encased in mucus as a gelatinous mass. See also illustration: Interphase.

Panacronematic (adj.)
See *Pantacronematic*.

Pankinetoplastic (adj.)
Pertaining to a morphology of kinetoplastids in which the kDNA is not localized in one or more discrete bodies but is irregularly distributed as stainable masses throughout the kinetoplast mitochondrion.

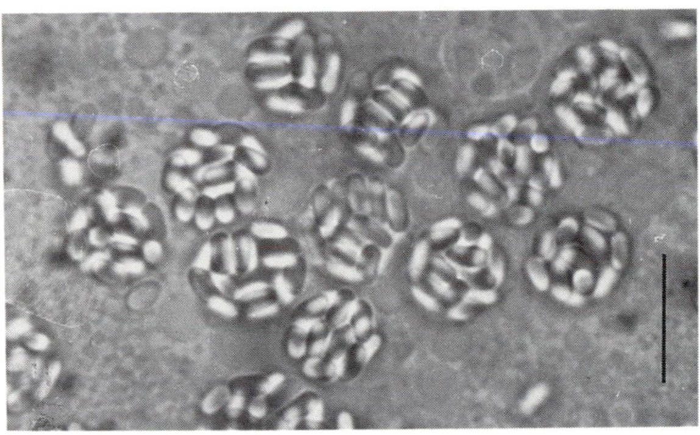

Pansporoblast
The microsporan *Pleistophora operophterae*. Spores in sporophorous vesicles surrounded by pansporoblast envelope. PC. Bar=10 μm.

Pansporoblast
Membrane surrounding multinucleate structure. Syncytium that undergoes cytokinesis to yield parasites. Gives rise to two sporoblasts contained within a single membrane in some apicomplexans. In actinosporean myxozoans, two-to-four-celled envelope containing groups of eight spores, and sometimes called a pansporocyst; in myxozoans, a thick envelope around one or more spores consisting of two degraded pansporoblast (or pericyte) cells; in microsporans, obsolete term for the subpersistent membrane of the sporophorous vesicle.

Pansporoblast envelope
See *Pansporoblast*. Illustration: Pansporoblast.

Pansporoblast membrane
See *Sporophorous vesicle*.

Pansporocyst
Pansporoblast, usu. of actinosporeans.

Pantacronematic (adj.)
Panacronematic. Referring to mastigotes having undulipodia with two rows of mastigonemes (flimmers, fibrils) and a terminal fiber. See *Acronematic*, *Pleuronematic*. Illustration: Mastigoneme.

Pantonematic (adj.)
Referring to mastigotes having undulipodia with two rows of mastigonemes but no terminal filament or fiber. See *Acronematic*, *Pantacronematic*. Illustration: Mastigoneme.

Papilla (pl. papillae; adj. papillate)
Small bump or projection. Specialized structure found on the periphery of mature sporangia that is enzymatically degraded at

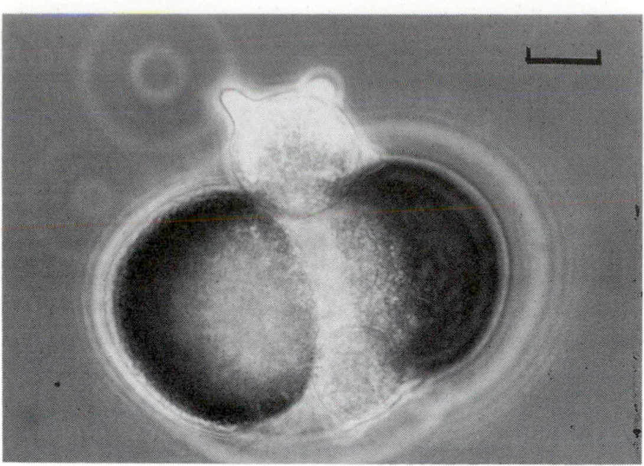

Papilla
A mature sporangium with three discharge papillae (top) of the chytridiomycote *Spizellomyces* sp. (Spizellomycetales) on pine pollen. Bar=1.0 µm.

the time of discharge of the sporangial contents, thereby allowing their escape (e.g., in the chytridiomycote *Blastocladiella*).

PAR
See *Paraxial rod.*

Parabasal apparatus
Parabasal body plus a single (or pair of) parabasal filament(s).

Parabasal body
Modified golgi apparatus in anterior of cell which defines the class Parabasalia. Located near the kinetosomes and their associated structures, the structure probably has a secretory function. Illustrations: Atractophore, Parabasalia.

Parabasal filaments
Microfibrillar, striated, often paired organelles of parabasalians arising from a complex kinetid and intimately associated with parabasal bodies.

Parabasal fold
Bent structure of golgi limited to members of the phylum Parabasalia. Illustration: Parabasalia.

Paracostal granules
See *Hydrogenosomes.*

Paracrystalline (adj.)
Pseudocrystalline. Pertaining to cellular inclusions of many types that exhibit a crystal-like organization as seen in the light or electron microscope.

Paradesmose
Cell structure that links two sets of polar kinetosomes during mitosis (e.g., paradesmose in some prasinophytes composed of a microtubular bundle).

Paraflagellar body (PFB)
See *Undulipodial swelling.*

Paraflagellar rod (PFR)
See *Paraxial rod.*

Paraflagellate
Opalinid; member of the class Opalina.

Parallel evolution
Convergent evolution. The independent development of similar structures or behaviors in populations that are not directly related but have been subjected to the same selection pressures (e.g., the evolution of cysts in response to desiccation).

Paralytic shellfish poisoning
PSP. Toxic response due to dinomastigote bloom in which the toxins do not kill many organisms, but are concentrated within the siphons or digestive glands of filter-feeding bivalve mollusks.

Paramylon
Cytoplasmic carbohydrate; the nutritional reserve of euglenids and prymnesiophytes; ß-1,3-glucose polymer, a glucan. Illustrations: Euglenida, Reservoir.

Paranuclear body
Cytoplasmic organelle found in the thraustochytrid labyrinthulomycotes located adjacent to the nuclei of developing thalli and consisting of a compact mass of inflated smooth endoplasmic reticulum cisternae containing a fine granular material.

Paranucleus (pl. paranuclei)
See *Nebenkörper.*

Paraphyletic group
Taxon. Group that includes a common ancestor and some but not all of its descendants.

Paraphysis
Structure of algae: sterile hair growing among reproductive structures (e.g., in *Fucus).*

Parapyla (pl. parapylae)
Secondary openings in the central capsule of phaeodarian actinopods. See *Astropyle.*

Parasexuality
Any process bypassing standard meiosis and fertilization that forms an offspring cell from more than a single parent (e.g., recovery of resistant recombinants in dictyostelids). See *Sex.*

Parasite (adj. parasitic)
Ecological term referring to organisms that live associated with members of different species as obligate or facultative symbiotrophs that tend toward necrotrophy. See *Mutualism, Necrotrophy, Pathogen, Symbiotrophy.* Illustrations: Ellobiopsida, Stachel.

Parasitemia
See *Parasitemia level.*

Parasitemia level
Parasitemia. Measure of parasites in the circulating blood of vertebrate hosts.

Parasitism
Ecological association between members of different species in which one partner (usu. the small form) is obligately or facultatively symbiotrophic and tends toward necrotrophy (see Table 2).

Parasitophorous vacuole
Membranous vacuole containing intracellular parasite; originally derived from the host plasma membrane during phagocytosis of the parasite (e.g., in the microsporan *Encephalitozoon* and in apicomplexans); its composition may subsequently be altered by the parasite. Illustrations: Macrogamont, Microgamete.

Parasomal sac
Structure found associated with each kinetid of the cortex of ciliates; small invagination of the plasma membrane adjacent to kinetosomes. Illustrations: Cortex, Kinetid.

Parasome
See *Nebenkörper*.

Parasporangium (pl. parasporangia)
Algal sporangium producing many spores.

Paratabulation
Numbering system for dinomastigote pellicle plates.

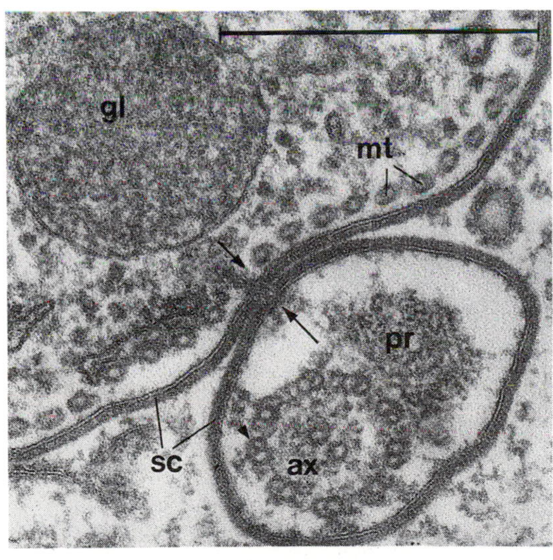

Paraxial rod
Transverse section through undulipodium of the metacyclic kinetoplastid zoomastiginid *Trypanosoma brucei*. Arrows=desmosome at discontinuity in microtubule layer; ax=axoneme; gl=glycosome; mt=pellicular microtubules; pr=paraxial rod; sc=surface coat of variable antigen on body and undulipodium. TEM. Bar=0.5 µm.

Paraxial rod
PAR; paraflagellar rod; PFR. Intraundulipodial structure in euglenids, dinomastigotes, and kinetoplastid zoomastiginids; elaborate cross-striated structure of unknown function that extends nearly the entire length of the undulipodium between the membrane and the axoneme. See also illustrations: Bodonidae, Dinomastigota.

Paraxostylar granules
See *Hydrogenosomes*.

Paraxostyle
Structure found alongside the axostyle (e.g., some zoomastiginids).

Parenchyma (adj. parenchymatous, parenchymous)
Tissue made of thin-walled cells that actively grow in any of three dimensions (e.g., thalli of large algae).

Parietal (adj.)
Referring to position of an organ or organelle: near or alongside a wall.

Paroral kinety
Row of kinetosomes around mouth region. In ciliates, zigzag row of kinetosomes on right side of mouth which form the paroral membrane. See *Undulating membrane*. Illustration: Oral region.

Parthenogenesis
Development of an unfertilized egg into an organism. Illustration: Acetabularian life history.

Parthenosporangium (pl. parthenosporangia)
Receptacle bearing parthenospores.

Parthenospore
Thick-walled spore developing from an unfertilized gamete (e.g., conjugating green algae); undulipodiated reproductive cell produced without conjugation (apomictically) (e.g., in phaeophytes); haploid dinomastigote that is morphologically similar or identical to planozygotes but is formed by mitosis instead of syngamy.

Pathogen (adj. pathogenic)
Ecological term referring to organism that is an obligate or facultative symbiotroph that tends toward necrotrophy and causes symptoms in its host. Disease-causing organism. See *Parasite*.

Pectin
Complex polysaccharide extractable from cell walls of plants and some algae.

Pedicel
Attachment stalk, holdfast (e.g., in some chonotrich ciliates and choanomastigotes); elongated protrusion from the posterior end of a cell; basal portion of a charophyte antheridium. Illustration next page.

Pedogamy
Paedogamy; autogamy. Fusion of two uninucleate sporoplasms and their haploid nuclei in myxozoans.

Peduncle (adj. pedunculate)
Holdfast, stalk, base, or stemlike structure; projection from sulcal region used to suck up food during heterotrophic feeding in

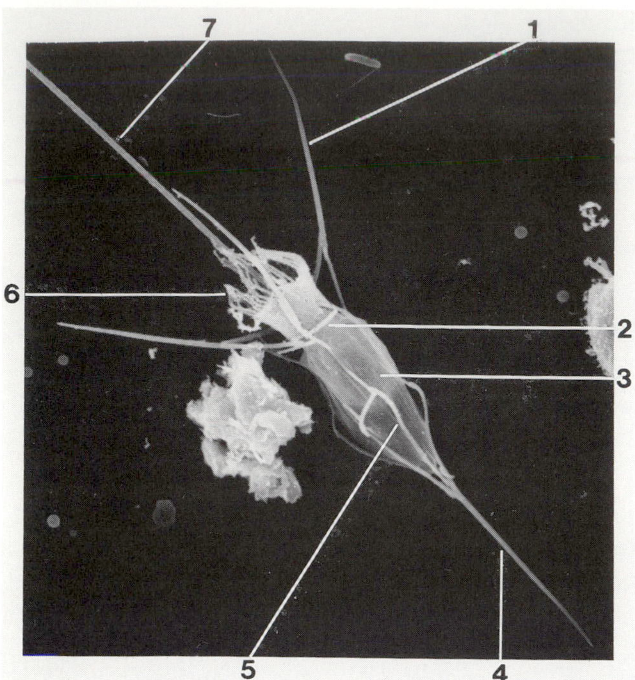

Pedicel
The choanomastigote zoomastiginid, *Calliacantha natans*. 1=anterior projection of lorica; 2=transverse costal strip; 3=choanomastigote cell; 4=pedicel; 5=longitudinal costal strip; 6=collar of tentacles; 7=undulipodium. SEM.

dinomastigotes (e.g., *Katodinium (Gymnodinium) fungiforme)*. Illustration: Heliozoa.

Pelagic (adj.)
Ecological term referring to organisms dwelling in open waters of the ocean (as opposed to benthic or neritic). Illustration: Habitat.

Pellicle (adj. pellicular, pelliculate)
Cortex. Outermost living layer of a protoctist, lying beneath any nonliving secreted material; pellicle contains the typical plasma membrane plus the pellicular alveoli or an underlying epiplasm or other membranes (in ciliates, dinomastigotes, and apicomplexans) and sometimes exhibits ridges, folds, or distinct crests; portion surrounding the cell after the theca is shed by ecdysis in armored dinomastigotes; proteinaceous ridged structure in euglenids. See also illustrations: Cortex, Lacunae, Oral region.

Pellicular folds
Wrinkles on surface; crenulations of pellicle. Illustration: Falx.

Pellicular lacunae system
System of flat membranous vesicles just beneath the pellicle of cells; micromorphological character typical of glaucocystophytes.

Pellicular microtubular armature
Microtubules located beneath the cell membrane that form a cytoskeleton involved in maintenance of the cell shape (e.g., chrysophytes); subpellicular microtubular cytoskeleton (e.g., euglenids). Illustrations: Pellicle, Retortamonadida, Zoomastigina.

Pellicular striae (sing. pellicular stria)
Striations, ridges, or striped markings in or on the pellicle. Illustrations: Euglenida, Reservoir.

Pelta (pl. peltae)
Crescent-shaped microtubular structure associated with the anterior portion of the axostyle (e.g., zoomastiginids such as pyrsonymphids or oxymonads). Illustration: Parabasalia.

Peneropliform (adj.)
Referring to a test that initially grows by adding chambers in a coiled single plane (planispirally) and then adds later chambers in a straight line (rectilinearly) (e.g., foraminifera such as *Peneroplis)*.

Pennate (adj.)
Pinnate. Morphological descriptive term for structure resembling a feather, esp. in having similar parts arranged on opposite sides of an axis like the barbs on the rachis of a feather; refers to shape of some diatoms. Illustration: Raphe.

Pentalith
Coccolith of five identical single calcite crystals; the cleavage plane of the crystals is in the plane of the pentalith (e.g., *Braarudosphaera)*. Illustration: Coccolith.

Peptide mapping
Technique used to compare a given protein from different organisms in which the protein is enzymatically cleaved, the resulting peptides are separated on a gel, and the peptides are identified by staining or reaction with a specific antibody. The similarity between the peptide patterns is related to amino acid sequence similarity.

Peptidoglycan
Glycan tetrapeptide. Rigid layer of bacterial cell walls con-

Pellicle
a. Euglenoid cell contracted in metaboly.
b. Cell surface of the euglenid *Distigma* showing interlocking pellicular strips and microtubules (mt).

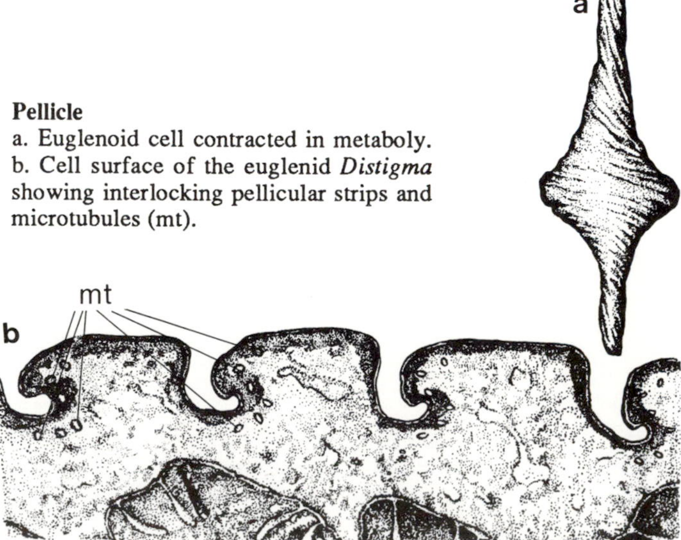

sisting of N-acetylglucosamine and N-acetylmurami acid attached to a few amino acid residues that form a repeating peptide.

PER
See *Plastid endoplasmic reticulum.*

Per os
Latin locution meaning orally.

Percurrent development
Proliferation through the sporangial septum of the sporangiophore, producing a new sporangium of similar dimensions; sporangia produced by limited internal renewal such that successive sporangial septa are formed at approximately the same point on the axis (e.g., oomycotes). See *Basipetal development, Internal renewal.*

Perennation
Overwintering in plants; in protoctists, survival of harsh conditions (e.g., seasonal desiccation).

Perennial (adj.)
Referring to an organism that lives for more than a year and produces a sexual phase annually or semiannually.

Perforatorium (pl. perforatoria)
Reinforced tip of mature microgamete that probably aids in penetration of macrogamete in coccidian apicomplexans. Illustration: Microgamete.

Periaxostylar (adj.)
Referring to material parallel to and surrounding the axostyle. Illustration: Parabasalia.

Pericarp
Sterile layer of cells that surrounds the carposporophyte in some rhodophytes.

Perichloroplastic compartment
See *Periplastidial compartment.*

Pericyte
Outer of two generative cells in myxosporean plasmodium that unite in pairs to produce a pansporoblast; stage in actinosporean myxozoans containing two nuclei that arises from the sporoplasm and that envelops the sporogonic cell.

Peridium (pl. peridia)
Structure of myxomycote sporophores consisting of a membranous surface layer.

Perioral kinety
Rows of cilia around the mouth derived from modified somatic kineties. Illustration: Oral region.

Periphyton
Aufwuchs community. Ecological term referring to microbial communities inhabiting interfaces such as the surfaces of other organisms attached to bottom substrates in lakes and other aquatic habitats. See *Microbenthos, Plankton, Seston.* Illustration: Habitat.

Periplasm
Peripheral cytoplasm. In prokaryotes, the space between the inner plasma membrane and the peptidoglycan layer of the cell wall.

Periplasmic cortex
Surface layer of the protoplasm in sexual organs remaining after differentiation of the sexual cells (e.g., peronosporacean oomycotes); outer pellicle in acantharian actinopods. Illustrations: Acantharia, Elastic junctions, Perispicular cone.

Periplast
Part of eukaryotic cell that lies external to the cell membrane; sometimes composed of elements such as scales, coccoliths, plates, etc., and including specialized structures such as cell walls, pellicles, thecae, etc.; a complex, ornamented plasma membrane. Illustration: Cryptophyta.

Periplastidial compartment
Perichloroplastic compartment. Space between the plastid membrane and the plastid endoplasmic reticulum. See also illustration: Chlorarachnida.

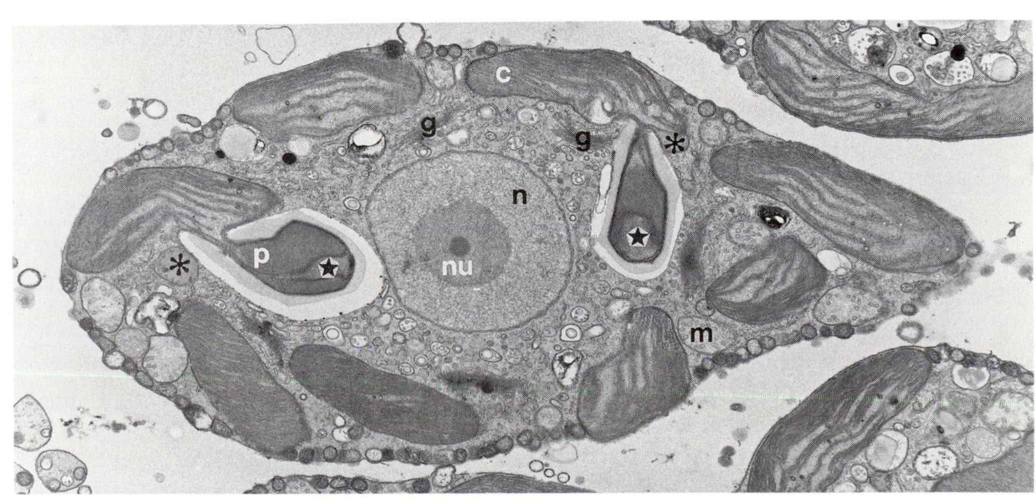

Periplastidial compartment
A cell of the chlorarachnid *Chlorarachnion reptans.* c=chloroplast; g=golgi body; m=mitochondrion; n=nucleus; nu=nucleolus; p=pyrenoid; asterisk=periplastidial compartment; star=invagination of periplastidial compartment into pyrenoid. TEM, x 12,500.

Periplastidial reticulum (pl. periplastidial reticula)

System of vesicles and tubules located in the periplastidial compartment; membranous reticulum in continuity with the inner membrane of the plastid endoplasmic reticulum, lying within the periplastidial compartment. Illustration: Phaeophyta.

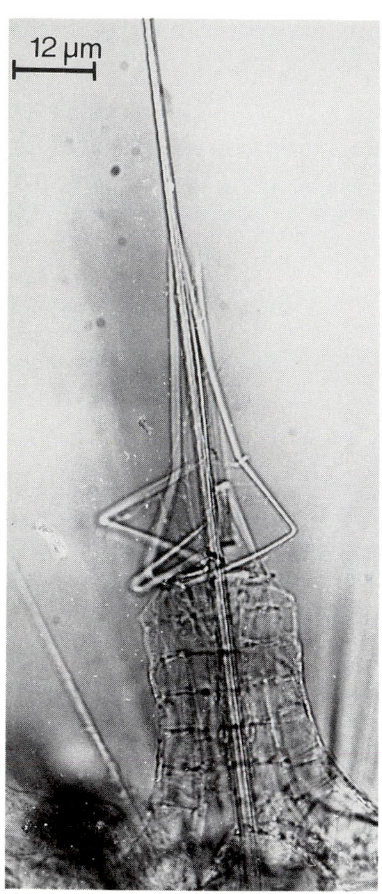

Perispicular cone
Spicule of the acantharian actinopod *Conacon foliaceus*. At the base, six myonemes are anchored to the periplasmic cortex. PC.

Perispicular cone

Region in acantharian actinopods in which the capsular wall is connected to the periplasmic cortex making a sleeve around each spicule. The sleeve defines a conical space containing the axoneme, dense granules, vesicles, and myonemes.

Perispicular vacuole

Spicular vacuole. Large vacuole in acantharian actinopods; structure in which the spicules are enclosed.

Peristome

See *Buccal cavity*.

Perithallium (pl. perithallia; adj. perithallial)

Portion of the growing thallus in which the cells or filaments are developed inwardly from the intercalary meristem (e.g., coralline rhodophytes).

Peritrichs (adj. peritrichous)

Ciliates of the subclass Peritrichia that bear an oral ring of ciliature; bacteria flagellated around their periphery.

Perizonium (pl. perizonia)

Outer membrane derived from the fertilization membrane after zygote (auxospore) formation in diatoms.

Perkinetal fission

Fission across or through the kineties or rows of cilia; most common type of homothetogenic fission; transverse fission of ciliates, as opposed to the longitudinal fission of mastigotes. Typical of ciliates. See *Interkinetal fission*.

Peroxisomes

Organelles containing enzymes, including catalase and peroxidase; site of the oxidation of a variety of substrates to form hydrogen peroxide (H_2O_2) using molecular oxygen as the oxidizing agent. Illustration: Organelle.

Petrographic thin section

Slice of rock polished and thin enough to allow light to pass through it; used to detect microfossils in a cryptocrystalline matrix (chert).

PFB

See *Paraflagellar body*.

PFR

See *Paraflagellar rod, Paraxial rod*.

pH

Scale for measuring the acidity of aqueous solutions; $pH=-\log[H^+]$; pure water has a pH of 7 (neutral); solutions having a pH greater than 7 are alkaline; less than 7 are acidic.

Phaeodium (pl. phaeodia)

Pigmented mass consisting primarily of waste products around the astropyle of the central capsule of phaeodarian actinopods.

Phaeophyte

Informal name of members of the phylum Phaeophyta. Illustrations: Mastigoneme, Phaeophyta.

Phaeoplast

Brown chlorophyll *c*-containing plastid; photosynthetic organelle of phaeophytes.

Phaeosome

Brown body, may be excretory products (e.g., those produced by many actinopods); surface-associated, ectosymbiotic, coccoid cyanobacteria, mostly *Synechococcus* spp. occurring in association with dinophysoid dinomastigotes.

Phaeosome chamber

Chamber of actinopods that harbor phaeosomes; chamberlike modification of the girdle of some complex dinophysoid genera (e.g., *Histioneis, Citharistes),* in which symbiotic cyanobacteria usu. occur.

Phage

Virus of bacteria. Illustration: Genophore.

Phagocytosis (adj. phagocytotic, phagocytic)

Mode of heterotrophic nutrition and immunological defense

involving ingestion, by a cell, of solid particles in which pseudopods flow over and engulf particulates.

Phagotrophy (n. phagotroph; adj. phagotrophic)
Mode of nutrition in which heterotrophic protoctists or tissue cells ingest solid food particles by phagocytosis (see Table 2). See *Carnivory, Histophagy, Holozoic, Osmotrophy.* Illustration: Euglenida.

Phaneroplasmodium (pl. phaneroplasmodia)
Largest and most conspicuous of the three types of plasmodia formed by myxomycotes (primarily of the order Physarales); plasmodium consisting of thin, fanlike advancing regions and a branching network of veins; veins consist of an outer gel zone of protoplasm and an inner fluid zone, in which protoplasmic streaming occurs. See *Aphanoplasmodium, Protoplasmodium.*

Phenetic taxonomy
Classification of organisms based on their visible, measurable (phenotypic) characteristics without regard to evolutionary (phylogenetic) relationships.

Phenological (adj.)
Ecological term referring to seasonal variation.

Pheromone
Ecological term referring to chemical substance that when released into the surroundings of organisms influences the behavior or development of other individuals of the same species. If produced by one sex and responded to by the other sex, the substance is called sex pheromone. See *Allelochemic.*

Phialine lip
Flask or cup-shaped outgrowth; in foraminifera, everted rim of aperture, common on neck.

Phialopore
Intercellular space in certain volvocalean chlorophytes through which the colony everts.

Phlebotominae
Dipteran insect family; sand flies.

Phoront
Stage in a polymorphic life cycle during which the protoctist is carried about (generally on or in the integument of) another (generally metazoan) organism. Stage typically preceded by a tomite and followed by a trophont (e.g., polymorphic apostome ciliates such as *Hyalophysa).* Illustration: Tomont.

Photic zone
See *Euphotic zone.*

Photoautotrophy
See *Phototrophy.*

Photoauxotrophy (n. photoauxotroph; adj. photoauxotrophic)
Mode of nutrition, usu. of algal mutants that grow phototrophically except for the requirement of a vitamin, amino acid, or other identifiable growth factor.

Photoheterotrophy (n. photoheterotroph; adj. photoheterotrophic)
Mode of nutrition, limited to bacteria, in which light is used as a source of energy (to generate ATP and osmotic gradients) but organic compounds are used as carbon sources (see Table 1).

Photoinhibition
Physiological response of algae or plants referring to the inhibition of photosynthesis at high light intensities.

Photokinesis
An effect of light intensity on speed of movement.

Photoperiodic response
Behavioral or growth response of an organism to changes in day length; a mechanism for measuring seasonal time.

Photoreceptor
Cell structure in which a specialized aggregate of pigments mediates a behavioral reaction to light stimuli (e.g., eustigmatophytes, euglenids, *Paramecium bursaria,* and many other protoctists). Illustration: Chrysophyta.

Photoresponse
Cell or organismal growth or behavioral response to light stimuli (e.g., positive and negative phototaxis; phototropism).

Photosensory transduction
Reaction chain of light-induced motor responses, i.e., the connecting link between photoreceptor and cell motility, consisting of stimulus transformation (conversion of one form of energy to another) and signal transmission (i.e., all steps in the reaction chain that cause signal transport).

Photosynthate
Any metabolic product of photosynthesis; total photosynthate contains sugars, amino acids, and organic acids and differs in exact composition in different phototrophs.

Photosynthesis
See *Phototrophy.*

Photosynthetic lamella (pl. photosynthetic lamellae)
See *Thylakoid.*

Photosystem
Functional light-trapping unit; an organized collection of chlorophyll and other pigments embedded in the thylakoids of plastids which trap photon energy and channel it in the form of energetic electrons to the thylakoid membrane.

Phototaxis (adj. phototactic)
Movement toward (positive) or away (negative) from a light source.

Phototrophy (n. phototroph; adj. phototrophic)
Photoautotrophy; photosynthesis. Mode of nutrition in which light provides the source of energy. An obligately photoautotrophic organism uses light energy to synthesize cell material from inorganic compounds (carbon dioxide, nitrogen salts) (see Table 1).

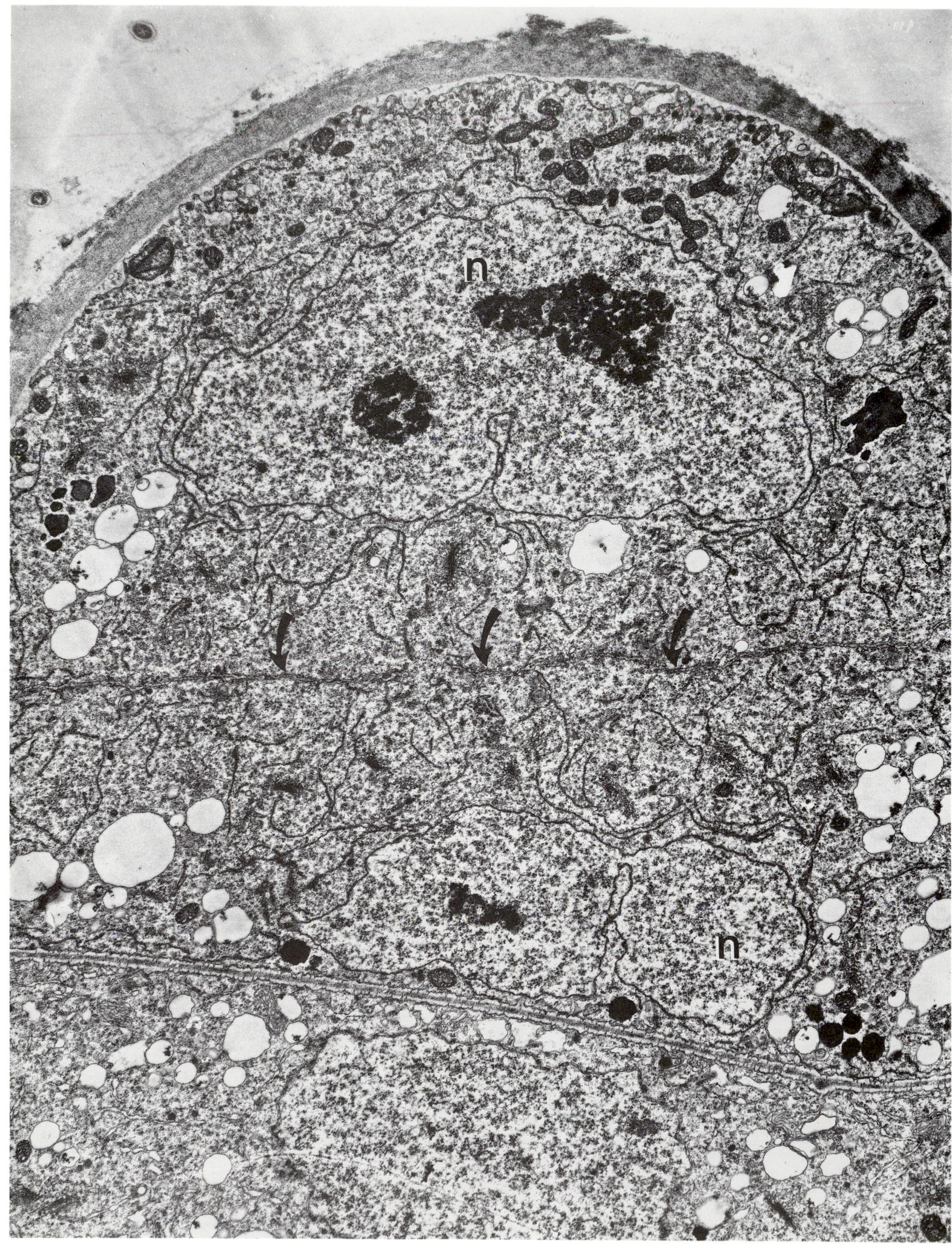
n
n

Phragmoplast
Cell plate. System of fusing vesicles guided by microtubules that form perpendicular to the spindle axis at telophase in the plane of division during cytokinesis (e.g., in plants and some chlorophytes). See *Phycoplast*.

Phycobilins
Class of protein-linked open tetrapyrrole pigments, water soluble, and generally bluish or red in color (e.g., in cyanelles of glaucocystophytes, plastids of rhodophytes, some cryptomonads, and thylakoids of cyanobacteria).

Phycobiliproteins
Biliproteins. Complex of phycobilins with protein found in cyanobacteria, rhodophytes, glaucocystophytes, and some cryptomonads. See *Phycobilins, Phycocyanin*.

Phycobilisome
Intracellular structure containing phycobilin pigments and arranged as protrusions on the surface of the thykaloids of cyanobacteria, rhodophytes, and glaucocystophytes but within the thykaloids (between membranous stacks) in the plastids of cryptomonads. Illustration: Plastid.

Phycobiont
Algal symbiotic partner of a lichen.

Phycocolloids
Complex polysaccharides produced by algae, the detailed structures of which are largely unknown (e.g., agarose, carrageenan).

Phycocyanin
Type of phycobiliprotein; water-soluble extract is blue; found in cyanobacteria, rhodophytes, and cryptomonads.

Phycoerythrin
Type of phycobiliprotein; water soluble extract is red; found in cyanobacteria, rhodophytes, and cryptomonads.

Phycology
Algology. Science of the study of algae.

Phycoma
Whole algal body; nonmotile, unicellular, spherical stage in the life history of some prasinophytes (*Pterosperma, Halosphaera*, and *Pachysphaera*) characterized by a thick, ornamented wall which may contain sporopollenin.

Phycomycetes
Lower fungi. Term for a class of fungi which is obsolete because it grouped zygomycotes with unrelated taxa (i.e., chytridiomycotes, oomycotes, and other "algalike" fungi).

Phragmoplast
Telophase in mitotically dividing cells in the chlorophyte *Chara preissii* (Charales). In the phragmoplast microtubules are oriented perpendicularly to the forming cell plate (arrows). n=nuclei. TEM.

Phycophage
Algal virus.

Phycoplast
System of fusing vesicles guided by microtubules that form parallel to the spindle axis at mitosis and in the plane of division in some algae. See *Phragmoplast*.

Phyllae (sing. phylla)
Flat ribbons of microtubules found in the oral region of some ciliates.

Phylogenetic tree
Graphic or diagrammatic representation of a partial phylogeny (e.g., ribosomal RNA or protein sequences) or complete phylogeny (e.g., family tree).

Phylogeny
Hypothesized sequence of ancestor/descendant relationships of groups of organisms as reflected by their evolutionary history. Illustration next page.

Physode
Fucosan vesicle. Small colorless vesicle occurring in cells of phaeophytes containing fucosan and certain tannins and terpenes. Illustrations: Phaeophyta, Sporelings.

Phytoalexins
Compounds of various kinds (some anti-microbial) induced by stress (infection, wound, etc.) in plants in direct response to injury; often are secondary metabolites.

Phytochrome
Pigment associated with the absorption of light found in plants and some algae (e.g., conjugating green algae); photoreceptor for red to far-red light; involved in the control of certain developmental processes.

Phytoflagellate
Mastigote alga; any swimming protist with at least one undulipodium and one plastid.

Phytophagy
Nutritional mode in which organisms feed on plants (and algae).

Phytopicoplankton
Ecological term referring to small photosynthetic microbes suspended in the water column, primarily in the ocean; cyanobacteria, chloroxybacteria, and the smallest plastidic protists (e.g., *Micromonas*).

Phytoplankton
Ecological term referring to aquatic free-floating algae and cyanobacteria (if motile they are unable to swim against the current). See *Picoplankton, Plankton*.

Phytotoxic (adj.)
Referring to chemical substances poisonous to plants.

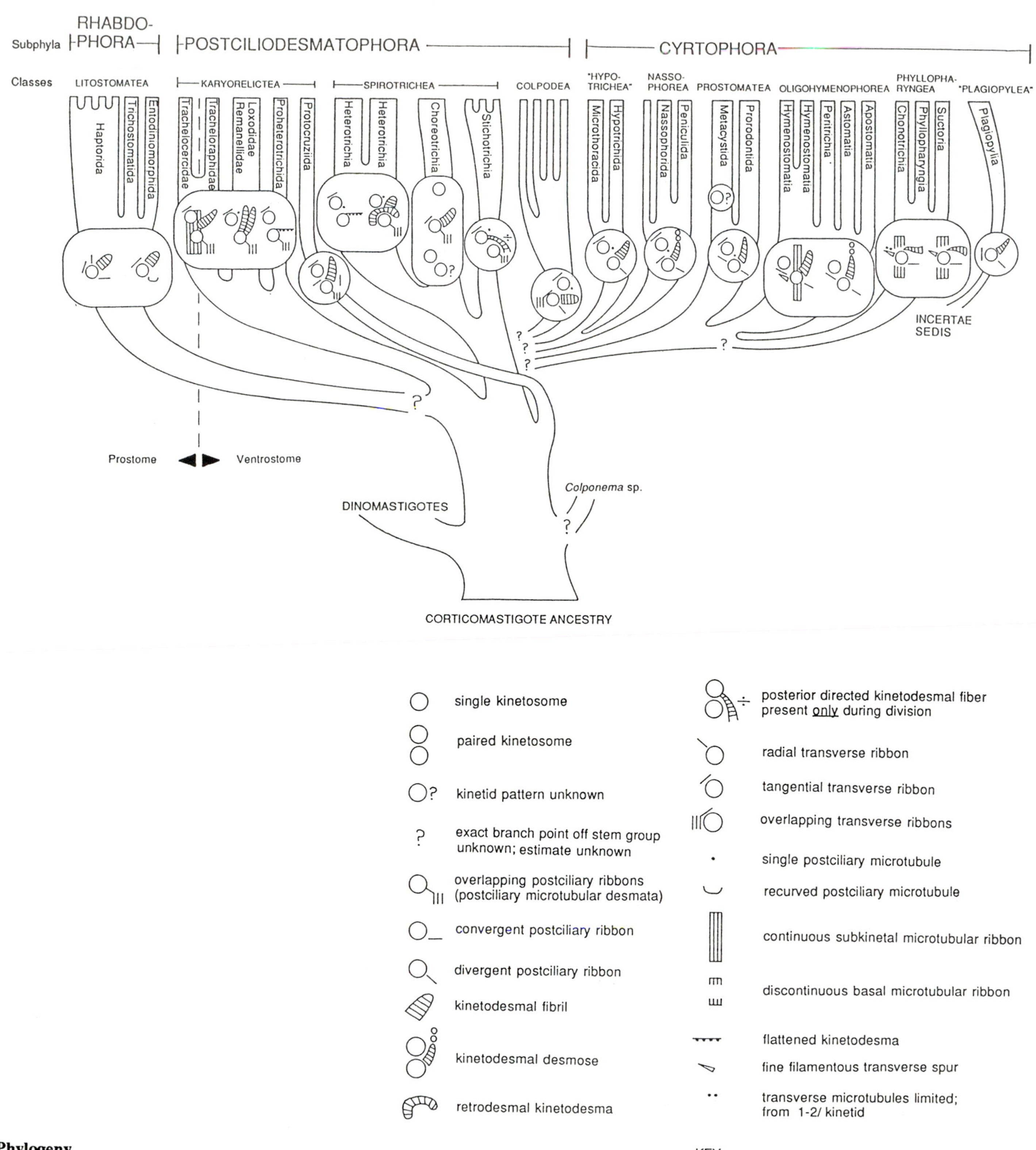

Phylogeny

Proposed phylogeny of the phylum Ciliophora. The phylum is divided into three subphyla: Rhabdophora, Postciliodesmatophora, and Cyrtophora. The classes and subclasses within each subphylum are distinguished by the structure of the somatic kinetids (diagrammed in cross section).

PI electrophoresis

Biochemical technique for separating proteins according to their electrically charged residues, i.e., their isoelectric point. A potential difference is applied across a system in which pH increases from anode to cathode. Proteins or peptides present in the system accumulate on a band in the region of the gradient corresponding to their isoelectric point, the point at which their total charge is neutralized.

Picoplankton

Ecological term referring to microorganisms found suspended in aquatic media, esp. the ocean. The planktonic cells in the 0.2-2.0 μm size range are dominated by prokaryotes but include small eukaryotes, both with and without plastids. Term refers to size, not to nutritional mode or cell structure. See *Phytopicoplankton, Phytoplankton, Plankton.*

Pinnate (adj.)

See *Pennate.*

Pinocytosis (adj. pinocytotic, pinocytic)

Type of eukaryotic intracellular motility process which uses microfibrils for cell "drinking"; endocytosis of liquid, dissolved solutes, and protein-sized particles through formation of membrane tunnels called pinocytotic vesicles. See *Phagocytosis.*

Pit areas

Areas of pit connections.

Pit connections

Pit plugs. Type of cell junction: protoplasmic connections joining cells by perforations in the cell wall which may or may not be plugged; typical of rhodophytes and fungi. For rhodophytes, pit connection is a misnomer because they are not connections between cells but rather plugs of proteinaceous material deposited in the pores that result from incomplete wall formation.

Pit field

Collection of plasmodesmata at the center of the cross wall between cells in certain chlorophytes (e.g., Trentepohliales).

Pit plugs

See *Pit connections.* Illustration: Florideophycidae.

Placoderm desmid

Kind of conjugating green alga; desmid composed of two semicells which are usu. joined by an isthmus and with pores usu. present in cell walls. Walls of the two semicells are of different ages. See *Saccoderm desmid.* Illustration: Conjugaphyta.

Placolith

Coccolith subtype with upper and lower "shields" composed of radial segments; heterococcolith composed of two plates or shields interconnected by a tube (e.g, *Coccolithus*). Illustration: Coccolith.

Plakea

Developmental stage in colonial volvocalean chlorophytes; curved plate of cells.

Planispiral (adj.)

Referring to test of foraminifera coiled in a single plane.

Plankton (adj. planktonic)

Ecological term referring to suspended, free-floating microscopic or small aquatic organisms in either marine or freshwater environments whose transport is subject to wave movements. Refers to size and passive motility, not to taxonomic affiliation. Illustration: Habitat.

Planont

Sporoplasm from a freshly germinated spore (e.g., myxosporean myxozoans).

Planozygote

Motile zygote of dinomastigotes; enlarged, undulipodiated, and sometimes thick-walled mastigote formed just after fusion. Illustration: Dinomastigote life history.

Plant

Multicellular, diploid organism that develops from an embryo supported by maternal tissue, generally photoautotrophic.

Plasma membrane

Cell envelope; cell membrane; cytoplasmic membrane; plasmalemma. Outer membrane, composed of lipids and proteins,

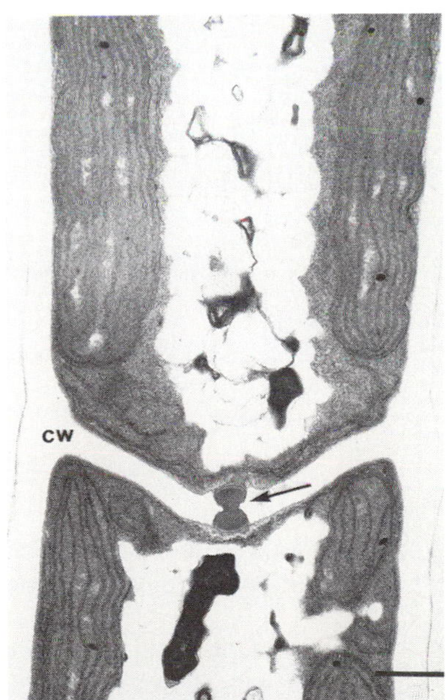

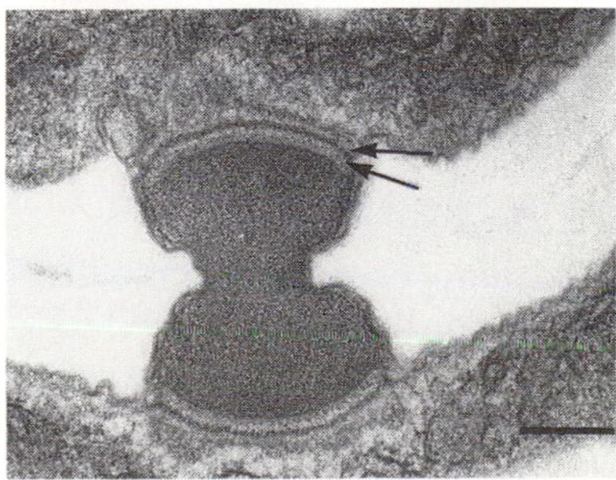

Pit connections
The rhodophyte *Audouinella hermannii.* a. Longitudinal section through adjacent vegetative cells. Arrow=pit plug; cw=cell walls. TEM. Bar=2.0 μm. b. Enlargement of pit plug. Two cap layers (arrows) overlie the more densely staining plug core. TEM. Bar=0.4 μm.

that surrounds a cell; regulates exchange of material between cell and environment; universal structure of cells. Illustration: Organelle.

Plasmalemma

See *Plasma membrane*. Illustrations: Conoid, Cortex, Raphidophyta.

Plasmid

Small piece of naked DNA; small replicon. Illustration: Genophore.

Plasmodesma (pl. plasmodesmata)

Cell junctions, i.e., the tiny cytoplasmic threads that extend through openings in cell walls and connect the protoplasts of adjacent living cells, esp. in algae (e.g., trentepohlialean chlorophytes) and plants.

Plasmodiocarp
The myxomycote plasmodial slime mold *Physarum serpula*.

Plasmodiocarp

Sporophore resembling thickened plasmodial veins or modifications of portions of veins in myxomycotes.

Plasmodium (pl. plasmodia; adj. plasmodial)

Coenocyte; syncytium. Multinucleate mass of cytoplasm lacking internal cell membranes or walls. Multinucleate cell generally has from two to over a dozen nuclei, while plasmodia have over a dozen and up to millions of nuclei per cell. Illustrations: Life cycle, Plasmodial Slime Molds, Plasmodiophoromycota.

Plasmogamy

Fusion of two cells or plasmodial cytoplasms without karyogamy (fusion of nuclei); cytoplasmic fusion that may be the first step in the fertilization process. Syngamy without karyogamy that may produce dikarya or heterokarya.

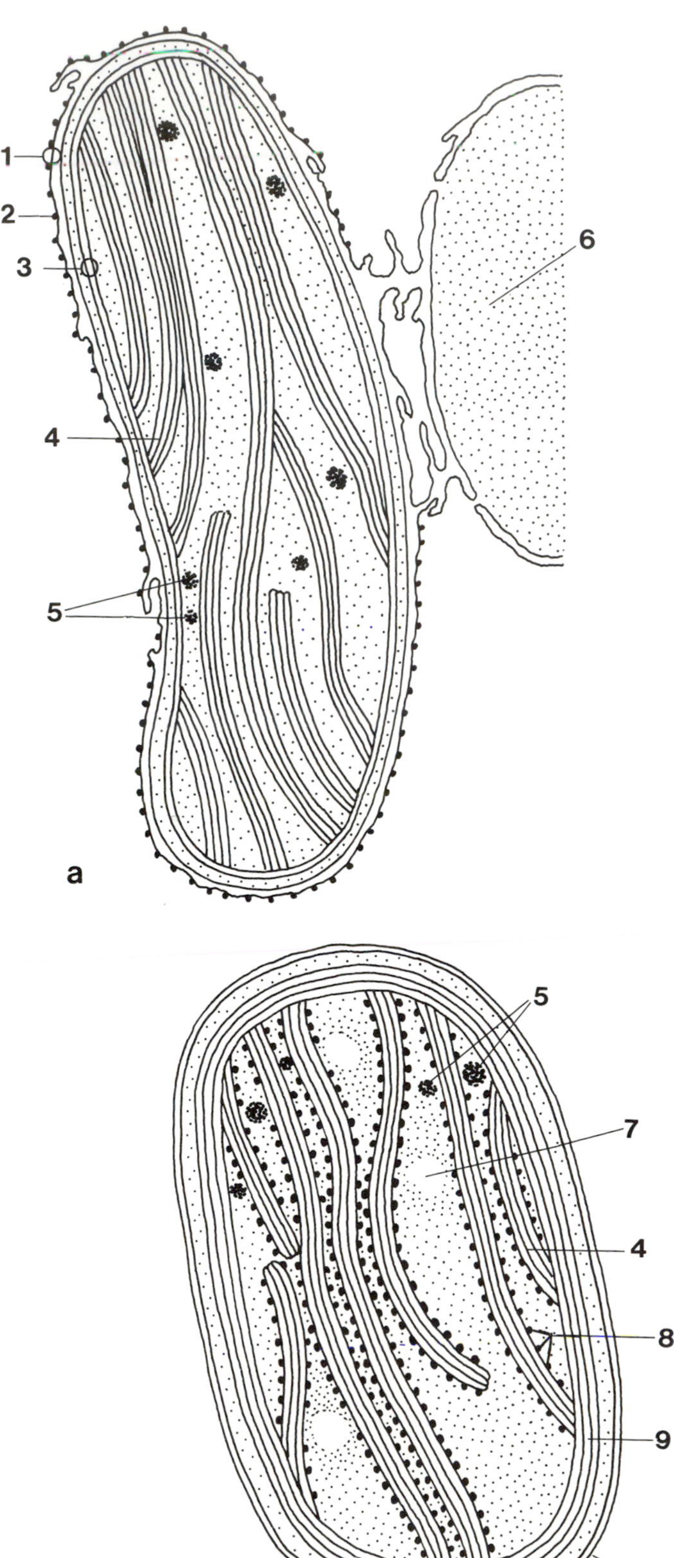

Plastid
Generalized plastids. a. Chloro-, Chryso-, Crypto-, and Xanthophyta types. b. Rhodophyta type. 1=plastid endoplasmic reticulum; 2=ribosome; 3=plastid envelope (2 membranes); 4=stacked thylakoids (grana); 5=plastoglobuli; 6=nucleus; 7=plastid DNA; 8=phycobilisomes; 9=peripheral encircling thylakoids; 10=outer membrane.

Plasmotomy
Form of binary or occasionally multiple fission of a plasmodium or multinucleate protoctistan cell; division of a plasmodium; characteristic of large, multinucleate amebas, opalinids, myxosporeans, and others, in which nuclei exhibit mitosis following (rather than during or immediately preceding) the process of somatic fission or in which nuclei may undergo divisions asynchronously. Some mitoses may be found at any time in the two (or more) separable multinucleate masses. Illustration: Life cycle.

Plastid
Generic term for photosynthetic organelle in plants and protoctists (all algae). Bounded by double membranes, plastids contain the enzymes and pigments for photosynthesis, ribosomes, nucleoids, and other structures. See *Chloroplast, Chrysoplast, Phaeoplast, Rhodoplast.* See also illustration: Organelle.

Plastid endoplasmic reticulum (pl. plastid endoplasmic reticula)
PER. Specialized layer of endoplasmic reticulum that closely surrounds the plastid and is usu. continuous with the nuclear membrane; ribosomes are present on the membrane facing the cytoplasm, but not the membrane facing the periplastidial compartment. See *Chloroplast endoplasmic reticulum.* Illustrations: Eyespot, Plastid.

Plastid matrix
Fluid contents of plastid.

Plastidic nanoplankton
Pnano. Nanoplanktonic algae; phototrophic nanoplankton; tiny phytoplankton; ecological term specifying certain small plankton; planktonic protists in the 2-20 µm size range that possess plastids. See *Phytoplankton, Picoplankton, Plankton.*

Plastidic protists
Unicellular algae.

Plastoglobuli (sing. plastoglobulus)
Lipid droplets usu. randomly distributed through the plastid matrix; sometimes seen concentrated at the periphery of the pyrenoid in xanthophytes. Illustration: Plastid.

Plate formula
System of labeling dinomastigote thecal plates. See also illustration: Thecal plates.

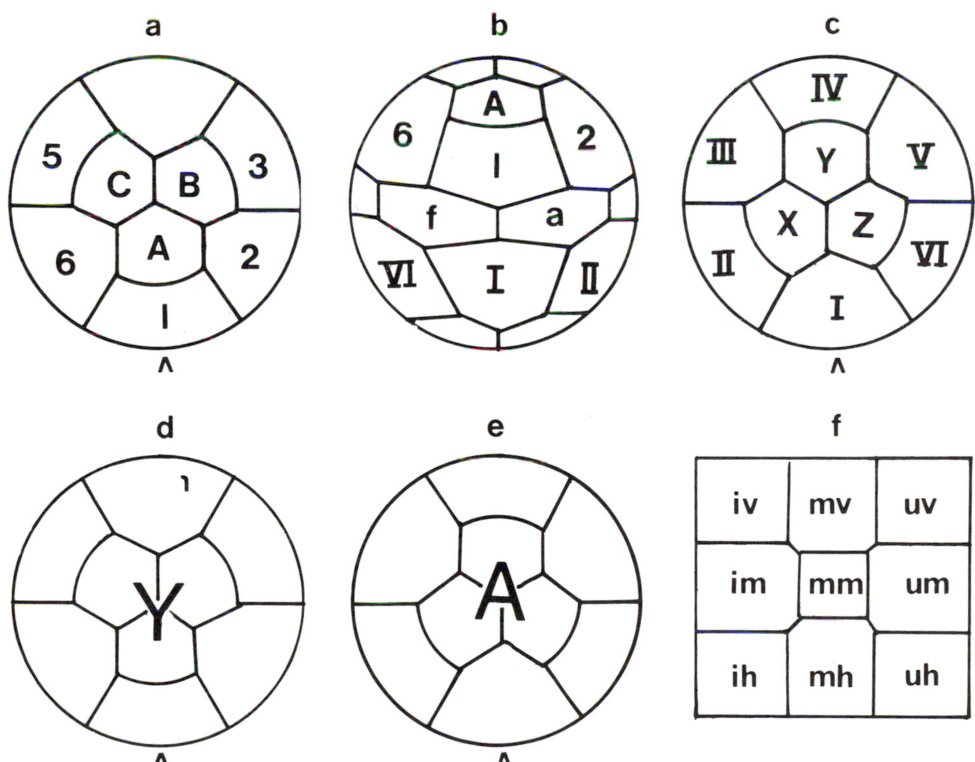

Plate formula
Dinomastigote thecal plate designation used in the Taylor homology system in which cell shape has been normalized to a sphere. a. Apical polar view. b. Ventral view. c. Antapical polar view. d. Polar view of the "Y" arrangement of polar plates relative to undulipodial insertion (on ventral side). e. The "A" arrangement. f. Designations for subdivision of a primary plate area (maximum subdivision), using Evitt's modification. 1 to 6 represent pre-equatorial sectors or plates; I to VI represent post-equatorial sectors or plates; X to Z represent hypothecal sectors or plates. h=hinter; i=initialis; m=medialis; u=ulter; v=vorner; ʌ=ventral side.

Plate scales
Oval or circular flat scales lacking superstructure (e.g., chrysophytes) as opposed to spiny scales (e.g., prymnesiophytes or prasinophytes). Illustration: Cryptophyta.

Playa (Spanish, meaning shore, beach)
Dry, barren area in the lowest part of an undrained basin (e.g., southwestern United States); also, small, sandy land area at the mouth of a stream or along a bay shore; beach. Illustration: Habitat.

Plectenchyma (adj. plectenchymatous, plectenchymous)
Interwoven tissues comprised of mycelial mass. Structural term designating mycelial tissues found in some heterotrophic protoctists and fungi.

Pleiomorphic (adj.)
See *Pleomorphic.*

Pleomorphic (adj.)
Pleiomorphic. Exhibiting several forms or shapes; many and variable expressions of shape in a genetically uniform population (e.g., organisms such as amebas or other protoctists that display changing form).

Plerotic (adj.)
Referring to oogenesis in oospores (e.g., oomycotes); clearly filling the oogonial cavity. See *Aplerotic.* Illustration: Oospore.

Plesiomorphy (adj. plesiomorphic)
Ancestral, generalized or primitive taxonomic character (seme) present in ancestor at the bifurcation of the lineage. See *Apomorphy, Symplesiomorphy, Synapomorphy.*

Plethysmothallus (pl. plethysmothalli)
Diploid microscopic life cycle phase of some phaeophytes in which reproduction is by zoospores that transform into diploid thalli (resembling *Ectocarpus* or *Streblonema*) capable of producing more zoospores (in the absence of sexual processes).

Pleura (pl. pleurae)
See *Girdle.* Illustration: Hypovalve.

Pleuromitosis
Cryptopleuromitosis. Closed mitosis (nuclear membrane remains intact) with an extranuclear spindle lateral to the nucleus, in which no equatorial plate forms; mitosis with a sharply asymmetrical intranuclear spindle. Illustration: Mitosis.

Pleuronematic (adj.)
Referring to mastigotes having an undulipodium with one or more rows of mastigonemes; may be pantacronematic, pantonematic, or stichonematic. Illustration: Mastigoneme.

Ploidy
The number of sets of chromosomes. See *Aneuploid, Diploid, Euploid, Haploid, Polyploid.*

Plurilocular sporangium (pl. plurilocular sporangia)
Sporangium composed of a multicellular structure in which each cell produces a single reproductive cell and spores are produced in several cavities. See *Unilocular sporangium.*

Pluriseriate (adj.)
See *Multiseriate.*

Pnano
See *Plastidic nanoplankton.*

Poikilothermic (adj.)
Referring to organisms whose body temperatures are very similar to those of their external environment, i.e., organisms unable to regulate their body temperature. Characteristic of all protoctists.

Polar body
One of two cells divided from ovum during maturation, before gametic nuclei fuse.

Polar cap
Chromophilic body beneath the anterior spore wall contained in the polar sac in myxosporean spores.

Polar capsule
Capsule. Apical, thick-walled vesicle of a myxozoan spore (one to seven per spore) containing spirally coiled, extrusible polar filament; in heliozoan actinopods, regions of dense cytoplasm at opposite sides of the nucleus during mitosis. Illustrations: Myxozoa, Sporoblasts.

Polar fenestrae (sing. polar fenestra)
Polar gaps. Gaps in the nuclear membrane associated with semiopen mitosis (mitosis in which the nuclear membrane dissolves only at the poles of the spindle).

Polar filament
Distally closed, tubelike structure coiled within the polar capsule of myxozoans. When everted it has a sticky surface and possibly serves to anchor the hatching spore to the surface of the intestine of its host; "hairpoint" on the terminus of protistan undulipodia. See *Anisofilar, Isofilar.* Illustrations: Myxozoa, Sporoblasts.

Polar gaps
See *Polar fenestrae.*

Polar ring
Part of apical complex of apicomplexans; typical of sporozoites and merozoites; probably a microtubule-organizing center. Illustrations: Conoid, Kinete.

Polar sac
Anchoring disc. Structure of microsporan spores that develops from a vesicle and into which the base of the polar tube is inserted (e.g., *Nosemoides vivieri*). Illustration: Microspora.

Polar tube
Tubular extrusome of microsporan spores serving for injection of the sporoplasm into the host cell. See *Anisofilar, Isofilar.* Illustration: Microspora.

Polarizing microscopy

Microscopy in which specimen is between a polarizer and an analyzer such that if regular features of that specimen lead to alterations in the path of polarized light they are detectable. Useful for analysis of petrographic thin sections and longitudinally aligned microtubules (e.g., of axopods) or microfibrils (e.g., cellulose walls of charophytes).

Polaroplast

Structure consisting of a series of flattened sacs and vesicles, thought to be involved in polar tube extrusion in microsporan spores.

Polycentric (adj.)

Referring to algal thallus radiating from many centers at which reproductive organs (sporangia or resting spores) are formed; descriptive of cells or organisms demonstrating a number of centers of growth and development and more than one reproductive structure (e.g., oomycotes, chytridiomycotes, hyphochytrids); descriptive of chromosomes or chromatids with more than one kinetochore, leading to parallel (rather than V-shaped) segregation of chromatids during anaphase. See *Monocentric.*

Polycomplexes

Structures formed by the fusion of components from synaptonemal complexes which have detached from diplotene chromosomes (e.g., insects, the haplosporidian *Minchinia louisiana*).

Polyeder

Polyhedral cell; angular cell formed by zoospores in some chlorophytes (e.g., *Pediastrum, Hydrodictyon).*

Polyenergid

Cell containing multiple genomes either within one nucleus or within several nuclei; state of having either multiple nuclei and/or multiple ploidy in a nucleus within a single cell (e.g., some radiolarian actinopods and ciliate macronuclei).

Polygenomic (adj.)

Having multiple genomes (e.g., as in an endosymbiotic association); may also refer to polyploidy.

Polyglucan granules

Storage bodies in the cytoplasm of some algal cells; dark-staining polymers of glucose resembling animal glycogen.

Polykinetid

See *Cirrus.* Illustration: Oral region.

Polykinetoplastic (adj.)

Referring to stage in trypanosome development in which the kDNA is present as several distinct kinetoplasts in the mitochondrion.

Polykinety (pl. polykineties)

Row of polykinetids (e.g., cirrus). Infraciliary bases, with or without their cilia, of the buccal membranelles *sensu lato* of certain groups of ciliates having more than two kinetosomes per unit kinetid (e.g., scuticociliates); oral membranelles of the peritrichous ciliates.

Polymorphism (adj. polymorphic)

Morphological or genetic differences seen in normal wild-type individuals that are members of the same species and same population. See *Dimorphism.*

Polyphyletic (adj.)

Referring to a trait or group of organisms derived by parallel (convergent) evolution from different ancestors. See *Monophyletic.*

Polyploid (adj.)

Referring to cells in which number of sets of chromosomes exceeds two; i.e., a multiple of the haploid number of chromosomes greater than diploid (e.g., triploid (3N) or hexaploid (6N)). See *Polygenomic.*

Polypodial ameba

Ameba that moves by means of several pseudopods that are extended simultaneously. See *Monopodial.* Illustration: Rhizopoda.

Polysaprobic (adj.)

Ecological term referring to an aquatic environment rich in dissolved organic material and low in dissolved oxygen. See *Oligotrophy.*

Polyseme

Evolutionary change in trait (seme) by repetition of that seme (e.g., segmentation in worms, increase in kinetids in ciliates). See *Seme.*

Polysiphonous (adj.)

Referring to an algal thallus composed of vertically aligned tubes composed of parallel cells (e.g., *Polysiphonia).*

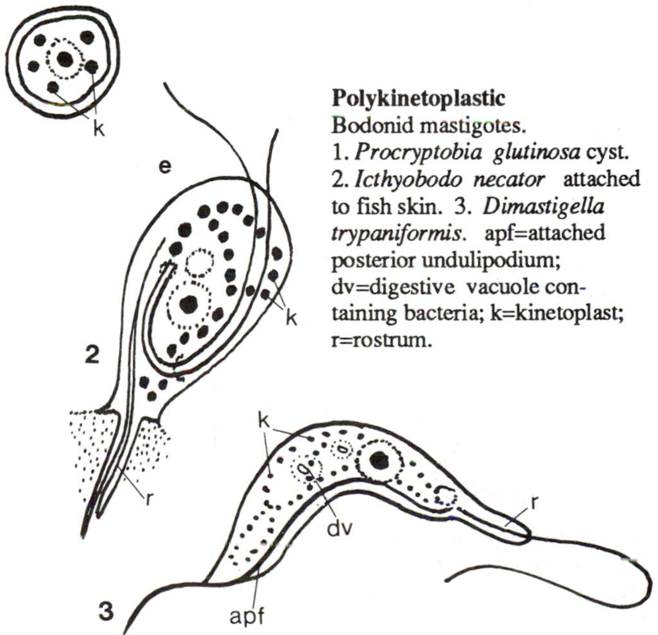

Polykinetoplastic
Bodonid mastigotes.
1. *Procryptobia glutinosa* cyst. 2. *Icthyobodo necator* attached to fish skin. 3. *Dimastigella trypaniformis.* apf=attached posterior undulipodium; dv=digestive vacuole containing bacteria; k=kinetoplast; r=rostrum.

Polyspore (adj. polysporous)
Gonospore. Germ cell; sex or reproductive cell. See *Somatic cell.*

Polystichous (adj.)
Referring to an algal thallus that has parenchymatous organization and hence is multicellular in cross section (e.g., phaeophytes).

Polystromatic (adj.)
Referring to an algal thallus composed of many cell layers.

Polyteny (adj. polytenic)
Condition in cells in which chromosomes have many times the normal (1X) quantity per length of DNA as a result of repeated replication without division so that the many (poly), threadlike (tenon) chromatids lie side-by-side. Whereas in polyploidy the number of chromosome sets augments, in polyteny the number of chromosomes stays constant but the quantity of DNA per set increases (from 2X to 10^3X). Polyteny is characteristic of certain stages of macronuclear maturation in hypotrichous ciliates such as *Stylonychia.*

Polythalamic (adj.)
Polythalamous. Referring to test of foraminifera having several chambers or cells; many-chambered, multilocular.

Polythalamous (adj.)
See *Polythalamic.*

Polytomic fission
Multiple fission; mode of reproduction involving division of a single individual into numerous offspring products. See *Merogony, Progressive cleavage, Schizogony.*

Polyxenic (adj.)
Pertaining to cultures containing more than one type of unknown (and undesired) organism; descriptive of a culture with many contaminants. See *Axenic, Monoxenic.*

Polyxenous parasite
Parasite requiring more than two different hosts for completion of its life cycle. See *Heteroxenous.*

Pond
A body of standing fresh water occupying a small surface depression, usually too small to sustain waves, i.e., smaller than a lake and larger than a puddle or pool. Illustration: Habitat.

Population
Individuals, members of the same species, found in the same place at the same time.

Porcellaneous test
Test that is white, opaque, or slightly translucent in reflected light (e.g., foraminifera). Illustration: Foraminiferan test.

Pore apparatus
Complex pore organ; openings through secondary wall in some desmids that consist of a lined pore channel and a web of fibrous material at the inner opening.

Pore plate
Structure of diatom frustule: fine plate of lightly silicified material with small pores that stretches across the areola of many diatoms. See *Rica, Velum.* Illustration: Velum.

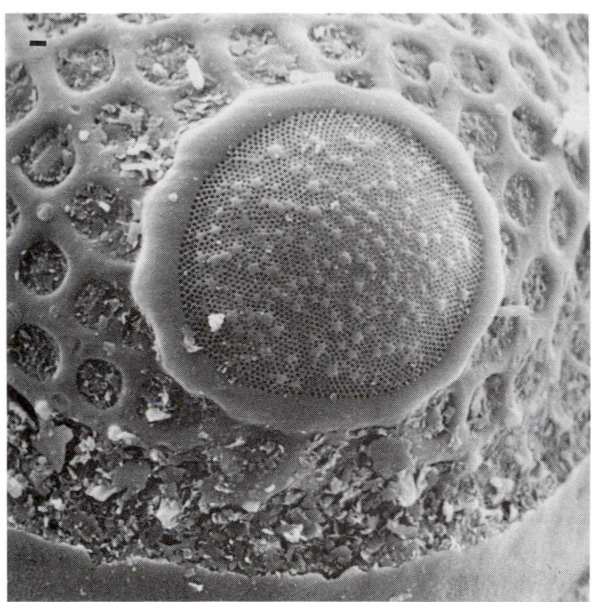

Porelli
Ocellus of the diatom *Odontella.* SEM. Bar (upper left)=1 µm.

Porelli (sing. porellus)
Small, regularly arranged pores in the ocellus of a diatom.

Pores
Openings. Minute rounded openings in the chamber wall, usu. covered by an internal membrane or sieve (e.g., in foraminifera).

Porphyran
Sulfated storage carbohydrate of *Porphyra,* composed of galactose units.

Porus (pl. pori)
Opening in chrysophyte stomatocysts (statospores) that is closed by a pectic plug at maturity. Illustration: Stomatocyst.

Postciliary microtubular ribbon
See *Postciliary ribbon.*

Postciliary ribbon
Postciliary microtubular ribbon. Part of kinetid structure in ciliates. Conspicuous in the subphylum Postciliodesmatophora; ribbon of microtubules associated with a kinetosome, originating in the right-posterior part at triplet 9 (by convention). See *Transverse ribbon.* Illustrations: Cortex, Transverse ribbon.

Postciliodesma (pl. postciliodesmata)
Bundle of overlapping postciliary ribbons found in a large

group of ciliates. The basis for classification at the level of subphylum (i.e., Postciliodesmatophora).

Postcingular (adj.)
Referring to cell-covering plates on hypotheca in contact with the cingulum in certain dinomastigotes. See *Precingular*. Illustration: Thecal plate.

Posterosome
Posterior vacuole formed from coalescence of golgi vacuoles and involved in polar tube formation in microsporans.

Preapical platelet
Small thecal plate that occurs between the first apical plate and the apical pore complex (APC) in some peridinioid dinomastigotes. Illustration: Thecal plate.

Precingular (adj.)
Referring to cell-covering plates on the epitheca in contact with cingulum in some dinomastigotes. See *Postcingular*. Illustration: Thecal plate.

Predation (n. predator)
Mode of nutrition in which an organism hunts, attacks, and digests other heterotrophic organisms for food (e.g., *Didinium* that seizes *Paramecium*).

Preoral crest
Part of the oral apparatus; ridge reinforced by band of microtubules (e.g., in kinetoplastid mastigotes). Illustration: Bodonidae.

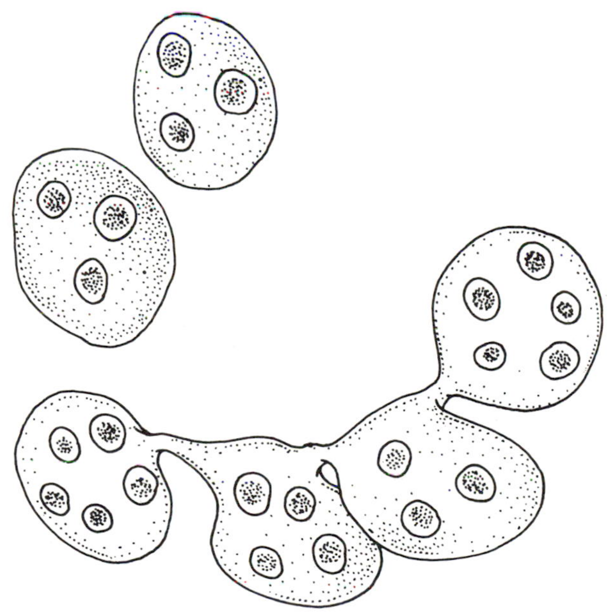

Prespore cell
Cleavage of plasmodium (right) into prespore cells (left) in the protostelid plasmodial slime mold *Ceratiomyxella tahitiensis*.

Prespore cell
Cell of a slug that will ultimately develop into spores in dictyostelids; earliest stage of sporocarp development in protostelids in which ameboid cells cleaved from plasmodium begin to round up and secrete a slime sheath before the stalk is produced. See also illustration: Protostelida.

Presporogonic (adj.)
Refers to part of a life history that precedes the sporogonic one, i.e., that during which spores are formed. See *Sporogonic*.

Primary cell
Pseudoplasmodium enclosing one generative cell (e.g., myxosporean myxozoans).

Primary metabolite
Organic compound produced metabolically and essential for completion of the life cycle of the organism that produces it (e.g., any of the 20 protein amino acids or nucleotides in RNA and DNA). Chemical component required for autopoiesis. See *Secondary metabolite*.

Primary nucleus (pl. primary nuclei)
Large diploid nucleus that undergoes meiosis to give rise to secondary nuclei in certain chlorophytes (e.g., Dasycladales).

Primary plasmodium (pl. primary plasmodia)
Sporangial plasmodium. Plasmodium that develops into thin-walled sporangium in plasmodiophorids. See *Secondary plasmodium*.

Primary production
Primary productivity; productivity. The production of reduced carbon (organic) compounds by autotrophs (see Tables 1 and 2).

Primary productivity
See *Primary production*.

Primary rhizoid
First rootlike protoplasmic extension (rhizoid) that develops from the encysted zoospore in chytridiomycotes.

Primary zoospore
Zoospore that germinates directly from a cyst (e.g., in plasmodiophorids). See *Secondary zoospore*.

Principal zoospore
First-formed zoospore that has laterally inserted undulipodia that are shed on encystment (e.g., the oomycote *Phytophthora*). See *Auxiliary zoospore*.

Proboscis
Emergent process on the anterior end of spermatozoids that contains a band of eight or nine microtubules originating near the kinetosomes (e.g., in the xanthophyte *Vaucheria*); structure thought to facilitate attachment to the egg in the phaeophyte *Fucus*; trunklike extension emerging from the oral area at anterior of certain ciliates (e.g., *Dileptus*).

Procentriole
Cell organelle; immature centriole (e.g., in *Labyrinthula*, electron-dense, granular structure 240 nm in diameter with a core of

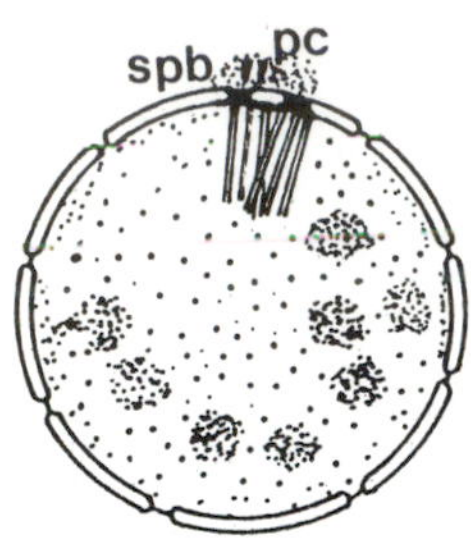

Procentriole
Early prophase of cell division in an acantharian actinopod. pc=procentriole; spb=spindle pole body.

nine radiating spokes and no microtubules at first; found in trophic cells, these structures arise *de novo* prior to each mitotic division). See also illustration: Kinetosome.

Prochlorophyta
See *Chloroxybacteria.*

Procyclic stage
Stage in life cycle that represents the beginning of development in an invertebrate host (e.g., trypanosomatid mastigotes).

Productivity
See *Primary production.*

Progamic fission
Binary fission occurring within a gamontocyst and resulting in the formation of two gamonts (e.g., heliozoan actinopods).

Progeny
See *Offspring.*

Progressive cleavage
Multiple fission; cytokinesis of multinucleate protoplasm to form uninucleate cells. Phycological term for process comparable to schizogony.

Prokaryote (procaryote)
Bacterium; member of the kingdom Monera (kingdom Procaryotae); cell or organism composed of cells with nucleoids (i.e., lacking a membrane-bounded nucleus). Illustration: Organelle.

Proloculum (pl. prolocula)
Proloculus. First chamber formed during development of the test of an adult gamont (e.g., foraminifera). Illustrations: Foraminiferan test, Granuloreticulosa.

Proloculus (pl. proloculi)
See *Proloculum.*

Promastigote
Stage in trypanosomatid development in which the kinetoplast lies in front of the nucleus and the associated undulipodium emerges laterally to form an undulating membrane along the anterior part of the body, usu. becoming free at its anterior end. Illustrations: Kinetoplastida, Trypomastigote.

Promitochondria (sing. promitochondrion)
Structures that develop into mature, cristate mitochondria. Illustration: Trypomastigote.

Promitosis (adj. promitotic)
Protomitosis. Mitosis in which the nuclear envelope remains closed throughout the process and the nucleolus pinches in two (e.g., in many amebomastigotes). See *Mesomitosis, Metamitosis.* Illustration: Closed mitosis.

Propagule
Generative structure; any unicellular or multicellular structure produced by organisms and capable of survival, dissemination, and further growth. Examples include cysts, spores, some kinds of eggs, seeds, akinetes, etc. Phycologists restrict the term to refer to hormogonia or other multicellular structures that function in asexual reproduction. Illustration: Cyst.

Prophase
First stage of mitosis or meiosis in which chromosomes condense and nucleolus and nuclear membrane may begin to disappear. In meiosis, prophase is broken down into five substages: leptonema, zygonema, pachynema, diplonema, and diakinesis. See *Meiosis, Mitosis.* Illustrations: Procentriole, Synaptonemal complex.

Prosporangium (pl. prosporangia)
Structure from which a sporangium develops.

Prostomial (adj.)
Referring to oral region that develops in the anterior of an organism (esp. ciliates).

Proter
Anterior offspring of transverse binary fission of the parental organism; it often retains the mouthparts of the parent (e.g., ciliates). See *Opisthe.*

Protists
Single-celled (or very-few-celled and, therefore, microscopic) protoctists.

Protocentriole
According to the serial endosymbiosis theory, free-living bacterial ancestor of the centriole. See *Protomitochondrion, Protoplastid, Serial Endosymbiosis Theory.*

Protoctists
Eukaryotic microorganisms (the single-celled protists and their multicellular descendants). All eukaryotic organisms with the exception of animals (developing from diploid blastulas), plants (developing from embryos supported by maternal tissue), and fungi (developing from zygo-, asco-, or basidiospores) are protoctists. Protoctists include two-kingdom system "protozoans" and all "fungi" with mastigote stages as well as all algae (including kelps), slime molds, slime nets, and other obscure eukaryotes.

Protomerite
Anterior part separated from the deuteromerite by a transverse septum (e.g., trophozoites of some gregarine apicomplexans).

Protomite
Separate form between the tomont and the tomite; a relatively rare stage in the polymorphic life cycle of a few ciliates (e.g., some apostomes). Illustration: Tomont.

Protomitochondrion (pl. protomitochondria)
According to the serial endosymbiosis theory, immediate free-living bacterial ancestor to mitochondria (e.g., oxygen respirer like *Paracoccus* or *Daptobacter*).

Protomitosis
See *Promitosis*.

Protomont
Separate form between the feeding trophont and the often encysted true tomont (dividing) stage; a relatively rare stage in the polymorphic life cycle of a few ciliates (e.g., some apostomes). Illustration: Tomont.

Protonema (pl. protonemata)
Thread-shaped structure developing from a spore (e.g., algae and plants) or the product of zygote germination (charophytes).

Protoplasm
Fluid contents of cells, i.e., cytoplasm and nucleoplasm.

Protoplasmic streaming
See *Cyclosis*.

Protoplasmodium (pl. protoplasmodia)
Smallest and simplest type of myxomycote plasmodium; microscopic in size, it lacks any system of veins or protoplasmic streaming and usually gives rise to a single sporophore. See *Aphanoplasmodium, Phaneroplasmodium*.

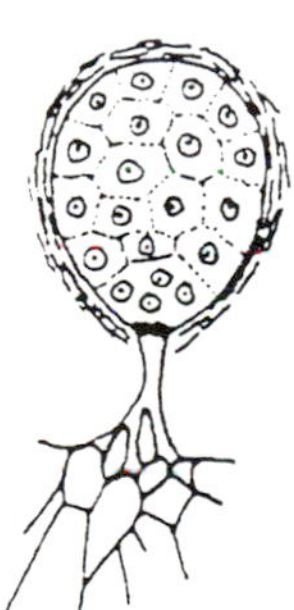

Protoplast
Cleaving protoplast in the sorus of the labyrinthulomycote *Thraustrochytrium* leads to formation of zoospores, which are later released.

Protoplast
Actively metabolizing membrane-bounded part of a cell as distinct from the cell wall. Cells that, after treatment to remove them, lack cell walls. See also illustration: Protostelida.

Protoplastid
According to the serial endosymbiosis theory, immediate free-living bacterial ancestor to chloroplast, rhodoplast, and other plastids (e.g., *Prochloron* or cyanobacterium).

Protoseptate (adj.)
Referring to rudimentary, incomplete, or partial internal walls (septa) separating successive growth stages (e.g., foraminifera, superfamily Astrorhizacea).

Protosphere
Phase in the life history of certain chlorophytes; the lobose cell developing from a germinating zygote that will next form a siphonous juvenile.

Protozoa
Obsolete term referring, in the two-kingdom classification, to a phylum in the Animal kingdom consisting of large numbers of primarily heterotrophic, microscopic eukaryotes. Traditionally the smaller heterotrophic protoctists and their immediate photosynthetic relatives (e.g., phytomastigotes). See *Metazoa*.

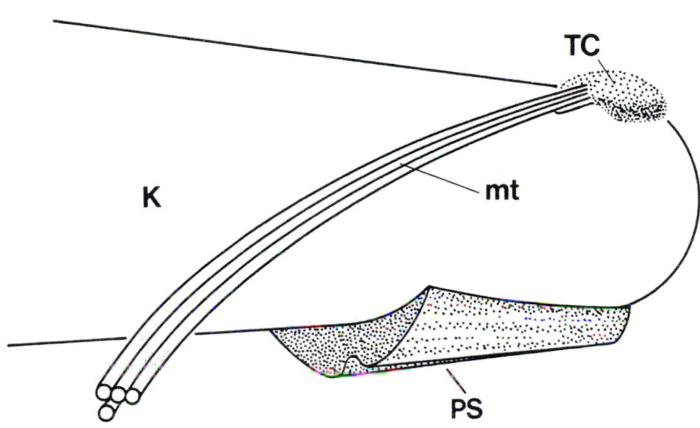

Proximal sheath
Bilobed proximal sheath (PS) attached to kinetosome (K) of motile reproductive cells of ulvophycean chlorophytes. mt=microtubules of rootlets; TC=terminal cap.

Proximal sheath
Wedge-shaped or bilobed component of kinetid associated with proximal end of uppermost kinetosome (e.g., ulvophycean chlorophytes). See also illustration: Terminal cap.

Psammolittoral (adj.)
Ecological term referring to the sandy environment along marine coasts.

Psammophile (adj. psammophilic)
Ecological term referring to organisms that live in sandy environments, esp. in the spaces between sand grains (e.g., many karyorelictid ciliates). Illustration: Habitat.

Pseudocapillitium (pl. pseudocapillitia)
Structure of myxomycote sporophores consisting of irregularly shaped thread- or platelike fragments dispersed among the spores (e.g., in the order Liceales).

Pseudocilium (pl. pseudocilia)
Type of mastigoneme in glaucocystophytes; protoplasmic protrusion of a cell containing microtubules, derived from the typical axoneme but immotile. See *Pseudoflagellum*.

Pseudocrystalline
See *Paracrystalline*.

Pseudoflagellum (pl. pseudoflagella)
Pseudocilium. Nonmotile undulipodium (e.g., of the chlorophyte *Tetraspora gelatinosa*).

Pseudogene
A nonfunctional gene closely resembling a known gene of a different locus.

Pseudointracellular (adj.)
Referring to position relative to a cell; appearing intracellular but topologically extracellular because of failure to cross the plasma membrane (e.g., symbiotrophs contained in parasitophorous vacuoles).

Pseudointratissular (adj.)
Referring to position relative to tissue; surrounded by tissue and appearing to be inside tissue but topologically external to the tissue because of failure to penetrate into or between cells. See *Intratissular*.

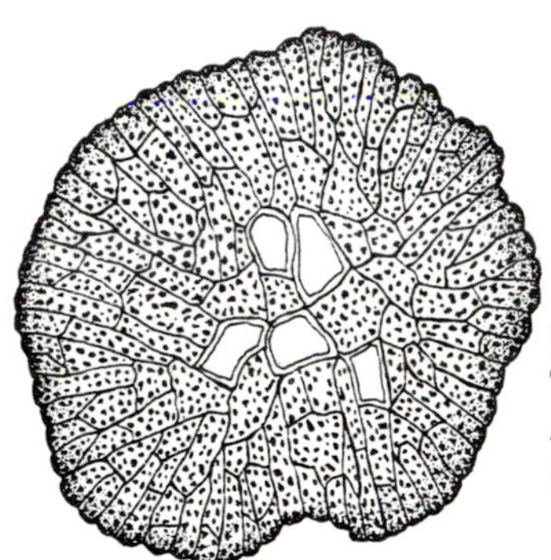

Pseudoparenchyma
Thallus of filamentous chlorophyte *Phycopeltis epiphyton* Millard (Trentepohliales).

Pseudoparenchyma (adj. pseudoparenchymatous, pseudoparenchymous)
Thallus construction; contiguous filaments rather than true parenchymatous cells capable of three-dimensional growth.

Pseudoplasmodium (pl. pseudoplasmodia)
Structure resembling a multinucleate plasmodium that has retained its cell membrane boundaries. An aggregate of amebas, esp. that constituting the initial stage of sorocarp formation in the cellular slime molds (dictyostelids, acrasids); uninucleate trophozoite cell containing one to several generative cells (myxosporean life cycle stage). See *Slug*. Illustration: Dictyostelida.

Pseudopod
See *Pseudopodium*.

Pseudopodium (pl. pseudopodia)
Pseudopod. Temporary cytoplasmic protrusion of an ameboid cell used for locomotion or phagocytotic feeding. Illustrations: Contractile vacuole, Rhizopoda.

Pseudospore
Nonmotile wall-less spore (e.g., in some acrasids).

Pseudostome
"False mouth"; aperture through which a testate ameba projects its pseudopods. Illustration: Rhizopoda.

PSP
See *Paralytic shellfish poisoning*.

Psychrophile
Cryophile. Organism that grows well and completes its life cycle at low temperatures (i.e., near 0°C).

Pulsed-field gradient electrophoresis
Technique for separating large DNA molecules, including small chromosomes. DNA from lysed organisms is subjected to alternate electrical pulses at right angles in agarose gels. The longer the linear molecule, the more time it takes to traverse the gel when the direction of the field changes, thus providing a basis for separation.

Punctum (pl. puncta)
Pore containing smaller pores (e.g., diatom wall markings).

Pustule
Blisterlike, frequently eruptive spot or spore mass (e.g., fungi, foraminifera).

Pusules
Fluid-filled intracellular sacs responsive to changes in pressure. Specialized vacuolelike organelles, presumably osmoregulatory. Usu. two per cell and consisting of two closely appressed membranes that bound a vesicle, they open by canals to the kinetosomes and thence to the outside of the cell. Illustration: Dinomastigota.

Pycnotic (adj.; n. pycnosis)
Referring to darkly staining chromosomes or nuclei; moribund nuclei in cells (e.g., nongenerative nuclei of foraminifera and degenerating fragments of ciliate macronuclei).

Pyrenoid
Proteinaceous structure associated with plastids serving as the center of starch formation or glucan deposits in some algae. Illustrations: Chlorophyta, Conjugaphyta, Cryptophyta, Periplastidial compartment, Phaeophyta.

Pyrenoid cap
Starchy structure surrounding specialized region of plastid (pyrenoid). Illustration: Phaeophyta.

Pyriform (adj.)
Referring to any structure with the form of a tear or pear (e.g., *Tetrahymena pyriformis).*

Q

Quadriflagellate (adj.)
See *Quadrimastigote.*

Quadrimastigote (adj.)
Quadriundulipodiated; referring to mastigote cell bearing four undulipodia (e.g., some trichomonads or chlorophytes).

Quinqueloculine (adj.)
Referring to foraminiferan test in which five chambers are visible and each chamber is angled 144° from previous chamber. Illustration: Foraminiferan test.

R

r
Potential maximal rate of increase of a population.

R-body
Body found inside the kappa particles in the cytoplasm of killer paramecia in some members of the *Paramecium aurelia* complex. Ribbon-shaped body of the kappa extrusome, viruslike in appearance.

Radial fibrils
Fibers arranged in a spokelike array such as those seen in many thin sections of axonemes.

Radial wall
Foraminiferan test wall composed of calcite or aragonite crystals oriented with their C-axis perpendicular to the surface.

Radiolarite
Rock made of chert (siliceous microcrystalline quartz) composed of radiolarian tests that have undergone diagenetic alteration.

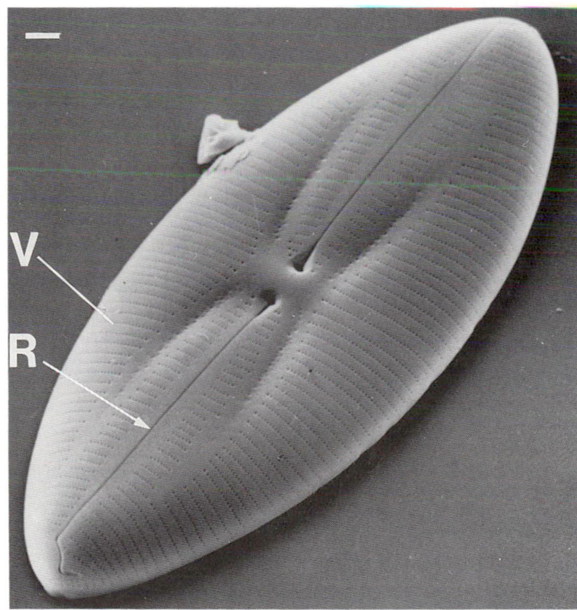

Raphe
Valves (V) of the pennate diatom *Navicula* showing the raphe (R). SEM. Bar=10 µm.

Raphe
Raphe slit; raphe system. The slit, elongate cleft, groove, or pair of grooves through the valve of most pennate diatoms that facilitates gliding cell motility.

Raphe fiber
Structure immediately below the forming raphe, thought to be responsible for the curve of the raphe slit in some diatoms (e.g., *Navicula* spp. and *Pinnularia* spp.).

Raphe slit
See *Raphe.*

Raphe system
See *Raphe.*

Receptacle
Swollen structure containing conceptacles on the thalli of phaeophytes on which reproductive organs (i.e., gametangia or sporangia) are borne. See *Conceptacle.*

Rectilinear test
Test in which chambers accumulate by growth in a straight line (e.g., foraminifera).

Recurrent flagellum (pl. recurrent flagella)
See *Recurrent undulipodium.*

Recurrent undulipodium (pl. recurrent undulipodia)
Recurrent flagellum. Undulipodium that does not lead an organism but adheres to it; trailing undulipodium of heterokont mastigotes. Illustrations: Bodonidae, Endocytosis, Funis.

Red tide
Seawater discolored by the presence of large numbers of dinomastigotes (esp. of the genera *Peridinium* and *Gymnodinium*); blooms of some chrysophytes, euglenids, and the ciliate *Mesodinium rubrum* have also been correlated with red tides.

Refringent (adj.)
Referring to the ability to refract (break up) (e.g., rays of light).

Regolith
Loose, rocky surface materials (boulders, gravel, silt, sand, etc.) covering a planet.

Replication
Process that augments the number of DNA or RNA molecules. Molecular duplication process requiring copying from a template.

Reproduction
Process that augments the number of individuals. A single parent is sufficient for the increase in numbers of individuals in asexual reproduction whereas two parents are required in sexual reproduction. Requires at least one autopoietic entity.

Reservoir
Holding structure or vestibule; deep part of the oral region of some protoctists; the base of the flask-shaped invagination of euglenids. Illustration next page.

Reservoir host
Ecological term, primarily used by parasitologists, for habitats of symbiotrophs in which infected species of animals serve as a source from which other species of animals can become infected (e.g., antelopes are reservoir hosts for *Trypanosoma rhodesiense*, the causative agent of African sleeping sickness in humans).

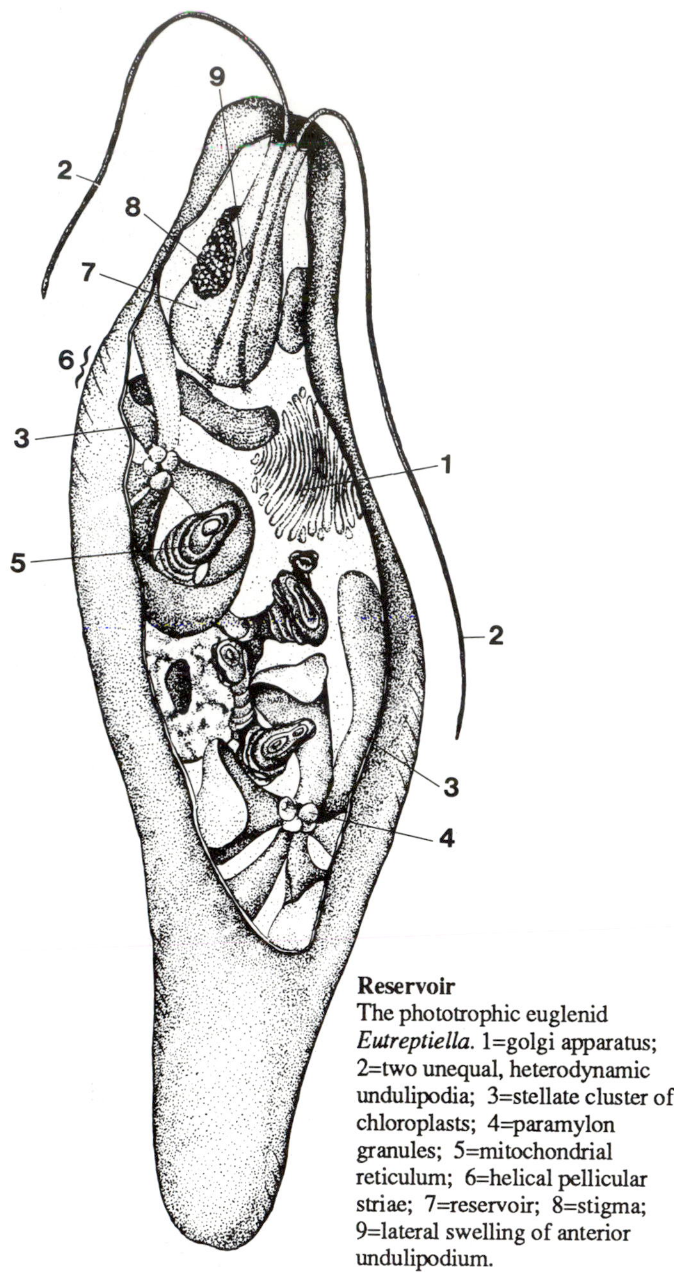

Reservoir
The phototrophic euglenid *Eutreptiella.* 1=golgi apparatus; 2=two unequal, heterodynamic undulipodia; 3=stellate cluster of chloroplasts; 4=paramylon granules; 5=mitochondrial reticulum; 6=helical pellicular striae; 7=reservoir; 8=stigma; 9=lateral swelling of anterior undulipodium.

Residual body

Residuum. That which exists after the formation of offspring cells (gametes or zoites); in dinomastigotes, dark brown body left in empty cyst; in apicomplexans, residual cytoplasm and nuclei of the parent cell. See also illustrations: Microgamete, Myxozoa.

Residuum (pl. residua)

See *Residual body.*

Resistant cyst

Resting cyst; resting spore; dormant propagule of many different kinds of protoctists, equivalent to protoctist spore. Stage surrounded by a wall protecting it from desiccation or other physical injuries; thick-walled, uni- or multinucleate cell that can remain dormant for periods of time under adverse environmental conditions. See *Spore, Statospore, Stomatocyst.* Illustration: Proteromonadida.

Resistant sporangium (pl. resistant sporangia)

Resting spore or covering of many spores (e.g., in chytridiomycotes, a zoosporangium with a thickened wall formed in response to desiccation and capable of extended survival). Illustration: Heterothallism.

Resting cyst

Resistant cyst or protoctist spore; dormant life cycle stage, equivalent to aplanospores or hypnospores of dinomastigotes and other protoctists. Illustration: Dinomastigote life history.

Resting spore

See *Resistant cyst, Resistant sporangium.* Illustration: Hypha.

Restriction enzyme digestion

The use of endonucleases, enzymes that cleave foreign DNA molecules at specific recognition sites, to generate DNA fragments.

Reticular body

Net-shaped body; any structure that is netlike or covered with netlike ridges. Illustration: Prymnesiophyta.

Reticulate (adj.)

Referring to any arrangement in a network; netted.

Reticulopod

See *Reticulopodium.*

Reticulopodial network

Network of cross-connected pseudopods through which a two-way flow of cytoplasm and food particles is detectable; functions more often in food capture than in locomotion (e.g., phylum Granuloreticulosa). Illustration: Granuloreticulosa.

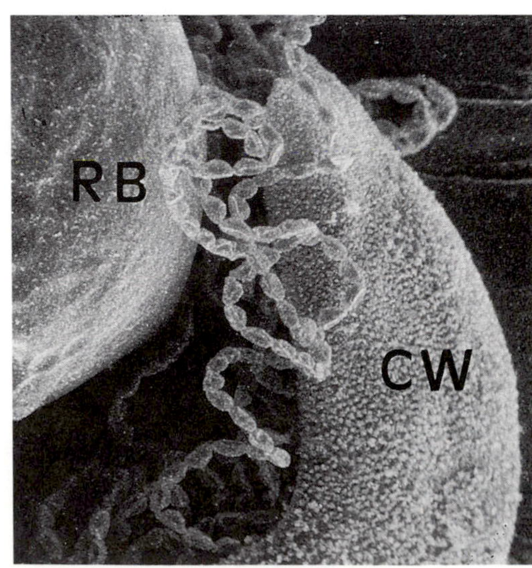

Residual body
Sporulating cyst and chain of spores of the apicomplexan *Stylocephalus* sp. CW=cyst wall; RB=residual body. SEM, x 250.

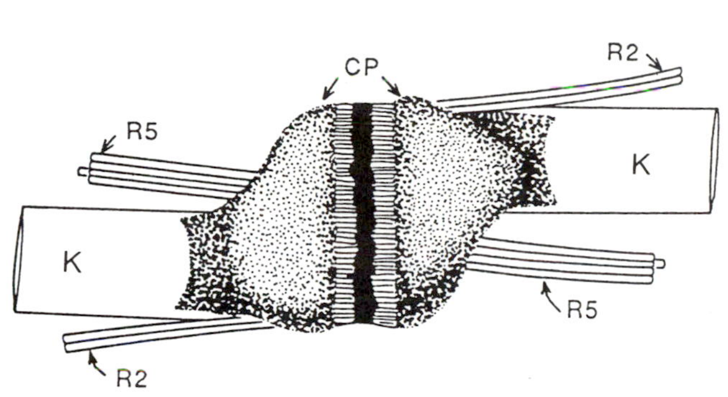

Rhizoplast
Kinetid of the gamete of the ulvophycean chlorophyte *Batophora oerstedii*. View from above (toward posterior end of cell) (above); longitudinal view (right). CP=striated distal fiber; F2=rhizoplasts; K=kinetosome; R2, R5=microtubular rootlets.

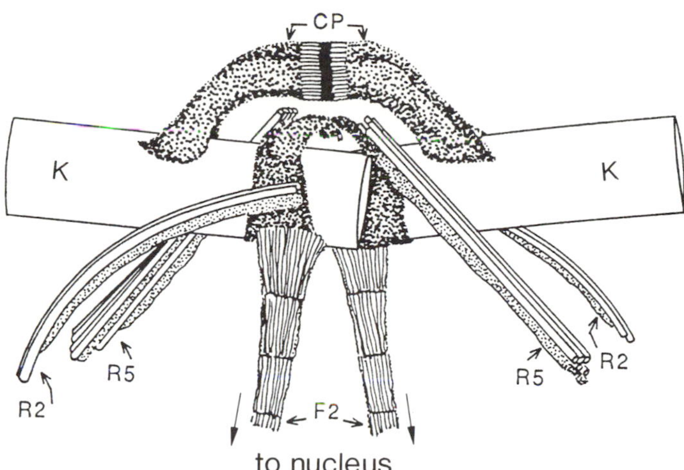

Reticulopodium (pl. reticulopodia)

Reticulopod; rhizopod; rhizopodium. Very slender, anastomosing pseudopod that is part of a reticulopodial network (e.g., phylum Granuloreticulosa). Illustrations: Chlorarachnida, Granuloreticulosa.

Retinoid

Light-sensing component located within the melanosome of the dinomastigote ocellus; pigment layer in dinomastigote cells that produce ocelli. Illustration: Ocellus.

Reversion

Change from mastigote to ameba form. Illustration: Amebomastigote.

Rhabde

Main branch of the siliceous skeleton of ebridians from which clades branch. Illustration: Ebridians.

Rhabdolith

Heterococcolith bearing a stem or club-shaped extension on its distal face (e.g., *Rhabdosphaera*). Illustration: Coccolith.

Rheoplasm

More fluid exterior of reticulopodia. See *Stereoplasm.*

Rheotaxis

Directed growth in response to flow of current (e.g., algae).

Rhizoid

Rootlike structure usu. with anucleate filaments that anchor and absorb (e.g., of chytridiomycotes). See *Rhizomycelium, Primary rhizoid.* Illustrations: Acetabularian life history, Chytridiomycota.

Rhizomycelium (pl. rhizomycelia; adj. rhizomycelial)

Delicate rootlike (rhizoidal) system extensive enough to resemble superficially the mycelia of fungi; nucleated rhizoids having the potential for unlimited growth under favorable environmental conditions (e.g., of chytridiomycotes). See *Rhizoid.*

Rhizoplast

Fibrillar rhizoplast. Cross-banded microtubular ribbon extending from the bases of kinetosomes and directed toward the nucleus or to cytoplasmic microtubule-organizing centers; in chytridiomycotes, fibrillar structure in the zoospore connecting the kinetosomes (at its proximal face) with the nuclear envelope. A rhizoplast is a type of kinetid. See also illustrations: Eustigmatophyta, Eyespot, Proteromonadida, Somatonemes.

Rhizopod

See *Reticulopodium.*

Rhizopodium (pl. rhizopodia)

See *Reticulopodium.*

Rhizosphere

Root zone of plants.

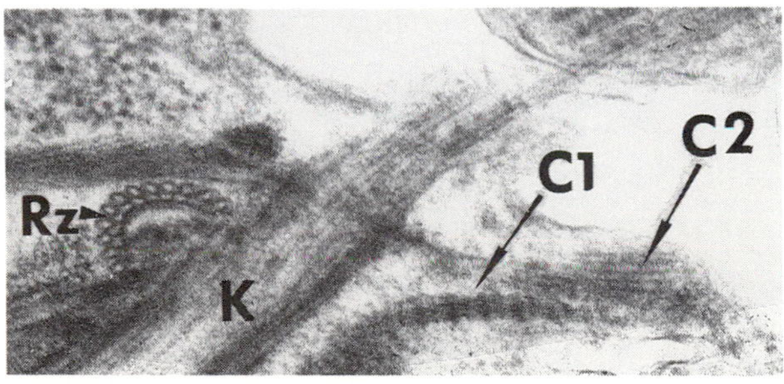

Rhizostyle
Kinetid of the cryptophyte *Chilomonas erosa*. C1=striated band of compound rootlet; C2=microtubule of compound rootlet; K=kinetosome; Rz=rhizostyle. TEM, x 64,000.

Rhizostyle

Kinetid of cryptomonads; posteriorly directed microtubular undulipodial rootlet; the microtubules have winglike projections in some species. See also illustrations: Lamella, Undulipodial rootlet.

Rhodomorphin

Hormone isolated from the red alga *Griffithsia* that can induce cell division and is thus involved in processes of cell repair.

Rhodoplast
Red plastid; photosynthetic membrane-bounded organelle of red algae containing chlorophyll *a* and phycobiliproteins.

Rhoptry
Part of the apical complex of some apicomplexans; dense body extending back from the anterior region of the zoite; may be tubular, saccular, or club-shaped (pedunculate); believed to release secretions facilitating entry of the zoite into its hosts' cells. Illustrations: Apicomplexa, Conoid.

Ribosome
Organelle composed of protein and ribonucleic acid; site of protein synthesis. Illustration: Organelle.

Rica (pl. ricae)
Structure of diatom frustules; thin closing plate of silica usu. with circular perforations across the areolae of some biraphid pennate diatoms; type of pore plate. See *Velum*.

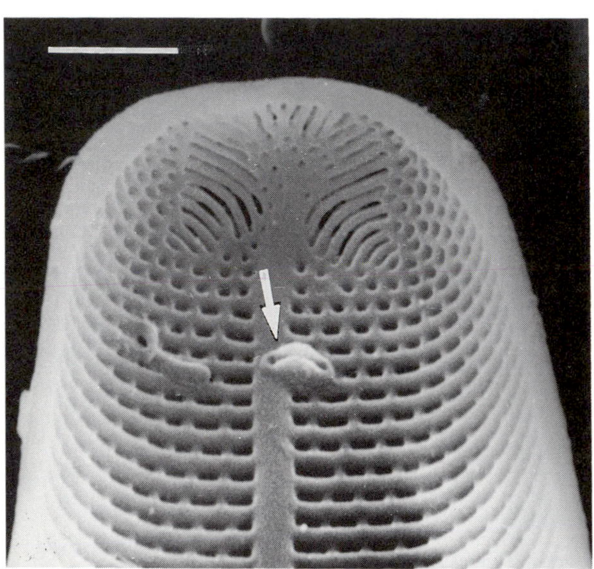

Rimoportule
Portion of the frustule of the diatom *Cyclophora* showing a rimoportule (arrow). SEM. Bar=1 μm.

Rimoportule
Diatom organelle forming a tubular passage through the siliceous wall, infrequently extended into an external tube. Internally a slitlike aperture surrounded by a stalked ridge, giving the appearance of two lips.

Rock
Any naturally formed, consolidated (lithified), loosely consolidated (friable), or unconsolidated (e.g., sand or gravel) material (but not soil) composed of two or more minerals or occasionally of one mineral and having some degree of chemical or mineralogic constancy. Illustration: Habitat.

Rohr (German, meaning pipe, tube)
Extracellular infection apparatus of plasmodiophorids; long, tubular cavity. See *Schlauch*, *Stachel*. Illustration: Stachel.

Root fiber system
Proximal portion of kinetid; portion of kinetid below kinetosome. See *Kinetid*.

Root microtubules
Rootlet microtubules; part of kinetid structure. Microtubules attached proximally to kinetosomes. See *Kinetid*.

Rootlet
Any small structure extending vertically (or proximally into cells) and resembling a tap root of plants (e.g., portion of kinetid, undulipodial (flagellar) rootlet). Illustrations: Rhizoplast, Rhizostyle.

Rostrum (pl. rostra)
General term describing the apical end of a cell when it is beak-shaped or when there is a protuberance (esp. ciliates or mastigotes). Head. Usu. less conspicuous than a proboscis. Illustrations: Parabasalia, Polykinetoplastic.

Ruderal (adj.)
Ecological term referring to the habitat of rubbish, waste, or disturbed places; an organism that grows in such a habitat.

Rumposome
Intracellular structure; honeycomblike organelle of unknown function consisting of regularly fenestrated cisternae in zoospores (chytridiomycote orders Chytridiales and Monoblepharidales). Illustration: Chytridiomycota.

S

S phase
Phase in the mitotic cell cycle of eukaryotes during which DNA synthesis occurs. See *Mitosis*.

Saccate (adj.)
Referring to any pouched or bag-shaped structure.

Saccoderm desmid
Conjugating green alga (desmid) lacking semicells and pitted walls. See *Placoderm desmid*.

Sagenogen
Bothrosome; sagenogenetosome. Organelle on cell surface limited to phylum Labyrinthulomycota (labyrinthulids and thraustochytrids) from which the ectoplasmic network arises. In the sagenogen, an electron-dense plug separates the cell cytoplasm from the matrix of the ectoplasmic network.

Sagenogenetosome
See *Sagenogen*.

Sagittal ring
Ring-shaped component of radiolarian skeletons that lies in a medial sagittal plane separating the skeleton into fragments.

Sagittal suture

Thecal plate boundary between left and right halves of many dinomastigotes (e.g., *Prorocentrum, Gymnodinium).*

Salt marsh

Flat, poorly drained land that is subject to periodic or occasional overflow by salt water, containing water that is brackish to strongly saline and usually covered with a thick mat of grassy halophytic plants. Illustration: Habitat.

Saltatory motion

Jumping motion, usu. intracellular motility (e.g., that exhibited by mitochondria and refractile granules in actinopod cytoplasm as a result of cyclosis).

Sand

A tract or region of rock fragments or detrital particles smaller than pebbles and larger than coarse silt; usually composed of silica but occasionally of carbonate, gypsum, or other composition. Illustration: Habitat.

Saprobe (adj. saprobic)

Saprophyte; saprotroph. Organism utilizing a type of heterotrophy in which it obtains food from dead organic matter; organism feeding by osmotrophy, the mode of nutrition involving the absorption of soluble organic nutrients.

Saprophyte (adj. saprophytic)

Saprobe. Heterotrophic organism living on and deriving its nutrition from dead organic matter. Obsolete term for bacteria and fungi (e.g., fungi living on dead animals). Term to be avoided meaning "plant feeding on dead matter"; refers to osmotrophy of bacteria and fungi.

Saprotrophy (n. saprotroph; adj. saprotrophic)

Mode of nutrition of a saprobe; heterotrophic nutrition obtained from a once-living, still recognizable organism (see Table 2). See *Autotrophy, Biotrophy.*

Sarcinoid (adj.)

Referring to a growth habit in which cubical cell packet arises because the component cells divide in successive perpendicular planes (e.g., bacteria, algae).

Saturation density

See *K*.

Saxicolous (adj.)

Epilithic; lithophilic. Referring to organisms dwelling on the surface of rocks (e.g., algae, cyanobacteria).

Saxitoxin complex

Group of toxins produced by the dinomastigotes *Protogonyaulax* and *Pyrodinium* that cause paralytic shellfish poisoning; includes saxitoxins, neosaxitoxin, gonyautoxins.

Scalariform conjugation

Sexual process in conjugating green algae involving exchange of gametes through conjugation tubes between cells of parallel filaments. The filaments and conjugation tubes form a ladderlike structure.

Scale reservoir

Invagination of cell surface harboring scales in scaly protoctists. Scales are deposited into the scale reservoir by exocytosis from the golgi where they are produced. Production is periodic and apparently synchronized with cell division (e.g., coccolithophorids).

Scales

Organic or mineralized structures of specific shape deposited on the cell surface. Organic or mineralized platelets forming part of a scaly envelope (or scale case) surrounding a cell; cell structures produced endogenously usu. within cisternae of the golgi apparatus and then deposited on the cell surface through vesicle exocytosis, usu. in ordered arrays; often with elaborate surface decoration and sometimes with an outer deposit of $CaCO_3$ (as in coccoliths). Scales may be disclike (plate scales) or elaborated to form cup scales, spine scales, or small dense bodies (knob scales). Illustrations: Coated vesicles, Coccolith, Knob scales.

Schizodeme

Strain or variety; ecological term referring to a population of kinetoplastids which display similarities in patterns of kDNA as determined by electrophoresis.

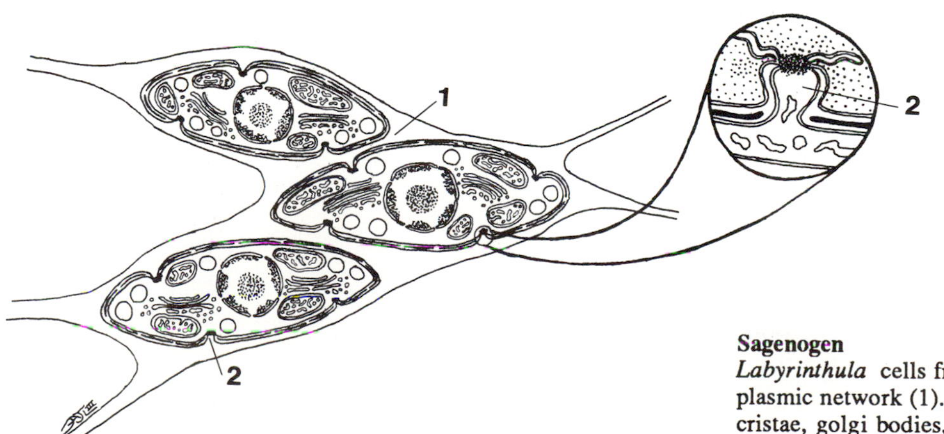

Sagenogen
Labyrinthula cells from a growing colony surrounded by the ectoplasmic network (1). A central nucleus, mitochondria with tubular cristae, golgi bodies, and several sagenogens (2, enlargement) are visible in each cell.

Schizogony

Type of multiple fission; formation of offspring cells in apicomplexans, microsporans, and myxozoans by multiple fission; if the products are merozoites, the process can be subtermed merogony, if gametes, gametogony, if sporozoites, sporogony; in the past, some workers have equated schizogony with merogony only. Process comparable to progressive cleavage of algal plasmodia. Illustrations: Coccidian life history, Foraminifera, Life cycle.

Schizont

Multinucleate organism that will undergo schizogony (e.g., apicomplexans). Illustrations: Life cycle, Sporocyst.

Schizozoite

See *Merozoite.*

Schlauch (German, meaning hose)

Narrow, open-ended extension of the rohr that is oriented toward the cytoplasm of the encysted zoospore of plasmodiophorids. See *Rohr, Stachel.* Illustration: Stachel.

Scintillons

Particles isolated from cell extracts of luminescent dinomastigotes that bioluminesce *in vitro.* See *Microsource.*

Sclerotium (pl. sclerotia)

Type of propagule; darkened amorphous cystlike material derived from desiccated plasmodia of myxomycotes, desiccation-resistant and capable of germination into viable slime mold. Illustration: Plasmodial Slime Molds.

Secondary cytoskeletal microtubules

Cytoplasmic microtubules originating at microtubule-organizing centers close to but not directly attached to kinetosomes (e.g., in ciliates, euglenids).

Secondary metabolite

Organic compound, produced metabolically, not essential for completion of the life cycle of the organism that produces it (e.g., alkaloids, flavonoids, and tannins). They seem primarily to play ecological roles; may serve as pheromones, phytoalexins. See *Primary metabolite.*

Secondary pit connection

Pit connection developed between two adjacent cells by the cutting off of a small cell from one of the pair of adjacent cells and the fusion of that small cell with the other member of the pair.

Secondary plasmodium (pl. secondary plasmodia)

Plasmodium of plasmodiophorids that develops into thick-walled resting cysts. See *Primary plasmodium.*

Secondary zoospore

Zoospore of zoosporangial origin (e.g., plasmodiophorids).

Sedimentation coefficient

Rate at which a given solute molecule suspended in a less dense solvent sediments in a field of centrifugal force; given in Svedberg units, abbreviated S (e.g., ribosomal subunits 23S, 16S; transfer RNA 5S, etc.).

Segregation

Movement to opposite poles of chromatids (mitosis) or chromosomes (meiosis).

Seirosporangia (sing. seirosporangium)

Algal sporangia produced in series at the termini of thalli. Such rows of sporangia may be either branched or unbranched (e.g., the rhodophyte *Seirospora seirosperma*).

Seirospore

Spore produced by a seirosporangium.

Seme

Complex trait of identifiable selective advantage, and therefore of evolutionary importance, resulting from evolution of an interacting set of genes. Unit of study by evolutionary biologists (e.g., nitrogen fixation, cell motility, eyes). See *Aposeme, Hyperseme, Hyposeme, Neoseme.*

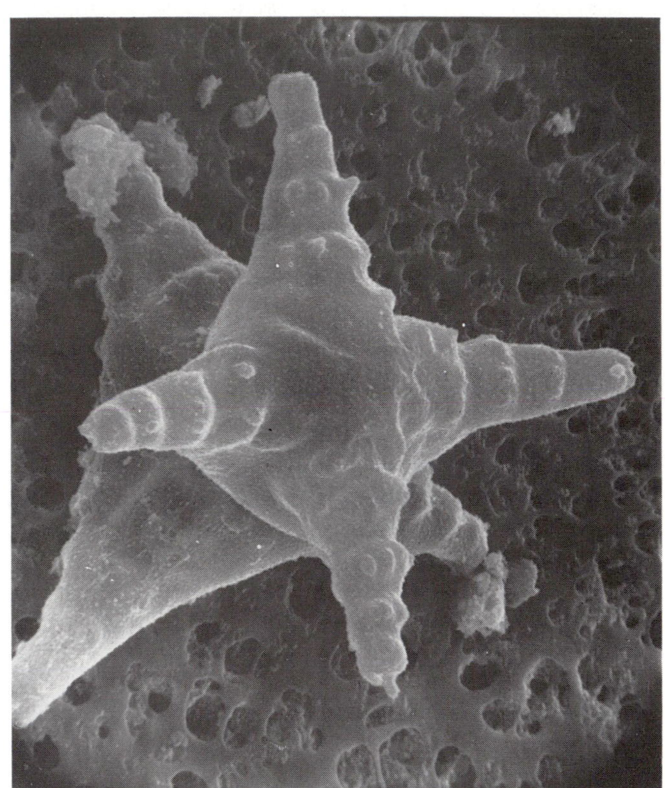

Semicell
The gamophyte desmid *Staurastrum pingue.* The semicell in the foreground is quadriradiate, the lower triradiate. SEM, x 1,430.

Semicell

One of a pair, usu. mirror-image halves, that form the cell of placoderm desmids. See also illustration: Conjugaphyta.

Semiconservative replication

Method of DNA replication in which the molecule splits, each half being conserved and acting as a template for the formation of a new strand.

Septate junction
Type of cell junction in animal tissues; specialized area of adjoining cell membranes showing partitions (i.e., epithelial cells).

Septum (pl. septa; adj. septate)
Partition; cell wall separating constituent cells of multicellular organisms. Illustration: Oomycota.

Sequential zoosporangium formation
Process in which zoosporangia are formed over a period of time on the same subtending hypha, either by means of regrowth through the sporangial septum, cymose renewal of the hypha below the base of the zoosporangial septum, or basipetal, retrogressive zoosprangium delimitation (e.g., oomycotes).

Serial endosymbiosis theory
SET. Theory that mitochondria, plastids, and undulipodia began as free-living bacteria that established symbioses with other bacterial hosts, i.e., that these organelles began as xenosomes.

Serodeme
Populations of e.g., trypanosomes (clones or strains) related by descent and capable of expressing the same variable antigen type repertoire.

Sessile (adj.)
Attached; referring to any organism not free to move about because of attachment to other organisms or to rocks. See *Vagile.*

Seston
Ecological term for microbial communities or populations of particulate matter (including organisms) suspended in the water column in aquatic environments.

SET
See *Serial endosymbiosis theory.*

Seta (pl. setae)
Stiff bristle, hair, or other elongate immotile process (e.g., mastigonemes); common in chlorophytes (e.g., *Coleochaete).* Hollow projection of the frustule that extends beyond the valve margin (e.g., diatoms). Illustration: Hypovalve.

Sex
Process of formation of new organism containing genetic material from more than a single parent. Minimally involves uptake of genetic material from solution and DNA recombination by at least one autopoietic entity; mode of reproduction involving the formation of haploid nuclei in eukaryotes (meiosis) and fertilization (karyogamy, syngamy) to form zygotes. Sexuality. See *Parasexuality.*

Sex cell
See *Germ cell.*

Sex pheromone
See *Pheromone.*

SGO
See *Spicule-generating organelle.*

Shadow casting
Technique used in transmission electron microscopy in which a coating of a heavy metal is deposited on a sample at an angle such that the metal builds up on one side, creating a shadow image. The shape and length of the shadow allows calculation of the dimensions of the sample.

Sheath
Mucopolysaccharide periplast; extracellular, noncellular matrix produced by cells; thought to protect cells from desiccation (e.g., made by pseudoplasmodia of dictyostelids, developing sorocarps of acrasids, sporocarps of protostelids, or by trichomes or coccoid cells of algae and cyanobacteria).

Shield cell
Wall cell of the antheridium in charophyte chlorophytes.

Shock reaction
Behavioral response to a sudden change in environmental conditions; in euglenids, the cell halts, spins, or turns *in situ* end-over-end for a second or more, then proceeds to swim in a random direction.

Shuttle streaming
Protoplasmic streaming in which there is a rapid flow of protoplasm in one direction, a gradual decrease in the flow rate until it ceases, and then a resumption of flow in the opposite direction (e.g., myxomycotes).

Side body complex
Collective name for cisterna, microbody, and lipid globules in the zoospores of the blastocladialean chytridiomycotes (e.g., *Blastocladiella*).

Sieve area
Field of pores lined by plasma membrane through which products of photosynthesis are translocated (e.g., in cells of large algae). The pores may be numerous and small (e.g., *Laminaria*) or few and large (e.g., *Macrocystis*).

Sieve element
Cells with sieve areas. Sieve elements may be randomly oriented or superimposed in longitudinal series constituting sieve tubes.

Sieve tubes
Longitidunal series of sieve elements that form tubes for translocation of photosynthate (e.g., in *Nereocystis* and *Macrocystis).*

Silicalemma
Intracellular membranous vesicle derived from golgi in silica-depositing algae (e.g., membrane upon which opaline silica of the diatom frustule is deposited). Silicalemma, to which silica adheres tightly, is found associated with microtubule-organizing center in central region between offspring cells just inside cell membrane.

Silicoflagellates
See *Silicomastigotes.*

Silicoflagellite
Chert rock composed of accumulated silicomastigote skeletons that were sedimented and diagenetically altered. See *Chert, Diagenesis.*

Silicomastigotes
Silicoflagellates. Undulipodiated photosynthetic marine protoctists with siliceous tests. Members of the phylum Chrysophyta, they are partially responsible for the depletion of dissolved silica from surface waters.

Silt
Clastic sediment composed of particles from 60-200 micrometers (grains are larger than clay and smaller than sand).

Sinus
Invaginated region at the isthmus in certain desmids (conjugating green algae).

Siphon
See *Tubular ingestion apparatus.* Illustration: Euglenida.

Siphonaceous
See *Siphonous.*

Siphonaxanthin
Carotenoid pigment of the chloroplasts of some chlorophytes (e.g., some Caulerpales, Siphonocladales).

Siphonein
Carotenoid pigment of chloroplasts of some chlorophytes, primarily members of the Caulerpaceae.

Siphoneous
See *Siphonous.*

Siphonous (adj.; n. siphon)
Siphonaceous; siphoneous. General term referring to cell or structure in the shape of a pipe or tube; in algae, multinucleate, without crosswalls, i.e., coenocytic, syncytial. See *Coenocyte, Plasmodium, Syncytium.*

Slime molds
See *Acellular slime molds, Cellular slime molds.*

Slime nets
Members of the phylum Labyrinthulomycota; labyrinthulids and thraustochytrids.

Slug
Grex. Pseudoplasmodium that, when mature, leaves a trail of slime and migrates toward a dry area before growing upward to form sporophore; stage in the life cycle of dictyostelids characterized by the appearance of a translucent migrating structure resembling a tiny shell-less snail. Illustration: Dictyostelida.

Soil
Regolith or loose, rocky, organic-rich surface cover of planet Earth; area of unconsolidated material over bedrock; usually supporting or capable of supporting growth of plants.

Somatic (adj.; n. soma)
General term referring to the body (soma) of an organism, esp. the parts not involved in reproduction or germination. Illustration: Oral region.

Somatic cell
Differentiated cell comprising the tissues of soma; any body cell except germ cells. See *Germ cell.*

Somatic kinetid
Body kinetid (e.g., kinetid of the ciliate cortex, usu. not of the oral region).

Somatic nucleus (pl. somatic nuclei)
See *Macronucleus.*

Somatic region
Body region (e.g., in ciliates, the body of the cell exclusive of the oral region). Illustration: Oral region.

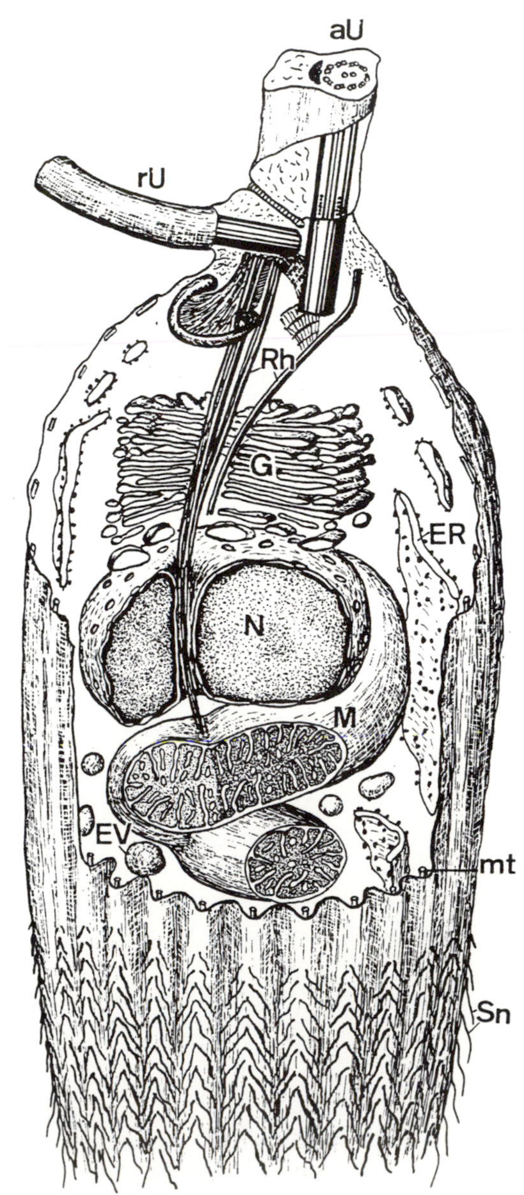

Somatonemes
Tubular hairs on the cell surface that are products of the golgi apparatus, associated with subpellicular microtubules (e.g., proteromonads). Illustration opposite page.

Sonication
Ultrasonication. Method for breaking cells open or homogenizing a mixture of particles by use of ultrahigh frequency vibration.

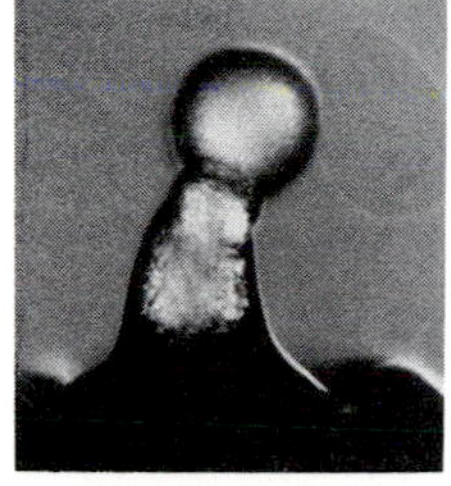

Sorocarp
The acrasid *Fonticula alba*. PC, x 100.

Sorocarp
Sorophore. Multicellular, aerial, stalked structure derived from the aggregation of many individual cells. Often called fructification or fruiting body; ambiguous botanical terms that should be avoided. Applies to dictyostelids, the ciliate *Sorogena*, and acrasids but not to protostelids. See *Sporocarp*. See also illustration: Dictyostelida.

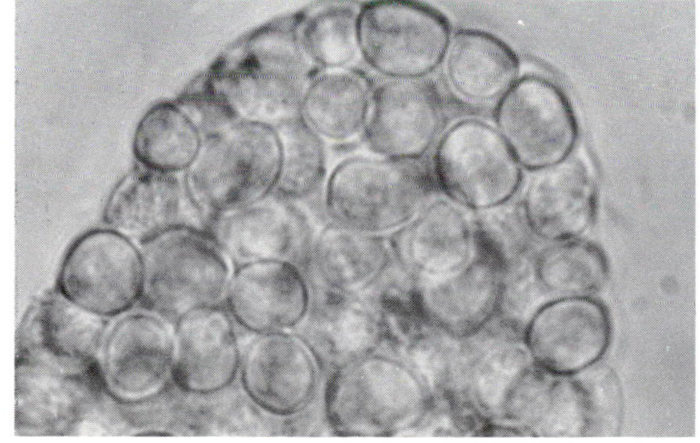

Sorocyst
Terminal area of a sorocarp of the acrasid *Copromyxella spicata*. PC, x 1,500.

Sorocyst
Cyst in sorus of cellular slime molds; sorocysts are virtually identical to ameba cysts, and can also be considered spores.

Sorogen
Culminating stage of cellular slime mold sorocarp.

Sorogenesis
Sorocarp development; formation of the stalked structure that bears the propagules.

Somatonemes
The proteromonad *Proteromonas*. aU=anteriorly directed undulipodium; ER=endoplasmic reticulum; EV=endocytotic vacuole; G=golgi apparatus; M=mitochondrion; mt=microtubules; N=nucleus; Rh=rhizoplast; rU=trailing undulipodium; Sn=somatonemes.

Sorophore
See *Sorocarp*.

Sorus (pl. sori)
Cluster of spores, sporangia, or similar structures in which spores are formed (e.g., in cellular slime molds). Illustration: Dictyostelida.

Sperm
Male gamete; motile and generally smaller than the female gamete. Zoosporelike structure requiring fertilization for further growth. Illustrations: Life cycle, Mitotic apparatus.

Spermary
Sperm storage organ. Illustration: Life cycle.

Spermatangium (pl. spermatangia)
Cell that produces spermatia (e.g., rhodophytes). Illustration: Florideophycidae.

Spermatium (pl. spermatia)
Minute, coccoid, colorless, male gamete released from a spermatangium; spermatia are never undulipodiated in rhodophytes. Illustrations: Florideophycidae, Life cycle, Rhodophyta.

Spermatozoid
Anisogamete, protoctist sperm; undulipodiated reproductive cell functioning as a sperm, i.e., gamete fertilizing a much larger nonmotile gamete (egg).

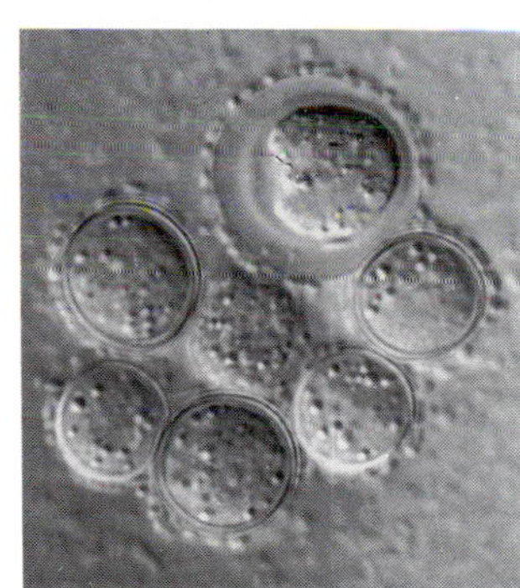

Sphaerocyst
The acrasid *Copromyxa protea*. DIC, x 820.

Sphaerocyst
Rough-walled, pigmented, spherical cyst that may result from cell fusions in the acrasid *Copromyxa protea*.

Sphaeromastigote
Rounded-up cell; developmental stage in kinetoplastids in which the anterior end cannot be identified, although an undulipodium is present.

Spherule
Prominent convoluted mass of cisternae at the anterior end of the sporoplasm in haplosporidians, possibly a modified golgi body. Macrocyst, i.e., dormant, usu. multinucleate, walled plasmodial segment in myxomycotes.

Spicular vacuole
Vacuole in actinopods in which spicules lie. See *Perispicular vacuole*.

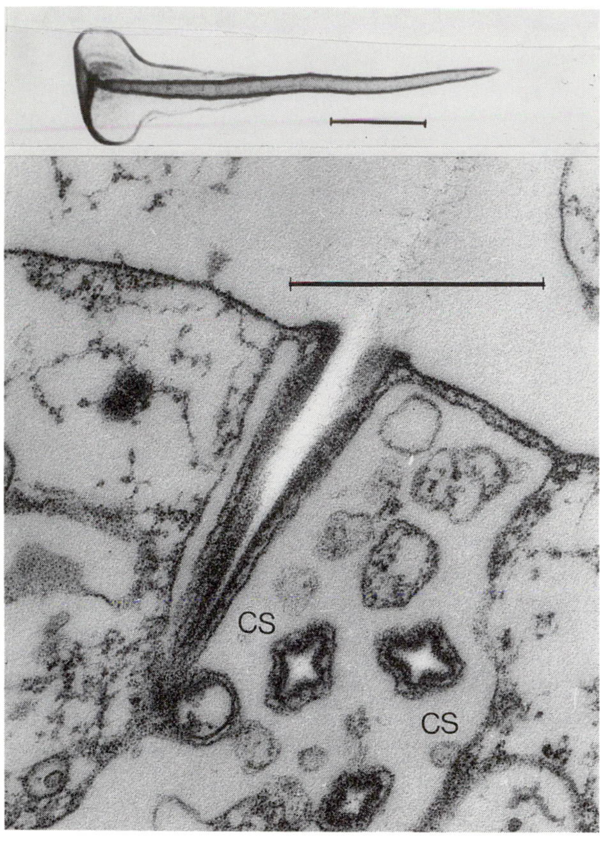

Spicule
Above: Silica spicule of heliozoan. Bar=1 μm.
Below: Cruciform spicules (CS) forming in the spicule-generating organelle (invagination of cell membrane where organic materials are deposited) of the heliozoan actinopod *Cienkowskya* (Heterophryidae). TEM. Bar=10 μm.

Spicule (adj. spicular)
Slender, typically needle-shaped process (e.g., biogenic crystals emerging from the siliceous tests of actinopods); small spine. See *Spines*. See also illustrations: Acantharia, Litholophus, Müller's Law.

Spicule-generating organelle
SGO. Invaginations in the cell membrane where organic substances are deposited to form the skeleton (e.g., heliozoan actinopods). Illustration: Spicule.

Spindle
See *Mitotic spindle.*

Spindle pole body
Nucleus-associated organelle (NAO). Granulofibrosal and microtubular material found at the poles of mitotic spindles; type of microtubule-organizing center. Many variations on NAOs exist in organisms that do not form [9(3)+0] kinetosomes (centrioles). See also illustrations: Mastigote division, Procentriole.

Spines
Slender needle-shaped protrusions (e.g., actinopods). Skeletal projections; defined differently by different authors as either a major rodlike projection from the skeleton or a minor barblike emanation on the skeleton; in the latter case the major projection is a spicule. Illustrations: Actinopoda, Pedicel.

Spirochete
Helically shaped bacterium with flagella in the periplasm. Illustrations: Parabasalia, Pyrsonymphida.

Sporangial plasmodium (pl. sporangial plasmodia)
See *Primary plasmodium.* Illustration: Plasmodiophoromycota.

Sporangial plug
Solid deposit of acellular cell-wall-like callous material that separates the sporangial protoplasm from the protoplasm of the rest of the thallus; expelled prior to sporangial release.

Sporangiogenesis
Formation of sporangium.

Sporangiophore
Subtending stalk to a sporangium.

Sporangium (pl. sporangia)
Hollow unicellular or multicellular structure in which propagules (cysts or spores) are produced and from which they are released. See *Gametangium.* See also illustrations: Heterothallism, Papilla, Plasmodial Slime Molds, Plasmodiophoromycota, Swarmer.

Spore
Type of propagule; small or microscopic agent of reproduction. Some are desiccation- and heat-resistant propagules capable of development into mature or active organisms. Spores are seldom homologous, sometimes even within a single taxon (e.g., coccidians). There is little, if any, difference between the spores of the acellular slime molds and the cysts of amebomastigotes. Yet the term spore is widely used for the nonresistant propagules developing from the sporangia of free-living myxomycote groups and the clearly nonhomologous resistant spores of all microsporan

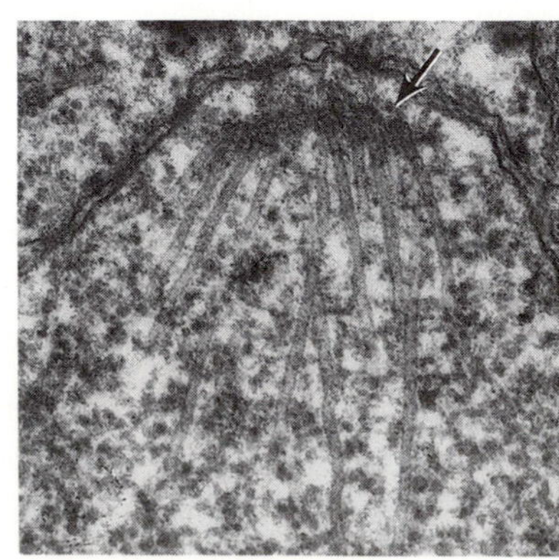

Spindle pole body
Mitotic nucleus with attached microtubules in the haplosporidian *Haplosporidium nelsoni* showing a spindle pole body (arrow). TEM, x 35,000.

and myxozoan groups. Controversial is the use of spore for the oocysts of gregarines or the sporocysts of coccidians, even though these stages are both resistant and infective; for some earlier authors, the sporozoites themselves were the "naked spores," similar as they are to the spore stage of myxomycotes (and various non-protoctist) species; even cyst (e.g., oocyst) and spore have sometimes been confounded; investigators working with apicomplexans suggest replacement of the term spore with specific terms in the life cycle stages of the organisms. Also called vegetative resting state, an ambiguous botanical term to be avoided. See *Aplanospore, Autospore, Auxospore, Azygospore, Ballistospore, Dinospore, Endospore, Epispore, Exospore, Hypnospore, Macrospore, Meiospore, Mesospore, Microspore, Oospore, Resting spore, Statospore, Zoospore, Zygospore.* Illustrations: Hilum, Life cycle, Pansporoblast, Residual body, Sporangium, Trophozoite.

Spore morphogenesis

Developmental process resulting in formation of a spore.

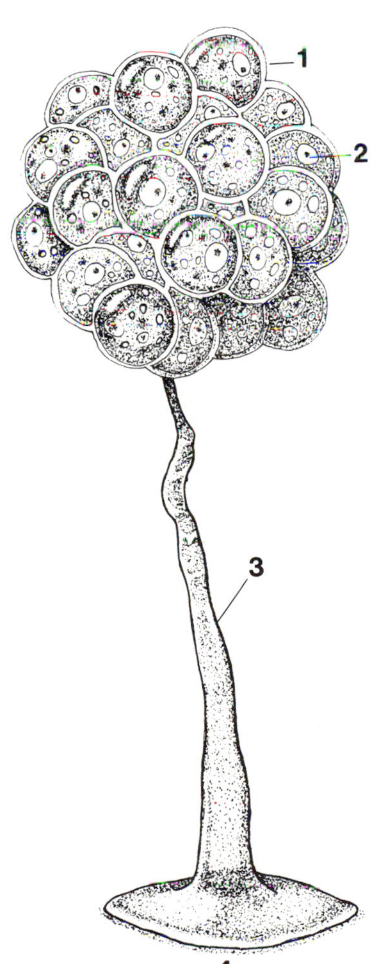

Sporangium
The plasmodial slime mold *Echinostelium minutum.* Entire structure has been called sporocarp or sporophore. 1=spore wall; 2=spore nucleus; 3=stalk; 4=basal disc.

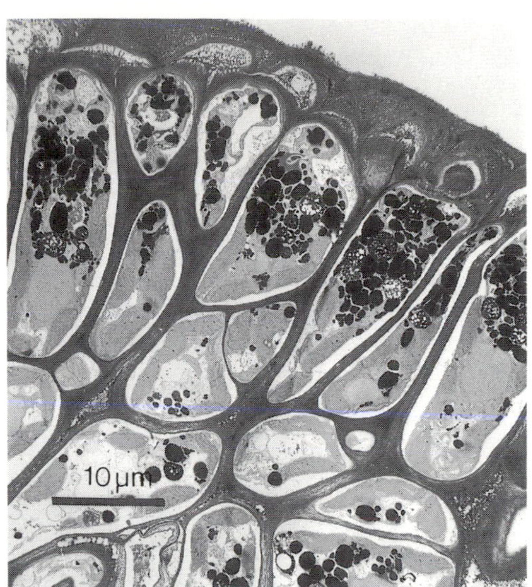

Sporelings
Section through a sporeling of the phaeophyte *Hormosira* showing large concentration of physodes (dark bodies) in outer cells. TEM.

Sporelings

Growths resulting from germinated spores.

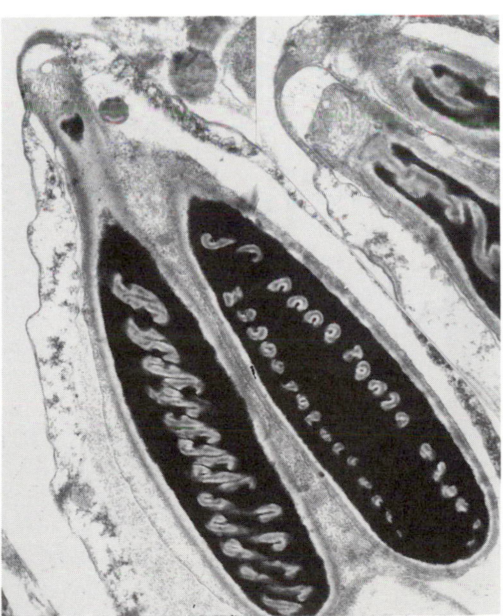

Sporoblasts
Mature polar capsule-containing spores in surrounding sporoblast of the myxozoan *Henneguya psorospermica.* The polar filaments of the capsule pair in the upper right are continuous with the capsule wall. TEM, x 11,500.

Sporoblasts (adj. sporoblastic)

Structures giving rise to spores (e.g., myxozoans). Elliptical, nucleated structures pointed at the ends, the result of a process of segmentation undergone by the protoplasm in apicomplexans.

Sporocarp

Usu. stalked spore-bearing structure in which one initial cell is the source of all the spores (e.g., myxomycotes and protostelids). Also called fruiting body, an ambiguous botanical term that should be avoided. See *Sorocarp*. Illustrations: Protostelida, Sporangium.

Sporocyst

Cyst formed within the divided oocyst that will contain the sporozoites (e.g., coccidians); cyst containing spores (e.g., microsporans); sometimes the oocyst itself in gregarine apicomplexans, which actually have no sporocyst stage. See also illustrations: Coccidian life history, Life cycle.

Sporogen

Stage of sporocarp development in which stalk is being formed; in protostelids, the stage of sporocarp development in which the cell that will ultimately differentiate into a spore or spores is rising off the substrate and depositing the microfibrillar stalk. Illustration: Protostelida.

Sporogenesis (adj. sporogenic, sporogenous)

Sporulation. Formation of spores; reproduction by spores. See *Presporogonic*. Illustration: Labyrinthulomycota.

Sporogonial plasmodium (pl. sporogonial plasmodia)

Structure that undergoes sporogony in apicomplexans.

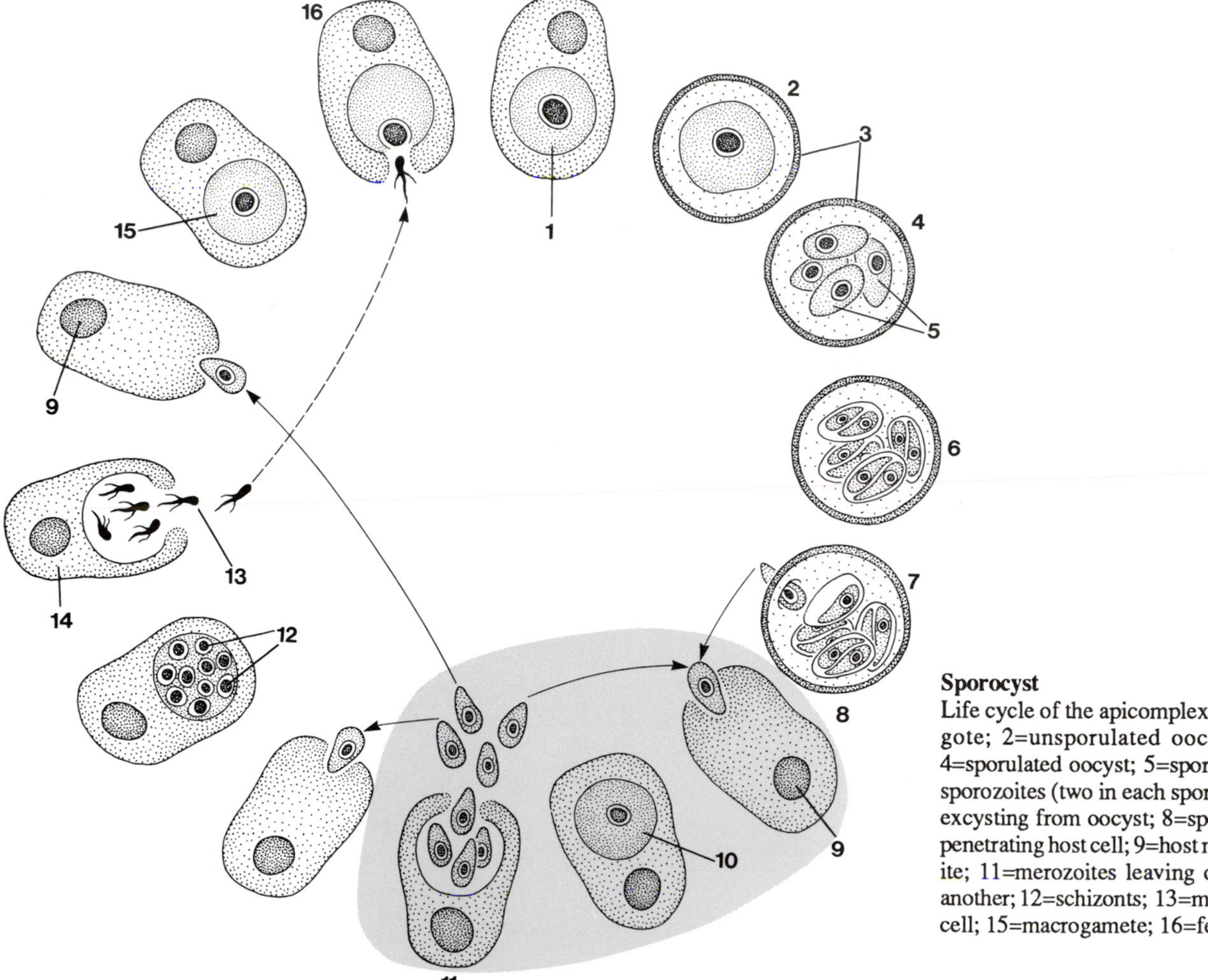

Sporocyst
Life cycle of the apicomplexan *Eimeria* sp. 1=zygote; 2=unsporulated oocyst; 3=oocyst wall; 4=sporulated oocyst; 5=sporocysts; 6=developing sporozoites (two in each sporocyst); 7=sporozoites excysting from oocyst; 8=sporozoite or merozoite penetrating host cell; 9=host nucleus; 10=trophozoite; 11=merozoites leaving one host cell to enter another; 12=schizonts; 13=microgametes; 14=host cell; 15=macrogamete; 16=fertilization.

Sporocyte

Diploid (2N) cell that undergoes meiosis to form haploid (1N) spores; aggregations of cells that divide to produce heterokont bimastigote zoospores (e.g., in the labyrinthulomycotes *Labyrinthula vitelline* and *L. algeriensis);* product of division of the gonocyte in parasitic dinomastigotes. See *Palisporogenesis.*

Sporoduct

Tubular expansion of a cyst wall allowing the escape of mature sporocysts in coccidians.

Sporogony (adj. sporogonic)

Kind of multiple fission; multiple mitoses of a spore or zygote without increase in cell size; zygotic production of haploid sporozoites; production of sporoblasts by schizogony. Illustrations: Coccidian life history, Life cycle.

Sporont

Stage in the life cycle that will form sporocysts (e.g, in coccidians, zygote within the oocyst wall), sporoblasts (haplosporidians), or spores (paramyxeans). Illustration: Paramyxea.

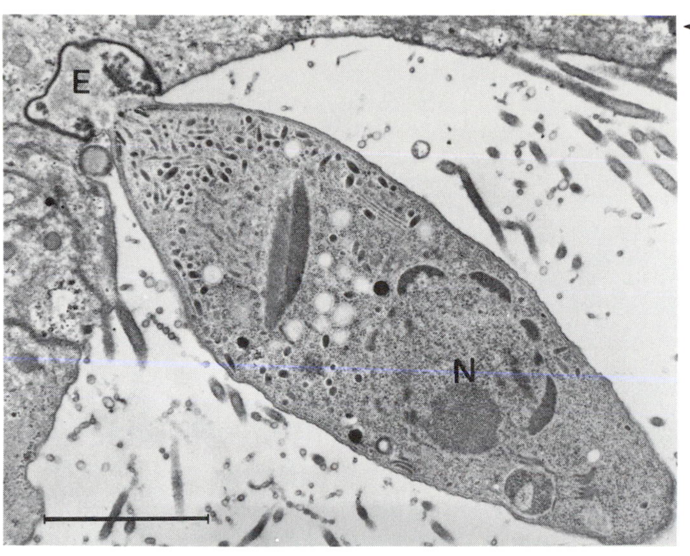

Sporozoite
The sporozoite of the gregarine apicomplexan *Stylocephalus* sp. differentiating an epimerite (E) in the intestinal epithelium of its host. N=nucleus. TEM. Bar=1 μm.

Sporophore
Any structure that bears spores, usu. a multicellular or noncellular stalked aerial structure bearing spores at the apex (e.g., myxomycotes); fruiting body, an ambiguous botanical term that should be avoided. Illustrations: Plasmodial Slime Molds, Sporangium.

Sporophorous vesicle
Pansporoblast membrane. Envelope laid down by sporont external to its plasma membrane in microsporans. Illustration: Pansporoblast.

Sporophyte generation
Life cycle stage in plants and algae: diploid generation that produces spores. The sporophyte is the thallus (body) composed of diploid cells. The sporophyte generation terminates with meiosis, usu. during sporogonic processes. See *Gametophyte generation*. Illustration: Life cycle.

Sporoplasm
Ameboid organism within a spore; infective body (e.g., in myxozoans). Illustrations: Myxozoa, Paramyxea.

Sporopollenin
Complex, extremely resistant, organic polymer that tends to survive diagenesis in the lithification process. Part of the organic geochemical record of life. Sporopollenin, complex heterogeneous material derived from carotenoids, is found in pollen and some algal cell walls; acid-hydrolysis-resistant material considered the diagenetic product of spore or cyst walls.

Sporozoa
Ambiguous former name for apicomplexans which also included the spore-forming parasites, myxozoans and microsporans.

Sporozoite
Life cycle stage of apicomplexans; motile product of multiple mitoses (sporogony) of zygote or spores; trophic stage which is usu. infective. See also illustrations: Coccidian life history, Life cycle, Sporocyst, Trophozoite.

Sporulation
Sporogenesis. Apicomplexan multiple fission; formation of spores that involves division of a large cell into small spores.

Stachel (German, meaning stinger or spine)
Bulletlike structure contained in the rohr whose pointed end is oriented toward the approsorium and the host cell wall in plasmodiophorids. See *Rohr, Schlauch.* Illustration next page.

Stalk
Peduncle; stipe; stem; basal process. Stalk tubes are tubular, microfibrillar components of stalk of the sporocarp in protostelids; outer layer of stalk laid down by prestalk cells in dictyostelids. Illustrations: Acetabularian life history, Heliozoa, Sporangium.

Stalkless migration
Aggregation and migration of slug stage not followed by sorocarp development in dictyostelid cellular slime molds; directional movement of the pseudoplasmodium (slug) in response to environmental stimuli (light, heat, pH, humidity).

Statospore
Stomatocyst. Resistant cyst that consists of two pieces in some algae (e.g., chrysophytes, xanthophytes); endogenously formed resting stage with a conspicuous plug (e.g., in chrysophytes).

Stem cell
Initial cell; cell giving rise by division to identifiable progeny. Ameboid cell located between host cells in which differentiation of the secondary cells occurs (e.g., paramyxeans). Illustration: Paramyxea.

Stenohaline (adj.)
Ecological term referring to the ability of organisms to tolerate only narrow ranges of salinity. See *Euryhaline.*

Stenothermal (adj.)
Stenothermic. Ecological term referring to the ability of organisms to tolerate only limited ranges of temperatures. See *Eurythermal.*

Stenothermic (adj.)
See *Stenothermal.*

Stephanokont (adj.)
Referring to a mastigote that bears an anterior ring or crown of undulipodia.

Stercomares
Masses, usually formed as strings, of stercomes lumped together in large numbers and covered by a thin membrane. Products of xenophyophores.

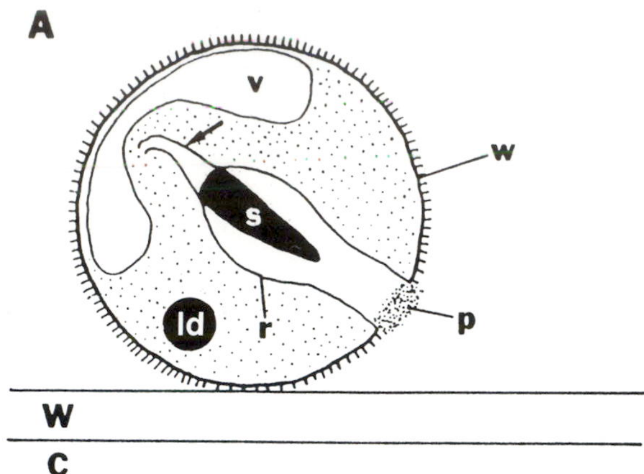

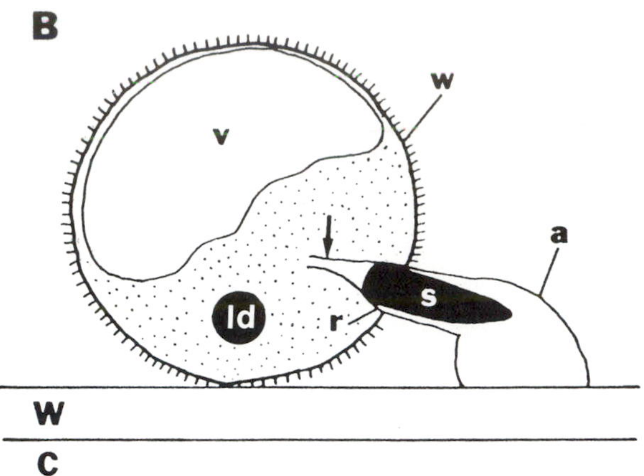

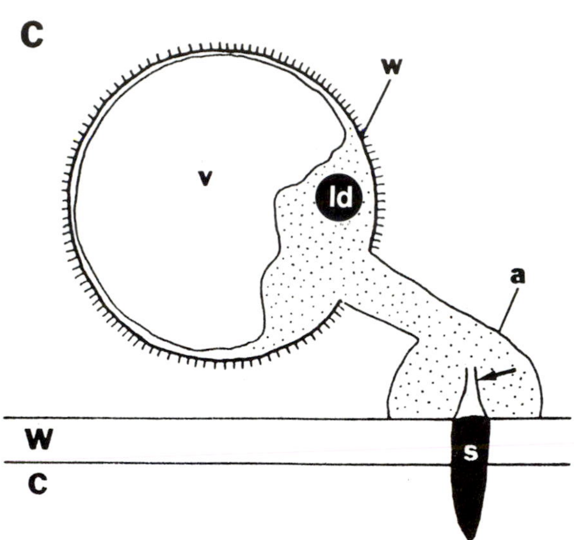

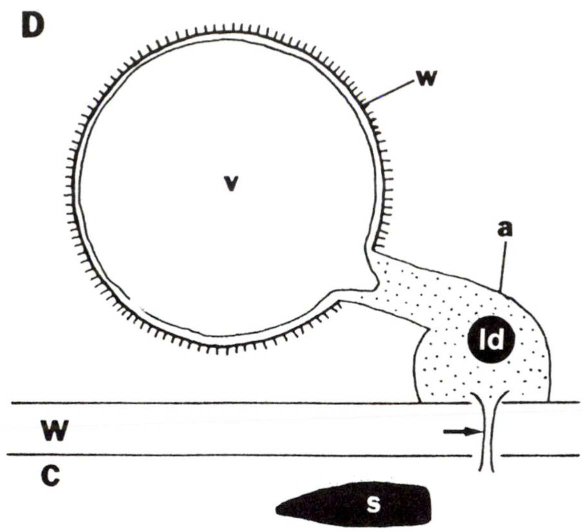

Stachel
Zoospore encystment and penetration of the plant host cell wall by a plasmodiophorid. a. encystment of zoospore. b. development of adhesorium. c. penetration of host wall by stachel. d. passage of plasmodiophorid cytoplasm into host. Host: c=cytoplasm; w=host cell wall. Parasite: arrows=schlauch; a=adhesorium; ld=lipid droplet; p=rohr plug; r=rohr well; s=stachel; v=cyst vacuole; w=cyst wall.

Stercomes
Fecal pellets of xenophyophores.

Stereoplasm
Solid axis of reticulopodia (e.g., foraminifera). See *Rheoplasm*. Illustration: Actinopod.

Stichidium (pl. stichidia)
Specialized branch in rhodophytes that bears tetrasporangia.

Stichonematic (adj.)
Referring to mastigote bearing an undulipodium with a single row of mastigonemes. See *Pleuronematic*. Illustration: Mastigoneme.

Stigma (pl. stigmata)
See *Eyespot*. Illustrations: Eustigmatophyta, Reservoir.

Stipe
General morphological term referring to slender stalk of an organ or organism.

Stipitate (adj.)
Stalked; with a stipe or little stalk.

Stolon system
Internal canal system; tubular structure connecting chambers; system of prolonged extensions (e.g., tests of foraminifera).

Stomatocyst
Statospore. Endogenous silicified resistant cyst produced by chrysophytes.

Stomatogenesis
Mouth formation, esp. in ciliates. In cyrtophoran ciliates, the process involves formation or replacement of all oral kineties, kinetosomes, and the infraciliature plus the associated openings, cavities, etc., in both the proter and opisthe during binary fission.

This resorption and reformation provides the basis for classifying the taxon (subphylum Cyrtophora).

Strain
Population of microorganisms under investigation in the field or taken into the laboratory. See *Isolate*.

Stratum (pl. strata)
Layer of sedimentary rock.

Streptospiral (adj.)
Coiled like a ball of wool (e.g., foraminiferan test in which axis of growth and plane of coiling change as it forms).

Stria (pl. striae)
Linear row of alveoli, areolae, or puncta (i.e., diatom frustules). Illustration: Hypovalve.

Striated disc
Part of kinetid of zoospores of Monoblepharidales (phylum Chytridiomycota); morphologically distinctive rootlet consisting of a flattened, often fan-shaped assemblage of microtubules and fibrils extending from the side of the kinetosome. See *Adhesive disc*. Illustration: Diplomonadida.

Striated (kinetodesmal) fiber
See *Kinetodesma*. Illustration: System I fiber.

Striker
Structure of ejectosome (taeniocyst) that contacts prey. Illustration: Taeniocyst.

Stroma (pl. stromata)
The fluid contents of an organelle (e.g., chloroplast).

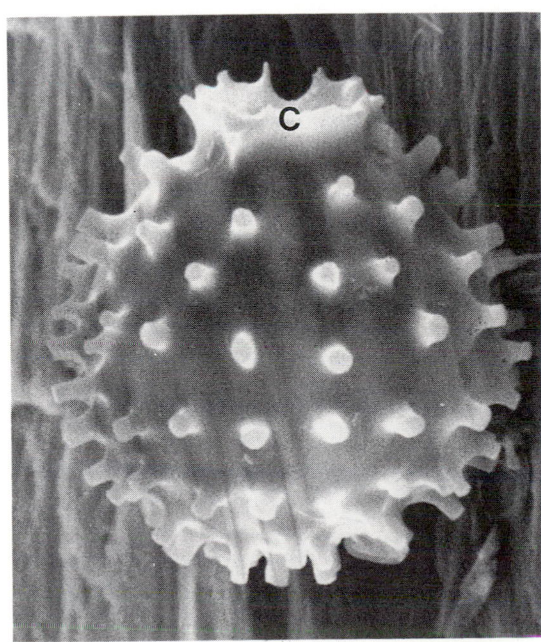

Stomatocyst
Stomatocyst of the chrysophyte *Mallomonas teilingii*. The collar (C) surrounding the porus. SEM, x 1,800.

Stummel (German, meaning little stump or butt)
Very short or reduced undulipodium in certain prymnesiophytes (haptophytes); the short bulbous haptonema found in some coccolithophorids.

Stylet
General morphological term for any of several rigid elongated organs or appendages.

Subaerial (adj.)
Ecological or geological term for processes occurring in the open air on Earth's surface (but not under water) (e.g., evaporation on an evaporite flat).

Subkinetal microtubules
Portion of cell cortex of ciliates composed of components derived from many linearly aligned kinetids (e.g., set of microtubules that arise from the base of kinetosomes and extend anteriorly or posteriorly beneath a kinety).

Sublittoral (adj.)
Ecological term referring to the environment lying below the level of low tide. Subtidal near the shore or just below the shore line or littoral zone. See *Littoral, Supralittoral*. Illustration: Habitat.

Submetacentric (adj.)
See *Mediocentric*. Illustration: Chromosome.

Subpseudopodium (pl. subpseudopodia)
Fine extension at the leading edge of a pseudopodium (e.g., amebas, foraminifera).

Subraphe costae (sing. subraphe costa)
Subraphe fibulae. Supporting bars in the form of flying buttresses running beneath and at a 90° angle to the raphe of pennate diatoms; they are continuations of the valve costae.

Subraphe fibulae (sing. subraphe fibula)
See *Subraphe costae*.

Substrate
Underlayer; carbon source, nitrogen source, food; stable surface to which organisms are attached (e.g., rocks); molecule that is acted upon by an enzyme.

Subtelocentric (adj.)
See *Acrocentric*. Illustration: Chromosome.

Succession
Ecological term referring to ecosystem change; the more-or-less regular phenomenon of community replacement through time.

Sucking disc
See *Adhesive disc*.

Sulcal groove
See *Sulcus*. Illustration: Epicone.

Sulcus (pl. sulci; adj. sulcate)
Groove running from the posterior end anteriorly in dinomastigotes; at the equatorial region it joins the transverse groove; the sulcus contains the insertion and often the proximal part of the longitudinal undulipodium. Illustration: Dinokont.

Supplementary aperture
Opening to the exterior; such an aperture is in addition to and independent of the primary aperture (i.e., tests of foraminifera).

Supralittoral (adj.)
Ecological term referring to the environment of the spray zone lying just above the shore line or littoral zone; i.e., above high tide. See *Littoral, Sublittoral*. Illustration: Habitat.

Surra
Disease of camels caused by *Trypanosoma evansi* and transmitted by biting flies.

Suture
General morphological term referring to a seam or furrow between adjacent parts (e.g., between thecal plates in armored dinomastigotes). Illustration: Thecal plate.

Suture line
Line of adhesion between the two to seven valves of myxozoan spore walls; contact area between adjacent plates that acts as a line of separation in dinomastigotes; region of discontinuity in the cortex, defined by the end of kineties terminating near or on each other in ciliates.

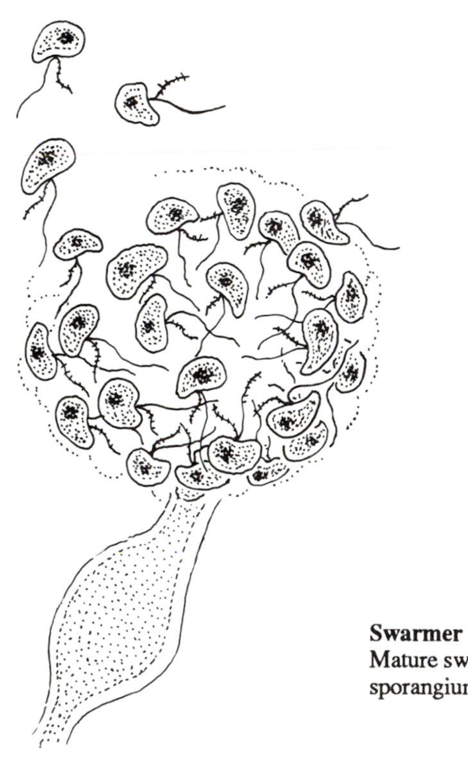

Swarmer
Mature swarmers leaving a sporangium of an oomycote.

Swarmer
Zoospore. Mastigote propagule; undulipodiated, dispersive form in the life cycle of protoctists of many different taxa; swarmer cell (e.g., some actinopod zoospores, rapidly produced motile cells of chytridiomycotes, dinomastigotes). See also illustration: Plasmodial Slime Molds.

Symbionts
Members of a symbiosis, i.e., organisms that have an intimate and protracted association with one or more organisms of a different species. Illustration: Pyrsonymphida.

Symbiosis
Prolonged physical association between two or more organisms belonging to different species. Levels of partner integration in symbioses may be behavioral, metabolic, gene product, or genic. For nutritional modes of symbionts, see Table 1.

Symbiotrophy (n. symbiotroph; adj. symbiotrophic)
Mode of nutrition involving a heterotrophic symbiont that derives both its carbon and its energy from a living partner (see Table 2). See *Necrotrophy*.

Symmetrogenic fission
Type of cell division, generally longitudinal, of a parent such that the two offspring are mirror images of one another with respect to principal structures (e.g., opalinids, pseudociliates). Typically occurs in nonciliate protoctists. See *Homothetogenic fission*.

Symplectic (adj.)
See *Metachronal waves*.

Symplesiomorphy (adj. symplesiomorphic)
Term derived from cladistics that refers to an ancestral, homologous trait (seme) that arose prior to the bifurcation of the lineages of organisms. See *Apomorphy, Plesiomorphy, Synapomorphy*.

Sympodial (adj.)
Referring to a mode of development in which the primary axis is continually replaced by lateral axes, which become dominant but are soon replaced by their own laterals (e.g., sympodial branching in phaeophytes, sympodial renewal in oomycotes).

Sympodial zoosporangium formation
Term describing morphogenesis in chytridiomycotes in which the zoosporangium forms on an apparent main axis derived from successive secondary axes.

Synapomorphy (adj. synapomorphic)
Homologous taxonomic character (seme) that arose in the ancestral species with the bifurcation of the lineage. See *Apomorphy, Plesiomorphy, Symplesiomorphy*.

Synaptonemal complex →
Complex proteinaceous, longitudinally aligned structure seen with the electron microscope that usu. unites homologous chromosomes during the prophase of meiosis. See also illustration: Plasmodiophoromycota.

Synchronous culture
Culture in which all cells or organisms are simultaneously in the same stage of growth or reproduction.

Syncytium (pl. syncytia; adj. syncytial)
See *Coenocyte, Plasmodium*.

Syngamy

Fertilization; gametogamy. Fusion of two cells, usu. gametes. The nuclear fusion process that often follows syngamy is called karyogamy. Illustrations: Acetabularian life history, Florideophycidae, Life cycle, Plasmodial Slime Molds.

Synkaryon (pl. synkarya)

Fusion nucleus; zygotic nucleus; product of fusion of two haploid gametic nuclei or pronuclei.

Synzoospore

Compound zoospore. Multiple zoospore with two to many sets of undulipodia and equivalent multiples of other organelles; usu. the result of incomplete cleavage during zoospore formation in multinucleate xanthophytes (e.g., *Botrydium*). Huge synzoospores, each forming hundreds of biundulipodiated zoospores, are characteristic of *Vaucheria*.

System I fiber

Part of kinetid structure; striated rootlet (not consisting of a bundle of 5-8 nm filaments) often associated with rootlet microtubules and exhibiting a narrow (25-35 nm) repeat of cross-striations (i.e., pedinomonadalean chlorophytes). See *System II fiber*.

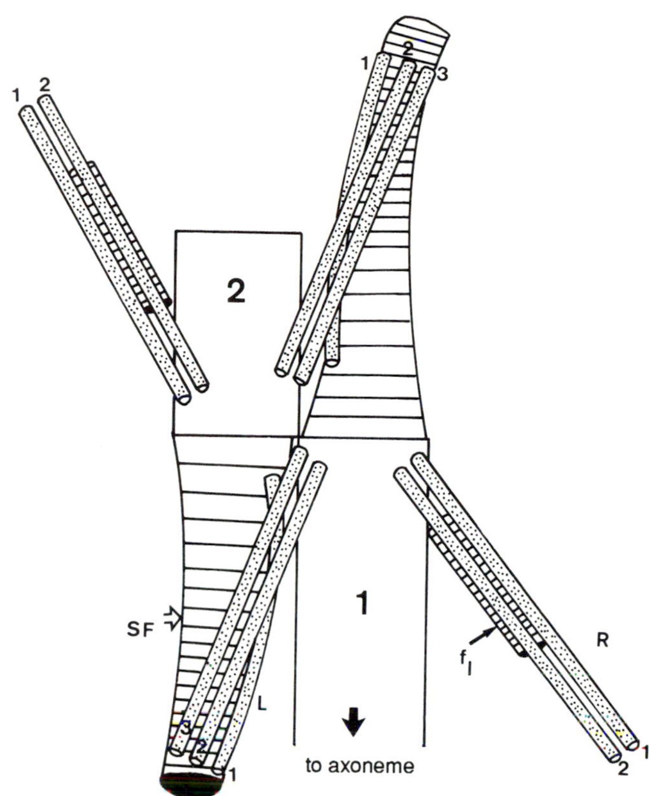

System I fiber
Kinetid in the chlorophyte *Pedinomonas* (Pedinomonadales). The structure is viewed from the outside of the cell. Each kinetosome has two microtubular roots. 1=kinetosome with axoneme (arrow); 2=naked kinetosome; f_1=system I fibers underlying each two-stranded root; L=left root with three microtubules; R=right root with two microtubules; SF=striated fiber underlying each three-stranded root.

System II fiber

Part of kinetid structure; rootlet consisting of a bundle of 5-8 nm filaments, often cross-striated (i.e., chlorophycean chlorophytes). See *System I fiber*.

Systematics

A biological science; that subfield of evolutionary science that deals with naming, classifying, and grouping organisms on the basis of their evolutionary relationships.

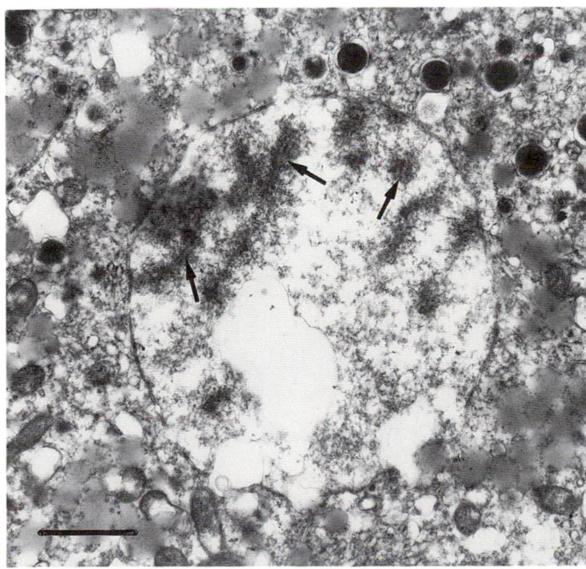

Synaptonemal complex
Pachynema of prophase I of meiosis in cystogenous plasmodium of the plasmodiophorid *Sorosphaera veronicae*, showing synaptonemal complexes (arrows). TEM. Bar =1.0 µm.

Syzygy

Association side-by-side or end-to-end (frontal syzygy or in caudo-frontal association) of gamonts (esp. of gregarine apicomplexans) prior to formation of gametocysts and gametes.

T

T bands

Morphological feature seen with an electron microscope in the myonemes of acantharian actinopods. Thin, dark transverse lines separating repeated clear areas known as L zones. Illustration: Myonemes.

T joints

Morphological feature of the loricae of choanomastigotes. Longitudinal costae joined midway along the anterior costal strips.

Tabular (adj.)

Referring to a laminar form, i.e., having a flat surface.

Tabulation

System of classifying dinomastigote envelope plates.

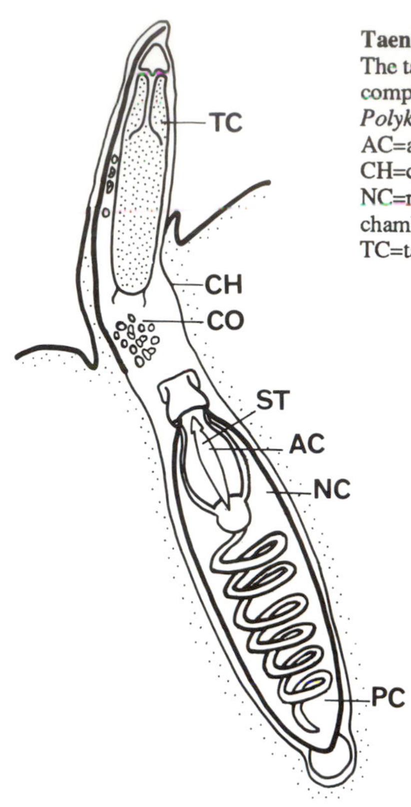

Taeniocyst
The taeniocyst-nematocyst complex of the dinomastigote *Polykrikos schwartzii.* AC=anterior chamber; CH=chute; CO=chute organelles; NC=nematocyst; PC=posterior chamber; ST=striker; TC=taeniocyst.

Taeniocyst
Extrusome with a complex structure characteristic of some dinomastigotes.

Taeniogene
Organelle that gives rise to the taeniocyst in dinomastigotes.

Tannins
Brown polyphenolic compounds that yield tannic acid on hydrolysis. Characteristic of phaeophytes and plants.

Taxis (adj. tactic)
Movement of an organism or organelle toward or away from a stimulus (e.g., geotaxis, phototaxis, magnetotaxis, thigmotaxis).

Taxon (pl. taxa)
Paraphyletic group. Unit in the hierarchy of biology that classifies all living organisms (e.g., in order of descending inclusiveness, taxa include kingdom, phylum, class, order, family, genus, and species).

Taylor-Evitt System
System of thecal plate or cyst paraplate designation used in dinomastigote taxonomy. Illustration: Plate formula.

Tectiform replication
Asexual reproduction in loricate choanomastigotes in which the offspring cell may have component costal strips when it departs from the parent lorica (e.g., Acanthoecidae). See *Nudiform replication.*

Tectin (adj. tectinous)
Complex of protein and mucopolysaccharides comprising some tests (e.g., foraminifera). See *Organic test.*

Telocentric (adj.)
Referring to chromosomes with centromeres (kinetochores) at the ends (telomeres) of the structure. Terminal (the very end) or subtelocentric chromosomes with very small quantities of chromatin lying distal to the centromere are termed acrocentric. See *Acrocentric.* Illustration: Chromosome.

Telomere
Chromosome end, usu. composed of highly repetitious DNA sequences.

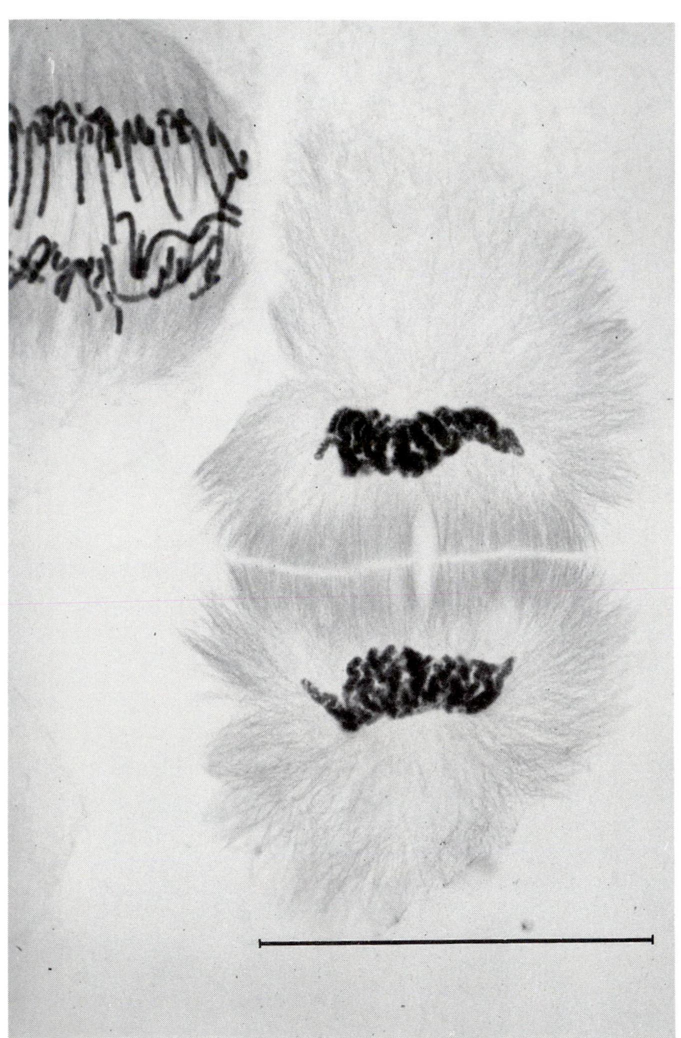

Telophase
Endosperm cell of the plant *Haemanthus katherinae* Bak. (African lily). Bar=30 μm.

Telophase
Stage in mitosis in which chromosomes are at opposite ends of the spindle, chromatin begins to uncoil, and cytokinesis occurs. Nucleolus and nuclear membrane often reform in telophase. See *Mitosis.* See also illustration: Phragmoplast.

Temporary cyst
Cyst produced directly and reversibly from trophic cell in rapid

response to feeding or unfavorable conditions (formed by amebomastigotes, some dinomastigotes, and ciliates such as *Colpoda*).

Tentacles
General term for long protrusions. In suctorian ciliates they are protoplasmic processes, underlain by microtubules, bearing missilelike projectiles (extrusive organelles) that attack prey. Tentacles are distinguished from undulipodia, haptonemes, stalks, and pseudopodia by their substructure and aggressive function. Illustration: Choanomastigota.

Teratological (adj.)
Monstrous; referring to the formation of abnormal growths (e.g., tumors).

Terete (adj.)
Cylindrical.

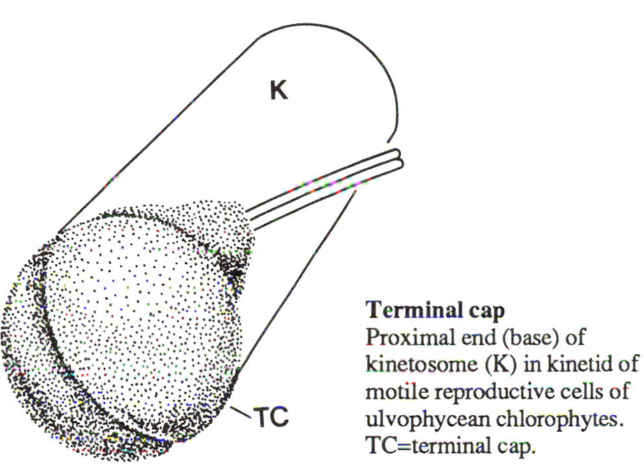

Terminal cap
Proximal end (base) of kinetosome (K) in kinetid of motile reproductive cells of ulvophycean chlorophytes. TC=terminal cap.

Terminal cap
Component of kinetid associated with the proximal end of the uppermost kinetid; more-or-less electron-dense flap at the anterior end of the kinetosome (ulvophycean and trentepohlialean chlorophytes). See also illustration: Proximal sheath.

Terminal nodule
Diatom valve structure; site of the terminal pore of raphe on a motile pennate diatom.

Terminal plate
Kinetid substructure (e.g., chytridiomycotes and hyphochytrids); structure just proximal to where the axoneme contacts the kinetosome and thus the cytoplasm of the rest of the zoospore. Illustrations: Chytridiomycota, Hyphochytriomycota.

Test
Cell covering; hardened, continuous periplast; general descriptive term for any of a large number of shells, hard coverings, valves, or thecae. See *Agglutinated test, Biserial, Calcareous, Evolute test, Hyaline, Megalospheric test, Microgranular test, Microspheric test, Organic test, Peneropliform, Planispiral, Porcellaneous test, Quinqueloculine, Rectilinear test, Streptospiral, Triserial, Trochospiral test, Uniserial, Valve.* Illustrations: Foraminiferan test, Granuloreticulosa.

Testes (sing. testis)
Sperm storage organs. Illustration: Life cycle.

Tethyan realm
The elongated east-west seaway that separated Eurasia from Gondwanaland from at least the early Paleozoic to late Cretaceous.

Tetrapyrroles
Class of carbon compounds formed from four heterocyclic pyrrole rings linked by single carbon bridges and often chelated with metal ions (e.g., Fe^{++} in heme, Mg^{++} in chlorophyll).

Tetrasporangium (pl. tetrasporangia)
Cell in which a diploid nucleus undergoes meiosis to form four haploid spores (tetraspores) in rhodophytes. Illustration: Florideophycidae.

Tetraspores
Spores formed in a tetrasporangium (e.g., rhodophytes). Illustration: Florideophycidae.

Tetrasporoblastic (adj.)
Referring to a life cycle in some rhodophytes in which carpospores germinate to produce a diploid tetrasporophyte that is borne on the gametophyte.

Tetrasporophyte
Diploid thallus in rhodophytes that produces tetrasporangia. Illustration: Florideophycidae.

Thallophytes
Literally "flat plants"; obsolete term for bacteria, fungi, and other nonvascular photosynthetic and heterotrophic organisms.

Thallus (pl. thalli)
General descriptive term, derived from botany, referring to body type in plants and algae. Thalli are flat, leaflike structures undifferentiated into organs and lacking vascular tissue characteristic of tracheophytes, i.e., lacking roots, stems, and leaves. Illustrations: Life cycle, Monosporangium, Pseudoparenchyma, Rhodophyta.

Thanosis
A process by which selected cells are programmed to die as a normal component of development. This loss of cells plays a role in sculpting the structure of an organism during morphogenesis. Thanosis is initiated by specific signals and requires *de novo* gene expression. (After the Greek god of death, Thanatos.)

Theca (pl. thecae; adj. thecate, thecal)
General descriptive term used for many unrelated structures; coat, periplast, test, valve, shell, hard covering, enveloping sheath, or case. Total cell wall, composed of many closely fitting cellulose plates (sometimes used equivalently to the amphiesma) in dinomastigotes. Illustrations: Dinomastigote life history, Plate formula.

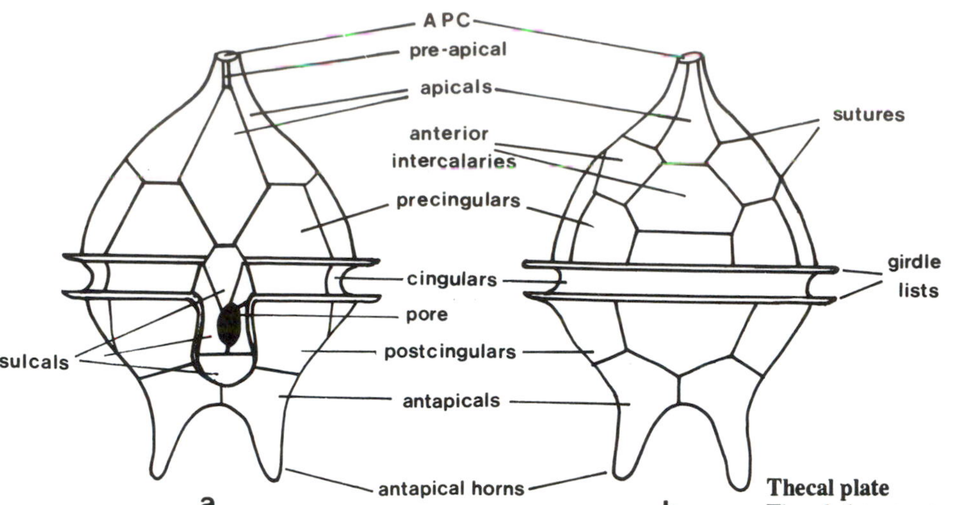

Thecal plate
Thecal plate structure of peridinioid or gonyaulacoid dinomastigotes. a. Ventral view. b. Dorsal view.

Thecal plate
Component of cell coat, or hardened structure, external to the outer plasma membrane (e.g., dinomastigotes). See also illustration: Plate formula.

Thermocline
Ecological term referring to a sharp temperature gradient; the zone of water in which temperature decreases rapidly with depth; in lakes, zone between the epilimnion and hypolimnion. Illustration: Habitat.

Thigmotactic (adj.; n. thigmotaxis)
Referring to organisms that are touch-sensitive or adherent. Thigmotaxis leads to production of structures functioning as holdfasts (e.g., certain somatic cilia of some epibiotic ciliates).

Thorotrast
Electron-dense substance that when added to a sample becomes trapped inside phagocytic vesicles; used in electron microscopy to identify such vesicles.

Thylakoid
Photosynthetic membrane, lamella, or sac; photosynthetic membrane bearing chlorophylls, carotenoids, and their associated proteins usu. stacked in layers; photosynthetic membranes in bacteria and in plastids. See also illustrations: Florideophycidae, Organelle.

Thylakoid doublet
Paired thylakoids, the outer surface of which in cyanobacteria and rhodophytes bears phycobilisomes.

Tight junction
Type of cell junction in animal tissues. Continuous bandlike junction between epithelial cells and, rarely, other cells.

Tinsel
See *Mastigoneme.*

Tinsel flagellum (pl. tinsel flagella)
See *Tinsel undulipodium.*

Tinsel undulipodium (pl. tinsel undulipodia)
Undulipodium bearing mastigonemes. See *Whiplash undulipodium.* Illustration: Mastigoneme.

Tomite
Stage in the polymorphic life cycle of histophagous ciliates in which organisms are small, free-swimming, and nonfeeding; one of two or more fission products of a tomont (or sometimes a protomite). Illustrations: Macrostome, Tomont.

Tomont
Prefission or dividing stage in the polymorphic life cycle of a number of histophagous ciliates (e.g., apostomes and some

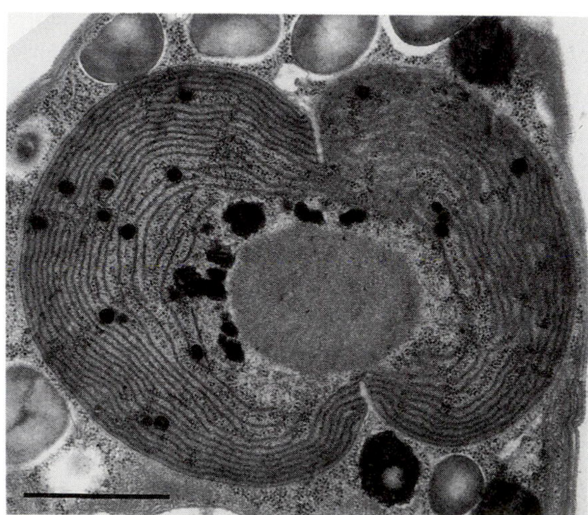

Thylakoid
Dividing cyanelle of the glaucocystophyte *Cyanophora paradoxa*, strain IABH 1555, showing concentric arrangement of thylakoids with lipid droplets between them. Dark granules surrounding the large central body are probably polyphosphates. Adjacent to the cyanelles are several starch grains lying freely in the cytoplasm. TEM. Bar = 1 μm.

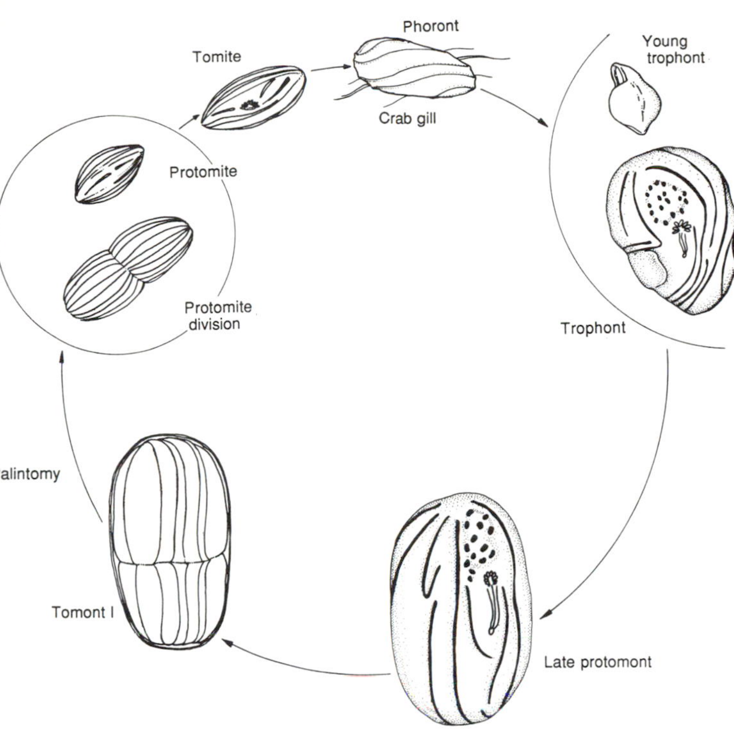

Tomont
Life cycle of the apostome ciliate *Hyalophysa*. The phoront stage is a resting or encysted stage on the crab gill.

hymenostomes). A large form, typically encysted. Tomont may undergo multiple fission (e.g., divide a number of times in quick succession to yield tomites). See also illustration: Macrostome.

Totipotency
Developmental term referring to propagule or growing cell that is capable of repeating all steps of development and giving rise to all cell types.

Toxicyst
Type of extrusome; slender tubular structure that probably contains both paralytic and proteolytic enzymes helping to penetrate, immobilize, and cytolyze prey.

Trace element
See *Micronutrient.*

Trace fossil
See *Ichnofossil.*

Transcription
Synthesis of messenger RNA from a DNA template with a sequence determined directly by the base pair sequence of the DNA template.

Transduction
The transfer of small replicons (e.g., viral or plasmid DNA) from an organelle or bacterium to another organelle or bacterium usu. mediated by bacteriophage. Change of energy from one form to another (e.g., light to chemical or mechanical energy to heat). Illustration: Genophore.

Transfection
Natural genetic change in bacteria and eukaryotic cells in culture induced by uptake of DNA from aqueous medium. Illustration: Genophore.

Transformation
The process of conversion of an ameba to a mastigote by the production of undulipodia or the reverse transformation of a mastigote to an ameba by active absorption of the undulipodia. Characteristic of amebomastigotes, myxomycotes, phaeophytes, some actinopods, and other organisms. The process is probably of evolutionary significance; whether it is monophyletic is unknown. Uptake, incorporation, and inheritance of exogenous genetic material (e.g., transforming principle DNA of *Hemophilus* bacteria). Illustrations: Amebomastigota, Genophore.

Transition fibers
Transition zone fibers; part of a kinetid; fine, fibrillar elements connecting the undulipodial membrane in the transition zone with the undulipodial axoneme at a point between the A- and B-tubules.

Transition region
See *Transition zone.*

Transition zone
Flagellar transition zone; transition region; transitional region; undulipodial transition region. Part of a kinetid; region of the undulipodium at its proximal (basal) end adjacent to the kinetosome displaying cytological characteristics of diagnostic and phylogenetic interest in the systematics of undulipodiated organisms. Illustrations: Coated vesicles, Eyespot, Mitotic apparatus, Multilayered structure.

Transitional helix
Coiled fiber. Helical structure, probably composed of ribonuclear protein, in transition zone of undulipodia of most heterokont groups (e.g., xanthophytes, eustigmatophytes, proteromonads, chrysophytes); called "Spiralkörper" in chrysophytes. Illustrations: Eustigmatophyta, Eyespot, Xanthophyta.

Transitional region
See *Transition zone.*

Translation
Synthesis of protein on ribosomes from activated amino acids using messenger RNA (mRNA) transcripts as templates.

Transverse fission
See *Homothetogenic fission, Perikinetal fission.*

Transverse flagellum (pl. transverse flagella)
See *Transverse undulipodium.*

Transverse microtubular ribbon
See *Transverse ribbon.*

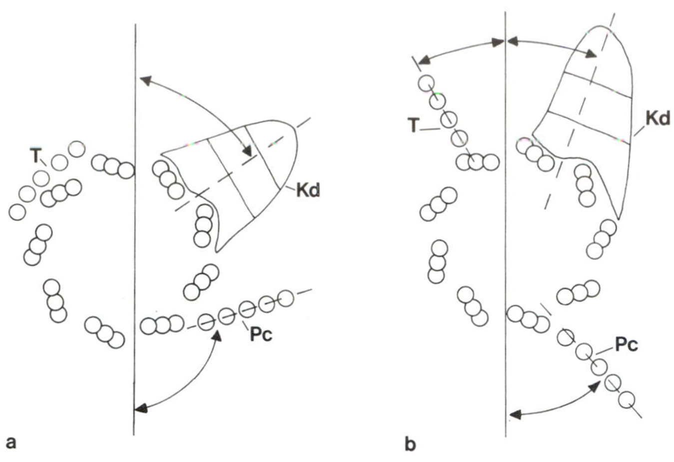

Transverse ribbon
Transverse sections of two ciliate kinetids. Vertical lines indicate the longitudinal axis of the kinety. Orientation of the transverse ribbon to the kinetosome: a. tangential; b. radial. Kd=kinetodesmal fibril; Pc=postciliary ribbon; T=transverse ribbon.

Transverse ribbon
Transverse microtubular ribbon. Transverse fiber; part of kinetid structure characteristic of ciliates; ribbon of microtubules associated with kinetosomes that originate near triplets 3, 4, and 5 and extend laterally. See *Postciliary ribbon*. See also illustration: Cortex.

Transverse undulipodium (pl. transverse undulipodia)
Undulipodium that wraps around the cell and lies in the equatorial groove in dinomastigotes.

Triaene
Arrangement in an ebridian skeleton in which the initial branching point is of four branches (e.g., in *Hermesinum*).

Triatomine bugs
Blood-sucking insects of the order Hemiptera, family Reduviidae (subfamily Triatominae) which defecate while feeding; they transmit *Trypanosoma cruzi*, infecting the host via their contaminated fecal material.

Trichocyst
Extrusome underlying the surface of many ciliates and some mastigotes; capable of sudden discharge to sting prey; probably nonhomologous structures (e.g., dinomastigotes, prasinophytes, raphidophytes). Illustration: Extrusome.

Trichocyst pore
Aperture in the thecal plate through which trichocysts are discharged in armored dinomastigotes. Illustration: Epicone.

Trichogyne
Receptive protuberance or threadlike elongation of a female gametangium to which male gametes become attached (e.g., rhodophytes and many fungi). Illustration: Florideophycidae.

Trichome
Morphological term referring to filamentous or threadlike shape (e.g., single row of cells of filament, exclusive of sheath, of cyanobacteria or algae).

Trichothallic growth
Mode of cell division in phaeophyte tissue in which active cell division occurs at the base of a filament or group of filaments.

Triode
Arrangement in an ebridian skeleton in which the initial branching point is of three branches (e.g., *Ebria*).

Triphasic life cycle
Three-part life history displaying three distinct types of morphology. Sequential polymorphism.

Triserial (adj.)
Triseriate. General morphological term for structures organized in three rows or series (e.g., tests of foraminifera).

Triseriate (adj.)
See *Triserial.*

Trisomy (adj. trisomic)
Karyotype (2N + 1) of a diploid organism with one extra chromosome. The extra chromosome is homologous with one of the existing pairs; one chromosome is present in triplicate.

Trochospiral test
Helicoid spiral test. Coiled test in which the pattern of growth involves the addition of chambers in a spiral coil; the hollow or depressed side of the cone-shaped test is the involute side; the higher opposite side is known as the evolute side (e.g., foraminifera). Illustration: Foraminiferan test.

Trophic cell
Trophic stage; trophont. General term for a heterotrophic cell that feeds and grows, common in the life cycle of many protoctists (e.g., apicomplexans, ciliates, etc.). Also called vegetative cell, an ambiguous botanical term that should be avoided. See *Trophont, Trophozoite.* Illustration: Labyrinthulomycota.

Trophic stage
See *Trophic cell, Trophont.*

Trophocyst
Enlarged cell capable of feeding by osmotrophy.

Trophocyte
Feeding cell (e.g., in multicellular parasitic dinomastigotes, the cell that attaches the host to the colony).

Trophomere
Proximal section of the body or thallus of ellobiopsids that carries terminal reproductive structures, the gonomeres. Illustrations: Ellobiopsida, Gonomere.

Trophont
Trophic stage. Trophic cell or organism; feeding and growing

stage. Adult stage in the life cycle in ciliates. An interfissional form; form that shows a preceding tomite and a succeeding tomont stage, as in the polymorphic life cycles of various symbiotrophic apostome and hymenostome species. Also called vegetative cell, an ambiguous botanical term that should be avoided. See *Trophozoite*. Illustrations: Opalinata, Tomont.

Trophozoite

Motile trophont stage of symbiotrophic protists (primarily apicomplexans, microsporans, and myxozoans). See also illustrations: Bodonidae, Diplomonadida, Life cycle, Sporocyst.

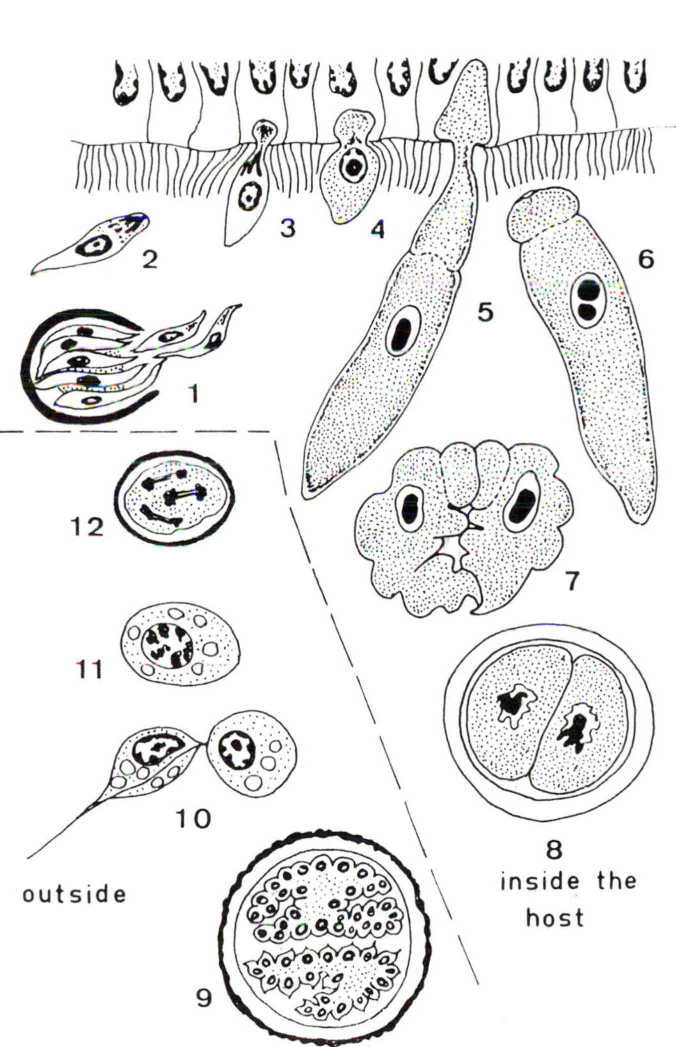

Trophozoite
Life cycle of the gregarine apicomplexan *Stylocephalus* sp. 1. sporozoites escaping from the spore ingested by the host, 2. free sporozoite. 3. penetration of the intestinal epithelium and differentiation of the epimerite. 4-5. growing gamonts (trophozoites). 6. mature gamont detached from the intestinal epithelium. 7. pairing of gamonts. 8. encystment. 9. differentiation of gametes. 10. fertilization. 11. the zygote. 12. meiosis. Stages 1 to 8 occur in the host intestine, 9 to 12 outside the host, and 10 to 12 within the cyst wall.

Tropism

Morphogenetic movement or growth toward or away from an external stimulus (e.g., phototropism, geotropism).

Trypanosomatid

Informal name of members of the trypanosome kinetoplastids. Illustration: Kinetoplastida.

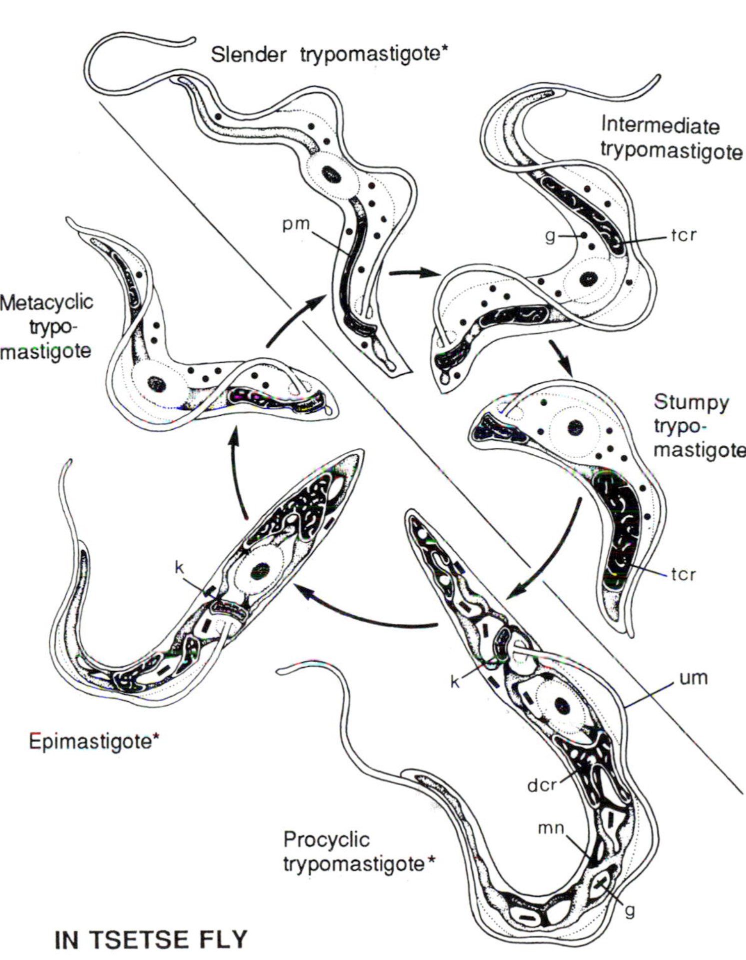

Trypomastigote
Life cycle of the kinetoplastid *Trypanosoma brucei*. Cell division occurs in stages marked with asterisks. dcr=discoid cristae; g=glycosomes; k=prenuclear kinetoplast; mn=mitochondrion network; pm=promitochondrion; tcr=tubular cristae; um=undulating membrane.

Trypomastigote

Stage in trypanosome development in which the kinetoplast lies behind the nucleus and the associated undulipodium emerges laterally to form an undulating membrane along the length of the body, usu. becoming free at its anterior end. See also illustration: Kinetoplastida.

Tubular cristae (sing. tubular crista)

Descriptive term for the morphology of mitochondrial membranes. Cristae that are finger-shaped, circular in transverse section, and round rather than flattened. Characteristic of ciliates, dinomastigotes, and other protoctists. See *Vermiform cristae, Vesicular cristae.* Illustrations: Cristae, Endoplasmic reticulum, Microbody, Sagenogen, Trypomastigote.

Tubular ingestion apparatus
Siphon. General descriptive term for an oral apparatus that has the form of a long tube.

Tubuli (sing. tubulus)
Small tubes (e.g., organic tubules that transverse the microgranular calcareous walls of certain foraminifera). See *Microtubule.*

Tufa
Porous, sedimentary rock composed of calcium carbonate formed by evaporation or by precipitation from spring water or seeps.

Tundra
Treeless area of arctic regions that has a permanently frozen subsoil (permafrost) and low-growing vegetation (e.g., lichens, mosses, and stunted shrubs). Illustration: Habitat.

Tychoplankton
Collective term for benthic organisms that become temporarily suspended in water column by turbulence or other disturbance.

U

Ultrasonication
See *Sonication.*

Ultrastructure
Fine structure. The appearance of the cell and/or cell organelles as seen in the transmission electron microscope. Illustrations: Diplomonadida, Somatonemes.

Umbilicus (pl. umbilici; adj. umbilical)
Morphological term generally meaning navel or button. Refers to a depressed region of a trochospirally coiled foraminifera surrounded by all the chambers of the last formed whorl.

Undefined medium (pl. undefined media)
Culture medium with one or more component, the exact chemical nature of which is unknown. See *Defined medium.*

Undulating membrane
Waving membrane; refers to several kinds of nonhomologous structures: (1) the paroral membrane, an organelle on the right side of the buccal cavity in ciliates with a compound ciliary apparatus; (2) in symbiotic mastigotes, an extension of the plasma membrane combined with the undulipodial membrane so that the axoneme of the undulipodium is attached to the body by a thin fold; or (3) a membranous fibrillar structure not underlain by undulipodia. Illustration: Trypomastigote.

Undulipodial apparatus
See *Kinetid.*

Undulipodial bracelet
Structure composed of intramembrane particles occurring at the junction between the undulipodium and the cell body; in green algae some consist of two or three closely associated rings of intramembrane particles; possibly homologous to ciliary necklace.

Undulipodial groove
Invagination of a cell from which undulipodia emerge. See *Gullet.*

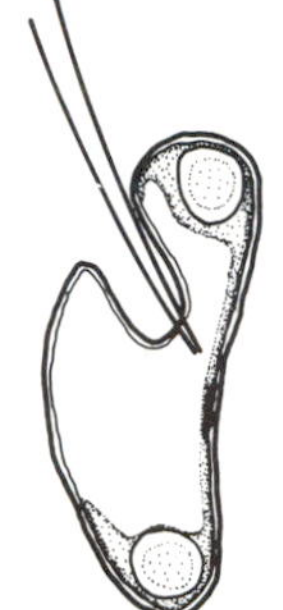

Undulipodial groove
The prasinophycean chlorophyte *Mesostigma viride.*

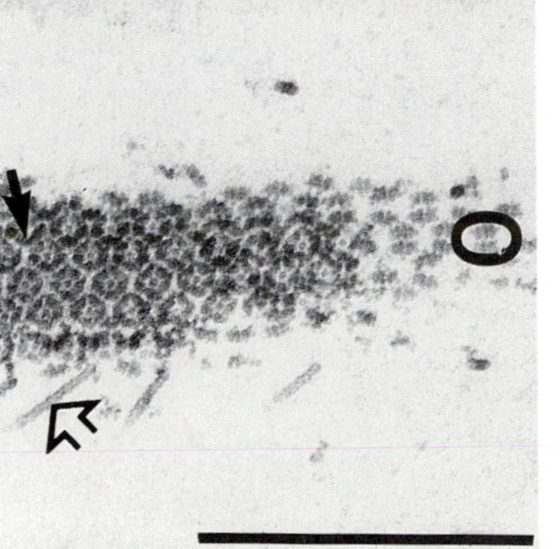

Undulipodial hairs
Tangential section through an undulipodium of the prasinophycean chlorophyte *Tetraselmis cordiformis* showing scales and hairs. Open arrow=undulipodial hair; small arrow=pentagonal scale; encircled area= rod-shaped "double scale." TEM. Bar=0.5 μm.

Undulipodial hairs
Filamentous appendages at right angles to the axoneme and arranged in one or more rows, associated with or coating the undulipodia of many phototrophic mastigotes (phytoflagellates). May be simple, nontubular structures or tubular hairs consisting of at least two distinct regions. See *Anisokont, Mastigoneme, Tinsel flagellum.* See also illustrations: Eustigmatophyta, Raphidophyta.

Undulipodial pocket
Invagination of the cell surface to form a pit or deep pocket from which the undulipodia emerge in euglenids and kinetoplastids. See *Undulipodial groove.* Illustrations: Bodonidae, Kinetoplastida, Zoomastigina.

Undulipodial pore
Opening through which undulipodium protrudes. Illustration: Thecal plate.

Undulipodial root
See *Undulipodial rootlet.*

Undulipodial rootlet
Undulipodial root. Portion of kinetid. Microtubular, fibrous, or amorphous structure originating at kinetosomes, extending proximally into the cell, and terminating somewhere in the cyto-

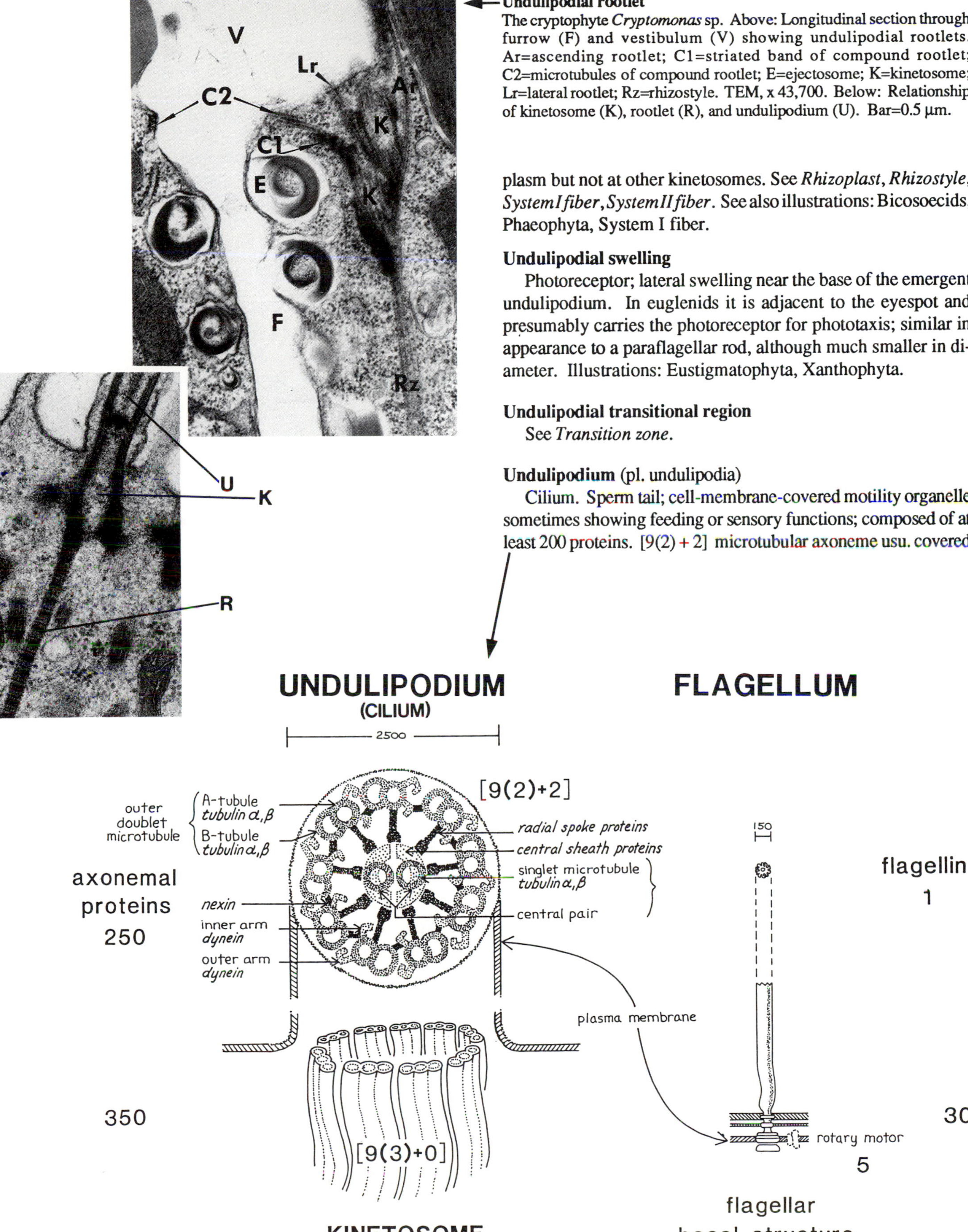

Undulipodial rootlet

The cryptophyte *Cryptomonas* sp. Above: Longitudinal section through furrow (F) and vestibulum (V) showing undulipodial rootlets. Ar=ascending rootlet; C1=striated band of compound rootlet; C2=microtubules of compound rootlet; E=ejectosome; K=kinetosome; Lr=lateral rootlet; Rz=rhizostyle. TEM, x 43,700. Below: Relationship of kinetosome (K), rootlet (R), and undulipodium (U). Bar=0.5 µm.

plasm but not at other kinetosomes. See *Rhizoplast*, *Rhizostyle*, *System I fiber*, *System II fiber*. See also illustrations: Bicosoecids, Phaeophyta, System I fiber.

Undulipodial swelling

Photoreceptor; lateral swelling near the base of the emergent undulipodium. In euglenids it is adjacent to the eyespot and presumably carries the photoreceptor for phototaxis; similar in appearance to a paraflagellar rod, although much smaller in diameter. Illustrations: Eustigmatophyta, Xanthophyta.

Undulipodial transitional region

See *Transition zone*.

Undulipodium (pl. undulipodia)

Cilium. Sperm tail; cell-membrane-covered motility organelle sometimes showing feeding or sensory functions; composed of at least 200 proteins. [9(2) + 2] microtubular axoneme usu. covered by plasma membrane; limited to eukaryotic cells. Includes cilia and eukaryotic "flagella." Each undulipodium invariably develops from its kinetosome. Contrasts in every way with the prokaryotic motility organelle or flagellum, a rigid structure composed of a single protein (which belongs to the class of proteins called flagellins). Undulipodia in the cell biological literature are often referred to by the outmoded term flagella or euflagella. See Introduction for discussion of these terms. Illustration previous page. See also illustrations: Mitotic apparatus, Organelle.

Unialgal culture
Culture containing only one species of algae; other protoctists, fungi, and/or bacteria may be present. See *Axenic, Monoxenic.*

Unikaryon (pl. unikarya; adj. unikaryotic)
Organism with a single nucleus.

Unilocular sporangium (pl. unilocular sporangia)
Sporangium in which all spores are produced in a single cavity. See *Plurilocular sporangium.*

Unimastigote
Cell with a single undulipodium.

Uniporate (adj.)
Referring to a structure with a single pore.

Uniserial (adj.)
Uniseriate. Referring to any of several structures arranged in or consisting of one series or row of structures; descriptive of cells characterized by such an arrangement (e.g., foraminiferan tests).

Uniseriate (adj.)
See *Uniserial.*

Uralga (pl. uralgae)
Hypothetical common ancestor of all algae (e.g., given the direct filiation (monosymbiotic) theory of the origin of the chloroplasts; the uralga is that common ancestral organism thought to combine the features of phototrophic bacteria (including cyanobacteria), algae, and plants).

Uroid
Descriptive morphological term for the tail-like protuberance at the posterior end of a moving lobose ameba. Structure is active in pinocytosis and possibly in defecation and water expulsion. Illustrations: Karyoblastea, Rhizopoda.

Uroid region
Region in the ameba opposite locomotory end (posterior); sometimes distinct from rest of body by constriction. See *Uroid.*

Utricle
General morphological term for "little bladder" (e.g., the swollen terminus of a filament of the green alga *Codium*).

V

Vacuole
A small space or cavity in the protoplasm of a cell containing fluid or air and surrounded by a membrane. Illustration: Organelle.

Vacuome
Morphological term referring to the complete system of vacuoles in a cell (analogous to genome or chondriome). Illustration: Dinomastigota.

Vagile (adj.; n. vagility)
Referring to behavior of a cell or organism; free to move about. See *Sessile.*

Valve
Opposite faces, or distal plates of a diatom frustule or dinomastigote theca, typically flattened or somewhat convex. Portion of myxozoan spore wall, formed by a specialized (valvogenic) cell during sporogenesis; two or more such valves adhere together along suture line, composing the spore wall (or shell). See *Epivalve, Hypovalve.* Illustrations: Myxozoa, Raphe.

Valve cell
Cell that forms part of the valve in myxozoans.

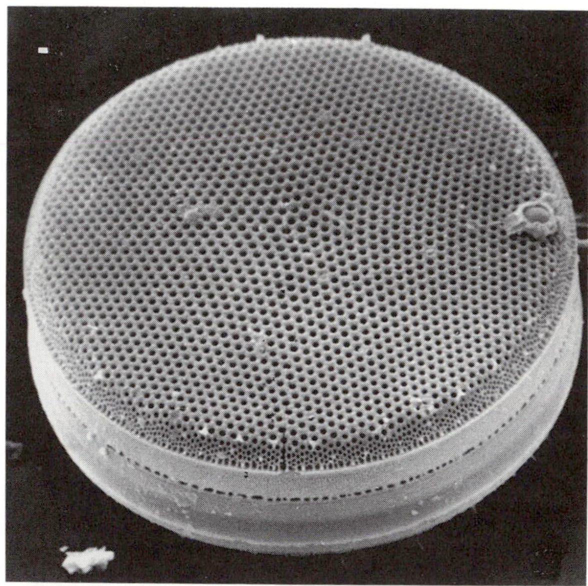

Valve face
The centric diatom *Coscinodiscus*. SEM. Bar=10 μm.

Valve face
Structure of diatoms; the surface of a valve.

Valve mantle
Structure of diatoms; marginal part of valve differentiated by slope, sometimes also by structure, from the valve face.

Valve view
Front view of a diatom valve.

Valvogenic cell
Cell of myxozoans in the sporoblast that gives rise to the valve of mature spores.

Variable antigen type
VAT. Antigen type (serotype) expressed by a trypanosome as a consequence of having a variant-specific glycoprotein on its surface. Illustration: Paraxial rod.

Variant surface glycoprotein
VSG. Surface macromolecules of trypanosomes: the glycoprotein present as a monomolecular layer on the surface of bloodstream trypanosomes and constituting the surface coat; its exposed epitope determines the variable antigen type of the organism.

VAT
See *Variable antigen type.*

Vector
Motile organism (e.g., insect, mammal) that transmits symbionts to other organisms. Parasitologists sometimes limit vector to mean an essential intermediate host in which a parasite undergoes a significant life cycle change.

Vegetative cell
Growing cell; trophont; trophic cell; trophozoite. The term vegetative, borrowed from growing plants, should be avoided. Illustration: Pit connections.

Vegetative resting state
See *Cyst, Spore.*

Vegetative state
Trophic state. See *Vegetative cell.*

Velum (pl. vela)
Veil; in diatoms, thin, perforated layer of silica over an areola, i.e., a type of pore plate; known as rica in many biraphid diatoms. In some species of Dasycladales (e.g., the chlorophyte *Acetabularia*), a protective covering over emergent lateral branches.

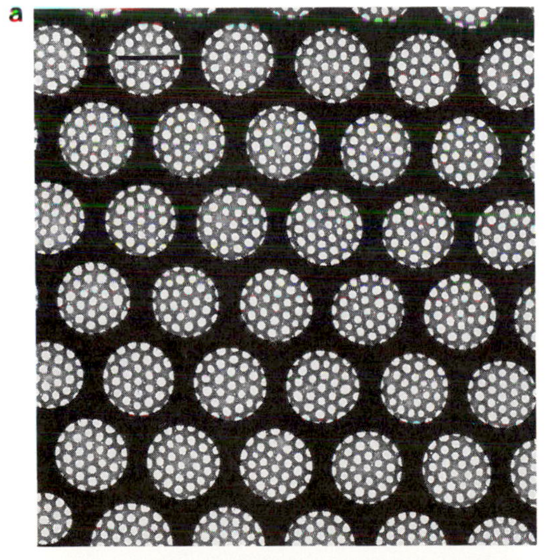

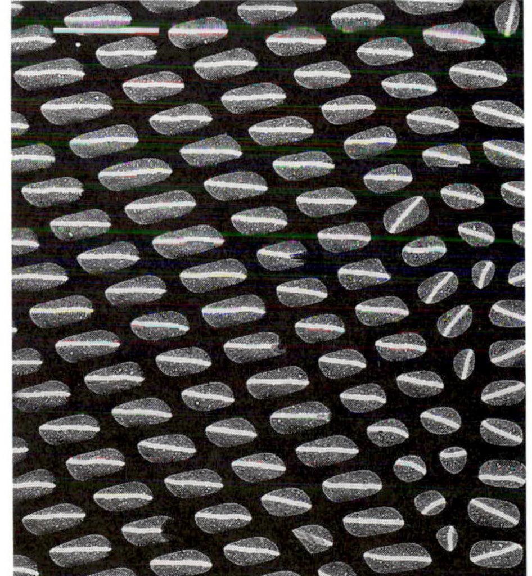

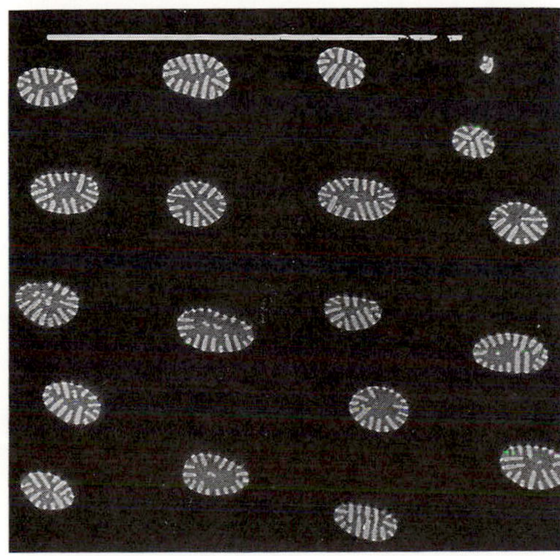

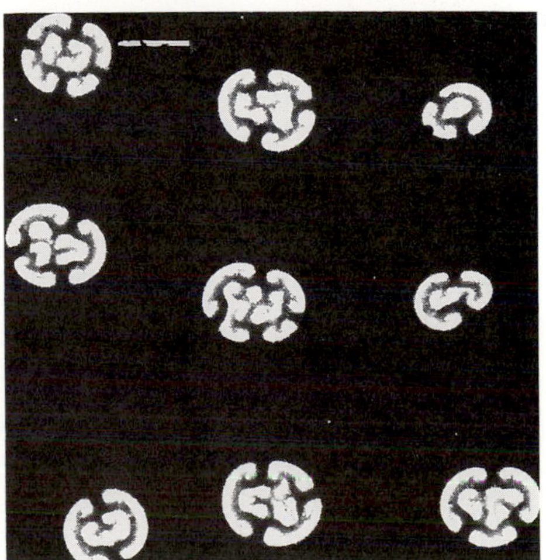

Velum
a. Each areola of *Roperia* has many small pores in a thin siliceous velum. b. The velum of *Rhizosolenia* has just one narrow slit. c. The areolae of *Cocconeis* are variable in size and shape and so is the pattern of slits in the velum. d. Vela of *Rhaphoneis* are interconnected projections from the side of the areolae. TEM. Bars=1 µm.

Ventral cortex
Portion of a ciliate that contains the oral region in cases where the oral region is not at the anterior pole. Illustration: Oral region.

Ventral disc
See *Adhesive disc.*

Ventral skid
Recurrent undulipodium; usu. nonmotile relative to the cell, which serves as a "runner" upon which a mastigote glides over the substrate.

Ventral sucker
See *Adhesive disc.*

Ventrostomial (adj.)
Morphological term referring to an area on the ciliate cortex; around the oral region, on the ventral side.

Vermiform cristae (sing. vermiform crista)
Descriptive term of intramitochondrial membranes; "worm-like" cristae; those occurring as a pancake-shaped flattened plate. See *Tubular cristae, Vesicular cristae.* Illustration: Cristae.

Vermifuge
Biologically active substance (e.g., antihelminthics) having the power or property of expelling worms or other parasites from the intestines of people or domestic animals (e.g., reported to occur in some ulvophytes).

Verrucae (sing. verruca; adj. verrucose)
General term referring to wartlike thickenings. Illustration: Cystosorus.

Vesicle
General structural term for a membranous sac. Usu. refers to a cell organelle in protoctists. Illustration: Microspora.

Vesicular cristae (sing. vesicular crista)
Descriptive term of intramitochondrial membranes; vesicles or sac-shaped cristae. See *Tubular cristae, Vermiform cristae.* Illustration: Cristae.

Vestibulum (pl. vestibula)
Morphological term referring to different oral structures in protoctists; subapical depression from which undulipodia emerge in cryptomonads; depression of the body at or near the apical end leading to the cytostome-cytopharyngeal complex and adorned with undulipodia in some ciliates; in other protoctists, intracellular compartment containing oral and lateral apertures for nutrient passage and waste disposal. Illustrations: Cryptophyta, Undulipodial rootlet.

Vitreous (adj.)
Glassy; referring to hyaline wall in which the crystals of calcite have their C-axes optically aligned normal to the surface of the shell (e.g., the Archaediscidae foraminifera).

VSG
See *Variant surface glycoprotein.*

W

Wall-forming bodies
Inclusions that give rise to the oocyst wall after fertilization in coccidian apicomplexans; wall-forming bodies I are more-or-less dense granules; wall-forming bodies II have a spongelike appearance. Illustration: Macrogamont.

Water molds
Common name for several unrelated groups of hyphae-forming organisms found in damp or aquatic environments. White rusts; downy mildews; symbiotrophic or osmotrophic funguslike protoctists most of which are members of the phylum Oomycota in the five-kingdom system.

Whiplash flagellum (pl. whiplash flagella)
See *Whiplash undulipodium.*

Whiplash undulipodium (pl. whiplash undulipodia)
Undulipodium lacking mastigonemes. See *Tinsel undulipodium.*

Whorl
General descriptive term for coiled form or radial structures emerging from a common axis (e.g., term is applied to a group of chambers which collectively make up a 360° turn of the test in coiled foraminifera and to the disposition of long cells around the nodal cells in charalean chlorophytes such as *Nitella*). Illustrations: Acanthopodium, Acetabularian life history.

Whorled vesicles
Intracellular membranous sacs disposed in a whorled conformation (e.g., arrangement characteristic of the contractile vacuole of paramecia). See *Whorl.*

Window
Opening (e.g., between the branches in the siliceous skeleton of ebridians).

Wrack
Tangled mass of fucalean seaweeds on the seashore.

X

Xanthophylls
Class of plastid pigments; oxygenated carotenoids.

Xanthosomes
Yellow bodies (e.g., reddish-brown or yellowish, rounded, and often aggregated bodies found between the stercomes in the stercomare of xenophyophores).

Xenogenous (adj.)
Of alien or foreign origin. Organism of a different species; heterospecific; heterogeneric.

Xenoma
Symbiotic aggregate formed by multiplying intracellular

symbiotrophs within their growing host cells, the whole structure increasing in size, as in the single-celled tumors formed by microsporans.

Xenophyae (sing. xenophya)
Foreign bodies of which the inorganic part of xenophyophoran tests is composed.

Xenosomes
Intracellular structures. Literally, "alien bodies," referring to micrometer-size bodies found in the cytoplasm and nuclei of protoctists of all kinds. Growth in the absence of the host provides the definitive proof that a structure is a xenosome. These may be foreign infective agents but are easily confused with natural components of the organism when their physiological and even genetic incorporation into the life of the host cell has occurred in the remote past. Endosymbiotic entities such as the (bacterial) kappa and omikron (and other Greek-letter) particles of *Paramecium, Euplotes*, as well as *Holospora*, etc., zooxanthellae, and cyanelles are xenosomes. Most of the Greek-letter particles (cytoplasmic genes of *Paramecium*) are now classified as Gram-negative bacteria in the genus *Caedibacter*. From an evolutionary point of view, the serial endosymbiosis theory claims that plastids and mitochondria began as xenosomes as well.

Xylan
Xylose polymer.

Z

Zerfall (German, meaning disintegration, breaking up)
Breakup of nuclear material prior to schizogony (e.g., agamont stage of the foraminiferan *Allogromia laticollaris*). Illustration: Foraminifera.

Zoid
See *Monad*.

Zoite
Endozoite. Trophic cell produced by multiple fission (e.g., the infective motile stage of apicomplexans whether of sexual or asexual origin). See *Merozoite, Sporozoite*. Illustration: Conoid.

Zonate (adj.)
Referring to structure that is zoned; marked with zones, bands, rings, or zones of color.

Zoobenthos
Ecological term referring to heterotrophic protoctists and animals that comprise the biota of the benthos.

Zoochlorella (pl. zoochlorellae)
Green photosynthetic symbionts found in protoctists and animals. Although many belong to the genus *Chlorella* (e.g., algae of *Coleps hirtus*, *Hydra viridis*, and *Paramecium bursaria*) others belong to the prasinophytes or other taxa; often symbionts are unidentified to genus.

Zoocyst
Undulipodiated propagule. Illustration: Protostelida.

Zooflagellate
Aplastidic mastigote; so-called "animal flagellates"; in two kingdom classifications, unicellular animals with undulipodia, members of the phylum Protozoa in the class Flagellata; in the five-kingdom classification, zoomastiginids, members of the phylum Zoomastigina, which includes the classes amebomastigotes, opalinids, pseudociliates, choanomastigotes, bicosoecids, kinetoplastids, diplomonads, pyrsonymphids, and parabasalians. Term should be avoided. See Introduction.

Zoonosis
Ecological term referring to specific protoctist (occasionally bacterial or viral) symbiotrophs that have animals including people as their hosts. Infection naturally transferable between animals other than humans.

Zoophagy
Mode of heterotrophic nutrition displayed by organisms that feed on animals.

Zoosporangium (pl. zoosporangia)
Sporangium that produces zoospores. Illustration: Heterothallism.

Zoospore
Swarmer. Mastigote propagule, undulipodiated motile reproductive cell capable of transformation into a different developmental stage but incapable of sexual fusion. Although spermlike in appearance, they are not sperm. See *Auxiliary zoospore, Macrozoospore, Microzoospore*. Illustrations: Heterothallism, Plasmodiophoromycota, Stachel.

Zoosporic fungi (sing. zoosporic fungus)
Protoctists, primarily osmotrophic, that are capable of forming hyphae and have undulipodiated stages in their life cycle. Outmoded term referring to members of the protoctist phyla Chytridiomycota, Hyphochytriomycota, Oomycota, and sometimes Plasmodiophoromycota.

Zoosporogenesis
Process by which zoospores are formed.

Zooxanthella (pl. zooxanthellae)
Yellowish or yellow-brown photosynthetic symbiont found in protoctists and animals. Although many belong to the dinomastigote group *Symbiodinium* (*Gymnodinium*), others belong to diatom or other taxa; often the symbionts are unidentified to genus. Illustration: Actinopoda.

Zygocyst
Encysted zygote (e.g., structure of opalinids usu. found in intestines or feces of anuran amphibians). Illustration: Opalinata.

Zygolith
Dome-shaped coccolith subtype (e.g., *Homozygosphaera*, *Periphyllophora*); holococcoliths with arched crossbow(s). Illustration: Coccolith.

Zygospore

Resistant structure formed by conjugation; thick-walled zygote of the conjugating green algae; large, multinucleate resting spore in zygomycote fungi. Illustration: Life cycle.

Zygote

Diploid (2N) nucleus or cell produced by the fusion of two haploid nuclei or cells. In animals, plants, and some protoctists (those undergoing gametic meiosis) the zygote is destined to develop into a new organism. In fungi and protoctists undergoing zygotic meiosis, the zygote stage is unstable and haploid nuclei or cells are formed as soon as the zygote resumes activity. Illustrations: Acetabularian life history, Coccidian life history, Foraminifera, Life cycle, Opalinata, Trophozoite.

Zygotene

Zygonema; stage in meiotic prophase I in which homologous chromosomes pair. See *Meiosis*.

Zygotic meiosis

Life cycle in which meiosis immediately follows zygote formation as in most fungi and some algae (e.g., conjugating green algae). See *Gametic meiosis*.

Zymodeme

Strain or variety; ecological term referring to a population of organisms members of which display similarities in patterns of isoenzymes as determined by electrophoresis, i.e., individuals having similar zymograms are said to belong to the same zymodeme.

Zymogram

Stained gel (starch or agarose) that shows isoenzyme banding patterns following electrophoresis of a cell lysate; method by which zymodemes are established.

Acantharia

1=contracted myoneme; 2=axopod; 3=ectoplasm; 4=food vacuole; 5=nucleus; 6=golgi; 7=mitochondrion; 8=photosynthetic symbiont; 9=endoplasmic reticulum; 10=cell membrane; 11=spicule; 12=capsular wall; 13=microbody; 14=endoplasm; 15=periplasmic cortex with elastic junctions; 16=central mass; 17=relaxed myoneme.

Organism Glossary

This comprehensive alphabetical list of organisms illustrates only one example of each higher taxon (phyla and some classes). Many others are depicted under specific General Glossary entries (e.g., apicomplexans under "conoid," ciliates under "oral region," *Giardia* under "karyomastigont system," and proteromonads under "somatonemes"). Table 7 presents this taxonomic information in as systematic an order as possible based on principles explained in the Introduction. Although representative genera have been retained, all species have been omitted. Table 6A summarizes the features common for members of each phylum. Abbreviations used in the figure legends are: DIC=Nomarski differential interference contrast light microscopy; PC=phase contrast light microscopy; SEM=scanning electron microscopy; TEM=transmission electron microscopy.

A

Abadehellidae
Family in phylum Granuloreticulosa.

Acanthamoebidae
Family in phylum Rhizopoda.

Acantharia
Class in phylum Actinopoda. Phagotrophic unicells in which strontium sulfate skeletons underlie the periplasmic cortex (outer layer of each cell). Each possesses numerous ribbonlike or cylindrical motile organelles called myonemes. May form mononucleate bimastigote reproductive cells. No sexuality known. Generally marine planktonic.

Acanthoceraceae
Family in phylum Bacillariophyta.

Acanthochiasmidae
Family in phylum Actinopoda.

Acanthocystidae
Family in phylum Actinopoda.

Acanthoecidae
Family in phylum Zoomastigina.

Acanthometridae
Family in phylum Actinopoda.

Acanthoplegmidae
Family in phylum Actinopoda.

Acanthopodina
Suborder in phylum Rhizopoda.

Acervulinacea
Superfamily in phylum Granuloreticulosa.

Acervulinidae
Family in phylum Granuloreticulosa.

Achnanthaceae
Family in phylum Bacillariophyta.

Achnanthales
Order in phylum Bacillariophyta.

Achnanthidiaceae
Family in phylum Bacillariophyta.

Aconchulinida
Order in phylum Rhizopoda.

Acrasea
Acrasids; phylum of cellular (pseudoplasmodial) slime molds. Phagotrophic, ameboid organisms formed by aggregation of amebas to directly produce multicellular aerial, spore-bearing struc-

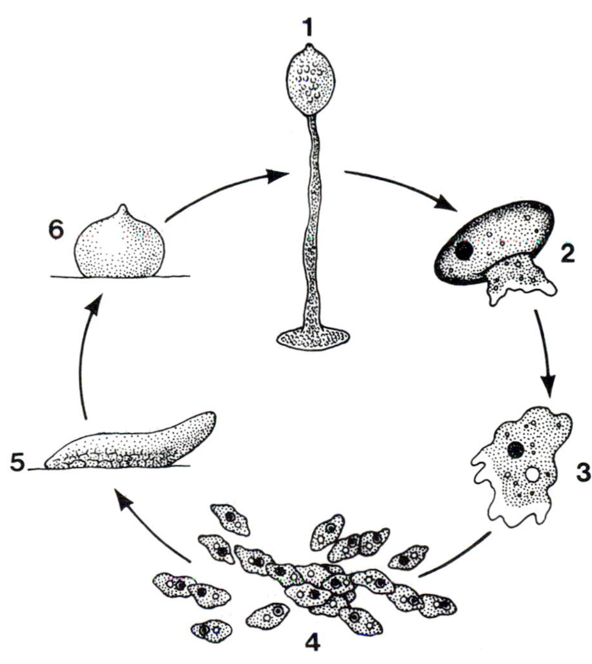

Acrasea
1=sorocarp; 2=excysting ameba; 3=trophic ameba; 4=aggregation; 5=migrating slug; 6=developing sorocarp.

tures (sorocarps). Damp soil habitats (e.g., dead plant parts, soil, or dung). Feed on bacteria. See also *Dictyostelida*, Table 6A.

Acrasida
Order in phylum Acrasea.

Acrasidae
Family in phylum Acrasea.

Acrochaetiales
Order in phylum Rhodophyta.

Acroseiraceae
Family in phylum Phaeophyta.

Acroseirales
Order in phylum Phaeophyta.

Acrosiphoniaceae
Family in phylum Chlorophyta.

Acrotrichaceae
Family in phylum Phaeophyta.

Actiniscaceae
Family in phylum Dinomastigota.

Actiniscales
Order in phylum Dinomastigota.

Actinocephalidae
Family in phylum Apicomplexa.

Actinomyxida
Order in phylum Myxozoa.

Actinophryida
Suborder in phylum Actinopoda.

Actinophryidae
Family in phylum Actinopoda.

Actinopoda
Phylum of protists, primarily large marine, heterotrophic unicells having long processes called axopods which develop from axoplasts. See *Acantharia*, *Heliozoa*, *Phaeodaria*, *Polycystina*, *Radiolaria*.

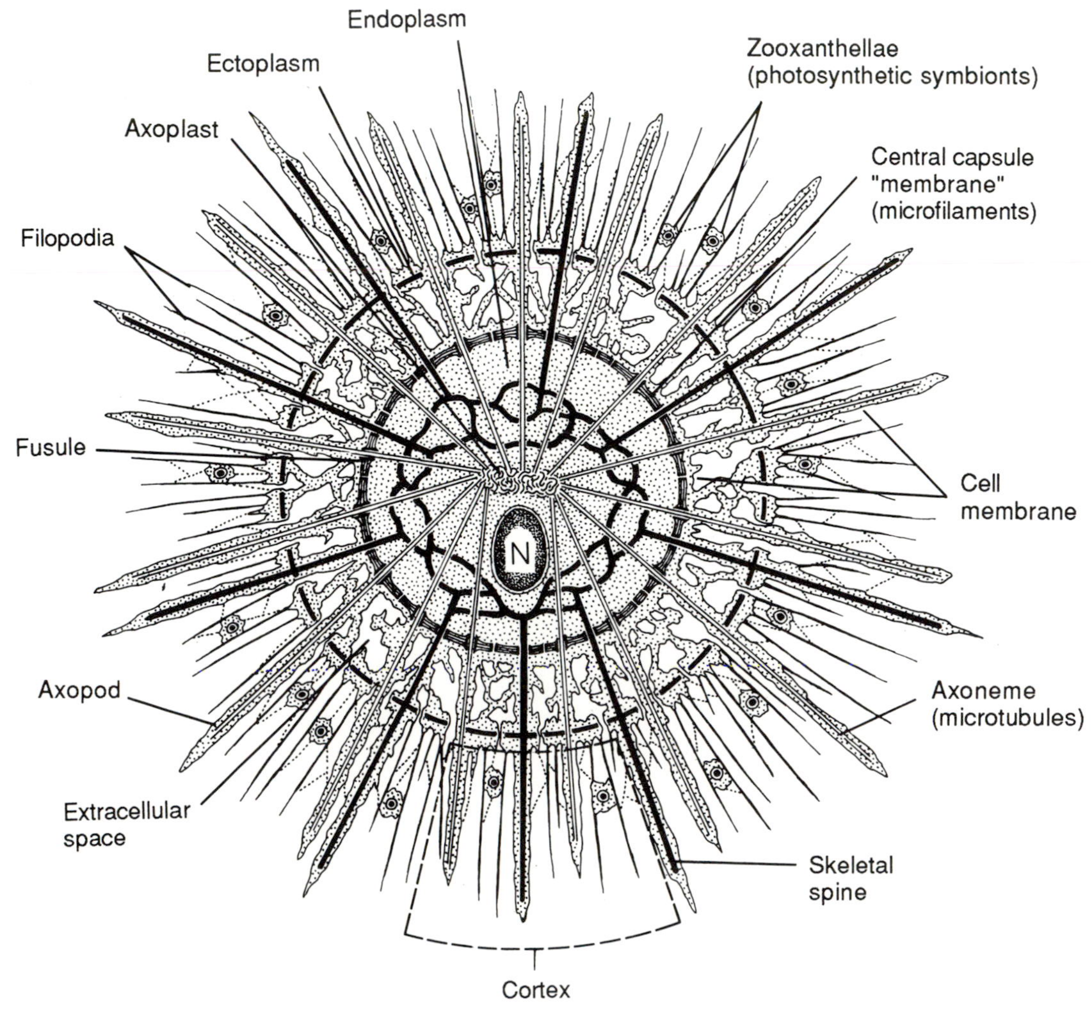

Actinopoda

Actinoptychaceae
Family in phylum Bacillariophyta.

Actinosporea
Class in phylum Myxozoa.

Acytosteliaceae
Family in phylum Dictyostelida.

Adeleida
Order in phylum Apicomplexa.

Adeleidae
Family in phylum Apicomplexa.

Adinomonadaceae
Family in phylum Dinomastigota.

Aggregatidae
Family in phylum Apicomplexa.

Alabaminidae
Family in phylum Granuloreticulosa.

Alariaceae
Family in phylum Phaeophyta.

Alatosporidae
Family in phylum Myxozoa.

Albuginaceae
Family in phylum Oomycota.

Alfredinidae
Family in phylum Granuloreticulosa.

Allogromida
Order in phylum Granuloreticulosa.

Allogromiidae
Family in phylum Granuloreticulosa.

Almaenidae
Family in phylum Granuloreticulosa.

Amblyosporidae
Family in phylum Microspora.

Amebomastigota
Class in phylum Zoomastigina consisting of heterotrophic, unicellular mastigotes that, during their life history, reversibly transform to monopodial, uninucleate or multinucleate amebas.

Ammodiscacea
Superfamily in phylum Granuloreticulosa.

Ammodiscidae
Family in phylum Granuloreticulosa.

Ammosphaeroidinidae
Family in phylum Granuloreticulosa.

Amoebida
Order in phylum Rhizopoda.

Amoebidae
Family in phylum Rhizopoda.

Amoebophryaceae
Family in phylum Dinomastigota.

Amphilithidae
Family in phylum Actinopoda.

Amphipleuraceae
Family in phylum Bacillariophyta.

Amphiroeae
Tribe in phylum Rhodophyta.

Amphiroideae
Subfamily in phylum Rhodophyta.

Amphisoleniaceae
Family in phylum Dinomastigota.

Amphisteginidae
Family in phylum Granuloreticulosa.

Amphoraceae
Family in phylum Bacillariophyta.

Amphorales
Order in phylum Bacillariophyta.

Anadyomenaceae
Family in phylum Chlorophyta.

Anaulaceae
Family in phylum Bacillariophyta.

Anaulales
Order in phylum Bacillariophyta.

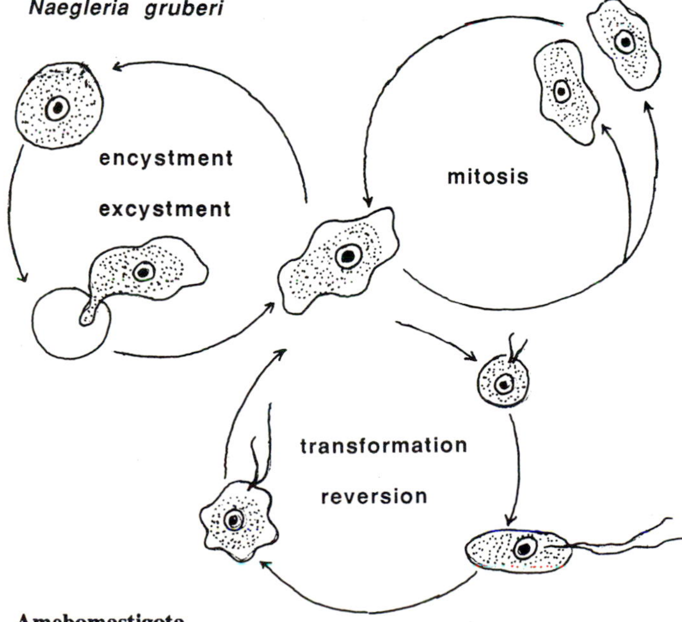

Amebomastigota
Life cycle of *Naegleria gruberi*.

Ancistrocomina
Suborder in phylum Ciliophora.

Angeiocystidae
Family in phylum Apicomplexa.

Anisolpidiaceae
Family in phylum Hyphochytriomycota.

Annulopatellinidae
Family in phylum Granuloreticulosa.

Anomoeoneidaceae
Family in phylum Bacillariophyta.

Apansporoblastina
Suborder in phylum Microspora.

Aphanochaetaceae
Family in phylum Chlorophyta.

Apicomplexa
Plasmodium. 1=apical complex; 2=nucleus; 3=mitochondrion; 4=nucleolus; 5=golgi body; 6=rhoptry; 7=conoid.

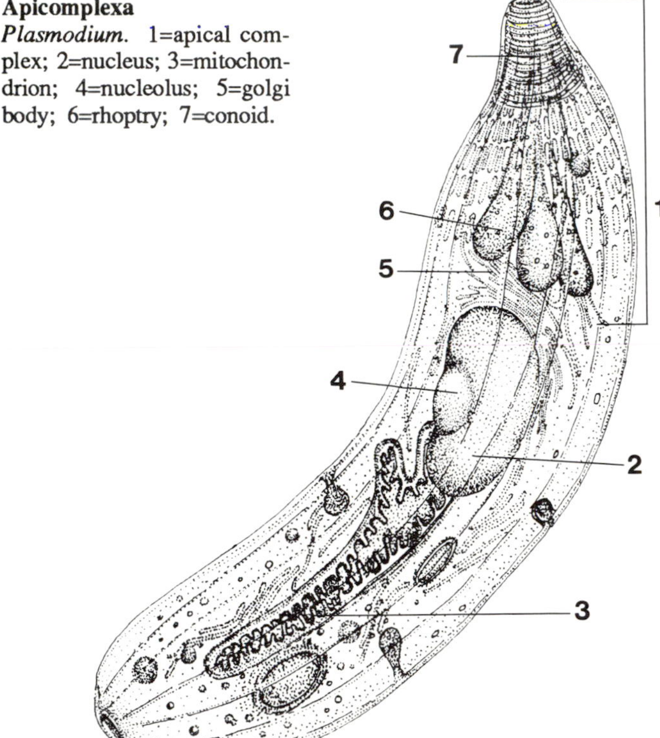

Apicomplexa
Phylum of protists parasitic on animals defined by a life history including a motile infective form (zoite) which possesses an apical complex. Life history generally has three phases: growth phase (by merogony or endogeny) during which the host is infected by the zoite, a sexual phase with gamete production and fertilization to form zygotes enclosed in oocysts, and a sporogenesis phase during which the sporoplasm within the oocysts divides successively to form sporozoites, the new infective forms. May be monoxenous or heteroxenous.

Apoaxoplastidiata
Superfamily in phylum Actinopoda.

Apodachlyellaceae
Family in phylum Oomycota.

Apodiniaceae
Family in phylum Dinomastigota.

Apostomatia
Subclass in phylum Ciliophora.

Apostomatida
Order in phylum Ciliophora.

Arachnoidiscaceae
Family in phylum Bacillariophyta.

Arachnoidiscales
Order in phylum Bacillariophyta.

Arcellidae
Family in phylum Rhizopoda.

Arcellinidae
Order in phylum Rhizopoda.

Archaediscacea
Superfamily in phylum Granuloreticulosa.

Archaediscidae
Family in phylum Granuloreticulosa.

Archigregarinida
Order in phylum Apicomplexa.

Archistomatina
Suborder in phylum Ciliophora.

Ardissoniaceae
Family in phylum Bacillariophyta.

Ardissoniales
Order in phylum Bacillariophyta.

Armophorida
Order in phylum Ciliophora.

Arnoldiellaceae
Family in phylum Chlorophyta.

Arthracanthida
Order in phylum Actinopoda.

Arthrocladiaceae
Family in phylum Phaeophyta.

Aschemonellidae
Family in phylum Granuloreticulosa.

Asterigerinacea
Superfamily in phylum Granuloreticulosa.

Asterigerinatidae
Family in phylum Granuloreticulosa.

Asterigerinidae
Family in phylum Granuloreticulosa.

Asterolampraceae
Family in phylum Bacillariophyta.

Asterolamprales
Order in phylum Bacillariophyta.

Astomatia
Subclass in phylum Ciliophora.

Astomatida
Order in phylum Ciliophora.

Astomatophorida
Order in phylum Ciliophora.

Astracanthidae
Family in phylum Actinopoda.

Astrephomenaceae
Family in phylum Chlorophyta.

Astrolithidae
Family in phylum Actinopoda.

Astrorhizacea
Superfamily in phylum Granuloreticulosa.

Astrorhizidae
Family in phylum Granuloreticulosa.

Asymmetrinidae
Family in phylum Granuloreticulosa.

Ataxophragmiacea
Superfamily in phylum Granuloreticulosa.

Ataxophragmiidae
Family in phylum Granuloreticulosa.

Athalamea
Class in phylum Granuloreticulosa.

Atlanticellidae
Family in phylum Actinopoda.

Attheyaceae
Family in phylum Bacillariophyta.

Auerbachiidae
Family in phylum Myxozoa.

Aulacanthidae
Family in phylum Actinopoda.

Aulacodiscaceae
Family in phylum Bacillariophyta.

Aulacosiraceae
Family in phylum Bacillariophyta.

Aulacosirales
Order in phylum Bacillariophyta.

Aulosphaeridae
Family in phylum Actinopoda.

Auriculaceae
Family in phylum Bacillariophyta.

Aurosphaeraceae
Family in phylum Chrysophyta.

Aveolinidae
Family in phylum Granuloreticulosa.

Axoplasthelida
Suborder in phylum Actinopoda.

B

Bacillariaceae
Family in phylum Bacillariophyta.

Bacillariales
Order in phylum Bacillariophyta.

Bacillariophyceae
Class in phylum Bacillariophyta.

Bacillariophycidae
Subclass in phylum Bacillariophyta.

Bacillariophyta
Diatoms. Phylum of diploid, sexual freshwater and marine algae. Cells enclosed by complex siliceous walls consisting of two valves. Centric, radially symmetric or pennate, bilaterally symmetric forms. Unicellular or colonial, diatoms reproduce by mitotic division with periodic formation of haploid, valveless gametes. Male gametes posteriorly undulipodiated in some centric genera. Cells nonmotile or motile by gliding accompanied by secretion through slits in cell walls and adhesion of the secreted material. Plastids with chlorophylls *a* and *c*, fucoxanthin, and other minor pigments. Extensive fossil forms, lower Cretaceous to Holocene.

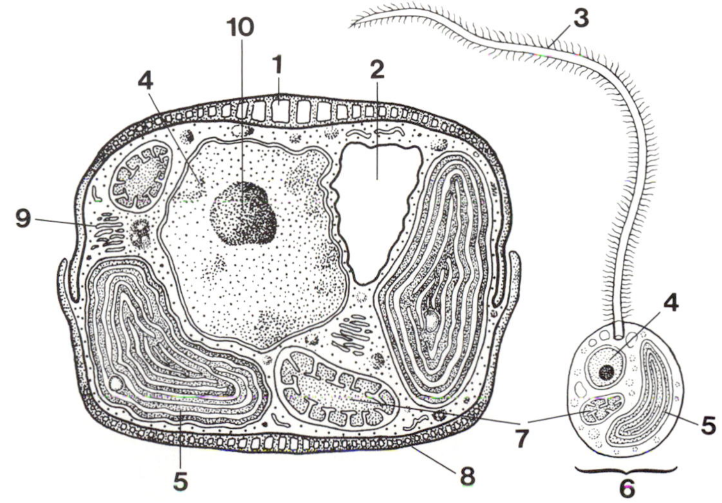

Bacillariophyta
1=hypovalve; 2=vacuole; 3=mastigoneme; 4=nucleus; 5=plastid; 6=gamete; 7=mitochondrion; 8=epivalve; 9=golgi body; 10=nucleolus.

Baculellidae
Family in phylum Granuloreticulosa.

Bagginidae
Family in phylum Granuloreticulosa.

Bangiaceae
Family in phylum Rhodophyta.

Bangiales
Order in phylum Rhodophyta.

Bangiophycidae
Subclass in phylum Rhodophyta.

Barkerinidae
Family in phylum Granuloreticulosa.

Bathysiphonidae
Family in phylum Granuloreticulosa.

Batrachospermales
Order in phylum Rhodophyta.

Bellerocheaceae
Family in phylum Bacillariophyta.

Berkeleyaceae
Family in phylum Bacillariophyta.

Bicosoecids (Bicoecids)
Class in phylum Zoomastigina. Free-living, planktonic, bimastigote, heterokont, heterotrophic, solitary or colonial cells, most contained in a vaselike shell or lorica composed of organic material.

Biddulphiaceae
Family in phylum Bacillariophyta.

Biddulphiales
Order in phylum Bacillariophyta.

Biddulphiophycidae
Subclass in phylum Bacillariophyta.

Biokovinidae
Family in phylum Granuloreticulosa.

Biomyxida
Order in phylum Granuloreticulosa.

Biseriamminidae
Family in phylum Granuloreticulosa.

Bivalvulida
Order in phylum Myxozoa.

Blastocladiaceae
Family in phylum Chytridiomycota.

Blastocladiales
Order in phylum Chytridiomycota.

Blastodiniaceae
Family in phylum Dinomastigota.

Blastodiniales
Order in phylum Dinomastigota.

Blastogregarinida
Order in phylum Apicomplexa.

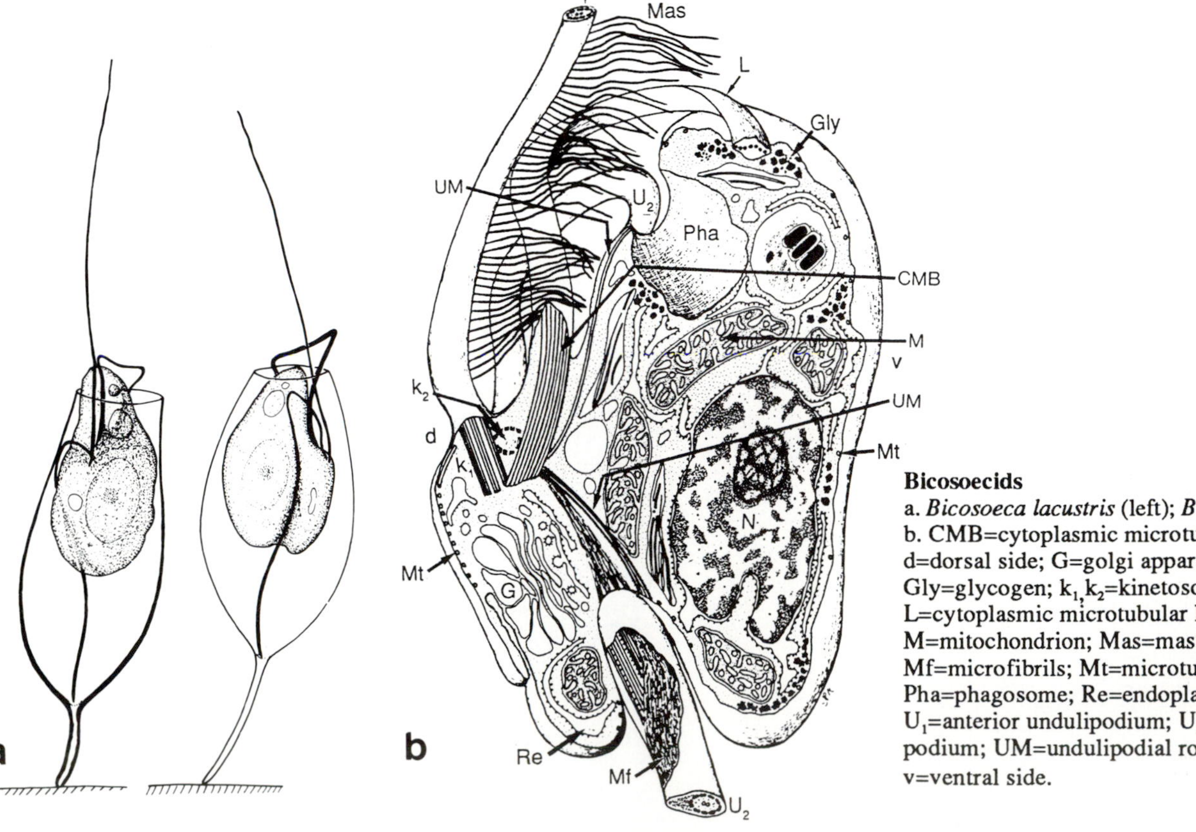

Bicosoecids
a. *Bicosoeca lacustris* (left); *B. kepneri* (right).
b. CMB=cytoplasmic microtubular band; d=dorsal side; G=golgi apparatus; Gly=glycogen; k_1,k_2=kinetosomes; L=cytoplasmic microtubular lip; M=mitochondrion; Mas=mastigonemes; Mf=microfibrils; Mt=microtubules; N=nucleus; Pha=phagosome; Re=endoplasmic reticulum; U_1=anterior undulipodium; U_2=recurrent undulipodium; UM=undulipodial root microtubule; v=ventral side.

Blepharocorythina
Suborder in phylum Ciliophora.

Bodonidae
Family in phylum Zoomastigina.

Bodonina
Suborder in phylum Zoomastigina.

Boldiaceae
Family in phylum Rhodophyta.

Bolivinellidae
Family in phylum Granuloreticulosa.

Bolivinidae
Family in phylum Granuloreticulosa.

Bolivinitidae
Family in phylum Granuloreticulosa.

Bolivinoididae
Family in phylum Granuloreticulosa.

Bonnemaisoniales
Order in phylum Rhodophyta.

Botrydiaceae
Family in phylum Xanthophyta.

Botrydiopsidaceae
Family in phylum Xanthophyta.

Botryochloridaceae
Family in phylum Xanthophyta.

Botryococcaceae
Family in phylum Chlorophyta.

Botryoidea
Family in phylum Actinopoda.

Brachydiniaceae
Family in phylum Dinomastigota.

Brachysiraceae
Family in phylum Bacillariophyta.

Bradyinidae
Family in phylum Granuloreticulosa.

Bronnimanniidae
Family in phylum Granuloreticulosa.

Bryophryida
Order in phylum Ciliophora.

Bryopsidaceae
Family in phylum Chlorophyta.

Bueningiidae
Family in phylum Granuloreticulosa.

Buffhamiaceae
Family in phylum Phaeophyta.

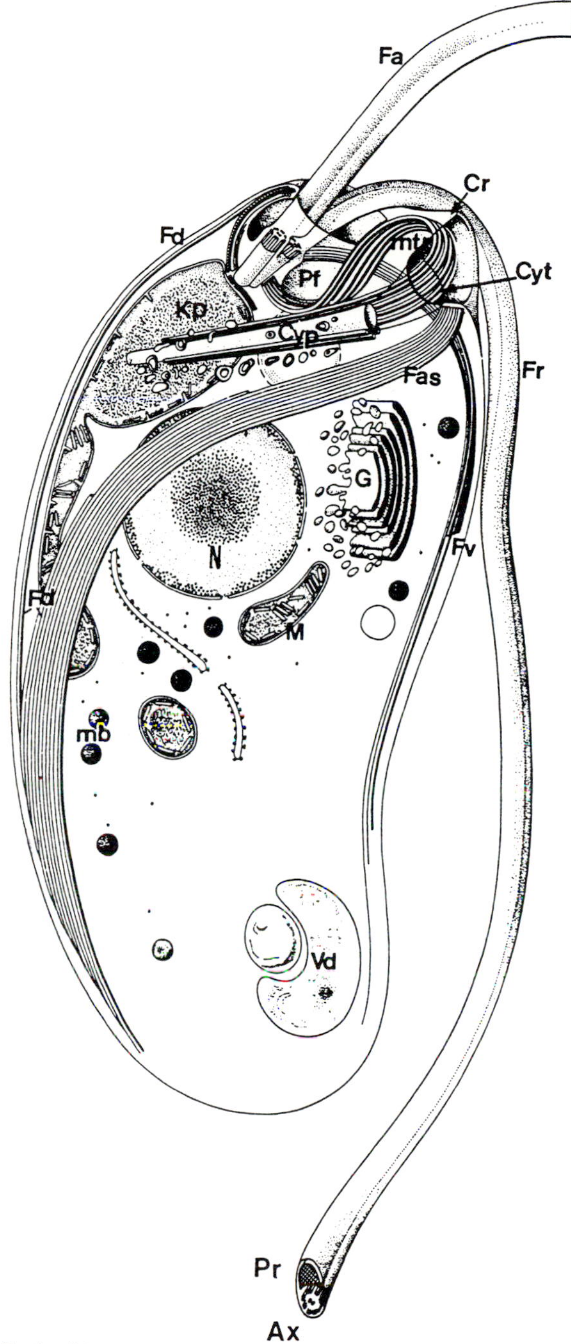

Bodonidae
Trophozoite of *Bodo caudatus*. Ax=axoneme; Cr=preoral crest; Cyp=cytopharynx; Cyt=cytostome; Fa=anterior undulipodium; Fas=band of microtubules; Fd=dorsal fiber; Fr=posterior (recurrent) undulipodium; Fv=ventral fiber; G=golgi apparatus; Kp=kinetoplast; M=mitochondrion; mb=microbody (glycosome); mt=band of microtubules; N=nucleus; Pf=undulipodial pocket; Pr=paraxial rod; Vd=digestive vacuole.

Buliminacea
Superfamily in phylum Granuloreticulosa.

Buliminellidae
Family in phylum Granuloreticulosa.

Buliminidae
Family in phylum Granuloreticulosa.

Buliminoidiae
Family in phylum Granuloreticulosa.

Burenellidae
Family in phylum Microspora.

Burkeidae
Family in phylum Microspora.

Bursariomorphida
Order in phylum Ciliophora.

Buxtehudeidae
Family in phylum Microspora.

C

Cachonellaceae
Family in phylum Dinomastigota.

Calcarinidae
Family in phylum Granuloreticulosa.

Caligellidae
Family in phylum Granuloreticulosa.

Calonymphidae
Family in phylum Zoomastigina.

Candeinidae
Family in phylum Granuloreticulosa.

Cannosphaeridae
Family in phylum Actinopoda.

Carteriaceae
Family in phylum Chlorophyta.

Carterinida
Order in phylum Granuloreticulosa.

Carterinidae
Family in phylum Granuloreticulosa.

Caryosporidae
Family in phylum Apicomplexa.

Caryotrophidae
Family in phylum Apicomplexa.

Cassidulinacea
Superfamily in phylum Granuloreticulosa.

Cassidulinidae
Family in phylum Granuloreticulosa.

Cassigerinellidae
Family in phylum Granuloreticulosa.

Castanellidae
Family in phylum Actinopoda.

Catapsydracidae
Family in phylum Granuloreticulosa.

Catenariaceae
Family in phylum Chytridiomycota.

Caucasinidae
Family in phylum Granuloreticulosa.

Caudosporidae
Family in phylum Microspora.

Caulerpaceae
Family in phylum Chlorophyta.

Caulerpales
Order in phylum Chlorophyta.

Caulochytriaceae
Family in phylum Chytridiomycota.

Cavosteliidae
Family in phylum Plasmodial Slime Molds.

Centritractaceae
Family in phylum Xanthophyta.

Centroaxoplastidiata
Superfamily in phylum Actinopoda.

Centrocollidae
Family in phylum Actinopoda.

Centroplasthelida
Suborder in phylum Actinopoda.

Centropyxidae
Family in phylum Rhizopoda.

Cephaloidophoridae
Family in phylum Apicomplexa.

Cephalolobidae
Family in phylum Apicomplexa.

Ceramiales
Order in phylum Rhodophyta.

Ceratiaceae
Family in phylum Dinomastigota.

Ceratiomyxaceae*
Family in phylum Plasmodial Slime Molds.

Ceratiomyxales*
Order in phylum Plasmodial Slime Molds.

Ceratiomyxidae*
Family in phylum Plasmodial Slime Molds.

Ceratiomyxomycetidae*
Subclass in phylum Plasmodial Slime Molds.

Ceratobuliminidae
Family in phylum Granuloreticulosa.

Ceratocoryaceae
Family in phylum Dinomastigota.

Ceratomyxidae
Family in phylum Myxozoa.

---

* See footnote p. 263.

Cerelasmidae
Family in phylum Xenophyophora.

Chaetocerophycidae
Subclass in phylum Bacillariophyta.

Chaetocerotaceae
Family in phylum Bacillariophyta.

Chaetocerotales
Order in phylum Bacillariophyta.

Chaetochloridaceae
Family in phylum Chlorophyta.

Chaetophoraceae
Family in phylum Chlorophyta.

Chaetophorales
Order in phylum Chlorophyta.

Chaetosiphonaceae
Family in phylum Chlorophyta.

Challengeriidae
Family in phylum Actinopoda.

Characeae
Family in phylum Chlorophyta.

Characidiopsidaceae
Family in phylum Xanthophyta.

Characiochloridaceae
Family in phylum Chlorophyta.

Characiopsidaceae
Family in phylum Xanthophyta.

Characiosiphonaceae
Family in phylum Chlorophyta.

Charophyceae
Class of green algae (phylum Chlorophyta) containing orders Chlorokybales, Klebsormidiales, Coleochaetales, and Charales. Members of first three orders produce undulipodiated swarmer cells with an intracellular multilayered structure associated with kinetids and typically are covered with small, square scales on the cell body. Since *Coleochaete* (and several other species) has a phragmoplast and forms bimastigote zoospores that resemble plant spermatozoids, some members of these groups are thought to resemble the ancestors of land plants. Members of the order Charales are large, submerged, phragmoplast-forming freshwater algae with multicellular sex organs. Egg cell enclosed within sterile (nondividing) tissue. Consist of large thalli with an erect main axis with regularly placed whorls of lateral branches with limited growth.

Chaunacanthida
Order in phylum Actinopoda.

Chiloguembelinidae
Family in phylum Granuloreticulosa.

Chilostomellacea
Superfamily in phylum Granuloreticulosa.

Chilostomellidae
Family in phylum Granuloreticulosa.

Chlamydodontina
Suborder in phylum Ciliophora.

Chlamydomonadaceae
Family in phylum Chlorophyta.

Chlamydomonadales
Order in phylum Chlorophyta.

Chloramoebaceae
Family in phylum Xanthophyta.

Chloramoebales
Order in phylum Xanthophyta.

Chlorarachnida
Phylum containing a phototrophic, marine organism in which an ameboid plasmodium contains individual green cells linked by a network of reticulopodia. Monospecific (one species: *Chlorarachnion reptans*). Cells contain plastids with chlorophylls *a* and *b*. Life history incompletely known but contains a spherical, walled stage and unimastigote zoospores. Extensive periplastidial compartment indicates origin from symbiosis between amebas and green algae.

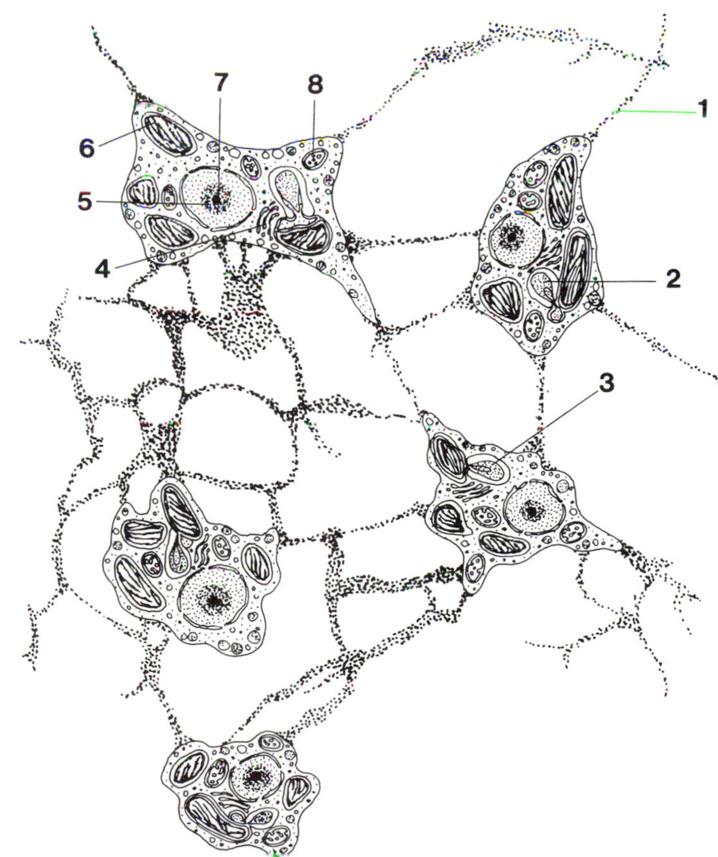

Chlorarachnida
1=reticulopodium; 2=pyrenoid; 3=periplastidial compartment; 4=golgi body; 5=nucleus; 6=chloroplast; 7=nucleolus; 8=mitochondrion.

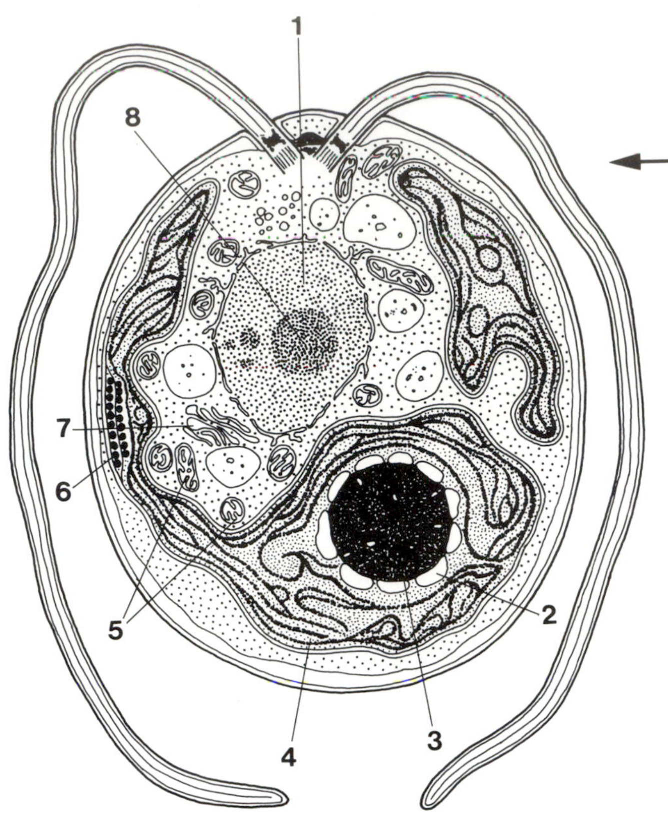

Chlorophyta
1=nucleus; 2=starch; 3=pyrenoid; 4=plastid; 5=mitochondria; 6=eyespot (lipid globules); 7=golgi body; 8=nucleolus.

Chlorarachniophyceae
Class in phylum Chlorarachnida.

Chlorobotryaceae
Family in phylum Eustigmatophyta.

Chlorochytriaceae
Family in phylum Chlorophyta.

Chlorococcaceae
Family in phylum Chlorophyta.

Chlorococcales
Order in phylum Chlorophyta.

Chlorodendraceae
Family in phylum Chlorophyta.

Chlorodendrales
Order in phylum Chlorophyta.

Chloromonads
See *Raphidophyta.*

Chloromyxidae
Family in phylum Myxozoa.

Chloropediaceae
Family in phylum Xanthophyta.

Chlorophyceae
Class of green algae (phylum Chlorophyta). Mainly freshwater; mastigotes covered by a cell wall (theca) or naked. Cell division characterized by phycoplast and a collapsing telophase spindle.

Chlorophyta
Green algae; phylum of cosmopolitan, unicellular or multicellular photosynthetic organisms that form mastigote stages as spores or gametes. Cells possess plastids surrounded by a double membrane. Thylakoid membranes contain chlorophylls *a* and *b*; primary storage material is starch. The colorless or amastigote immediate descendants of these algae included in phylum. See *Charophyceae, Chlorophyceae, Microthamniales, Pedinomonadales, Prasinophyceae, Prasiolales, Trentepohliales, Ulvophyceae.*

Choanomastigota
Choanoflagellates; class of marine heterotrophic mastigotes or sessile colonial organisms in phylum Zoomastigina. Cells enclosed by an organic (theca) or siliceous (lorica) structure with collars of tentacles; also, in class Kinetoplastida, phylum Zoomastigina, term for a stage in development of trypanosomatid mastigotes in which the kinetoplast lies anterior to the nucleus and the associated undulipodium emerges at the anterior extremity by way of an expanded undulipodial pocket.

Chonotrichia
Subclass in phylum Ciliophora.

Chordaceae
Family in phylum Phaeophyta.

Chordariaceae
Family in phylum Phaeophyta.

Chordariales
Order in phylum Phaeophyta.

Choreotrichia
Subclass in phylum Ciliophora.

Choreotrichida
Order in phylum Ciliophora.

Choristocarpaceae
Family in phylum Phaeophyta.

Chromulinaceae
Family in phylum Chrysophyta.

Chrysalidinidae
Family in phylum Granuloreticulosa.

Chrysamoebaceae
Family in phylum Chrysophyta.

Chrysamoebales
Order in phylum Chrysophyta.

Chrysanthemodiscaceae
Family in phylum Bacillariophyta.

Chrysanthemodiscales
Order in phylum Bacillariophyta.

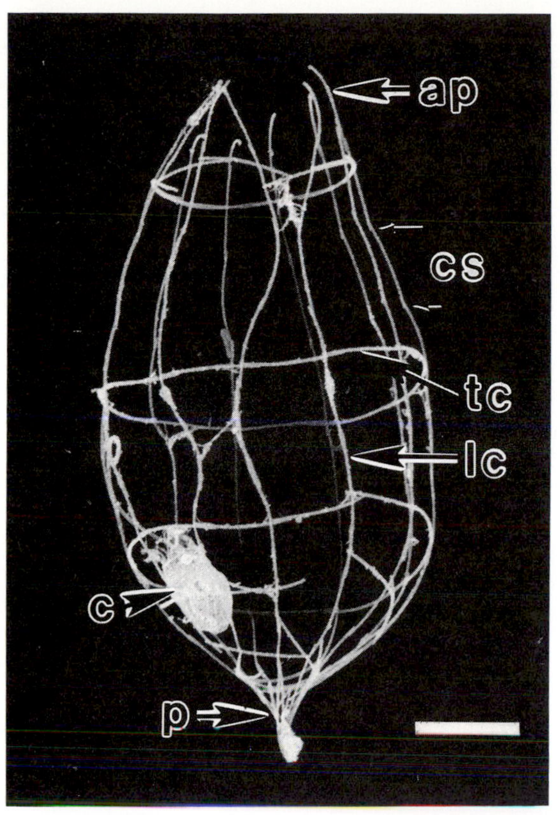

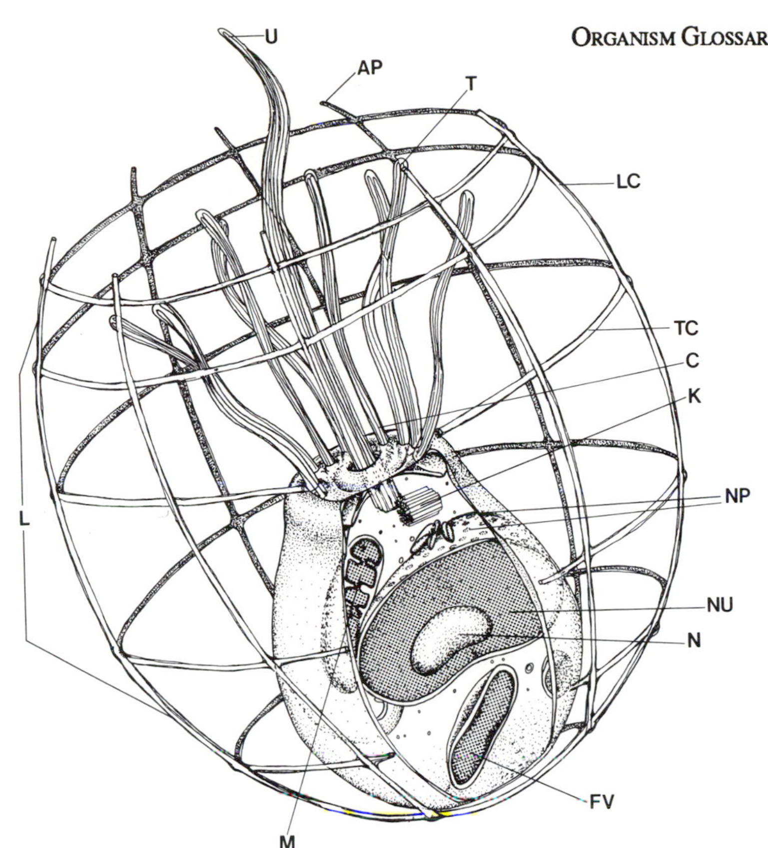

Choanomastigota
a. *Diaphanoeca multiannulata* from Antarctic sea ice. ap=anterior projection; c=cell with short collar tentacles subtending its anterior end; cs=single costal strip (delimited by small arrows); lc=longitudinal costa; p=pedicel; tc=transverse costa. Bar=5 μm. b. AP=anterior projections; C=collar; FV=food vacuole; K=kinetosome; L=lorica; LC=longitudinal costae; M=mitochondrion; N=nucleolus; NP=nuclear pores; NU=nucleus; T=tentacles; TC=transverse costa; U=undulipodium.

Chrysapiaceae
Family in phylum Chrysophyta.

Chrysocapsaceae
Family in phylum Chrysophyta.

Chrysocapsales
Order in phylum Chrysophyta.

Chrysochaetaceae
Family in phylum Chrysophyta.

Chrysococcaceae
Family in phylum Chrysophyta.

Chrysomeridaceae
Family in phylum Chrysophyta.

Chrysophyceae
Class in phylum Chrysophyta.

Chrysophyta
Golden yellow algae. Phylum of photosynthetic and related colorless organisms, single cells or colonial, primarily freshwater plankton. Includes the class Dictyochophyceae (silicoflagellates or silicomastigotes). Plastids contain chlorophylls *a* and *c*; chrysolaminarin as storage product. Form swarmers with

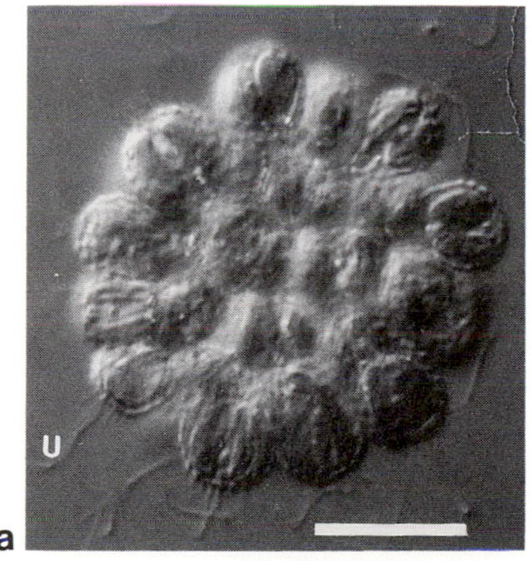

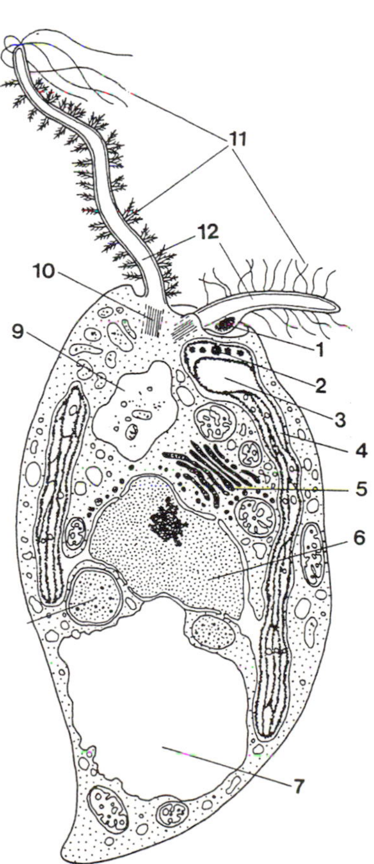

Chrysophyta
a. *Synura,* a motile, colonial chrysophyte. U=undulipodium. DIC. Bar=10 μm. b. 1=photoreceptor; 2=eyespot; 3=chrysoplast; 4=chrysoplast membranes; 5=golgi body; 6=nucleus; 7=chrysolaminarin storage vacuole; 8=lipid drop; 9=vacuole; 10=kinetosome; 11=mastigonemes; 12=undulipodia.

heterokont undulipodia. Fossil silicified cysts (stomatocysts) of class Chrysophyceae common, from Upper Cretaceous to Holocene.

Chrysosaccaceae
Family in phylum Chrysophyta.

Chrysosphaeraceae
Family in phylum Chrysophyta.

Chrysosphaerales
Order in phylum Chrysophyta.

Chytridiaceae
Family in phylum Chytridiomycota.

Chytridiales
Order in phylum Chytridiomycota.

Chytridiomycetes
Only class in phylum Chytridiomycota.

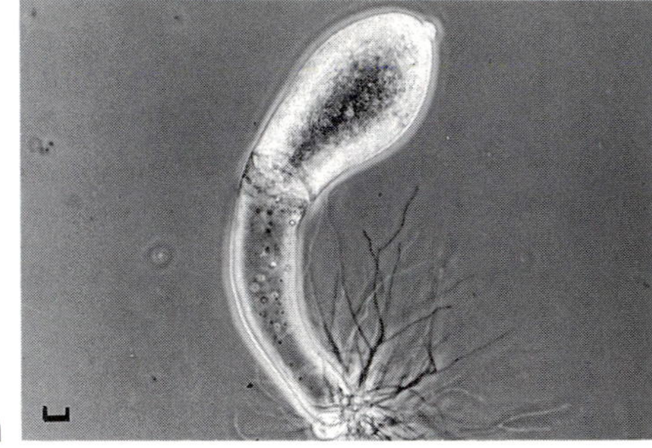

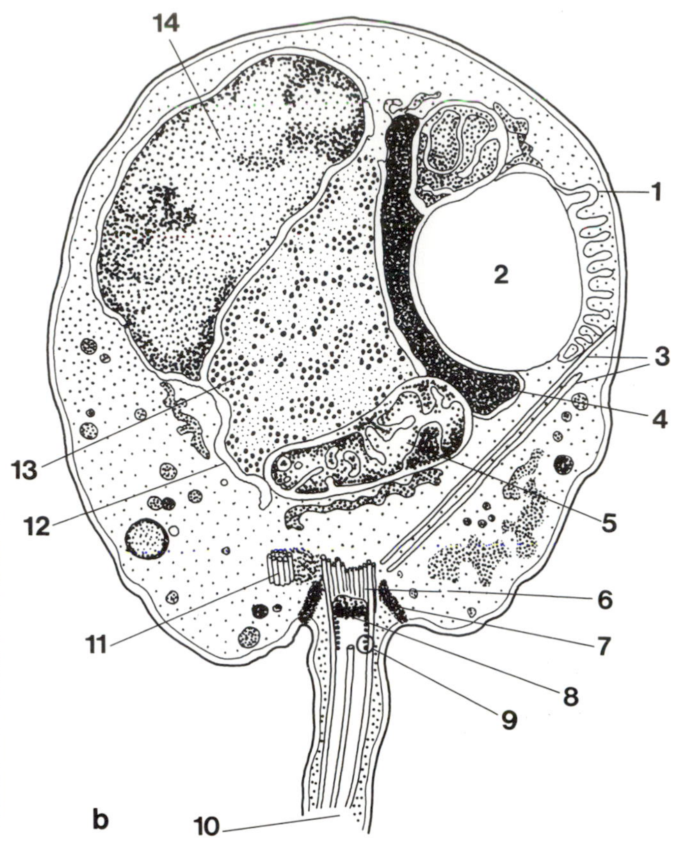

Chytridiomycota
a. Sporangium of *Blastocladiella emersonii* (Blastocladiales) with rhizoids at base. PC. Bar=1.0 μm. b. Posteriorly undulipodiated zoospore. 1=rumposome; 2=lipid globule; 3=microtubules; 4=microbody; 5=mitochondrion; 6=kinetosome; 7=kinetosome props; 8=terminal plate; 9=concentric fibers; 10=undulipodium; 11=barren kinetosome; 12=endoplasmic reticulum; 13=ribosomes; 14=nucleus.

Chytridiomycota
Phylum of chitinous-walled, heterotrophic, aquatic and soil protoctists which form undulipodiated zoospores and display absorptive nutrition. Filamentous or thalloid organisms that form sporangia that release undulipodiated propagules (zoospores), some of which may behave as gametes and fuse. Zoospores may transform into or fuse with the developing sporangium. Cells contain microbody-lipid globule complex (MLC). Some are necrotrophs in plants.

Chytridiopsidae
Family in phylum Microspora.

Chytriodiniaceae
Family in phylum Dinomastigota.

Chytriodiniales
Order in phylum Dinomastigota.

Cibicididae
Family in phylum Granuloreticulosa.

Ciliophora →
Phylum of dikaryotic, heterotrophic ciliates, primarily single motile cells with dimorphic nuclei (at least one macro- and one micronucleus, but often more) and complex cortices. The cortex, approximately 1 μm at the outer surface of the ciliate wall, is composed of precisely patterned tubules, fibers, and membranes. Files of kinetosomes, known as kineties, from which cilia extend, comprise a major portion of the cortex. Physiologically active macronucleus divides amitotically, whereas smaller, inactive, diploid mitotic micronucleus undergoes meiosis and reciprocal transfer in sexuality. Synkaryotes formed in conjugation and autogamy. Both types of nuclei lack centrioles and divide by closed karyokinesis (no nuclear membrane breakdown). Most are phagotrophic on bacteria or other protists. Cosmopolitan in aqueous habitats. Some are secondarily photosynthetic by acquisition of algae or plastids, others heterotrophic symbiotrophs which display dimorphic life cycles.

Ciliophryida
Suborder in phylum Actinopoda.

Ciliophryidae
Family in phylum Actinopoda.

Circoporidae
Family in phylum Actinopoda.

Cladochytriaceae
Family in phylum Chytridiomycota.

Cladophoraceae
Family in phylum Chlorophyta.

Cladopyxidaceae
Family in phylum Dinomastigota.

Cladostephaceae
Family in phylum Phaeophyta.

Clastodermataceae
Family in phylum Plasmodial Slime Molds.

Clathrulinidae
Family in phylum Actinopoda.

Clevelandellida
Order in phylum Ciliophora.

Climacospheniaceae
Family in phylum Bacillariophyta.

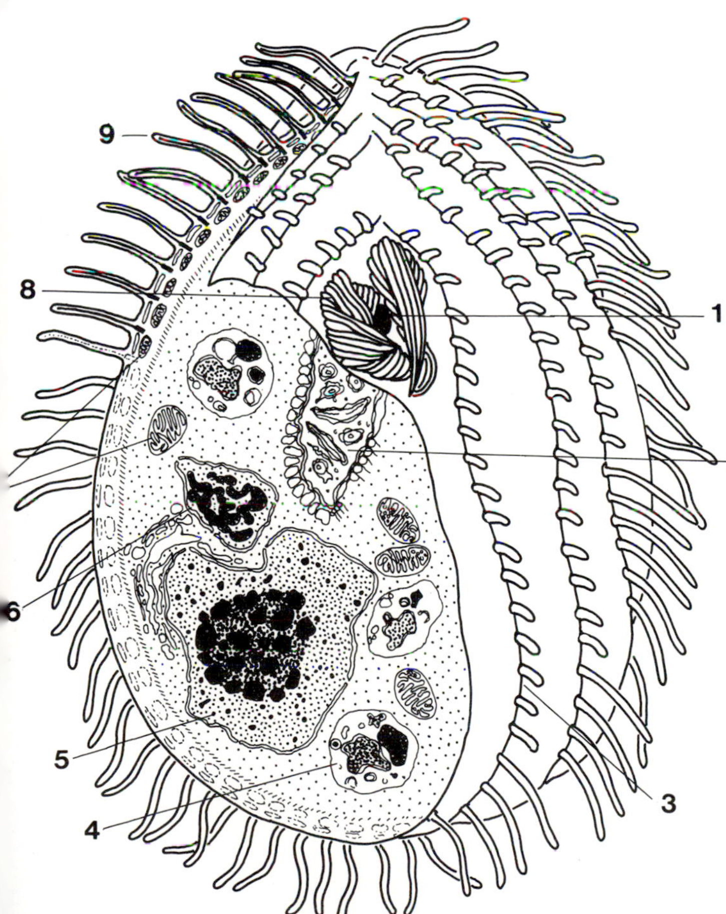

Ciliophora
1=cytostome (oral opening); 2=cytopharynx; 3=kinety; 4=food vacuole; 5=macronucleus with condensed chromatin; 6=micronucleus; 7=mitochondria; 8=oral cilia of buccal (oral) cavity; 9=ciliary axoneme.

Climacospheniales
Order in phylum Bacillariophyta.

Coccidia
Class in phylum Apicomplexa.

Coccidiniaceae
Family in phylum Dinomastigota.

Coccomyxaceae
Family in phylum Chlorophyta.

Cocconeidaceae
Family in phylum Bacillariophyta.

Coccosphaerales
Order in phylum Prymnesiophyta.

Codiaceae
Family in phylum Chlorophyta.

Codonosigidae
Family in phylum Zoomastigina.

Coelocladiaceae
Family in phylum Phaeophyta.

Coelodendriae
Family in phylum Actinopoda.

Coelomomycetaceae
Family in phylum Chytridiomycota.

Coelotrophiida
Order in phylum Apicomplexa.

Coelotrophiidae
Family in phylum Apicomplexa.

Coilodesmaceae
Family in phylum Phaeophyta.

Colaniellacea
Superfamily in phylum Granuloreticulosa.

Colaniellidae
Family in phylum Granuloreticulosa.

Coleitidae
Family in phylum Granuloreticulosa.

Coliphorina
Suborder in phylum Ciliophora.

Colpodea
Class in phylum Ciliophora.

Colpodida
Order in phylum Ciliophora.

Compsopogonaceae
Family in phylum Rhodophyta.

Compsopogonales
Order in phylum Rhodophyta.

Conaconidae
Family in phylum Actinopoda.

Concharidae
Family in phylum Actinopoda.

Conjugaphyta
Phylum of primarily freshwater, filamentous, zygonemalean (conjugacean) and desmid green algae distinguished from other chlorophytes by their isogamontous conjugating sexuality and their lack of undulipodia at all stages of development. Reproduction by mitotic division, sexuality by conjugation involving fusion of ameboid gametes to form synkaryon that develops into resistant spores. Also called Gamophyta.

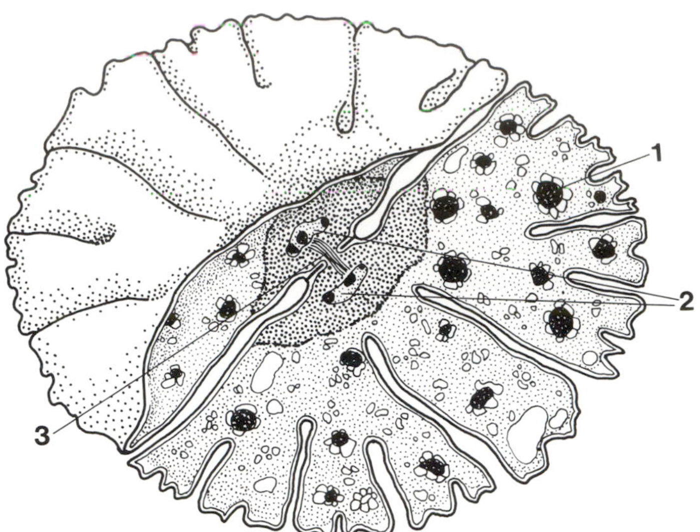

Conjugaphyta
Dividing cell at telophase. 1=pyrenoid surrounded by starch; 2=nuclei; 3=mitotic spindle.

Conjugatophyceae
Only class in phylum Conjugaphyta.

Conopodina
Suborder in phylum Rhizopoda.

Conorbinidae
Family in phylum Granuloreticulosa.

Conorboididae
Family in phylum Granuloreticulosa.

Copromyxidae
Family in phylum Acrasea.

Corallinales
Order in phylum Rhodophyta.

Corallineae
Tribe in phylum Rhodophyta.

Corallinoideae
Subfamily in phylum Rhodophyta.

Corethraceae
Family in phylum Bacillariophyta.

Corethrales
Order in phylum Bacillariophyta.

Corethronophycidae
Subclass in phylum Bacillariophyta.

Cornuspiracea
Superfamily in phylum Granuloreticulosa.

Cornuspiridae
Family in phylum Granuloreticulosa.

Coscinodiscaceae
Family in phylum Bacillariophyta.

Coscinodiscales
Order in phylum Bacillariophyta.

Coscinodiscophyceae
Class in phylum Bacillariophyta.

Coscinodiscophycidae
Subclass in phylum Bacillariophyta.

Coscinophragmatacea
Superfamily in phylum Granuloreticulosa.

Coscinophragmatidae
Family in phylum Granuloreticulosa.

Coskinolinidae
Family in phylum Granuloreticulosa.

Cougourdellidae
Family in phylum Microspora.

Cribrariaceae
Family in phylum Plasmodial Slime Molds.

Cribratinidae
Family in phylum Granuloreticulosa.

Cryptaxohelida
Order in phylum Actinopoda.

Crypthecodiniaceae
Family in phylum Dinomastigota.

Cryptoaxoplastidiata
Superfamily in phylum Actinopoda.

Cryptogemmida
Order in phylum Ciliophora.

Cryptonemiales
Order in phylum Rhodophyta.

Cryptophyceae
Only class in phylum Cryptophyta.

Cryptophyta
Cryptomonads. Phylum of asymmetric, flattened, mastigote algae with distinctive swimming motion or derived palmelloid forms. Vestibular depression from which undulipodia emerge is

anterior portion of crypt lined with refractile ejectosomes. Periplast formed by organic plates internal to the plasma membrane, rather than external cell wall. Plastids contain chlorophyll *c* and phycobilins. Based on presence of nucleomorph, phylum is thought to have evolved from symbiosis between heterotrophic mastigotes and red algae that retain remnant nuclei.

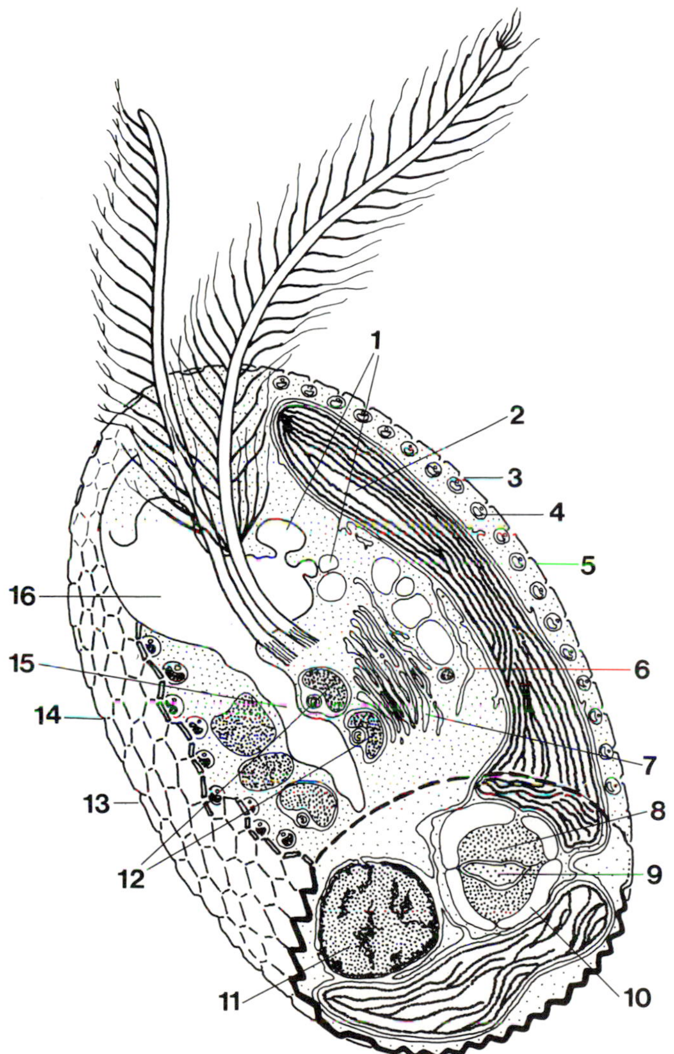

Cryptophyta (Cryptomonads)
1=vacuoles; 2=plastid; 3=pore; 4=small ejectosome; 5=plate; 6=plastid endoplasmic reticulum; 7=golgi body; 8=pyrenoid; 9=nucleomorph; 10=starch; 11=nucleus; 12=large ejectosomes; 13=periplast; 14=exterior plates; 15=gullet; 16=vestibulum.

Cryptosporidae
Family in phylum Apicomplexa.

Culicosporidae
Family in phylum Microspora.

Cuneolinidae
Family in phylum Granuloreticulosa.

Cutleriaceae
Family in phylum Phaeophyta.

Cutleriales
Order in phylum Phaeophyta.

Cyanophoraceae
Family in phylum Glaucocystophyta.

Cyanophorales
Order in phylum Glaucocystophyta.

Cyclamminidae
Family in phylum Granuloreticulosa.

Cyclolinacea
Superfamily in phylum Granuloreticulosa.

Cyclolinidae
Family in phylum Granuloreticulosa.

Cyclosporidae
Family in phylum Apicomplexa.

Cyclotellaceae
Family in phylum Bacillariophyta.

Cymatosiraceae
Family in phylum Bacillariophyta.

Cymatosirales
Order in phylum Bacillariophyta.

Cymatosirophycidae
Subclass in phylum Bacillariophyta.

Cymbaloporidae
Family in phylum Granuloreticulosa.

Cymbellaceae
Family in phylum Bacillariophyta.

Cymbellales
Order in phylum Bacillariophyta.

Cyrtoidea
Family in phylum Actinopoda.

Cyrtolophosidida
Order in phylum Ciliophora.

Cyrtophora
Subphylum in phylum Ciliophora.

Cyrtophorida
Order in phylum Ciliophora.

Cystoseiraceae
Family in phylum Phaeophyta.

D

Dactylophoridae
Family in phylum Apicomplexa.

Dariopsidae
Family in phylum Granuloreticulosa.

Dasycladaceae
Family in phylum Chlorophyta.

Dasycladales
Order in phylum Chlorophyta.

Delamareaceae
Family in phylum Phaeophyta.

Delosinacea
Superfamily in phylum Granuloreticulosa.

Delosinidae
Family in phylum Granuloreticulosa.

Dermatolitheae
Tribe in phylum Rhodophyta.

Desmarestiaceae
Family in phylum Phaeophyta.

Desmarestiales
Order in phylum Phaeophyta.

Desmidiaceae
Family in phylum Conjugaphyta.

Desmocapsaceae
Family in phylum Dinomastigota.

Desmocapsales
Order in phylum Dinomastigota.

Desmomonadaceae
Family in phylum Dinomastigota.

Desmomonadales
Order in phylum Dinomastigota.

Desmothoracida
Suborder in phylum Actinopoda.

Devescoviidae
Family in phylum Zoomastigina.

Diadesmidaceae
Family in phylum Bacillariophyta.

Dianemaceae
Family in phylum Plasmodial Slime Molds.

Diatoms
Any member of the phylum Bacillariophyta; unicellular and colonial aquatic protoctists renowned for their two-valved siliceous tests (frustules). See *Bacillariophyta*.

Dichotomosiphonaceae
Family in phylum Chlorophyta.

Dictyacanthidae
Family in phylum Actinopoda.

Dictyochaceae
Family in phylum Chrysophyta.

Dictyochales
Order in phylum Chrysophyta.

Dictyochophyceae
Class in phylum Chrysophyta. Silicoflagellates.

Dictyoneidaceae
Family in phylum Bacillariophyta.

Dictyoneidales
Order in phylum Bacillariophyta.

Dictyopsellidae
Family in phylum Granuloreticulosa.

Dictyosiphonaceae
Family in phylum Phaeophyta.

Dictyosiphonales
Order in phylum Phaeophyta.

Dictyosphaeriaceae
Family in phylum Chlorophyta.

Dictyostelia
Class in phylum Dictyostelida.

Dictyosteliaceae
Family in phylum Dictyostelida.

Dictyostelida
Dictyostelids; phylum of cellular (pseudoplasmodial) slime molds. Ameboid amastigote cells aggregate to form sorocarps. Damp soil, freshwater habitats. Differentiated from acrasids by cytology of the myxameba (slime ameba), production of well-differentiated stalk and spore cells, formation of more complex sorocarps and alignment of aggregating myxamebas into streams that form motile pseudoplasmodia (e.g., "slugs" or "Mexican hat stage," cf. *Dictyostelium discoideum*). Sexual fusion of compatible myxamebas occurs in plasmodium formation.

Dictyotaceae
Family in phylum Phaeophyta.

Dictyotales
Order in phylum Phaeophyta.

Dictyotopsidaceae
Family in phylum Phaeophyta.

Dicyclinidae
Family in phylum Granuloreticulosa.

Didymiaceae
Family in phylum Plasmodial Slime Molds.

Difflugiidae
Family in phylum Rhizopoda.

Diffusilinidae
Family in phylum Granuloreticulosa.

Dimorphyidae
Family in phylum Actinopoda.

Dinobryaceae
Family in phylum Chrysophyta.

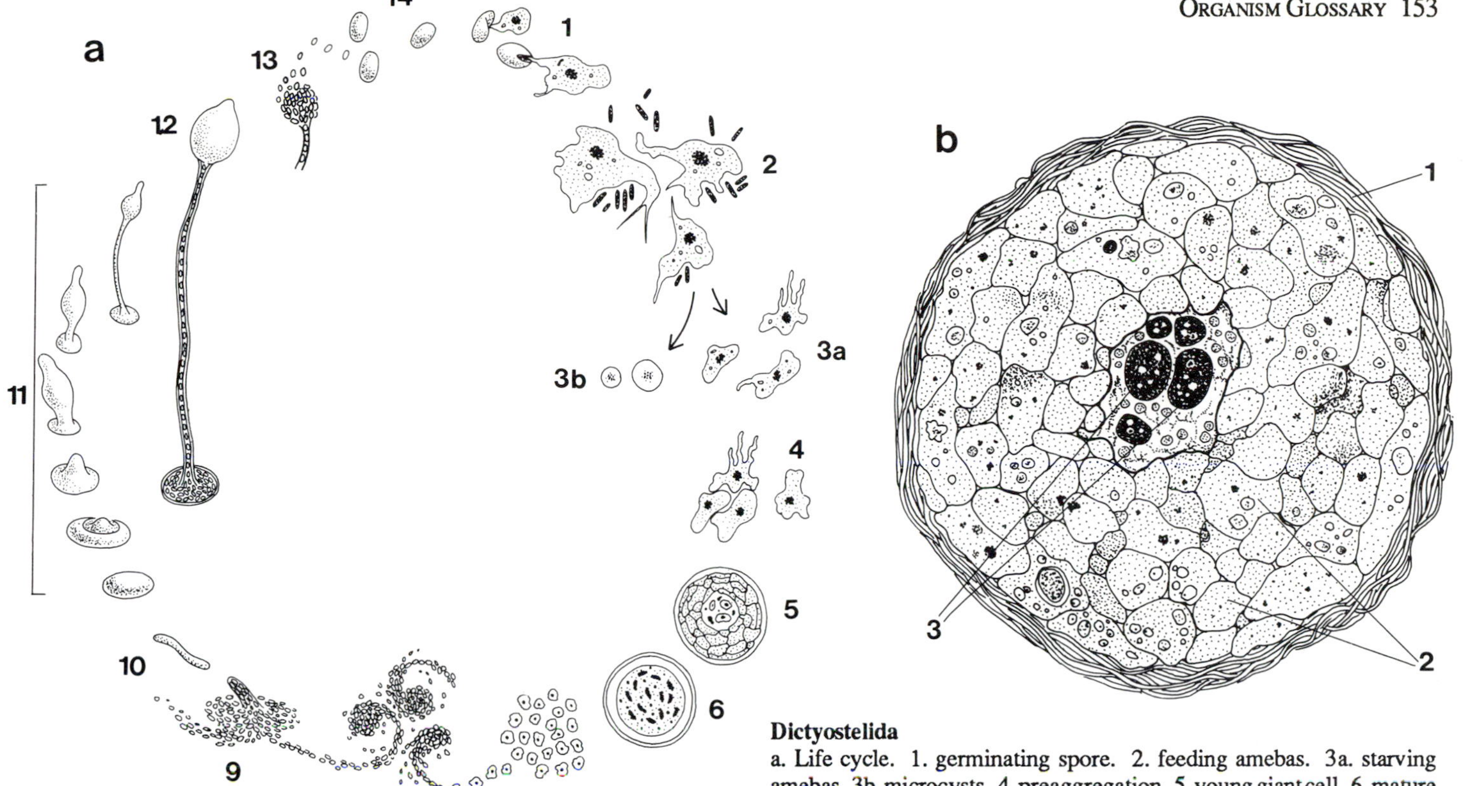

Dictyostelida

a. Life cycle. 1. germinating spore. 2. feeding amebas. 3a. starving amebas. 3b. microcysts. 4. preaggregation. 5. young giant cell. 6. mature giant cell/macrocyst. 7. division of giant cell. 8. defined aggregation centers. 9. slug forming from aggregation center. 10. slug. 11. development of sorocarp from slug. 12. mature sorocarp. 13. spores released from sorus. 14. spores. b. Developing macrocyst. 1=fibrillar sheath; 2=myxamebas; 3=ingested amebas.

Dinomastigota

Phylum of bimastigotes, usu. with one girdle and one transverse undulipodium. Amphiesmal plates form intramembranous wall-like structures; some lack walls. Primarily marine plankton, solitary or colonial cells. Distinctive chromatin organization: nucleus (dinokaryon or mesokaryon) has permanently condensed and visible chromosomes lacking nucleosomes and the histones which comprise them. Substitution of much thymine in DNA by 5-hydroxymethyl uracil. Photosynthetic forms contain plastids with chlorophylls *a* and *c2* and a unique xanthophyll, peridinin; many lack plastids. Resistant cysts fossilize as hystrichospheres.

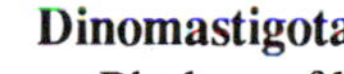

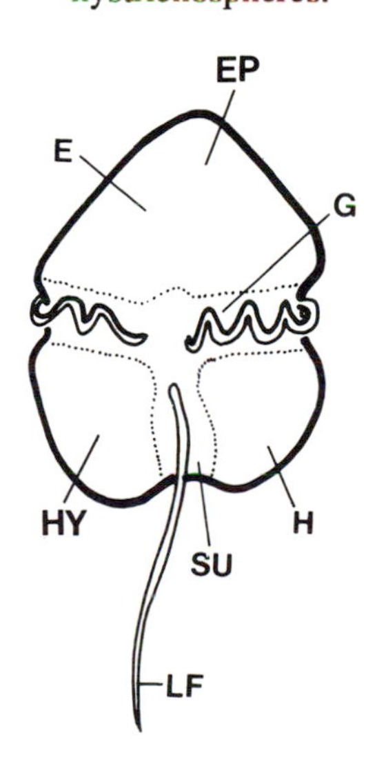

Dinomastigota

AM=amphiesma (outer surface); AV=amphiesmal vesicle; AX=axoneme; E=episome (epicone); EP=epitheca; G=girdle (cingulum); H=hyposome (hypocone); HY=hypotheca; LF=longitudinal undulipodium; MT=mitochondrion; NU=mesokaryotic nucleus; PC=collecting pusule; PL=plastid; PS= pusule (sac); PY=pyrenoid; SS=striated strand (paraxial rod); SU=sulcus; V=vacuome.

Dinophysiaceae
Family in phylum Dinomastigota.

Dinophysiales
Order in phylum Dinomastigota.

Diploconidae
Family in phylum Actinopoda.

Diplocystidae
Family in phylum Apicomplexa.

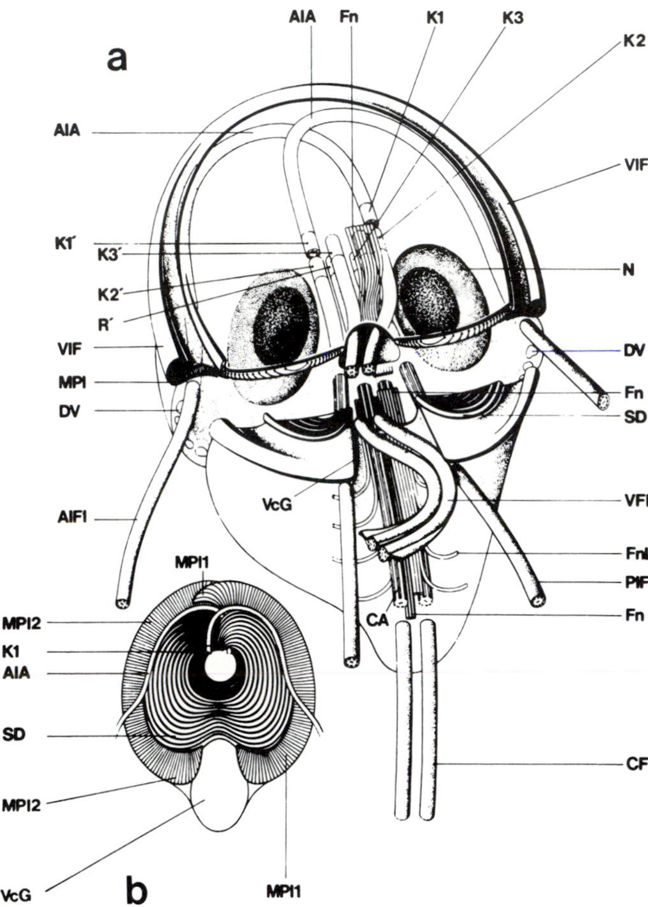

Diplomonadida
Trophozoite of *Giardia muris*. a. Ventral view. b. Dorsal view. AlA=intracellular axonemes of anterolateral undulipodia; AlFl=anterolateral undulipodia; CA=intracellular axonemes of caudal undulipodia; CFl=caudal undulipodia; DV=digestive vacuoles; Fn=funis; FnL=microtubules; K1, K2, K3, K1', K2', K3', R'=kinetosomes of two sets of undulipodia; MPl, MPl1, MPl2=marginal plates; N=nucleus; PlFl=posterolateral undulipodia; SD=striated disc; VcG=ventrocaudal groove; VFl=ventral undulipodia; VlF=ventrolateral flange.

Diplomonadida
Class in phylum Zoomastigina. Mastigotes with distinctive karyomastigont systems, lacking mitochondria and golgi apparatus. All heterotrophic; sexuality unknown. Free-living freshwater or symbiotic, including necrotrophic forms. Cysts formed only by symbiotrophic species.

Diploneidaceae
Family in phylum Bacillariophyta.

Diploneidineae
Suborder in phylum Bacillariophyta.

Diplosporidae
Family in phylum Apicomplexa.

Discamminidae
Family in phylum Granuloreticulosa.

Discocephalina
Suborder in phylum Ciliophora.

Discocyclinidae
Family in phylum Granuloreticulosa.

Discoidae
Family in phylum Actinopoda.

Discorbacea
Superfamily in phylum Granuloreticulosa.

Discorbidae
Family in phylum Granuloreticulosa.

Discorbinellidae
Family in phylum Granuloreticulosa.

Discospirinidae
Family in phylum Granuloreticulosa.

Dobelliidae
Family in phylum Apicomplexa.

Dorataspidae
Family in phylum Actinopoda.

Dorisiellidae
Family in phylum Apicomplexa.

Dorothiidae
Family in phylum Granuloreticulosa.

Dryorhizopsidae
Family in phylum Granuloreticulosa.

Duboscqiidae
Family in phylum Microspora.

Duboscquellaceae
Family in phylum Dinomastigota.

Dunaliellales
Order in phylum Chlorophyta.

Duostominacea
Superfamily in phylum Granuloreticulosa.

Duostominidae
Family in phylum Granuloreticulosa.

Durvillaeaceae
Family in phylum Phaeophyta.

Durvillaeales
Order in phylum Phaeophyta.

Dusenburyinidae
Family in phylum Granuloreticulosa.

Dysteriina
Suborder in phylum Ciliophora.

E

Earlandiacea
Superfamily in phylum Granuloreticulosa.

Earlandiidae
Family in phylum Granuloreticulosa.

Earlandinitidae
Family in phylum Granuloreticulosa.

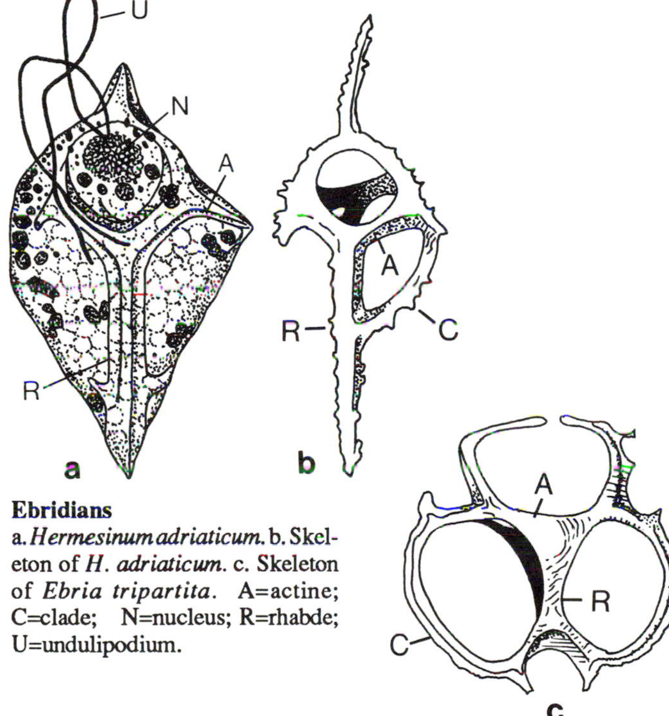

Ebridians
a. *Hermesinum adriaticum.* b. Skeleton of *H. adriaticum.* c. Skeleton of *Ebria tripartita.* A=actine; C=clade; N=nucleus; R=rhabde; U=undulipodium.

Ebridians
Coastal marine, free-living, bimastigote solitary cells with basketlike internal skeletons consisting of siliceous rods. Reproduction by simple fission; sexuality unknown; taxonomy *incertae sedis.* Fossil record from lower Cenozoic to present with greatest diversity in the Miocene.

Echinamoebidae
Family in phylum Rhizopoda.

Echinosteliaceae
Family in phylum Plasmodial Slime Molds.

Echinosteliales
Order in phylum Plasmodial Slime Molds.

Echinosteliopsidae
Family in phylum Plasmodial Slime Molds.

Ectocarpaceae
Family in phylum Phaeophyta.

Ectocarpales
Order in phylum Phaeophyta.

Eggerellidae
Family in phylum Granuloreticulosa.

Eimeriida
Order in phylum Apicomplexa.

Eimeriidae
Family in phylum Apicomplexa.

Elachistaceae
Family in phylum Phaeophyta.

Elaeomyxaceae
Family in phylum Plasmodial Slime Molds.

Elhasaellidae
Family in phylum Granuloreticulosa.

Ellobiopsida
Heterotrophic, coenocytic symbiotrophs including necrotrophs, esp. of planktonic marine arthropods. Larger members arborescent

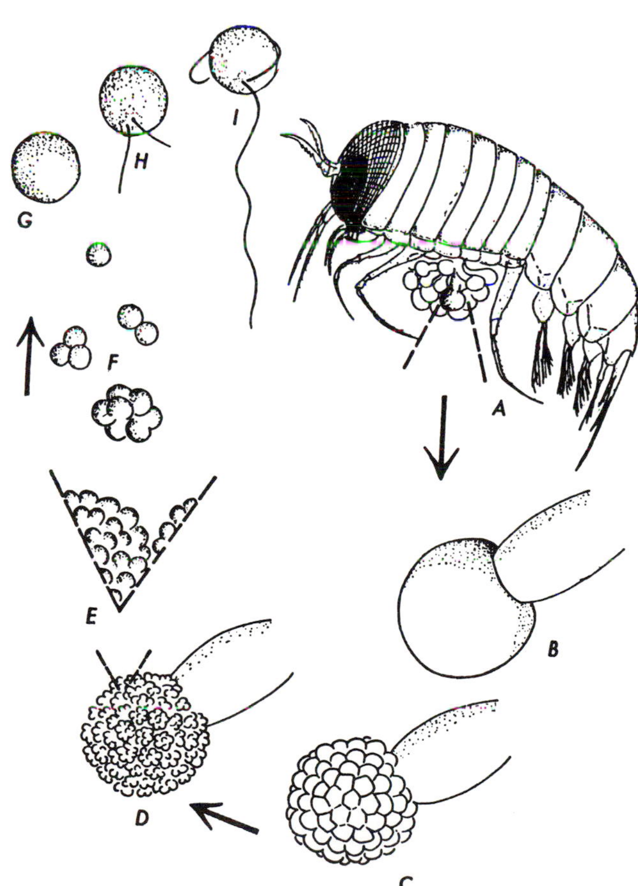

Ellobiopsida
Life history of *Thalassomyces marsupii.* A. parasite attached to host amphipod. B. mature gonomere distally attached to trophomere. C. initial faceting of gonomere. D. gonomere after spore cleavage is nearly completed. E. enlarged section of D showing partial cleavage. F. groups of detached spores. G. single uninucleate spore. H. early undulipodial formation of spore. I. fully developed swimming spore.

with their absorptive base anchored in the host nerve tissue. The trunk breaches the cuticle and then divides dichotomously into branches (trophomeres) carrying terminal reproductive segments (gonomeres) that form bimastigote zoospores; taxonomy *incertae sedis*.

Elphidiidae
Family in phylum Granuloreticulosa.

Endictyaceae
Family in phylum Bacillariophyta.

Endochytriaceae
Family in phylum Chytridiomycota.

Endogenida
Order in phylum Ciliophora.

Endonucleoaxoplasthelida
Suborder in phylum Actinopoda.

Endothyracea
Superfamily in phylum Granuloreticulosa.

Endothyridae
Family in phylum Granuloreticulosa.

Entamoebidae
Family in phylum Rhizopoda.

Enteridiaceae
Family in phylum Plasmodial Slime Molds.

Enterocystidae
Family in phylum Apicomplexa.

Enteromonadidae
Family in phylum Zoomastigina.

Entodiniomorphida
Order in phylum Ciliophora.

Entodiniomorphina
Suborder in phylum Ciliophora.

Entomoneidaceae
Family in phylum Bacillariophyta.

Entopylaceae
Family in phylum Bacillariophyta.

Entopylales
Order in phylum Bacillariophyta.

Eocristellariidae
Family in phylum Granuloreticulosa.

Eoglobigerinidae
Family in phylum Granuloreticulosa.

Eouvigerinacea
Superfamily in phylum Granuloreticulosa.

Eouvigerinidae
Family in phylum Granuloreticulosa.

Epistomariidae
Family in phylum Granuloreticulosa.

Epistominidae
Family in phylum Granuloreticulosa.

Epithemiaceae
Family in phylum Bacillariophyta.

Epithemiales
Order in phylum Bacillariophyta.

Eponididae
Family in phylum Granuloreticulosa.

Eremosphaeraceae
Family in phylum Chlorophyta.

Erythropeltidaceae
Family in phylum Rhodophyta.

Ethmodiscaceae
Family in phylum Bacillariophyta.

Ethmodiscales
Order in phylum Bacillariophyta.

Eucomonymphidae
Family in phylum Zoomastigina.

Euglenales
Order in phylum Euglenida.

Euglenamorphales
Order in phylum Euglenida.

Euglenida →
Euglenids; phylum of mastigotes with one or two anterior undulipodia. Unilateral hairs present on the emergent portion of the locomotory undulipodium; paramylon as storage material. Most euglenids are freshwater or soil phagotrophs or osmotrophs. Approximately one-third are photosynthetic with plastids that contain chlorophylls *a* and *b* in which photosynthesis supplements heterotrophy; no fully autotrophic species known. Stigma is outside the chloroplasts. Many have a flexible pellicle and move by metaboly. Cells solitary or colonial, display characteristic type of closed mitosis.

Euglenophyceae
Only class in phylum Euglenida.

Euglyphidae
Family in phylum Rhizopoda.

Eugregarinida
Order in phylum Apicomplexa.

Eunotiaceae
Family in phylum Bacillariophyta.

Eunotiales
Order in phylum Bacillariophyta.

Eunotiophycidae
Subclass in phylum Bacillariophyta.

Euplotida
Order in phylum Ciliophora.

Euplotina
Suborder in phylum Ciliophora.

Eupodiscaceae
Family in phylum Bacillariophyta.

Eupodiscales
Order in phylum Bacillariophyta.

Eustigmataceae
Family in phylum Eustigmatophyta.

Eustigmatales
Only order in phylum Eustigmatophyta.

Eustigmatophyceae
Only class in phylum Eustigmatophyta.

Eustigmatophyta
Eustigmatophytes; phylum of mastigote algae that form zoospores with prominent red eyespot at the extreme anterior end

Euglenida
Anisonema, a colorless phagotrophic euglenid with one short anterior undulipodium and one posteriorly trailing undulipodium that lies in a ventral groove. 1=undulipodia; 2=subapical canal; 3=pocket (reservoir) in which undulipodia are situated; 4=striated pellicle; 5=golgi apparatus; 6=nucleus; 7=paramylon granules; 8=mitochondrial reticulum; 9=tubelike ingestion apparatus.

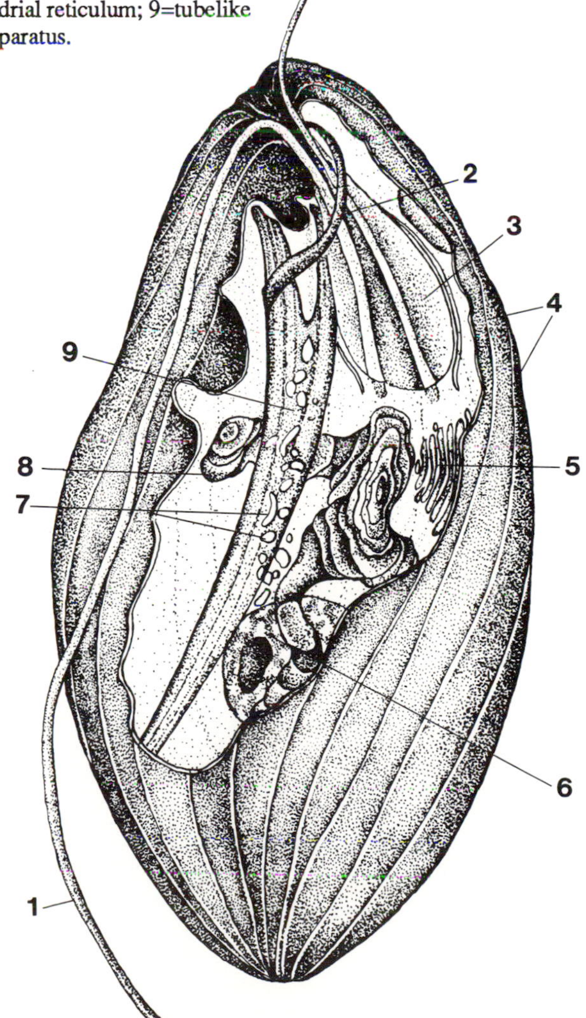

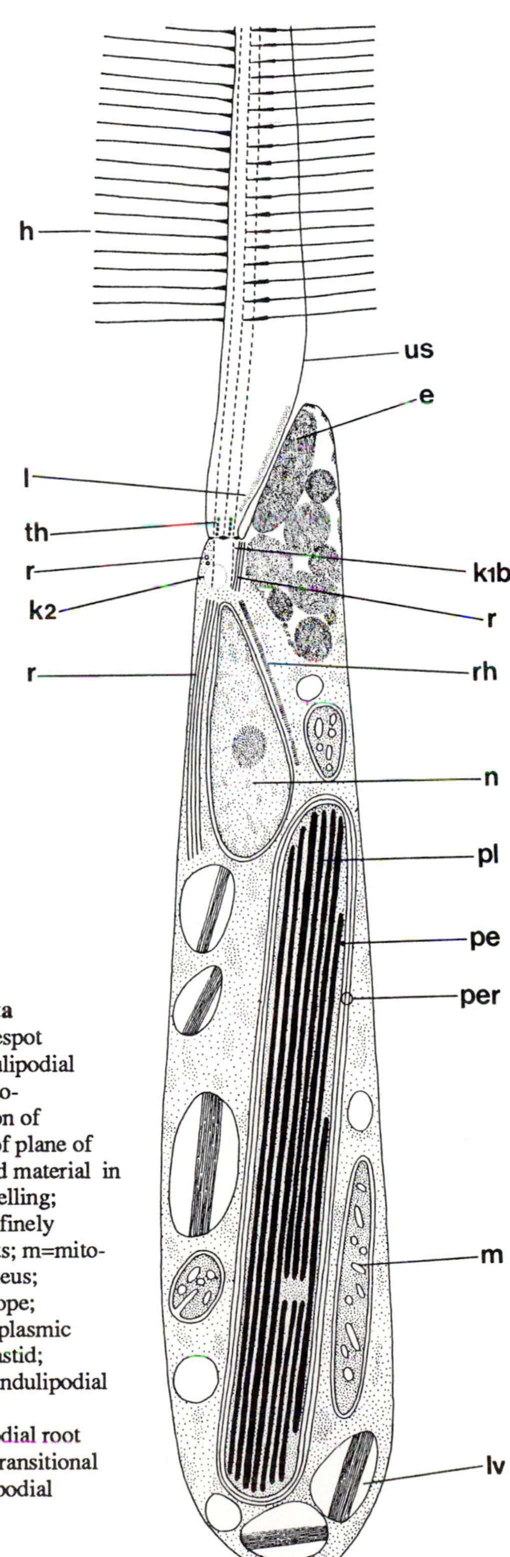

Eustigmatophyta
Zoospore. e=eyespot (stigma); h=undulipodial hairs; k1b=kinetosome; k2=position of kinetosome out of plane of section; l=layered material in undulipodial swelling; lv=vesicles with finely lamellate contents; m=mitochondria; n=nucleus; pe=plastid envelope; per=plastid endoplasmic reticulum; pl=plastid; r=microtubular undulipodial roots; rh=microfibrillar undulipodial root (rhizoplast); th=transitional helix; us=undulipodial swelling.

of their single anteriorly-inserted undulipodium. Reproduction by autospores or zoospores; no sexuality known. Plastids contain chlorophyll *a*, ß-carotene, and violoxanthin.

Eutreptiales
Order in phylum Euglenida.

Evaginogenida
Order in phylum Ciliophora.

Exoaxoplastidiata
Superfamily in phylum Actinopoda.

Exocryptoaxoplastidiata
Superfamily in phylum Actinopoda.

Exogemmida
Order in phylum Ciliophora.

Exogenida
Order in phylum Ciliophora.

Exonucleoaxoplasthelida
Suborder in phylum Actinopoda.

F

Fabesporidae
Family in phylum Myxozoa.

Fabulariidae
Family in phylum Granuloreticulosa.

Favusellidae
Family in phylum Granuloreticulosa.

Filosea
Class in phylum Rhizopoda.

Fischerinidae
Family in phylum Granuloreticulosa.

Flabellina
Suborder in phylum Rhizopoda.

Flabellulidae
Family in phylum Rhizopoda.

Florideophycidae
Subclass in phylum Rhodophyta.

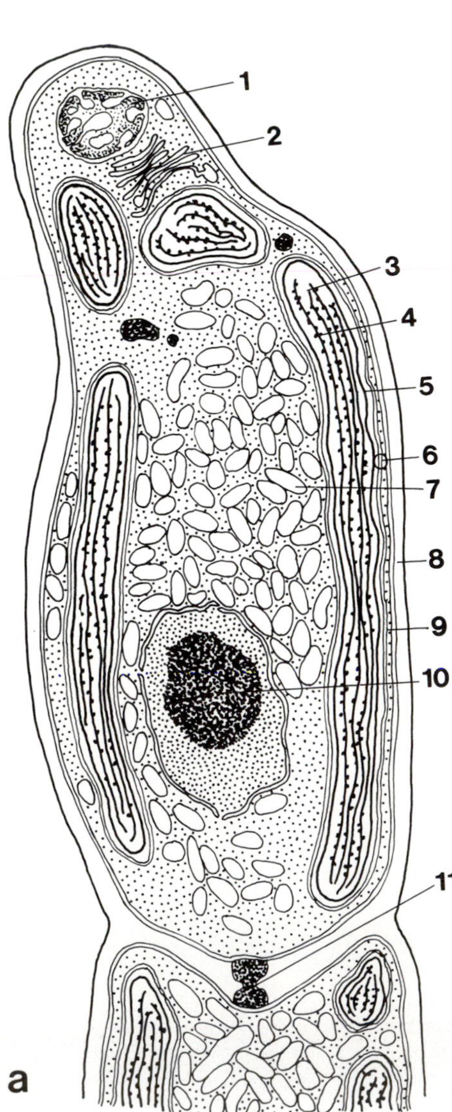

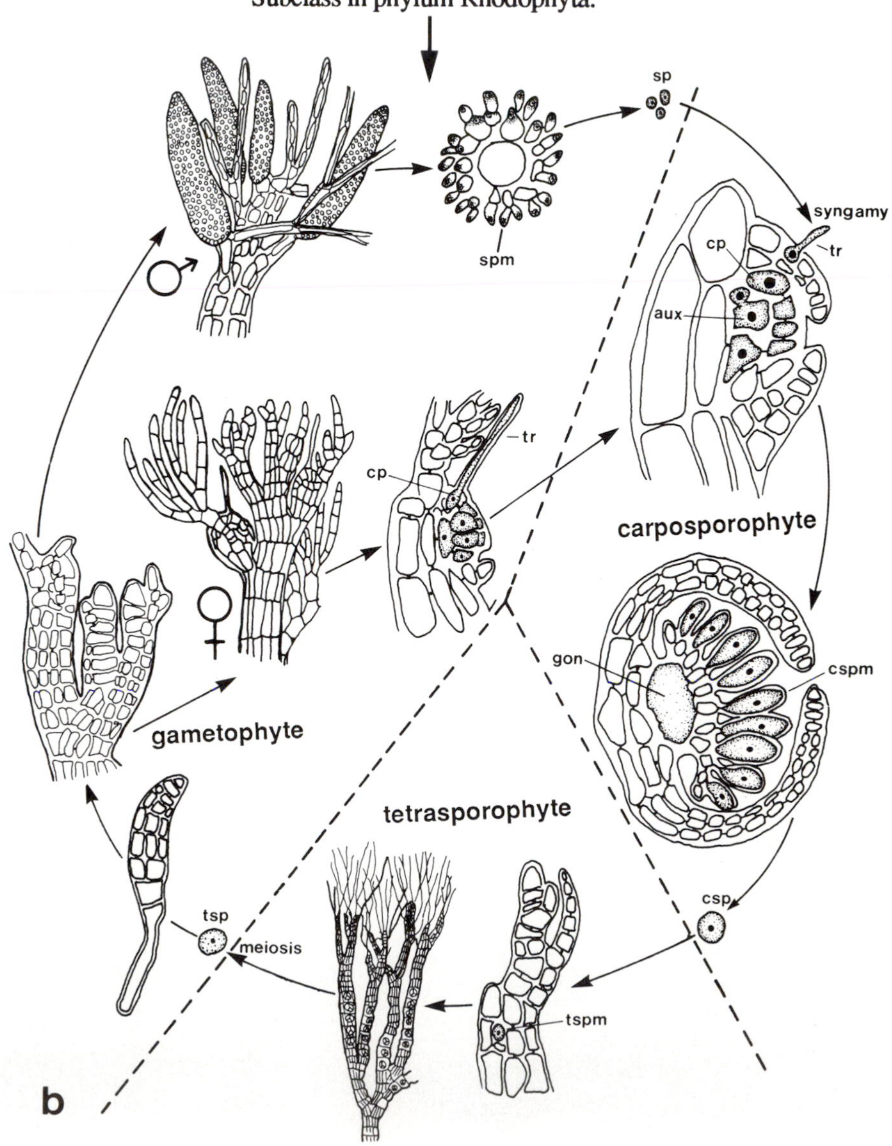

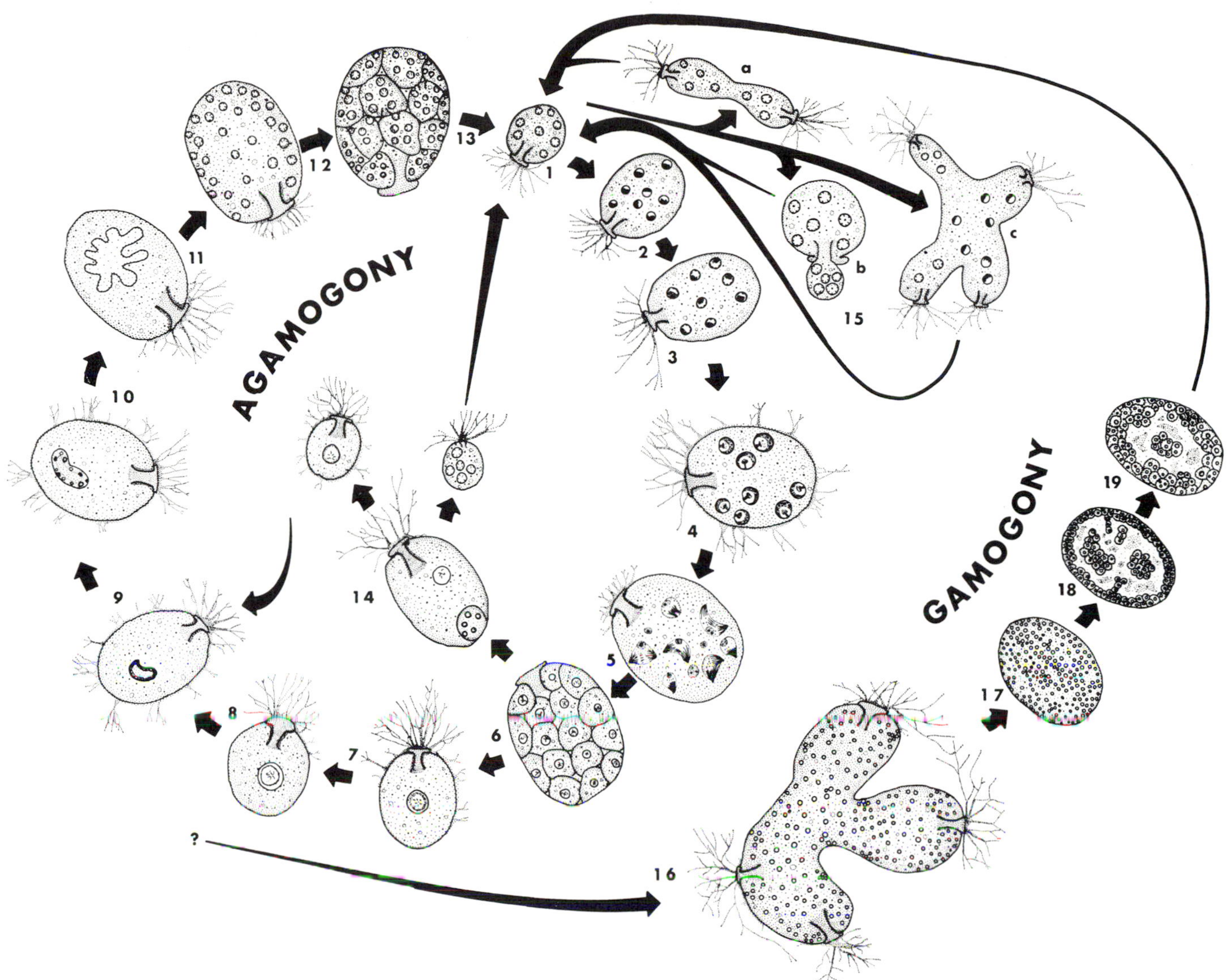

Fonticulidae
Family in phylum Acrasea, but with uncertain affinity.

Foraminifera
Shelled members of phylum Granuloreticulosa. →

Foraminiferea
Class in phylum Granuloreticulosa.

Fragilariaceae
Family in phylum Bacillariophyta.

Fragilariales
Order in phylum Bacillariophyta.

Fragilariophyceae
Class in phylum Bacillariophyta.

←

Florideophycidae
a. 1=mitochondrion; 2=golgi apparatus; 3=thylakoid; 4=phycobilisome; 5=peripheral encircling thylakoid; 6=chloroplast envelope; 7=starch grains; 8=cell wall; 9=plastid; 10=nucleus; 11=pit plug. b. Life history of *Polysiphonia*. aux=auxiliary cell; cp=carpogonium; csp=carpospore; cspm=carposporangium; gon=gonimoblast; sp=spermatium; spm=spermatangium; tr=trichogyne; tsp=tetraspore; tspm=tetrasporangium.

Foraminifera
Life cycle of *Allogromia laticollaris*. 1-6, 15. multinucleate agamont. 7-12. uninucleate agamont. 1. juvenile agamont. 2. young agamont. 3. growing agamont. 4. mature agamont. 5. karyokinesis. 6. cytokinesis (schizogony). 7. young agamont. 8. growing agamont. 9. mature agamont. 10. early "ameba-form" nucleus. 11. giant ameba-form nucleus. 12. breakup of giant nucleus (Zerfall). 13. schizogony. 14. relatively uncommon life cycle alternate pathway in which budding gives rise to a multinucleate agamont. 15. alternate life cycle pathways include: (a) binary fission; (b) budding; (c) cytotomy. 16. giant gametocytotomont. 17. multinucleate gamont prior to the formation of gametes. 18. gamont filled with gametes. 19. gamont with some gametes and zygotes.

Fragilariophycidae
Subclass in phylum Bacillariophyta.

Frontoniina
Suborder in phylum Ciliophora.

Fucaceae
Family in phylum Phaeophyta.

Fucales
Order in phylum Phaeophyta.

Fursenkoinacea
Superfamily in phylum Granuloreticulosa.

Fursenkoinidae
Family in phylum Granuloreticulosa.

Fusulinacea
Superfamily in phylum Granuloreticulosa.

Fusulinida
Order in phylum Granuloreticulosa.

Fusulinidae
Family in phylum Granuloreticulosa.

G

Ganymedidae
Family in phylum Apicomplexa.

Gavelinellidae
Family in phylum Granuloreticulosa.

Geinitzinacea
Superfamily in phylum Granuloreticulosa.

Geinitzinidae
Family in phylum Granuloreticulosa.

Gelidiales
Order in phylum Rhodophyta.

Giardiinae
Family in phylum Zoomastigina.

Gigartaconidae
Family in phylum Actinopoda.

Gigartinales
Order in phylum Rhodophyta.

Giraudiaceae
Family in phylum Phaeophyta.

Glabratellacea
Superfamily in phylum Granuloreticulosa.

Glabratellidae
Family in phylum Granuloreticulosa.

Glandulinidae
Family in phylum Granuloreticulosa.

Glaucocystaceae
Family in phylum Glaucocystophyta.

Glaucocystales
Order in phylum Glaucocystophyta.

Glaucocystophyceae
Only class in phylum Glaucocystophyta.

Glaucocystophyta
Phylum of miscellaneous blue-green nucleated algae. Photosynthetic, freshwater organisms containing cyanelles, intracellular organelles interpreted to be modified cyanobacterial symbionts (with chlorophyll *a* and phycobiliproteins) that retain remnants of cell walls.

Glaucophyta
Synonym for phylum Glaucocystophyta.

Glaucosphaeraceae
Family in phylum Glaucocystophyta.

Globanomalinidae
Family in phylum Granuloreticulosa.

Globigerinacea
Superfamily in phylum Granuloreticulosa.

Globigerinelloididae
Family in phylum Granuloreticulosa.

Globigerinida
Order in phylum Granuloreticulosa.

Globigerinidae
Family in phylum Granuloreticulosa.

Globigerinitidae
Family in phylum Granuloreticulosa.

Globorotaliacea
Superfamily in phylum Granuloreticulosa.

Globorotaliidae
Family in phylum Granuloreticulosa.

Globorotalitidae
Family in phylum Granuloreticulosa.

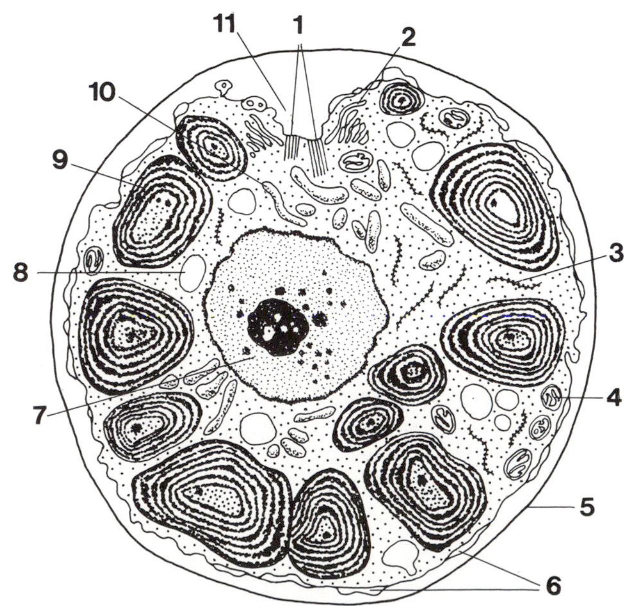

Glaucocystophyta
1=kinetosomes (prokinetosome left); 2=golgi; 3=endoplasmic reticulum; 4=mitochondrion; 5=plasma membrane; 6=cortical alveoli; 7=nucleus; 8=vacuole; 9=cyanelle; 10=starch grains; 11=apical depression.

Globotextulariidae
Family in phylum Granuloreticulosa.

Globotruncanacea
Superfamily in phylum Granuloreticulosa.

Globotruncanidae
Family in phylum Granuloreticulosa.

Gloeobotrydaceae
Family in phylum Xanthophyta.

Gloeochaetaceae
Family in phylum Glaucocystophyta.

Gloeochaetales
Order in phylum Glaucocystophyta.

Gloeodiniaceae
Family in phylum Dinomastigota.

Gloeopodiaceae
Family in phylum Xanthophyta.

Glugeidae
Family in phylum Microspora.

Gomphonemataceae
Family in phylum Bacillariophyta.

Gonapodyaceae
Family in phylum Chytridiomycota.

Gonyaulacaceae
Family in phylum Dinomastigota.

Gonyaulacales
Order in phylum Dinomastigota.

Gossleriellaceae
Family in phylum Bacillariophyta.

Granuloreticulosa
Phylum of marine protists having granular reticulopods that form anastomosing networks with distinctive two-way streaming. Most are enclosed by calcareous or agglutinated tests characteristic of the major class, Foraminiferea. Naked forms in class Athalamea. Possess single, dimorphic, or many nuclei. Many contain photosynthetic symbionts. Some have complex sexual life cycles, some have undulipodiated gametes; diploid asexual reproducing phase (agamont) alternating with a haploid sexual reproducing phase (gamont), or with only one phase (apogamic or apoagamic). Extremely useful as statigraphic markers because of abundance and diversity in Paleozoic and more recent marine sediment.

Gregarinia
Class in phylum Apicomplexa.

Gregarinidae
Family in phylum Apicomplexa.

Gurleyidae
Family in phylum Microspora.

Guttulinopsidae
Family in phylum Acrasea.

Gymnamoebia
Subclass in phylum Rhizopoda.

Gymnidae
Family in phylum Actinopoda.

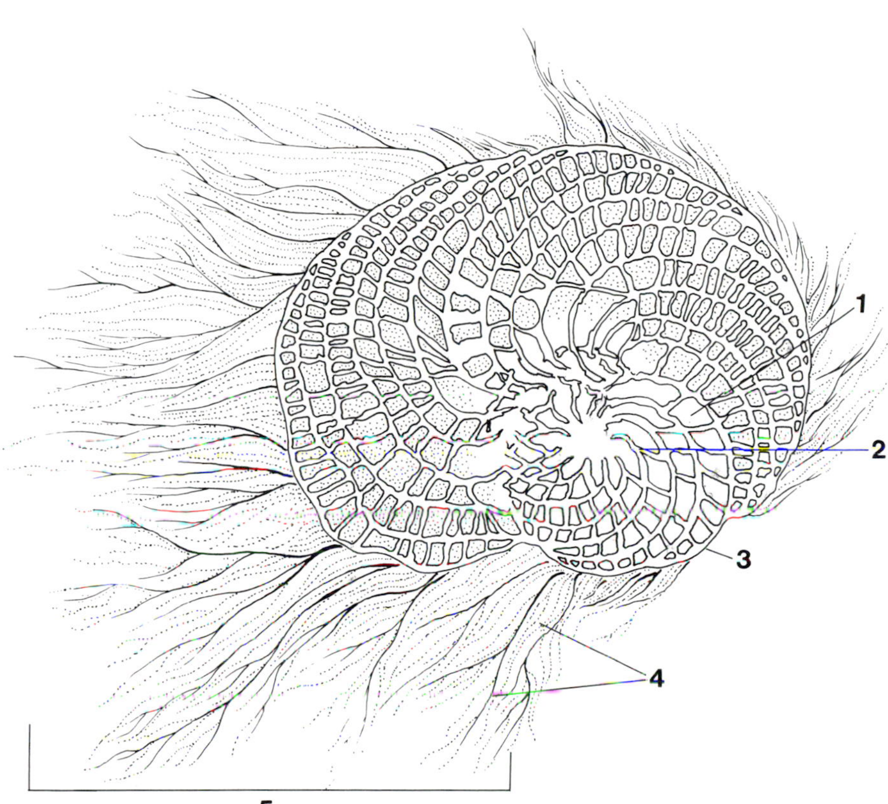

Granuloreticulosa
1=chambers; 2=proloculum; 3=test; 4=reticulopodia; 5=reticulopodial network.

Gymnodiniaceae
Family in phylum Dinomastigota.

Gymnodiniales
Order in phylum Dinomastigota.

H

Haddoniidae
Family in phylum Granuloreticulosa.

Haematococcaceae
Family in phylum Chlorophyta.

Haemogregarinidae
Family in phylum Apicomplexa.

Haemosporida
Order in phylum Apicomplexa.

Haliommatidae
Family in phylum Actinopoda.

Hantkeninacea
Superfamily in phylum Granuloreticulosa.

Hantkeninidae
Family in phylum Granuloreticulosa.

Haplophragmiidae
Family in phylum Granuloreticulosa.

Haplophragmoididae
Family in phylum Granuloreticulosa.

Haplosporea
Class in phylum Haplosporidia.

Haplosporida
Order in phylum Haplosporidia.

Haplosporidia
Phylum of unicellular amastigote symbiotrophs including necrotrophs (pathogens) primarily histozoic or coelozoic in marine animals. Form plasmodia with dense organelles called haplosporosomes in host tissue and produce unicellular, typically uninucleate propagules ("spores" that lack polar capsules and polar filaments).

Haplosporidiidae
Family in phylum Haplosporidia.

Haplozoaceae
Family in phylum Dinomastigota.

Haptoria
Subclass in phylum Ciliophora.

Haptorida
Order in phylum Ciliophora.

Harpochytriaceae
Family in phylum Chytridiomycota.

Hartmannellidae
Family in phylum Rhizopoda.

Hastigerinidae
Family in phylum Granuloreticulosa.

Hedbergellidae
Family in phylum Granuloreticulosa.

Hedraiophryidae
Family in phylum Actinopoda.

Heleochloridaceae
Family in phylum Chlorophyta.

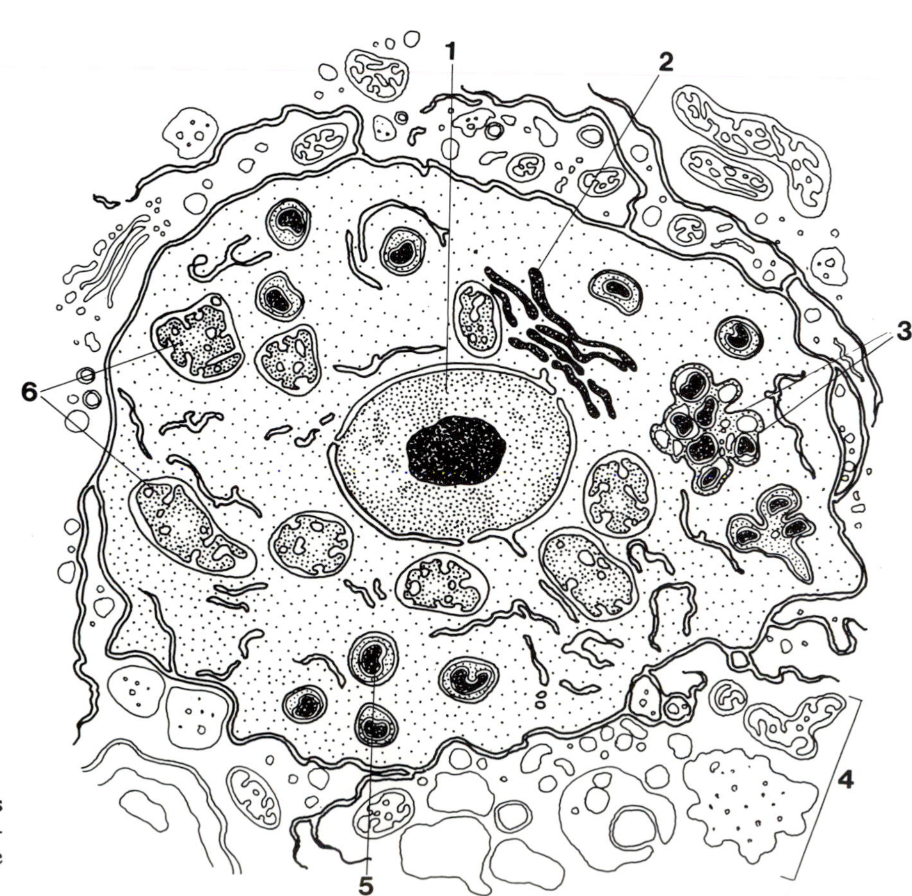

Haplosporidia
1=nucleus; 2=golgi; 3=membrane-bounded regions where haplosporosomes are formed; 4=host cell organelles; 5=haplosporosome with limiting membrane and internal membrane; 6=mitochondria.

Heliozoa

Class in phylum Actinopoda. Spherical, free-living, heterotrophic, primarily freshwater unicells that lack central capsules. Axopods, used for locomotion or predaceous feeding, radiate from naked, siliceous coated bodies. Some species also produce pseudopods, filopods, or undulipodia. Autogamy reported in two species.

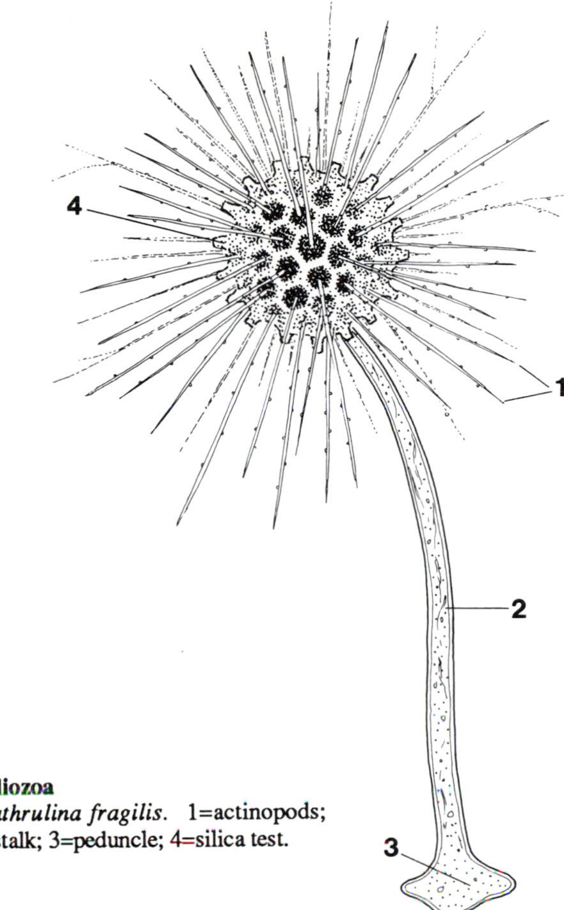

Heliozoa
Clathrulina fragilis. 1=actinopods; 2=stalk; 3=peduncle; 4=silica test.

Hematozoa
Class in phylum Apicomplexa.

Hemiaulaceae
Family in phylum Bacillariophyta.

Hemiaulales
Order in phylum Bacillariophyta.

Hemidiscaceae
Family in phylum Bacillariophyta.

Hemisphaeramminidae
Family in phylum Granuloreticulosa.

Hesseidae
Family in phylum Microspora.

Heterochordariaceae
Family in phylum Phaeophyta.

Heterodendraceae
Family in phylum Xanthophyta.

Heterodiniaceae
Family in phylum Dinomastigota.

Heterogloeaceae
Family in phylum Xanthophyta.

Heterogloeales
Order in phylum Xanthophyta.

Heterohelicacea
Superfamily in phylum Granuloreticulosa.

Heterohelicidae
Family in phylum Granuloreticulosa.

Heteronematales
Order in phylum Euglenida.

Heteropediaceae
Family in phylum Xanthophyta.

Heterophryidae
Family in phylum Actinopoda.

Heterotrichia
Subclass in phylum Ciliophora.

Heterotrichida
Order in phylum Ciliophora.

Heterotrichina
Suborder in phylum Ciliophora.

Hexacapsulidae
Family in phylum Myxozoa.

Hexactinomyxidae
Family in phylum Myxozoa.

Hexalaspidae
Family in phylum Actinopoda.

Hexamitidae
Family in phylum Zoomastigina.

Hexamitinae
Subfamily in phylum Zoomastigina.

Hildebrandiales
Order in phylum Rhodophyta.

Himanthaliaceae
Family in phylum Phaeophyta.

Hippocrepinellidae
Family in phylum Granuloreticulosa.

Hirmocystidae
Family in phylum Apicomplexa.

Holacanthida
Order in phylum Actinopoda.

Holomastigotidae
Family in phylum Zoomastigina.

Homotrematidae
Family in phylum Granuloreticulosa.

Hoplonymphidae
Family in phylum Zoomastigina.

Hormosinacea
Superfamily in phylum Granuloreticulosa.

Hormosinidae
Family in phylum Granuloreticulosa.

Hormosiraceae
Family in phylum Phaeophyta.

Hormotilaceae
Family in phylum Chlorophyta.

Hospitellidae
Family in phylum Granuloreticulosa.

Hyalodiscaceae
Family in phylum Bacillariophyta.

Hydrodictyaceae
Family in phylum Chlorophyta.

Hydruraceae
Family in phylum Chrysophyta.

Hymenostomatida
Order in phylum Ciliophora.

Hymenostomia
Subclass in phylum Ciliophora.

Hyperamminacea
Superfamily in phylum Granuloreticulosa.

Hyperamminidae
Family in phylum Granuloreticulosa.

Hyperamminoididae
Family in phylum Granuloreticulosa.

Hypermastigida
Order in phylum Zoomastigina. Illustration: Parabasalia.

Hyphochytriales
Order in phylum Hyphochytriomycota.

Hyphochytriomycetes
Class in phylum Hyphochytriomycota.

Hyphochytriomycota
Phylum of osmotrophic or necrotrophic soil and water organisms that reproduce by zoospores. Zoospores, motile by a single, anteriorly-directed undulipodium with mastigonemes, form from a multinucleate thallus by reduction of cleavage vesicles. Growth as heterotrophic thallus follows germination of an encysted zoospore. Autogamy reported in one species, *Anisolpidium ectocarpii*. "Funguslike protoctists."

Hypocomatina
Suborder in phylum Ciliophora.

Hypotrichia
Subclass in phylum Ciliophora.

I

Involutinida
Order in phylum Granuloreticulosa.

Involutinidae
Family in phylum Granuloreticulosa.

Ishigeaceae
Family in phylum Phaeophyta.

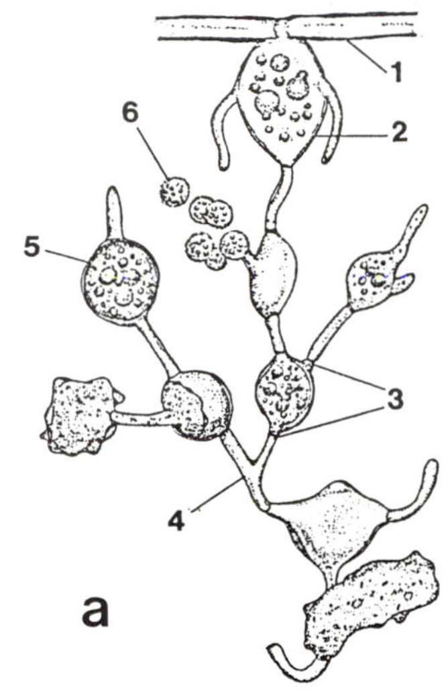

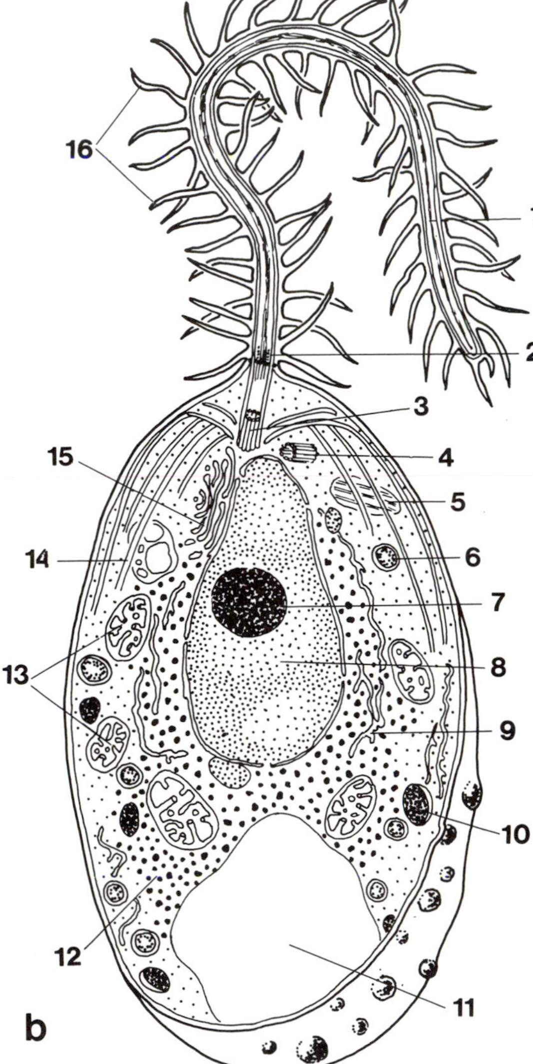

Hyphochytriomycota
a. 1=host cell wall; 2=sporangium attached to host; 3=septa; 4=hypha; 5=sporangium; 6=differentiating zoospores. b. Zoospore. 1=undulipodium; 2=terminal plate; 3=kinetosome; 4=centriole (barren kinetosome); 5=fibrillar inclusion; 6=osmiophilic body; 7=nucleolus; 8=nucleus; 9=endoplasmic reticulum; 10=electron-opaque body; 11=lipid; 12=ribosomal region; 13=mitochondria; 14=undulipodial rootlet; 15=golgi; 16=mastigonemes.

Islandiellidae
Family in phylum Granuloreticulosa.

Isochrysidales
Order in phylum Prymnesiophyta.

J

Janiae
Tribe in phylum Rhodophyta.

Joeniidae
Family in phylum Zoomastigina.

K

Karreriddae
Family in phylum Granuloreticulosa.

Karyoblastea
Phylum of the giant free-living, microaerophilic, multinucleate, algivorous, freshwater amebas. Monospecific: *Pelomyxa palustris*. Each ameba harbors three different morphotypes of endosymbiotic bacteria in proportions that change with conditions; at least one type is methanogenic and a second perinuclear. Lack mitochondria and possibly golgi bodies; have nonmotile surface projections that seem to be extreme variations on standard axonemal morphology.

Karyorelictea
Class in phylum Ciliophora.

Keramosphaeridae
Family in phylum Granuloreticulosa.

Kinetoplastida
Class in phylum Zoomastigina. Free-living or symbiotrophic mastigotes with one or two undulipodia associated with a conspicuous intracellular stainable structure: the kinetoplast. Masses of small and large circular DNA including mitochondria DNA sequences, within a single differentiating and dedifferentiating mitochondrion render the kinetoplast nearly as large and just as stainable as nucleic acids of the nucleus. Illustration next page.

Klossiidae
Family in phylum Apicomplexa.

Kofoidiidae
Family of hypermastigotes in phylum Zoomastigina.

Kofoidiniaceae
Family in phylum Dinomastigota.

Kolkwitziellaceae
Family in phylum Dinomastigota.

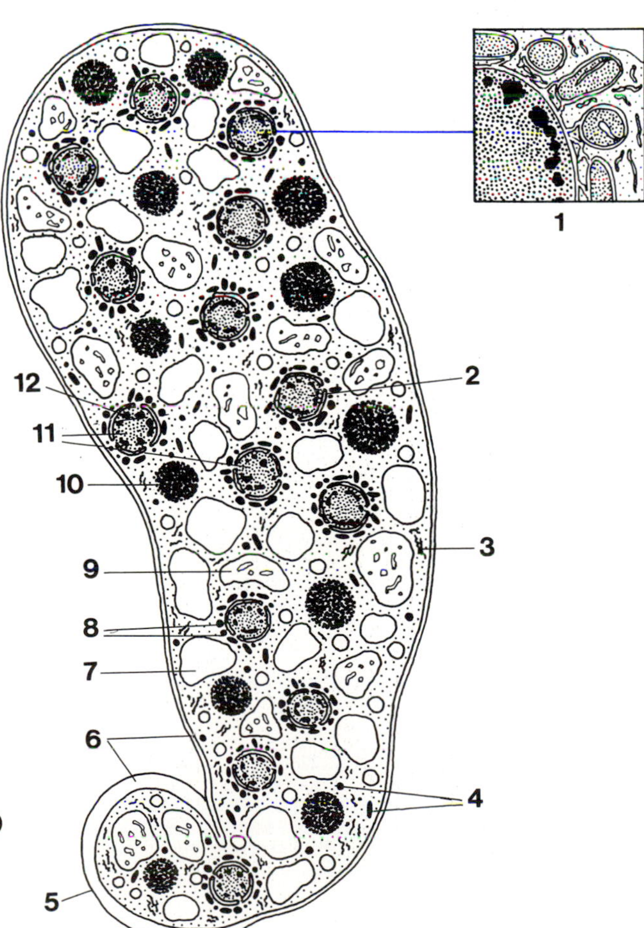

Karyoblastea
a. *Pelomyxa palustris*. b. 1=close-up of nucleus with bacterial symbionts attached; 2=nuclear membrane; 3=cytoplasmic membranes; 4=symbiotic bacteria; 5=uroid; 6=glycocalyx; 7=vacuole; 8=perinuclear symbiotic bacteria; 9=food vacuole; 10=glycogen body; 11=nuclei; 12=chromatin granules.

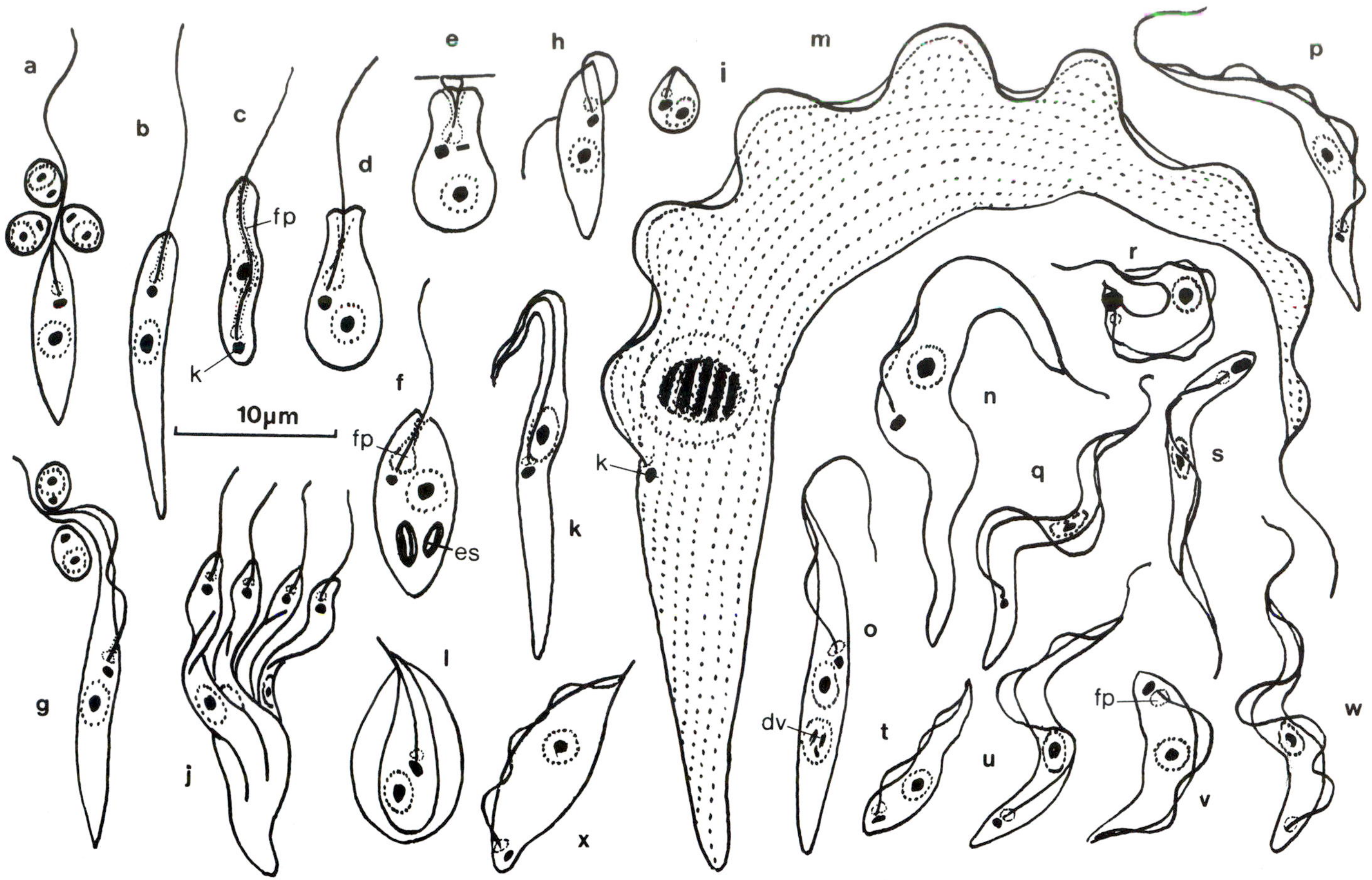

Kinetoplastida
Morphology of trypanosomatid mastigotes. a. *Leptomonas oncopelti* (PM) with "strap hanger" cysts. b. *Herpetomonas muscarum* (PM). c. *H. muscarum* (OPM). d. *Crithidia fasciculata* (CHM, nectomonad). e. *C. fasciculata* (CHM, haptomonad). f. *C. oncopelti* (CHM) with endosymbionts. g. *Blastocrithidia familiaris* (EPM) with cysts. h. *Leishmania major* (PM). i. *L. major* (AM). j. *Phytomonas elmassiani* (PM), multiple fission stage in plant latex. k. *Rhynchoidomonas drosophilae* (TPM). 1. *Endotrypanum schaudinni* (EPM) in sloth red blood cell. m. *Trypanosoma grayi* (TPM) from crocodile blood. n. *T. (Megatrypanum) cyclops* (TPM) from blood of Macaque. o. *T. cyclops* (EPM, with pigment in digestive vacuole) from culture. p. *T. (Herpetosoma) musculi* (TPM) from mouse blood. q. *T. (Tejeraia) rangeli* (TPM) from human blood. r. *T. (Schizotrypanum) dionisii* (TPM) from pipistrelle bat. s. *T. (Duttonella) vivax* and t. *T. (Nannomonas) congolense* (TPM), both from cattle blood. u. *T. brucei* (TPM, slender bloodstream form). v. *T. brucei* (TPM, short stumpy form). w. *T. evansi* (TPM, dyskinetoplastic) from camel. x. *T. (Pycnomonas) suis* (TPM) from pig blood. AM=amastigote; CHM=choanomastigote; EPM=epimastigote; OPM=opisthomastigote; PM=promastigote; TPM=trypomastigote. dv=digestive vacuole containing bacteria; es=endosymbionts; fp=undulipodial pocket; k=kinetoplast.

Kolkwitziellales
Order in phylum Dinomastigota.

Komokiacea
Superfamily in phylum Granuloreticulosa.

Komokiidae
Family in phylum Granuloreticulosa.

Kudoidae
Family in phylum Myxozoa.

Kybotiaceae
Family in phylum Chrysophyta.

L

Labyrinthulea
Only class in phylum Labyrinthulomycota.

Labyrinthulida
Only order in phylum Labyrinthulomycota.

Labyrinthulidae
Family in phylum Labyrinthulomycota.

Labyrinthulomycota

Labyrinthulids, slime nets, and thraustochytrids. Phylum of heterotrophic protoctists that produce an extracellular matrix (a wall-less ectoplasmic network), called a slime network, which absorbs nutrients and attaches the cells within it to surfaces. Ectoplasmic networks are devoid of cytoplasmic constituents; they are produced by cell organelles called sagenogens. Cells divide within the network; in some genera cells show gliding motility. Reproduction by breakup of the net or by propagules (heterokont bimastigote zoospores). The slime net in thraustochytrids is reduced, and the extracellular material is hardened into a structure that resembles superficially a chytrid thallus. Meiotic sexuality observed in at least one species. Saprotrophic to weakly symbiotrophic. Found in marine and estuarine environments.

Lacosteinidae
Family in phylum Granuloreticulosa.

Lagenida
Order in phylum Granuloreticulosa.

Lagenidiaceae
Family of phylum Oomycota.

Lagynidae
Family in phylum Granuloreticulosa.

Lagyniina
Suborder in phylum Granuloreticulosa.

Laminariaceae
Family in phylum Phaeophyta.

Laminariales
Order in phylum Phaeophyta.

Lankesterellidae
Family in phylum Apicomplexa.

Larcoidae
Family in phylum Actinopoda.

Lasiodiscidae
Family in phylum Granuloreticulosa.

Lauderiaceae
Family in phylum Bacillariophyta.

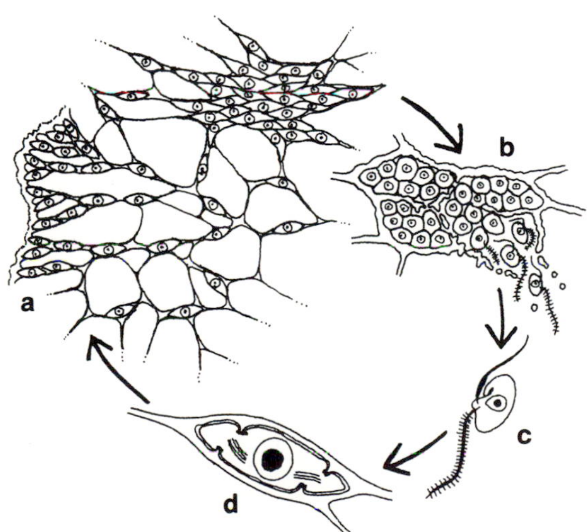

Labyrinthulomycota
Life cycle of *Labyrinthula*. Spindle-shaped trophic cells (a) aggregate to form sporogenous masses (b); enlarged cells undergo meiosis and release zoospores (c), which probably develop into new trophic cells (d).

Leathesiaceae
Family in phylum Phaeophyta.

Lecudinidae
Family in phylum Apicomplexa.

Legerellidae
Family in phylum Apicomplexa.

Lepidocyclinidae
Family in phylum Granuloreticulosa.

Lepidorbitoididae
Family in phylum Granuloreticulosa.

Leptocylindraceae
Family in phylum Bacillariophyta.

Leptocylindrales
Order in phylum Bacillariophyta.

Leptodiscaceae
Family in phylum Dinomastigota.

Leptolegniellaceae
Family in phylum Oomycota.

Leptomitaceae
Family in phylum Oomycota.

Leptomitales
Order in phylum Oomycota.

Lessoniaceae
Family in phylum Phaeophyta.

Liceaceae
Family in phylum Plasmodial Slime Molds.

Liceales
Order in phylum Plasmodial Slime Molds.

Licmophoraceae
Family in phylum Bacillariophyta.

Licnophorida
Order in phylum Ciliophora.

Linderinidae
Family in phylum Granuloreticulosa.

Lithodesmiaceae
Family in phylum Bacillariophyta.

Lithodesmiales
Order in phylum Bacillariophyta.

Lithodesmiophycidae
Subclass in phylum Bacillariophyta.

Lithophylleae
Tribe in phylum Rhodophyta.

Lithophylloideae
Subfamily in phylum Rhodophyta.

Lithopteridae
Family in phylum Actinopoda.

Lithothamnieae
Tribe in phylum Rhodophyta.

Lithothamnioideae
Subfamily in phylum Rhodophyta.

Lithotricheae
Tribe in phylum Rhodophyta.

Litostomatea
Class in phylum Ciliophora.

Lituolacea
Superfamily in phylum Granuloreticulosa.

Lituolidae
Family in phylum Granuloreticulosa.

Lituoliporidae
Family in phylum Granuloreticulosa.

Lituotubidae
Family in phylum Granuloreticulosa.

Lobosea
Class in phylum Rhizopoda.

Loeblichiidae
Family in phylum Granuloreticulosa.

Loftusiacea
Superfamily in phylum Granuloreticulosa.

Loftusiidae
Family in phylum Granuloreticulosa.

Lophodiniaceae
Family in phylum Dinomastigota.

Lophomonadidae
Family in phylum Zoomastigina.

Loxodida
Order in phylum Ciliophora.

Loxostomatidae
Family in phylum Granuloreticulosa.

Lyrellaceae
Family in phylum Bacillariophyta.

Lyrellales
Order in phylum Bacillariophyta.

M

Mackinnoniidae
Family in phylum Apicomplexa.

Mallodendraceae
Family in phylum Xanthophyta.

Mallomonadaceae
Family in phylum Chrysophyta.

Mallomonadales
Order in phylum Chrysophyta.

Mamiellaceae
Family in phylum Chlorophyta.

Mamiellales
Order in phylum Chlorophyta.

Mantonellidae
Family in phylum Apicomplexa.

Marteiliidea
Class in phylum Paramyxea.

Mastogloiaceae
Family in phylum Bacillariophyta.

Mastogloiales
Order in phylum Bacillariophyta.

Mastophoreae
Tribe in phylum Rhodophyta.

Mastophoroideae
Subfamily in phylum Rhodophyta.

Maylisoriidae
Family in phylum Granuloreticulosa.

Meandropsinidae
Family in phylum Granuloreticulosa.

Medusettidae
Family in phylum Actinopoda.

Melobesioideae
Subfamily in phylum Rhodophyta.

Melonidae
Family in phylum Granuloreticulosa.

Melosiraceae
Family in phylum Bacillariophyta.

Melosirales
Order in phylum Bacillariophyta.

Merocystidae
Family in phylum Apicomplexa.

Merogregarinidae
Family in phylum Apicomplexa.

Mesostigmataceae
Family in phylum Chlorophyta.

Mesotaeniaceae
Family in phylum Conjugaphyta.

Metchnikovellida
Order in phylum Microspora.

Metchnikovellidae
Family in phylum Microspora.

Micractiniaceae
Family in phylum Chlorophyta.

Micromonadaceae
Family in phylum Chlorophyta.

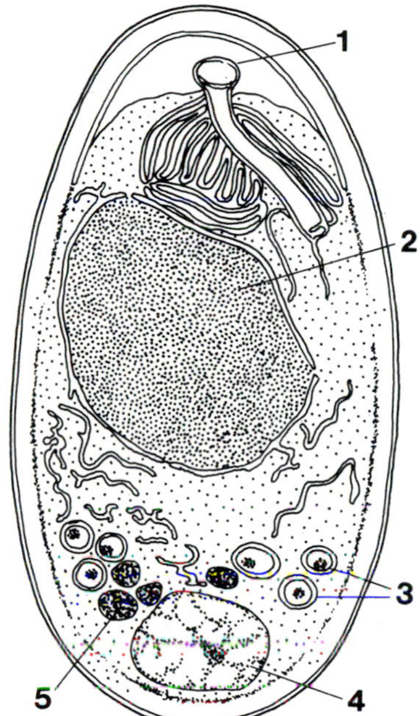

Microspora
Developing spore. 1=polar sac; 2=nucleus; 3=polar tubes; 4=posterior vacuole; 5=vesicle.

Microspora
Phylum of minute unicellular symbiotrophs causing single-cell tumors in a vast array of insects and other animals. All lack mitochondria. Propagules are spores that produce a polar tube deployed in the inoculation of the host with no damage to the host cell membrane. Penetration of the animal tissue is through this unprecedented inoculation device. Sexual fusion reported for some species.

Microsporea
Class in phylum Microspora.

Microsporida
Order in phylum Microspora.

Microthamniales
Order in phylum Chlorophyta of exclusively freshwater chlorophytes including several common phycobionts of lichens. Occur as solitary cell packets or branched filaments. Propagate by autospores, aplanospores (esp. symbiotic taxa), or naked, bimastigote zoospores. Sexuality unknown.

Microthoracida
Order in phylum Ciliophora.

Miliolacea
Superfamily in phylum Granuloreticulosa.

Miliolida
Order in phylum Granuloreticulosa.

Miliolidae
Family in phylum Granuloreticulosa.

Milioliporidae
Family in phylum Granuloreticulosa.

Millettiidae
Family in phylum Granuloreticulosa.

Minisporida
Order in phylum Microspora.

Miogypsinidae
Family in phylum Granuloreticulosa.

Mischococcaceae
Family in phylum Xanthophyta.

Mischococcales
Order in phylum Xanthophyta.

Mississippinidae
Family in phylum Granuloreticulosa.

Mobilida
Order in phylum Ciliophora.

Monoblepharidaceae
Family in phylum Chytridiomycota.

Monoblepharidales
Order in phylum Chytridiomycota.

Monocercomonae
Family in phylum Zoomastigina.

Monocystidae
Family in phylum Apicomplexa.

Monodopsidaceae
Family in phylum Eustigmatophyta.

Monoductidae
Family in phylum Apicomplexa.

Monostromataceae
Family in phylum Chlorophyta.

Moravamminacea
Superfamily in phylum Granuloreticulosa.

Moravamminidae
Family in phylum Granuloreticulosa.

Mrazekiidae
Family in phylum Microspora.

Multivalvulida
Order in phylum Myxozoa.

Myrionemataceae
Family in phylum Phaeophyta.

Myriosporidae
Family in phylum Apicomplexa.

Myriotrichiaceae
Family in phylum Phaeophyta.

Myxidiidae
Family in phylum Myxozoa.

Myxobolidae
Family in phylum Myxozoa.

Myxochloridaceae
Family in phylum Xanthophyta.

Myxochrysidaceae
Family in phylum Chrysophyta.

Myxogastromycetidae
Subclass in phylum Plasmodial Slime Molds.

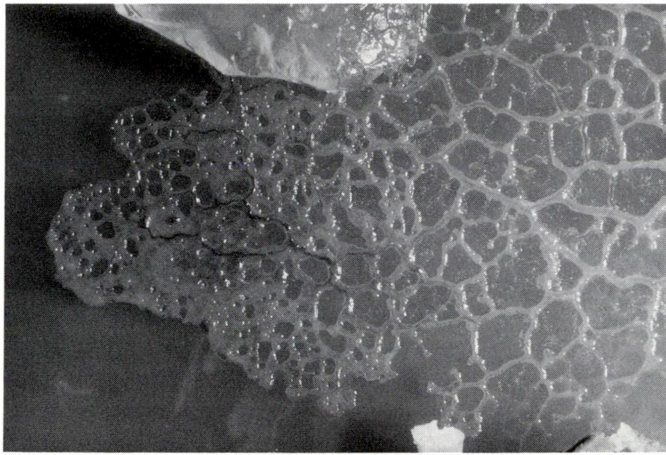

Myxomycota
Plasmodium of *Didymium iridis*.

Myxomycota
Largest and most diverse class in phylum Plasmodial Slime Molds. Phagotrophic bactivorous organisms form plasmodia. Propagate by spores shed by sporophore (stalked spore-bearing structure). Spores germinate to form ameboid (myxameba) or undulipodiated cells (mastigote swarmers); each type is a potential gamete or can develop into plasmodium, in many species by synchronous division of plasmodial nuclei. Plasmodium has a reversible type of protoplasmic streaming and ability to increase in size by coalescing with other compatible plasmodia.

Myxosporea
Class in phylum Myxozoa.

Myxozoa
Phylum of obligate symbiotrophs that produce multicellular spores, deploy polar capsules that penetrate and attach to animal tissue (e.g., oligochaetes, sipunculids, fish, and other vertebrates). Ameboid cells are released through valves. Symbiotrophic (including necrotrophic) heterotrophy is by ameboid cells or plasmodia. Plasmodia are formed by buds (internal or external) or by binary or multiple karyokinesis. Two classes are Myxosporea and Actinosporea.

N

Naegeliellaceae
Family in phylum Chrysophyta.

Nassellarida
Order in phylum Actinopoda.

Nassophorea
Class in phylum Ciliophora.

Nassophoria
Subclass in phylum Ciliophora.

Nassulida
Order in phylum Ciliophora.

Nassulina
Suborder in phylum Ciliophora.

Nautococcaceae
Family in phylum Chlorophyta.

Naviculaceae
Family in phylum Bacillariophyta.

Naviculales
Order in phylum Bacillariophyta.

Naviculineae
Suborder in phylum Bacillariophyta.

Neidiaceae
Family in phylum Bacillariophyta.

Neidiineae
Suborder in phylum Bacillariophyta.

Nemaliales
Order in phylum Rhodophyta.

Nematochrysidaceae
Family in phylum Chrysophyta.

Neocallimasticaceae
Family in phylum Chytridiomycota.

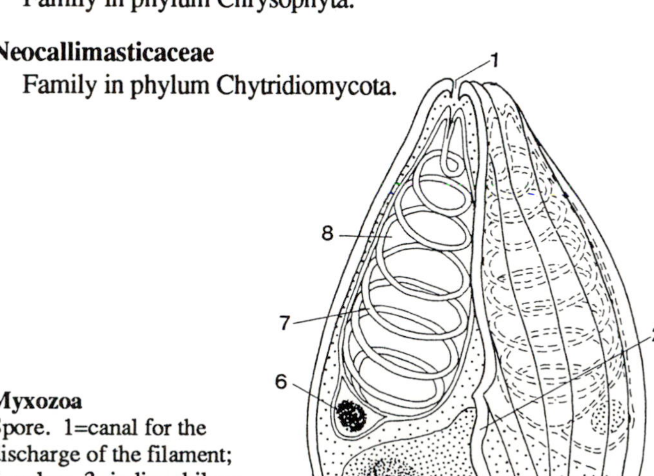

Myxozoa
Spore. 1=canal for the discharge of the filament; 2=valve; 3=iodinophilous vacuole; 4=sporoplasm; 5=nuclei; 6=residual nucleus of the capsulogenic cell; 7=polar filament; 8=polar capsule.

Neogoniolithoneae
Tribe in phylum Rhodophyta.

Neogregarinida
Order in phylum Apicomplexa.

Neonemataceae
Family in phylum Xanthophyta.

Neoschwagerinidae
Family in phylum Granuloreticulosa.

Nephroselmidaceae
Family in phylum Chlorophyta.

Nezzazatidae
Family in phylum Granuloreticulosa.

Nivalidae
Family in phylum Actinopoda.

Noctilucaceae
Family in phylum Dinomastigota.

Noctilucales
Order in phylum Dinomastigota.

Nodosariacea
Superfamily in phylum Granuloreticulosa.

Nodosariidae
Family in phylum Granuloreticulosa.

Nodosinellacea
Superfamily in phylum Granuloreticulosa.

Nodosinellidae
Family in phylum Granuloreticulosa.

Nonionacea
Superfamily in phylum Granuloreticulosa.

Nonionidae
Family in phylum Granuloreticulosa.

Nosematidae
Family in phylum Microspora.

Notheiaceae
Family in phylum Phaeophyta.

Notodendrodidae
Family in phylum Granuloreticulosa.

Nouriidae
Family in phylum Granuloreticulosa.

Nubeculariidae
Family in phylum Granuloreticulosa.

Nummulitacea
Superfamily in phylum Granuloreticulosa.

Nummulitidae
Family in phylum Granuloreticulosa.

O

Oberhauserellidae
Family in phylum Granuloreticulosa.

Ochromonadaceae
Family in phylum Chrysophyta.

Ochromonadales
Order in phylum Chrysophyta.

Odontostomatida
Order in phylum Ciliophora.

Oedogoniomycetaceae
Family in phylum Chytridiomycota.

Oligohymenophorea
Class in phylum Ciliophora.

Oligotrichida
Order in phylum Ciliophora.

Olpidiaceae
Family in phylum Chytridiomycota.

Oocystaceae
Family in phylum Chlorophyta.

Oodiniaceae
Family in phylum Dinomastigota.

Oomycota
Phylum of conjugating anisogamontous protoctists. Heterotrophic or osmotrophic in freshwater environments, or symbiotrophic on plants. Uninucleate or coenocytic with haplomitotic B ploidy cycle. Undulipodiated heterokonts are zoospores, not gametes; sexuality is by conjugation of nonmotile differentiated (male and female) hyphae.

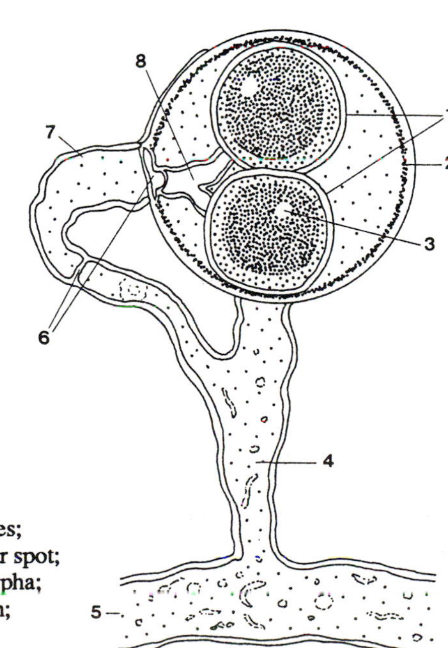

Oomycota
Fertilization. 1=oospores; 2=oogonium; 3=nuclear spot; 4=hyphal branch; 5=hypha; 6=septa; 7=antheridium; 8=fertilization tube.

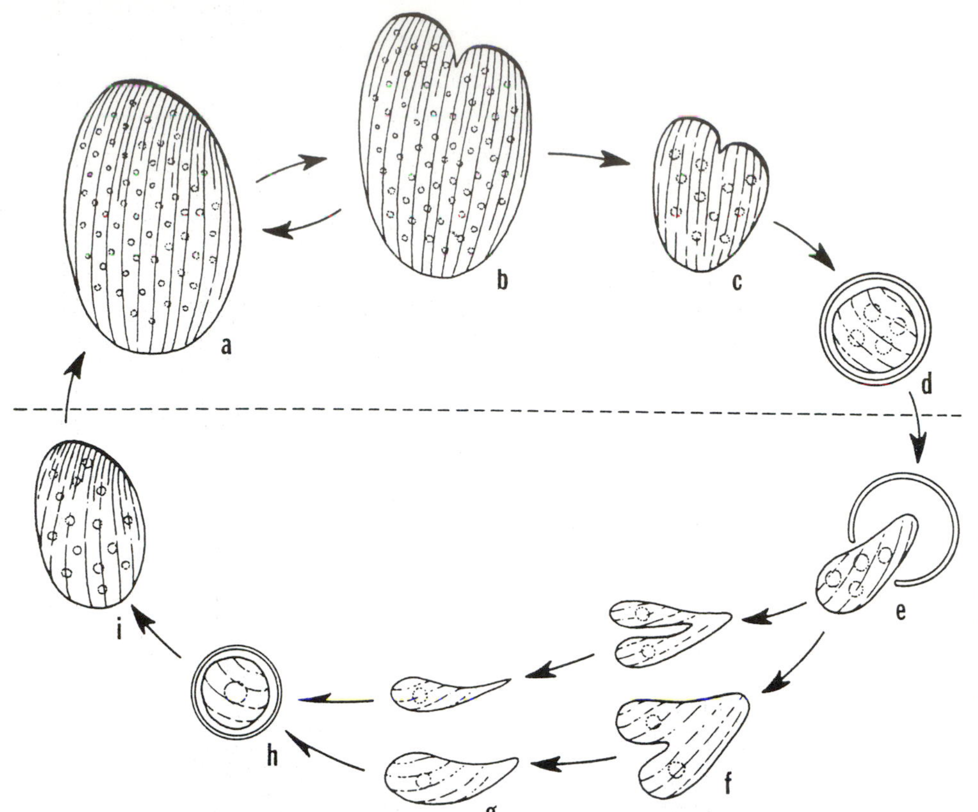

Opalinata
Partial life cycle of *Opalina*. a. trophont. b. mitotic division. c. palinotomic division. d. infective cyst. e. form emerging from cyst. f. gamonts. g. uninucleate gametes. h. encysted zygote (zygocyst). i. young trophont. Stages a to d occur in adult frogs; d passes into water; e to i occur in tadpoles.

Opalinata
Opalinids; class in phylum Zoomastigina. Large heterotrophic protists, mastigotes motile by numerous surface undulipodia. Symbiotrophic in the digestive system (cloaca) of poikilotherm vertebrates, mostly anuran amphibians. Opalinid cells contain two or many homokaryotic nuclei, lack cytostomes, and display fertilization and cyst formation.

Opalinida
Order in phylum Zoomastigina.

Opalinidae
Family in phylum Zoomastigina.

Ophiocytaceae
Sciadiaceae. Family in phylum Xanthophyta.

Ophryoglenina
Suborder in phylum Ciliophora.

Ophthalmidiidae
Family in phylum Granuloreticulosa.

Orbitoclypeidae
Family in phylum Granuloreticulosa.

Orbitoidacea
Superfamily in phylum Granuloreticulosa.

Orbitoididae
Family in phylum Granuloreticulosa.

Orbitolinacea
Superfamily in phylum Granuloreticulosa.

Orbitolinidae
Family in phylum Granuloreticulosa.

Orbitopsellidae
Family in phylum Granuloreticulosa.

Oridorsalidae
Family in phylum Granuloreticulosa.

Ormieractinomyxidae
Family in phylum Myxozoa.

Ortholineidae
Family in phylum Myxozoa.

Osangulariidae
Family in phylum Granuloreticulosa.

Ostreobiaceae
Family in phylum Chlorophyta.

Ostreopsidaceae
Family in phylum Dinomastigota.

Oxinoxisidae
Family in phylum Granuloreticulosa.

Oxyphysaceae
Family in phylum Dinomastigota.

Oxyrrhinaceae
Family in phylum Dinomastigota.

Oxyrrhinales
Order in phylum Dinomastigota.

Oxytoxaceae
Family in phylum Dinomastigota.

Ozawainellidae
Family in phylum Granuloreticulosa.

P

Pachyphloiidae
Family in phylum Granuloreticulosa.

Palaeospiroplectamminidae
Family in phylum Granuloreticulosa.

Palaeotextulariacea
Superfamily in phylum Granuloreticulosa.

Palaeotextulariidae
Family in phylum Granuloreticulosa.

Palmariales
Order in phylum Rhodophyta.

Palmellopsidaceae
Family in phylum Chlorophyta.

Palmodictyaceae
Family in phylum Chlorophyta.

Pannellainidae
Family in phylum Granuloreticulosa.

Pansporoblastina
Suborder in phylum Microspora.

Parabasalia
Parabasalians; class in phylum Zoomastigina. Uninucleate cells symbiotrophic in animals, containing one or more parabasal bodies (modified golgi apparatus) usu. associated with nuclei and kinetids. Heterotrophic mastigotes with few to hundreds of thousands of undulipodia. All lack mitochondria. Some have microtubular axostyles and distinctive undulating membranes. Includes three orders: trichomonads, polymonads, and hypermastigotes.

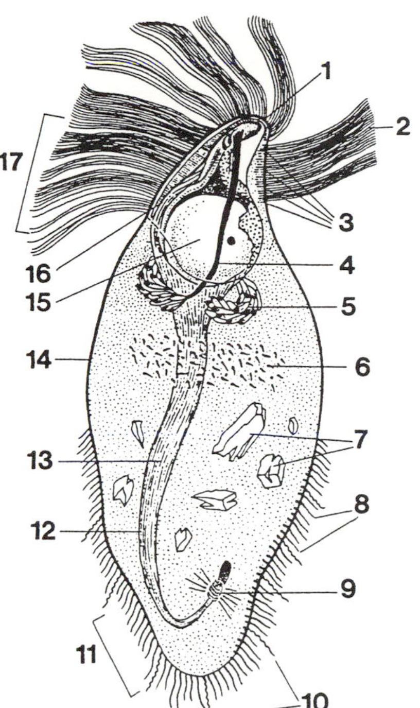

Parabasalia
Joenia annectens. 1=pelta; 2=undulipodia; 3=kinetosomes; 4=parabasal fold; 5=parabasal body (golgi body); 6=periaxostylar bacteria; 7=fragments of wood; 8=epibiotic bacteria; 9=periaxostylar ring; 10=epibiotic spirochetes; 11=ingestive zone; 12=axostyle; 13=axostylar microtubule; 14=cell membrane; 15=nucleus; 16=axostylar cap; 17=rostrum (anterior end).

Parahymenostomatina
Suborder in phylum Ciliophora.

Paraliaceae
Family in phylum Bacillariophyta.

Paraliales
Order in phylum Bacillariophyta.

Parameciina
Suborder in phylum Ciliophora.

Paramoebidae
Family in phylum Rhizopoda.

Paramyxea
Phylum of amastigote unicellular symbiotrophs of marine animals. Form propagules (spores) consisting of several cells enclosed inside each other arising by a process of internal cleavage or endogenous budding within an ameboid stem cell.

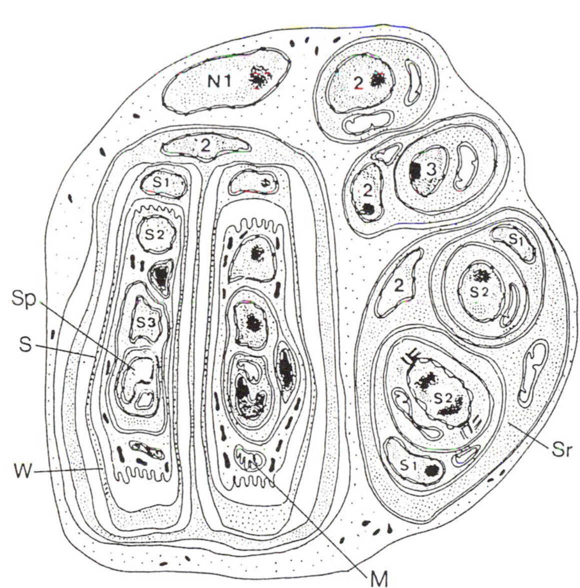

Paramyxea
The development of *Paramyxa paradoxa* (clockwise) in host cytoplasm. Only two of the four spores are shown in the young sporont (bottom right) and in the mature one (bottom left). 2=nucleus of secondary (stem) cell; 3=tertiary cell nucleus; M=mitochondrion; N1=stem cell nucleus; S1, S2, S3=nuclei of sporal cells 1, 2, 3; S=spore; Sp=sporoplasm nucleus; Sr=sporont; W=spore wall.

Paramyxidea
Class in phylum Paramyxea.

Paraphysomonadaceae
Family in phylum Chrysophyta.

Parathuramminacea
Superfamily in phylum Granuloreticulosa.

Parathuramminidae
Family in phylum Granuloreticulosa.

Parathuramminiina
Suborder in phylum Granuloreticulosa.

Paratikhinellidae
Family in phylum Granuloreticulosa.

Parrelloididae
Family in phylum Granuloreticulosa.

Partisaniidae
Family in phylum Granuloreticulosa.

Parvicapsulidae
Family in phylum Myxozoa.

Patellinidae
Family in phylum Granuloreticulosa.

Pavlovales
Order in phylum Prymnesiophyta.

Pavoninidae
Family in phylum Granuloreticulosa.

Pedinellaceae
Family in phylum Chrysophyta.

Pedinellales
Order in phylum Chrysophyta.

Pedinellophyceae
Class in phylum Chrysophyta.

Pedinomonadales
Order in phylum Chlorophyta. Small, naked mastigotes composed of flattened and asymmetric cells with one undulipodium laterally to subapically inserted, a nonfunctional kinetosome, and chloroplasts. Marine or freshwater; two species symbiotic. Sexuality unknown.

Pegidiidae
Family in phylum Granuloreticulosa.

Peneroplidae
Family in phylum Granuloreticulosa.

Peniculida
Order in phylum Ciliophora.

Pentacapsulidae
Family in phylum Myxozoa.

Pereziidae
Family in phylum Microspora.

Periaxoplastidiata
Superfamily in phylum Actinopoda.

Peridiniaceae
Family in phylum Dinomastigota.

Peridiniales
Order in phylum Dinomastigota.

Peritrichia
Subclass in phylum Ciliophora.

Peroniaceae
Family in phylum Bacillariophyta.

Peronosporaceae
Family in phylum Oomycota.

Peronosporales
Order in phylum Oomycota.

Peronosporomycetidae
Class (subclass) in phylum Oomycota.

Perryaceae
Family in phylum Bacillariophyta.

Pfeifferinellidae
Family in phylum Apicomplexa.

Pfenderinidae
Family in phylum Granuloreticulosa.

Phacodiniida
Order in phylum Ciliophora.

Phacotaceae
Family in phylum Chlorophyta.

Phaeocalpida
Order in phylum Actinopoda.

Phaeoconchia
Order in phylum Actinopoda.

Phaeocystida
Order in phylum Actinopoda.

Phaeodaria
Class in phylum Actinopoda. Large spherical solitary cells with siliceous skeletons consisting of isolated pieces or numerous hollow tubes. Some lack skeletons. Spheres in ectoplasm develop into polynucleated ameboids and eventually lead to bimastigote propagule formation. Marine planktonic radiolarialike protists. Illustration: Actinopoda.

Phaeodendrida
Order in phylum Actinopoda.

Phaeodermatiaceae
Family in phylum Chrysophyta.

Phaeodinidae
Family in phylum Actinopoda.

Phaeogromida
Order in phylum Actinopoda.

Phaeogymnocellida
Order in phylum Actinopoda.

Phaeophyceae
Only class in phylum Phaeophyta.

Phaeophyta
Brown algae; brown seaweeds. Phylum containing some of the largest multicellular protoctists. Algae are reproduced from heterokont mastigotes or zygotes formed by fusion of eggs with heterokont male gametes. Exclusively marine in subtidal and intertidal zones. May alternate diploid and haploid generations. Plastids contain chlorophylls *a*, *c*, and *c1* and fucoxanthin. Laminarin as storage material.

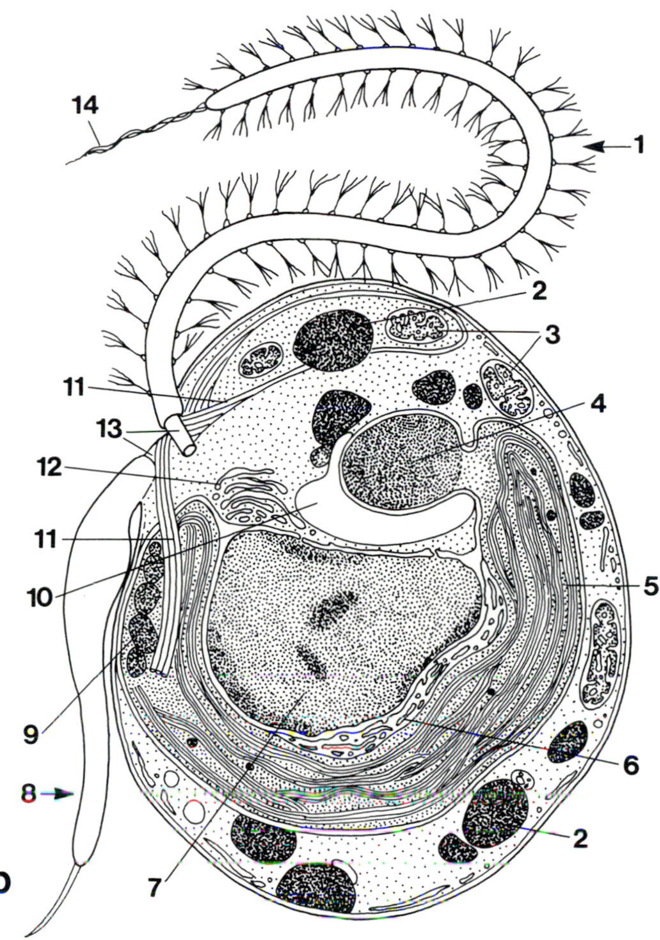

Phaeophyta
a. *Fucus vesiculosis*. b. Gamete. 1=anterior undulipodium; 2=physode; 3=mitochondria; 4=pyrenoid; 5=plastid; 6=periplastidial reticulum; 7=nucleus; 8=posterior undulipodium; 9=eyespot; 10=pyrenoid cap; 11=undulipodial root; 12=golgi; 13=kinetosomes; 14=terminal acroneme.

Phaeoplacaceae
Family in phylum Chrysophyta.

Phaeosacciaceae
Family in phylum Chrysophyta.

Phaeosphaerida
Order in phylum Actinopoda.

Phaeosphaeridae
Family in phylum Actinopoda.

Phaeothamniaceae
Family in phylum Chrysophyta.

Phaeothamniales
Order in phylum Chrysophyta.

Phaneraxohelida
Order in phylum Actinopoda.

Pharactopeltidae
Family in phylum Actinopoda.

Pharyngophorida
Order in phylum Ciliophora.

Philasterina
Suborder in phylum Ciliophora.

Phragmonemataceae
Family in phylum Rhodophyta.

Phthanotrochidae
Family in phylum Granuloreticulosa.

Phyllacantha
Suborder in phylum Actinopoda.

Phyllopharyngea
Class in phylum Ciliophora.

Phyllopharyngia
Subclass in phylum Ciliophora.

Phyllostauridae
Family in phylum Actinopoda.

Phymatolitheae
Tribe in phylum Rhodophyta.

Physaraceae
Family in phylum Plasmodial Slime Molds.

Physarales
Order in phylum Plasmodial Slime Molds.

Physematidae
Family in phylum Actinopoda.

Physodermataceae
Family in phylum Chytridiomycota.

Phytodiniaceae
Family in phylum Dinomastigota.

Phytodiniales
Order in phylum Dinomastigota.

Pilisuctorida
Order in phylum Ciliophora.

Pinnulariaceae
Family in phylum Bacillariophyta.

Piroplasmida
Order in phylum Apicomplexa.

Placentulinidae
Family in phylum Granuloreticulosa.

Placopsilinidae
Family in phylum Granuloreticulosa.

Plagiogrammaceae
Family in phylum Bacillariophyta.

Plagiogrammales
Order in phylum Bacillariophyta.

Plagiopylia
Subclass in phylum Ciliophora.

Plagiopylida
Order in phylum Ciliophora.

Plagiotomida
Order in phylum Ciliophora.

Plagiotropidaceae
Family in phylum Bacillariophyta.

Planomalinacea
Superfamily in phylum Granuloreticulosa.

Planomalinidae
Family in phylum Granuloreticulosa.

Planorbulinacea
Superfamily in phylum Granuloreticulosa.

Planorbulinidae
Family in phylum Granuloreticulosa.

Planulinidae
Family in phylum Granuloreticulosa.

Planulinoididae
Family in phylum Granuloreticulosa.

Plasmodial Slime Molds →
Phylum of phagotrophic, bactivorous, soil-, dung-, and plant debris-dwelling organisms that develop from spores borne in sporophores. Spores germinate to form amebas that develop into plasmodia (rate of karyokinesis exceeds that of cytokinesis). Conspicuous cyclosis in plasmodium. Can form mastigote and ameba stages as well as sclerotia (dry propagules). See *Myxomycota, Protostelida.*

Plasmodiophoraceae
Only family in phylum Plasmodiophoromycota.

Plasmodiophorales
Only order in phylum Plasmodiophoromycota.

Plasmodiophoromycetes
Only class in phylum Plasmodiophoromycota.

Plasmodiophoromycota
Phylum of soil and freshwater, obligate symbiotrophs (including necrotrophs) of many plants, fungi, and other protoctists. Multinucleate unwalled protoplasts (plasmodia) develop either into sporangia, which produce zoospores with two anteriorly-directed whiplash undulipodia, or cystosori, which form resting bodies that are aggregations of thick-walled, uninucleate cells. Cells show cruciform division. Meiosis thought to occur based on presence of synaptonemal complexes in some species. Illustration page 178.

Platysporina
Suborder in phylum Myxozoa.

Plectoidea
Family in phylum Actinopoda.

Plectorecurvoididae
Family in phylum Granuloreticulosa.

Pleurochloridaceae
Family in phylum Xanthophyta.

Pleurochloridellaceae
Family in phylum Xanthophyta.

Pleuronematina
Suborder in phylum Ciliophora.

Pleurosigmataceae
Family in phylum Bacillariophyta.

Pleurostomatida
Order in phylum Ciliophora.

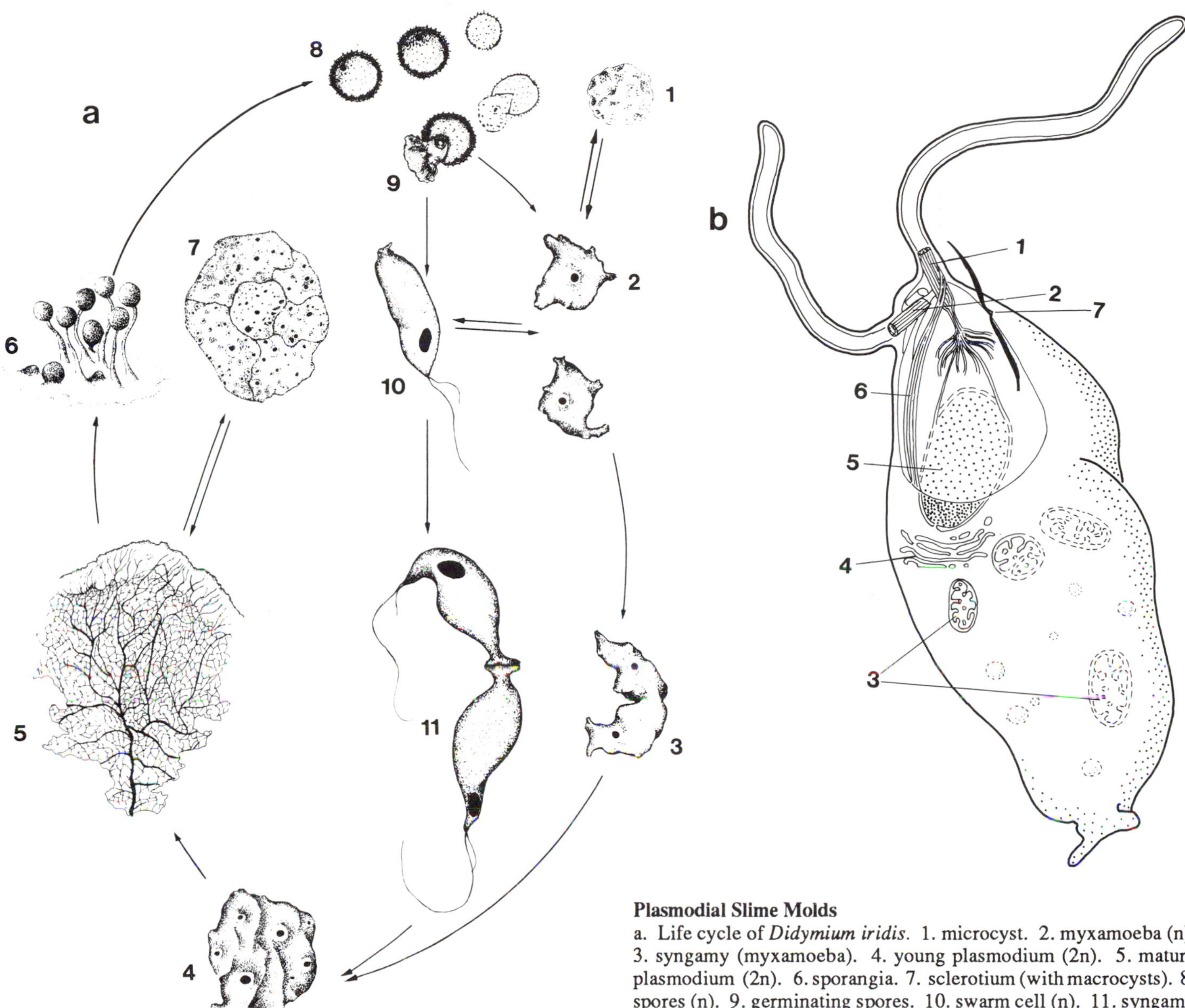

Plasmodial Slime Molds

a. Life cycle of *Didymium iridis*. 1. microcyst. 2. myxamoeba (n). 3. syngamy (myxamoeba). 4. young plasmodium (2n). 5. mature plasmodium (2n). 6. sporangia. 7. sclerotium (with macrocysts). 8. spores (n). 9. germinating spores. 10. swarm cell (n). 11. syngamy (swarm cells). b. Swarm cell. 1=anterior kinetosome; 2=posterior kinetosome; 3=mitochondria; 4=golgi; 5=nucleus; 6=outer microtubule array of rhizoplast; 7=rhizoplast.

Pleurostomellacea
Superfamily in phylum Granuloreticulosa.

Pleurostomellidae
Family in phylum Granuloreticulosa.

Podolampaceae
Family in phylum Dinomastigota.

Polycystina
Radiolaria. Class in phylum Actinopoda. Large solitary cells having regularly perforated silica skeletons with radial axopods emerging among fine ramified pseudopods. Mastigote propagules formed having two undulipodia, one emergent, and characterized by intracellular strontium sulfate crystal. No sexuality known. Extant organisms are marine; fossil record of polycystines dates from late Proterozoic Eon. Their skeletal debris may be the basis of formations of extensive marine silica deposits (radiolarite). Illustration: Actinopoda.

Polycyttaria
Family in phylum Actinopoda.

Polykrikaceae
Family in phylum Dinomastigota.

Polymonadida
Order in phylum Zoomastigina.

Polymorphinidae
Family in phylum Granuloreticulosa.

Polypyramidae
Family in phylum Actinopoda.

Polysaccamminidae
Family in phylum Granuloreticulosa.

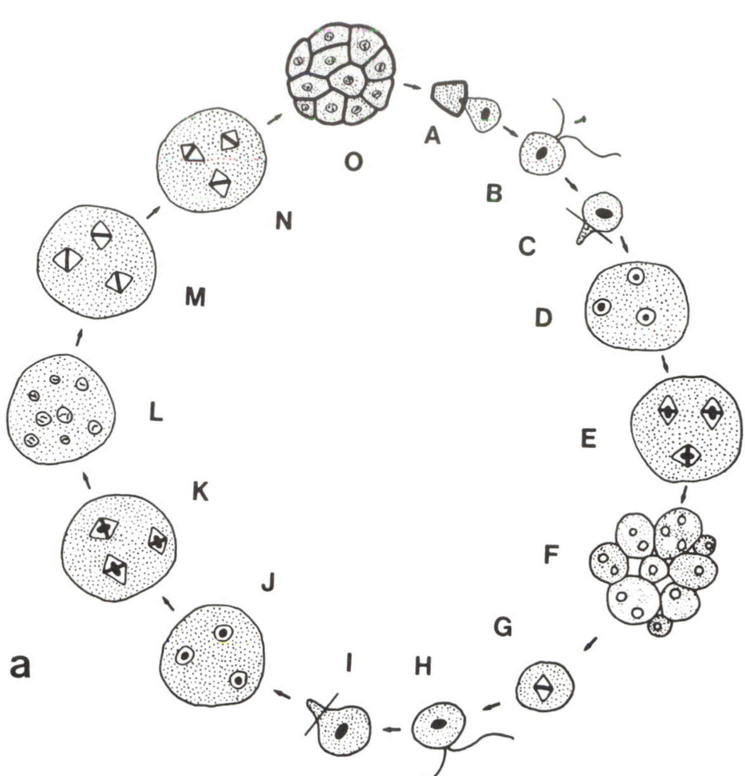

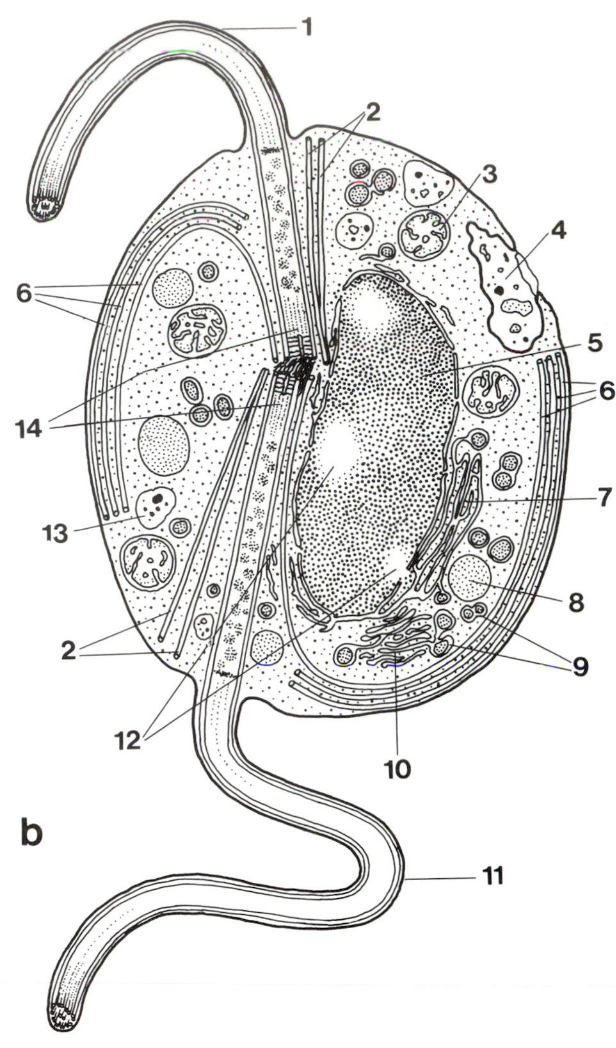

Plasmodiophoromycota

a.Plasmodiophorid life cycle. A. cyst germination. B. primary zoospore. C. infection. D. sporangial plasmodium. E. cruciform nuclear division in sporangial plasmodium. F. multilobed sporangium. G. mitosis in sporangial lobe. H. secondary zoospore. I. infection. J. cystosoral plasmodium. K. cruciform division in cystosoral plasmodium. L. pachynema of meiosis I (nuclei contain synaptonemal complexes). M. metaphase of meiosis I. N. metaphase of meiosis II. O. cystosorus composed of uninucleate cysts. b. Zoospore. 1=anterior undulipodium; 2=α-rootlets (kinetid microtubules); 3=mitochondria; 4=expulsion vacuole; 5=nucleus; 6=β-rootlets (kinetid microtubules); 7=endoplasmic reticulum; 8=lipid globule; 9=coated vesicles; 10=golgi; 11=posterior undulipodium; 12=electron-lucent areas in nucleoplasm; 13=vacuole; 14=kinetosomes.

Porospathidae
Family in phylum Actinopoda.

Porosporidae
Family in phylum Apicomplexa.

Porphyridiaceae
Family in phylum Rhodophyta.

Porphyridiales
Order in phylum Rhodophyta.

Postciliodesmatophora
Subphylum in phylum Ciliophora.

Praebuliminidae
Family in phylum Granuloreticulosa.

Prasinophyceae
Class in phylum Chlorophyta. Motile solitary green algae with cell bodies and undulipodia covered by nonmineralized organic scales. Undulipodia originate from a groove and golgi apparatus is in a parabasal position. Reproduction by binary division; no sexuality known.

Prasiolaceae
Family in phylum Chlorophyta.

Prasiolales
Order in phylum Chlorophyta. Multicellular flattened algae composed of walled, uninucleate cells. Reproduction primarily by aplanospores. Oogamous sexual reproduction known in a few species in which a bimastigote sperm has one undulipodium absorbed by the egg resulting in the formation of posteriorly unimastigote planozygote. Marine or freshwater.

Proaxoplastidiata
Superfamily in phylum Actinopoda.

Progonoiaceae
Family in phylum Bacillariophyta.

Proheterotrichida
Order in phylum Ciliophora.

Propeniculida
Order in phylum Ciliophora.

Prorocentraceae
Family in phylum Dinomastigota.

Prorocentrales
Order in phylum Dinomastigota.

Prorodontida
Order in phylum Ciliophora.

Proschkiniaceae
Family in phylum Bacillariophyta.

Prostomatea
Class in phylum Ciliophora.

Prostomatida
Order in phylum Ciliophora.

Proteromonadida
Proteromonads; class in phylum Zoomastigina. Small, symbiotrophic, nonmastigonemate, heterokont mastigotes in which a rhizoplast is associated with the golgi apparatus and the nucleus. Reproduction by multiple fission occurs in some species. Form resistant fecal cysts in the intestinal tract of many amphibians, reptiles, and mammals.

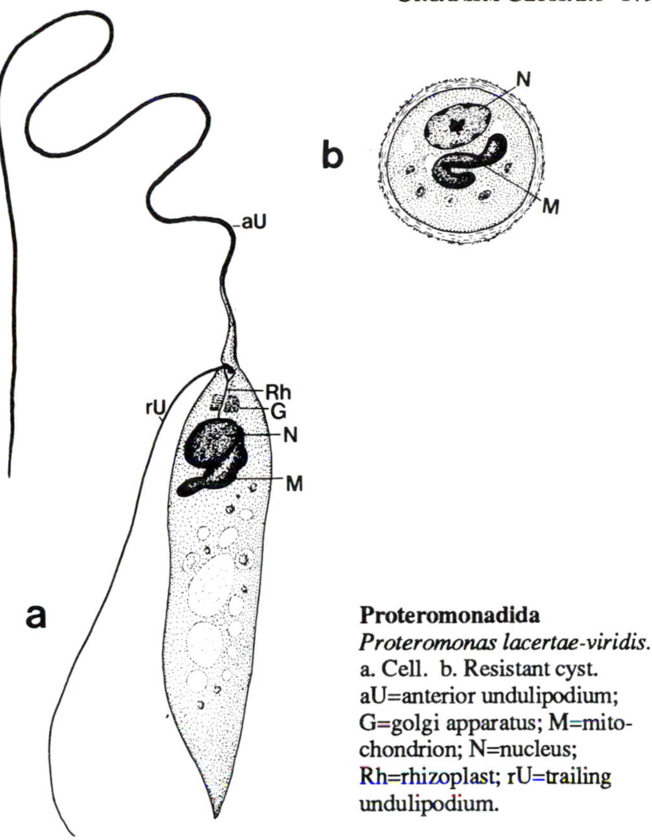

Proteromonadida
Proteromonas lacertae-viridis.
a. Cell. b. Resistant cyst. aU=anterior undulipodium; G=golgi apparatus; M=mitochondrion; N=nucleus; Rh=rhizoplast; rU=trailing undulipodium.

Proteromonadidae
Family in phylum Zoomastigina.

Protocruziida
Order in phylum Ciliophora.

Protoodiniaceae
Family in phylum Dinomastigota.

Protosiphonaceae
Family in phylum Chlorophyta.

Protostelida
Class in phylum Plasmodial Slime Molds. Sporocarp consists of a small delicate stalk bearing one to four spores. Growing stage ameboid; may also possess mastigote and plasmodial stages. Life history may be simple with one type of trophic cell, or complex, with several types. Found worldwide in soil, dung, or on living or dead plant parts.

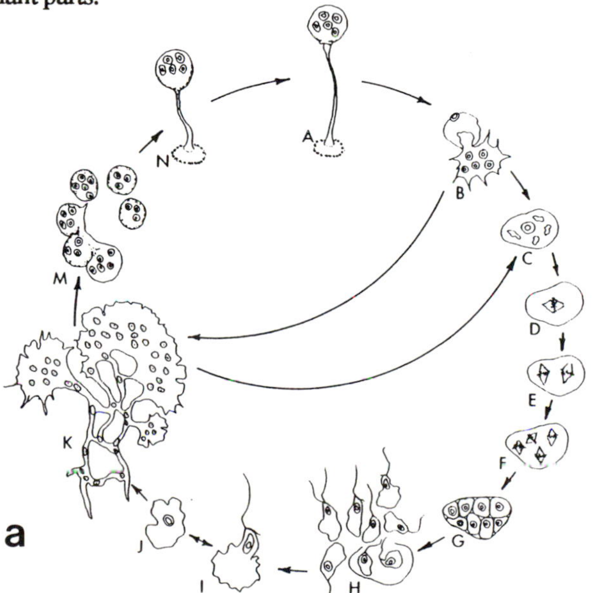

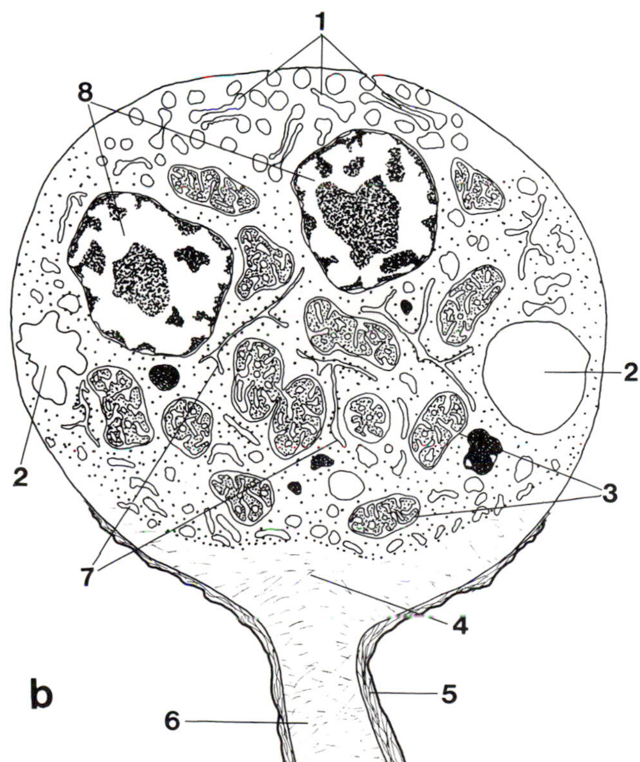

Protostelida
Ceratiomyxella tahitiensis. a. Life cycle: A. sporocarp. B. germinating protoplast. C. all but one of the nuclei degenerating in protoplast, or portion of plasmodium, which is converting into zoocyst. D-F. three nuclear divisions in zoocyst. G-H. eight or fewer mastigote cells cleaving and germinating from zoocyst. I-J. amebomastigote stage. K. plasmodium. M. plasmodium cleaving into prespore cells. N. rising sporogen. b. Rising sporogen. 1=secretory region at apex of sporogen; 2=contractile vacuole; 3=mitochondria; 4=actin-rich cortex at contractile base of sporogen; 5=stalk tube; 6=cytoplasmic plug of stalk (stalk plug); 7=rough endoplasmic reticulum; 8=nuclei.

Protosteliida
Family in phylum Plasmodial Slime Molds.

Protostomatida
Order in phylum Ciliophora.

Prunoidae
Family in phylum Actinopoda.

Prymnesiales
Order in phylum Prymnesiophyta.

Prymnesiophyceae
Class in phylum Prymnesiophyta.

Psammodiscaceae
Family in phylum Bacillariophyta.

Psammosphaeridae
Family in phylum Granuloreticulosa.

Pseudoammodiscidae
Family in phylum Granuloreticulosa.

Pseudobolivinidae
Family in phylum Granuloreticulosa.

Pseudocharaciopsidaceae
Family in phylum Eustigmatophyta.

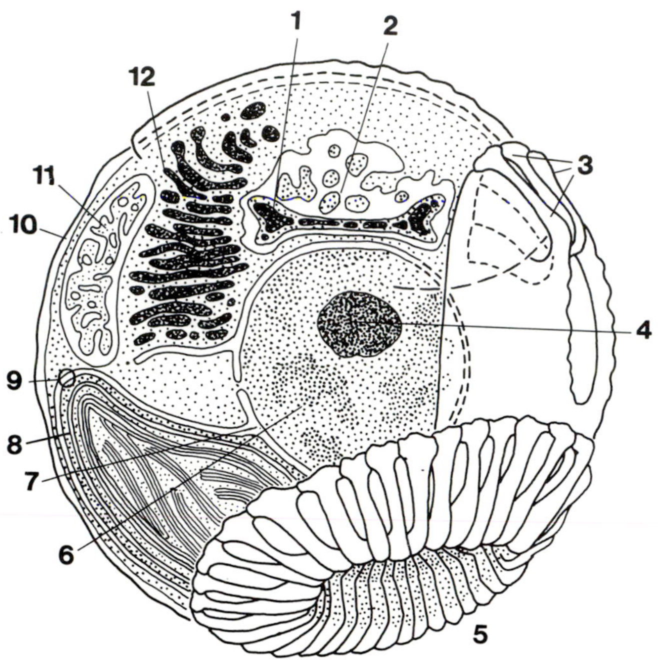

Prymnesiophyta
1=vacuole containing immature coccolith; 2=reticular body; 3=overlapping coccoliths; 4=nucleolus; 5=exterior of coccolithophorids; 6=chromatin; 7=nuclear envelope; 8=chloroplast; 9=chloroplast endoplasmic reticulum; 10=cell membrane; 11=mitochondrion; 12=golgi.

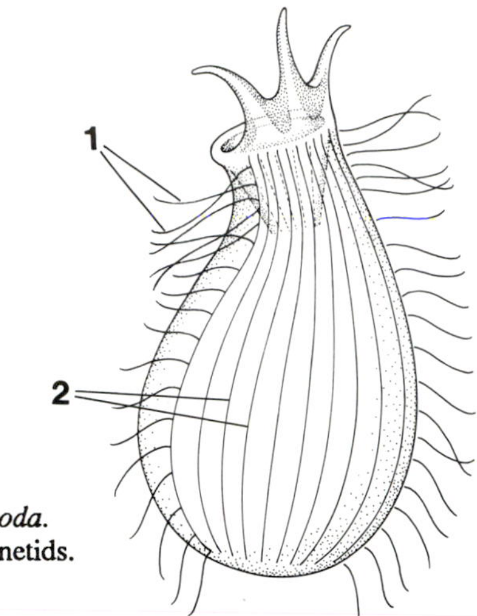

Pseudociliata
Stephanopogon colpoda.
1=undulipodia; 2=kinetids.

Prymnesiophyta
Phylum of yellow-brown algae, many covered with scales of varying degrees of complexity that may be unmineralized or calcified. Includes coccolithophorids, unicellular organisms with calcified plates (coccoliths). Many possess haptonemata, typically a filiform structure associated with the undulipodia. Generally marine; fossil coccolithophorids dating to Jurassic.

Psammettidae
Family in phylum Xenophyophora.

Psamminida
Class in phylum Xenophyophora.

Psamminidae
Family in phylum Xenophyophora.

Pseudociliata
Class in phylum Zoomastigina. Marine, benthic organisms with 2-16 homokaryotic nuclei and distinctive kinetids, kinetosomes connected by a desmose; cytostome- (mouth) cytopharyngeal apparatus supported by complex fibrillar system for active phagocytosis; reproduction by multiple cell division inside cyst. Formerly classified as ciliates, but lack the infraciliary features characteristic of ciliates. Feed on diatoms, small mastigotes, and bacteria. Monogeneric; four species of *Stephanopogon.*

Pseudoendothyridae
Family in phylum Granuloreticulosa.

Pseudohimantidiaceae
Family in phylum Bacillariophyta.

Pseudohimantidiales
Order in phylum Bacillariophyta.

Pseudolithidae
Family in phylum Actinopoda.

Pseudoparrellidae
Family in phylum Granuloreticulosa.

Pseudopleistophoridae
Family in phylum Microspora.

Pseudorbitoididae
Family in phylum Granuloreticulosa.

Pseudoscourfieldiaceae
Family in phylum Chlorophyta.

Pseudoscourfieldiales
Order in phylum Chlorophyta.

Pseudotaxidae
Family in phylum Granuloreticulosa.

Psycheneidaceae
Family in phylum Bacillariophyta.

Psyedoklossiidae
Family in phylum Apicomplexa.

Pterospermataceae
Family in phylum Chlorophyta.

Ptychocladiacea
Superfamily in phylum Granuloreticulosa.

Ptychocladiidae
Family in phylum Granuloreticulosa.

Pulleniatinidae
Family in phylum Granuloreticulosa.

Punctariaceae
Family in phylum Phaeophyta.

Pyramimonadaceae
Family in phylum Chlorophyta.

Pyramimonadales
Order in phylum Chlorophyta.

Pyrobotryaceae
Family in phylum Chlorophyta.

Pyrocystaceae
Family in phylum Dinomastigota.

Pyrocystales
Order in phylum Dinomastigota.

Pyrophacaceae
Family in phylum Dinomastigota.

Pyrsonymphida
Pyrsonymphids; class in phylum Zoomastigina. Heterotrophic mastigotes symbiotrophic in the hindguts of wood-eating cockroaches and termites; 4, 8, or 12 undulipodia and an intrinsically motile longitudinally aligned axostyle composed of laterally connected microtubules. Since all lack mitochondria, the group is presumed anaerobic.

Pyrsonymphidae
Family in phylum Zoomastigina.

Pythiaceae
Family in phylum Oomycota.

Pythiales
Order in phylum Oomycota.

Q

Quadrimorphinidae
Family in phylum Granuloreticulosa.

R

Radiolaria
Common name of polycystine and phaeodarian marine actinopods. See *Phaeodaria, Polycystina.*

Ralfsiaceae
Family in phylum Phaeophyta.

Raphidiophryidae
Family in phylum Actinopoda.

Raphidophyceae
Class in phylum Raphidcphyta.

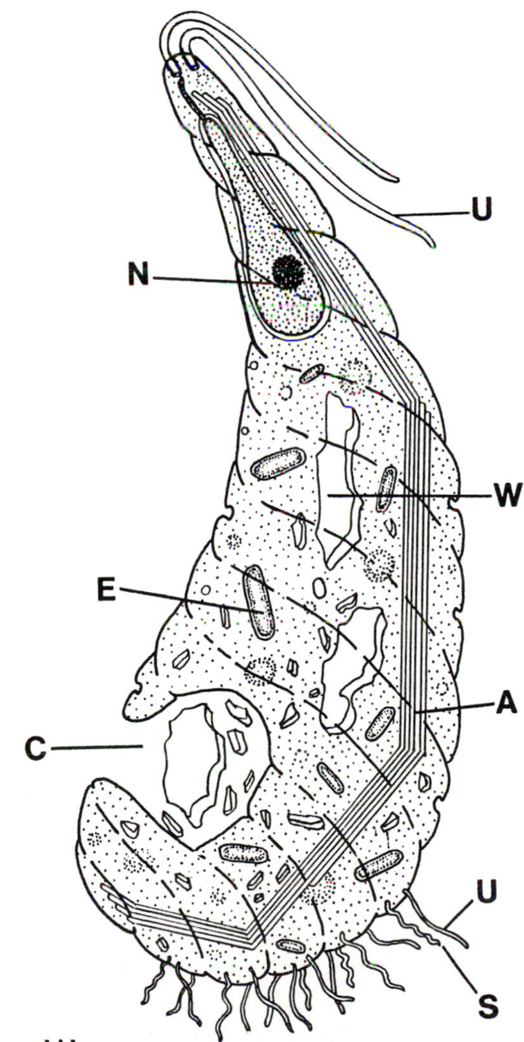

Pyrsonymphida
Pyrsonympha vertens. A=axostyle; C=cytostome; E=endosymbiotic bacteria; N=nucleus; S=symbiotic spirochetes; U=undulipodia; W=wood particle.

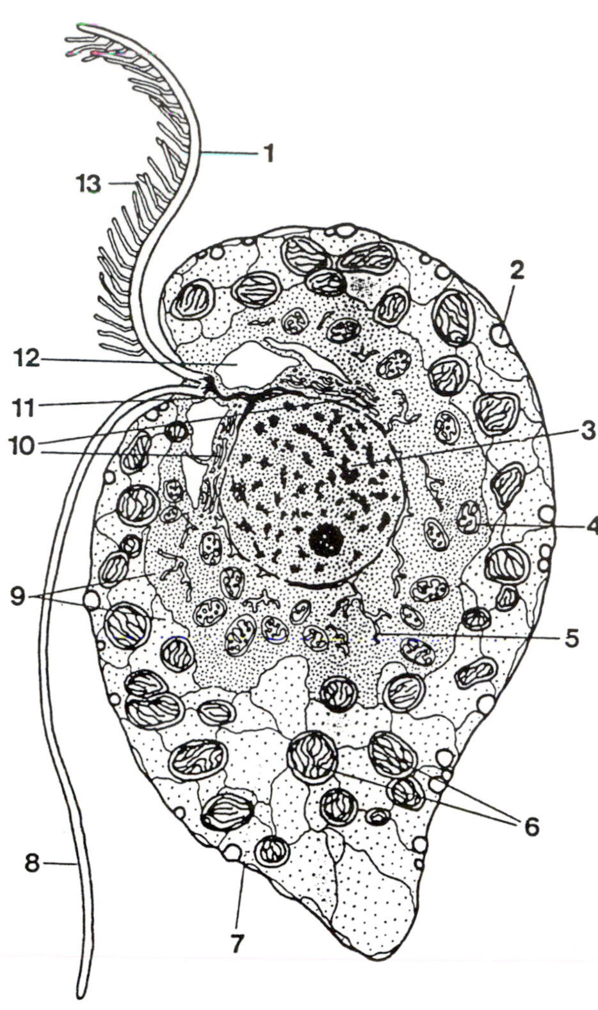

Raphidophyta
1=anterior undulipodium; 2=mucocyst; 3=nucleus; 4=mitochondrion; 5=endoplasmic reticulum; 6=plastids; 7=plasmalemma; 8=trailing (recurrent) undulipodium; 9=cytoplasm; 10=golgi; 11=undulipodial root (of rhizoplast); 12=contractile vacuole; 13=undulipodial hairs.

Raphidophyta
Chloromonads; phylum of wall-less heterokont mastigote algae. Solitary cells distinguished by large golgi apparatus extending over the anterior surface of the single nucleus. Plastids contain chlorophylls *a* and *c*. Found as motile or palmelloid cells in freshwater and marine habitats; sexuality unknown.

Remaneicidae
Family in phylum Granuloreticulosa.

Retortamonadida
Retortamonads; class in phylum Zoomastigina. Small, symbiotrophic mastigotes with twisted cell bodies bearing a cytostome in which a trailing undulipodium beats. Symbiotrophic usu. in digestive tract of insects, amphibians, reptiles, rodents, and other animals. Lack mitochondria and golgi, presumed anaerobes.

Retortamonadidae
Family in phylum Zoomastigina.

Reussellidae
Family in phylum Granuloreticulosa.

Rhabdomonadales
Order in phylum Euglenida.

Rhabdonemataceae
Family in phylum Bacillariophyta.

Rhabdonematales
Order in phylum Bacillariophyta.

Rhabdophora
Subphylum in phylum Ciliophora.

Rhaphoneidaceae
Family in phylum Bacillariophyta.

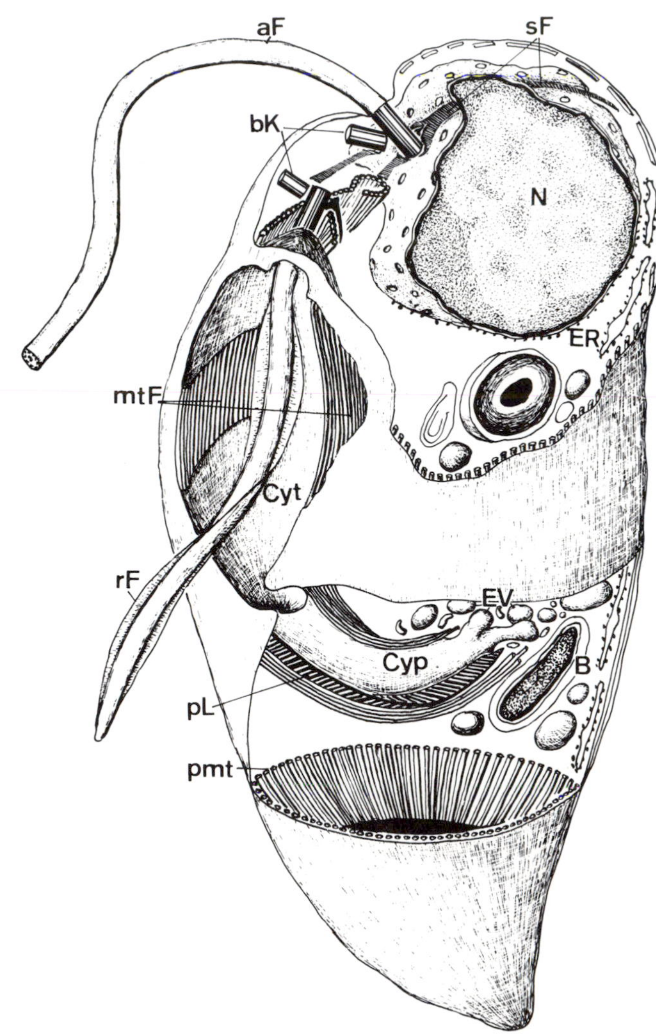

Retortamonadida
Retortamonas. aF=anterior undulipodium; B=bacterium; bK=barren kinetosomes; Cyp=cytopharyngeal region; Cyt=cytostomal ventral pouch; ER=endoplasmic reticulum; EV=endocytotic vesicles; mtF=microtubular fiber; N=nucleus; pL=striped lamina; pmt=subpellicular microtubules; rF=trailing undulipodium; sF=striated kinetosome fiber.

Rhaphoneidales
Order in phylum Bacillariophyta.

Rhapydioninidae
Family in phylum Granuloreticulosa.

Rhipidiaceae
Family in phylum Oomycota.

Rhipidiales
Order in phylum Oomycota.

Rhizamminideae
Family in phylum Granuloreticulosa.

Rhizidiomycetaceae
Family in phylum Hyphochytriomycota.

Rhizochloridaceae
Family in phylum Xanthophyta.

Rhizochloridales
Order in phylum Xanthophyta.

Rhizochrysidaceae
Family in phylum Chrysophyta.

Rhizonymphidae
Family in phylum Zoomastigina.

Rhizopoda
Phylum of amastigote soil, freshwater, and marine amebas. Typically single-celled uninucleate organisms motile by pseudopods, feeding by phagotrophy. Body naked or bears tests of silica, carbonate sand grains, or organic materials; many form resistant cysts. Some have two or more nuclei; reproduction by binary fission only. Sexuality unknown. Cosmopolitan distribution in aquatic or terrestrial habitats; some symbiotrophic to necrotrophic.

Rhizosoleniaceae
Family in phylum Bacillariophyta.

Rhizosoleniales
Order in phylum Bacillariophyta.

Rhizosoleniophycidae
Subclass in phylum Bacillariophyta.

Rhodochaetaceae
Family in phylum Rhodophyta.

Rhodochaetales
Order in phylum Rhodophyta.

Rhodochytriaceae
Family in phylum Chlorophyta.

Rhodophyceae
Class in phylum Rhodophyta.

Rhodophyta
Rhodophytes; red algae; phylum of primarily marine, photosynthetic protoctists. Life history involves alternation of generations which may include two free-living generations and a third, dependent generation. Sexuality via nonmotile male gametes that penetrate female tissue. Plastids contain chlorophyll *a* and the water-soluble accessory pigments allophycocyanin, phycocyanin, and phycoerythrin localized in phycobilisomes; thylakoids present as single lamellae. Undulipodia absent at all stages. Illustration next page.

Rhodymeniales
Order in phylum Rhodophyta.

Rhoicospheniaceae
Family in phylum Bacillariophyta.

Rhynchodida
Order in phylum Ciliophora.

Rhynchodina
Suborder in phylum Ciliophora.

Riveroinidae
Family in phylum Granuloreticulosa.

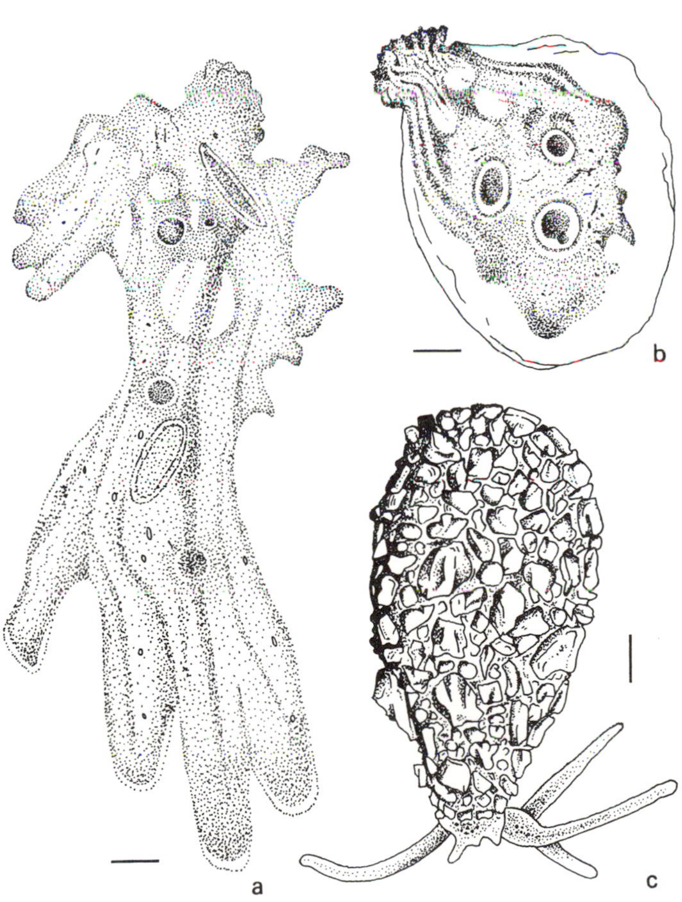

Rhizopoda
Types of large lobose amebas. a. *Amoeba proteus,* possessing several pseudopods. Bar=30 µm. b. *Thecamoeba terricola,* with clear ectoplasmic, indeterminate pseudopod. The uroid, or tail, is at the upper left of the organism. Bar=10 µm. c. *Difflugia oblonga.* Testate ameba with agglutinated test composed of small grains of sand. Pseudopods project through an aperture or pseudostome. Bar=10 µm.

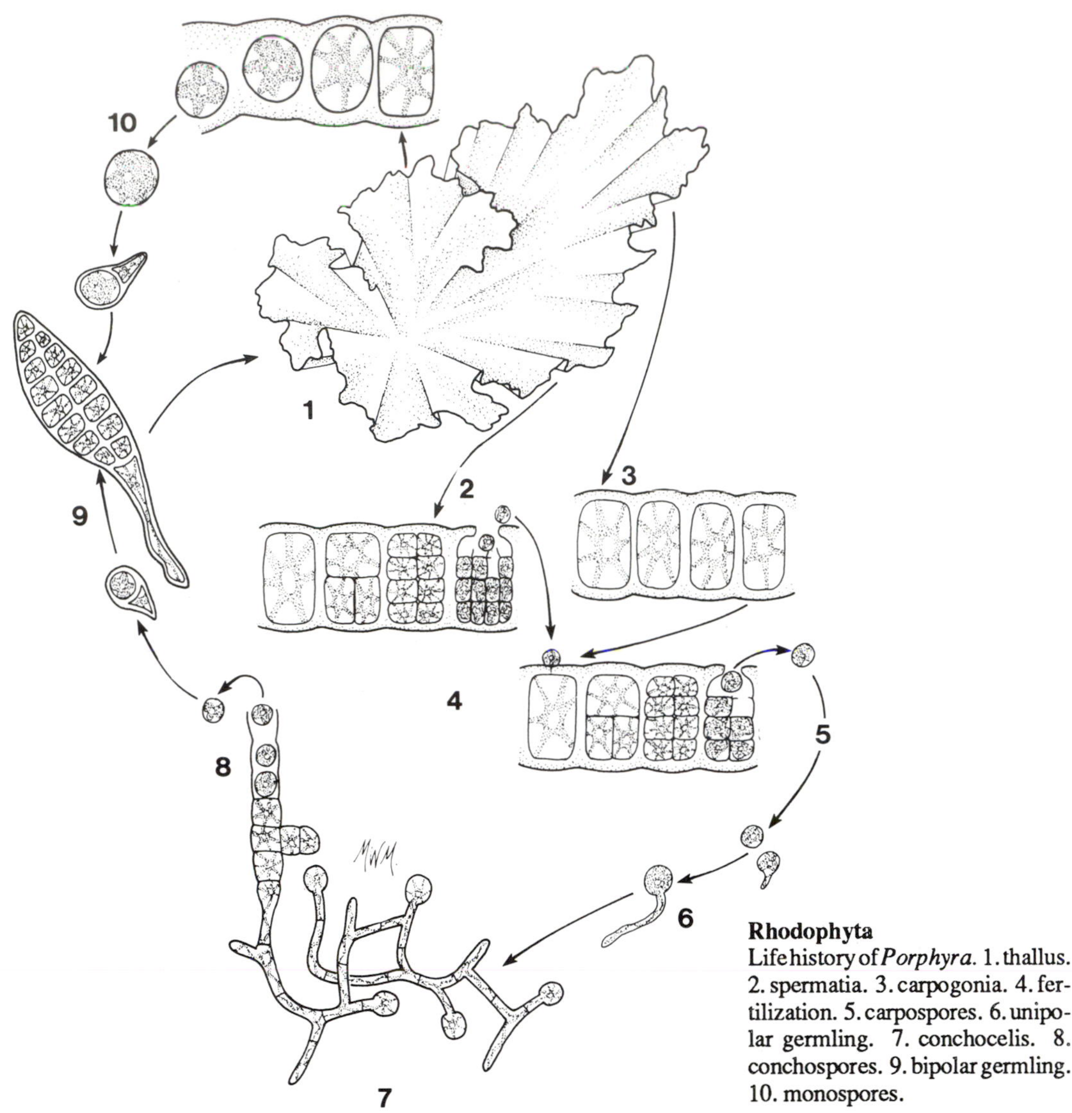

Rhodophyta
Life history of *Porphyra*. 1. thallus. 2. spermatia. 3. carpogonia. 4. fertilization. 5. carpospores. 6. unipolar germling. 7. conchocelis. 8. conchospores. 9. bipolar germling. 10. monospores.

Robertinacea
Superfamily in phylum Granuloreticulosa.

Robertinida
Order in phylum Granuloreticulosa.

Robertinidae
Family in phylum Granuloreticulosa.

Robuloidacea
Superfamily in phylum Granuloreticulosa.

Robuloididae
Family in phylum Granuloreticulosa.

Rotaliacea
Superfamily in phylum Granuloreticulosa.

Rotaliellidae
Family in phylum Granuloreticulosa.

Rotaliida
Order in phylum Granuloreticulosa.

Rotaliidae
Family in phylum Granuloreticulosa.

Rotaliporacea
Superfamily in phylum Granuloreticulosa.

Rotaliporidae
Family in phylum Granuloreticulosa.

Rudimicrosporea
Class in phylum Microspora.

Rugoglobigerinidae
Family in phylum Granuloreticulosa.

Rzehakinacea
Superfamily in phylum Granuloreticulosa.

Rzehakinidae
Family in phylum Granuloreticulosa.

S

Saccamminidae
Family in phylum Granuloreticulosa.

Sagosphaeridae
Family in phylum Actinopoda.

Salpingoecidae
Family in phylum Zoomastigina.

Saprolegniales
Order in phylum Oomycota.

Saprolegniomycetidae
Class (subclass) in phylum Oomycota.

Sarcinochrysidaceae
Family in phylum Chrysophyta.

Sarcinochrysidales
Order in phylum Chrysophyta.

Sarcocystidae
Family in phylum Apicomplexa.

Sargassaceae
Family in phylum Phaeophyta.

Sceletonemataceae
Family in phylum Bacillariophyta.

Scenedesmaceae
Family in phylum Chlorophyta.

Schackoinidae
Family in phylum Granuloreticulosa.

Schizamminidae
Family in phylum Granuloreticulosa.

Schizomeridaceae
Family in phylum Chlorophyta.

Schnellaceae
Family in phylum Plasmodial Slime Molds.

Schubertellidae
Family in phylum Granuloreticulosa.

Schwagerinidae
Family in phylum Granuloreticulosa.

Sciadiaceae
See *Ophiocytaceae*.

Sclerosporaceae
Family in phylum Oomycota.

Sclerosporales
Order in phylum Oomycota.

Scolioneidaceae
Family in phylum Bacillariophyta.

Scoliotropidaceae
Family in phylum Bacillariophyta.

Scuticociliatida
Order in phylum Ciliophora.

Scytosiphonaceae
Family in phylum Phaeophyta.

Scytosiphonales
Order in phylum Phaeophyta.

Seirococcaceae
Family in phylum Phaeophyta.

Selenidiidae
Family in phylum Apicomplexa.

Sellaphoraceae
Family in phylum Bacillariophyta.

Sellaphorineae
Suborder in phylum Bacillariophyta.

Semitextulariidae
Family in phylum Granuloreticulosa.

Septemcapsulidae
Family in phylum Myxozoa.

Sessilida
Order in phylum Ciliophora.

Silicoloculinida
Order in phylum Granuloreticulosa.

Silicoloculinidae
Family in phylum Granuloreticulosa.

Silicotubidae
Family in phylum Granuloreticulosa.

Sinuolineidae
Family in phylum Myxozoa.

Siphogenerinoididae
Family in phylum Granuloreticulosa.

Siphoninacea
Superfamily in phylum Granuloreticulosa.

Siphoninidae
Family in phylum Granuloreticulosa.

Siphonocladaceae
Family in phylum Chlorophyta.

Siphonocladales
Order in phylum Chlorophyta.

Soritacea
Superfamily in phylum Granuloreticulosa.

Soritidae
Family in phylum Granuloreticulosa.

Sorocarpaceae
Family in phylum Phaeophyta.

Spermatochnaceae
Family in phylum Phaeophyta.

Sphacelariaceae
Family in phylum Phaeophyta.

Sphacelariales
Order in phylum Phaeophyta.

Sphaenacantha
Suborder in phylum Actinopoda.

Sphaeractinomyxidae
Family in phylum Myxozoa.

Sphaeramminidae
Family in phylum Granuloreticulosa.

Sphaerellarina
Suborder in phylum Actinopoda.

Sphaeridiothricaceae
Family in phylum Chrysophyta.

Sphaeriparaceae
Family in phylum Dinomastigota.

Sphaerocollina
Suborder in phylum Actinopoda.

Sphaeroidinidae
Family in phylum Granuloreticulosa.

Sphaeromyxidae
Family in phylum Myxozoa.

Sphaeromyxina
Suborder in phylum Myxozoa.

Sphaerosporidae
Family in phylum Myxozoa.

Sphenomonadales
Order in phylum Euglenida.

Spirillinida
Order in phylum Granuloreticulosa.

Spirillinidae
Family in phylum Granuloreticulosa.

Spirocyclinidae
Family in phylum Granuloreticulosa.

Spiroplectamminacea
Superfamily in phylum Granuloreticulosa.

Spiroplectamminidae
Family in phylum Granuloreticulosa.

Spirotectinidae
Family in phylum Granuloreticulosa.

Spirotrichea
Class in phylum Ciliophora.

Spirotrichonymphidae
Family in phylum Zoomastigina.

Spirotrichosomidae
Family in phylum Zoomastigina.

Spizellomycetaceae
Family in phylum Chytridiomycota.

Spizellomycetales
Order in phylum Chytridiomycota.

Splachnidiaceae
Family in phylum Phaeophyta.

Sporadotrichina
Suborder in phylum Ciliophora.

Sporochnaceae
Family in phylum Phaeophyta.

Sporochnales
Order in phylum Phaeophyta.

Sporolitheae
Tribe in phylum Rhodophyta.

Sporolithoideae
Subfamily in phylum Rhodophyta.

Spraguidae
Family in phylum Microspora.

Spumellarida
Order in phylum Actinopoda.

Spyroidea
Family in phylum Actinopoda.

Squamulinacea
Superfamily in phylum Granuloreticulosa.

Squamulinidae
Family in phylum Granuloreticulosa.

Staffellidae
Family in phylum Granuloreticulosa.

Stainforthiidae
Family in phylum Granuloreticulosa.

Stannomida
Class in phylum Xenophyophora.

Stannomidae
Family in phylum Xenophyophora.

Stauraconidae
Family in phylum Actinopoda.

Staurojoenidae
Family in phylum Zoomastigina.

Stauroneidaceae
Family in phylum Bacillariophyta.

Stemonitaceae
Family in phylum Plasmodial Slime Molds.

Stemonitales
Order in phylum Plasmodial Slime Molds.

Stemonitomycetidae
Subclass in phylum Plasmodial Slime Molds.

Stenophoridae
Family in phylum Apicomplexa.

Stephanopyxidaceae
Family in phylum Bacillariophyta.

Stephoidea
Family in phylum Actinopoda.

Stichogloeaceae
Family in phylum Chrysophyta.

Sticholonchidae
Family in phylum Actinopoda.

Stichotrichia
Subclass in phylum Ciliophora.

Stichotrichida
Order in phylum Ciliophora.

Stichotrichina
Suborder in phylum Ciliophora.

Stictocyclaceae
Family in phylum Bacillariophyta.

Stictocyclales
Order in phylum Bacillariophyta.

Stilostomellidae
Family in phylum Granuloreticulosa.

Stipitococcaceae
Family in phylum Xanthophyta.

Striariaceae
Family in phylum Phaeophyta.

Strobilidiina
Suborder in phylum Ciliophora.

Strombidinopsina
Suborder in phylum Ciliophora.

Stylocephalidae
Family in phylum Granuloreticulosa.

Stylococcaceae
Family in phylum Chrysophyta.

Stypocaulaceae
Family in phylum Phaeophyta.

Suctoria
Subclass in phylum Ciliophora.

Surirellaceae
Family in phylum Bacillariophyta.

Surirellales
Order in phylum Bacillariophyta.

Symphiacanthida
Order in phylum Actinopoda.

Synactinomyxidae
Family in phylum Myxozoa.

Synchytriaceae
Family in phylum Chytridiomycota.

Syndiniaceae
Family in phylum Dinomastigota.

Syndiniales
Order in phylum Dinomastigota.

Synhymeniida
Order in phylum Ciliophora.

Syringamminidae
Family in phylum Xenophyophora.

Syringodermataceae
Family in phylum Phaeophyta.

Syringodermatales
Order in phylum Phaeophyta.

Syzraniidae
Family in phylum Granuloreticulosa.

T

Tabellariaceae
Family in phylum Bacillariophyta.

Tawitawiacea
Superfamily in phylum Granuloreticulosa.

Tawitawiidae
Family in phylum Granuloreticulosa.

Taxopodida
Suborder in phylum Actinopoda.

Telomyxidae
Family in phylum Microspora.

Teratonymphidae
Family in phylum Zoomastigina.

Testaceafilosida
Order in phylum Rhizopoda.

Testacealobosa
Subclass in phylum Rhizopoda.

Tetractinomyxidae
Family in phylum Myxozoa.

Tetradimorphyidae
Family in phylum Actinopoda.

Tetrahymenina
Suborder in phylum Ciliophora.

Tetrasporaceae
Family in phylum Chlorophyta.

Tetrasporales
Order in phylum Chlorophyta.

Tetrataxacea
Superfamily in phylum Granuloreticulosa.

Tetrataxidae
Family in phylum Granuloreticulosa.

Textulariacea
Superfamily in phylum Granuloreticulosa.

Textulariellidae
Family in phylum Granuloreticulosa.

Textulariida
Order in phylum Granuloreticulosa.

Textulariidae
Family in phylum Granuloreticulosa.

Textulariopsidae
Family in phylum Granuloreticulosa.

Thalassicollidae
Family in phylum Actinopoda.

Thalassionemataceae
Family in phylum Bacillariophyta.

Thalassionematales
Order in phylum Bacillariophyta.

Thalassiophysaceae
Family in phylum Bacillariophyta.

Thalassiosiraceae
Family in phylum Bacillariophyta.

Thalassiosirales
Order in phylum Bacillariophyta.

Thalassiosirophycidae
Subclass in phylum Bacillariophyta.

Thalassophysidae
Family in phylum Actinopoda.

Thalicolidae
Family in phylum Apicomplexa.

Thecadiniaceae
Family in phylum Dinomastigota.

Thecamoebidae
Family in phylum Rhizopoda.

Thecina
Suborder in phylum Rhizopoda.

Thelohaniidae
Family in phylum Microspora.

Thigmotrichina
Suborder in phylum Ciliophora.

Thomasinellidae
Family in phylum Granuloreticulosa.

Thoracosphaeraceae
Family in phylum Dinomastigota.

Thoracosphaerales
Order in phylum Dinomastigota.

Thraustochytriidae
Family in phylum Labyrinthulomycota.

Tilopteridaceae
Family in phylum Phaeophyta.

Tilopteridales
Order in phylum Phaeophyta.

Tintinnina
Suborder in phylum Ciliophora.

Tournayellacea
Superfamily in phylum Granuloreticulosa.

Tournayellidae
Family in phylum Granuloreticulosa.

Toxariaceae
Family in phylum Bacillariophyta.

Toxariales
Order in phylum Bacillariophyta.

Tremachoridae
Family in phylum Granuloreticulosa.

Trentepohliaceae
Family in phylum Chlorophyta.

Trentepohliales
Order in phylum Chlorophyta. Microscopic, branched filamentous chlorophytes, usu. with differentiated reproductive cells: quadrimastigote zoospores and bimastigote isogametes. Cells walled, usu. uninucleate. Often occur in subaerial habitats. Some are plant parasites, and at least one species is lichen phycobiont.

Treubariaceae
Family in phylum Chlorophyta.

Triactinomyxidae
Family in phylum Myxozoa.

Triadiniaceae
Family in phylum Dinomastigota.

Tribonemataceae
Family in phylum Xanthophyta.

Tribonematales
Order in phylum Xanthophyta.

Trichiaceae
Family in phylum Plasmodial Slime Molds.

Trichiales
Order in phylum Plasmodial Slime Molds.

Trichohyalidae
Family in phylum Granuloreticulosa.

Trichomonadida
Order in phylum Zoomastigina.

Trichomonadidae
Family in phylum Zoomastigina.

Trichonymphidae
Family in phylum Zoomastigina.

Trichosida
Order in phylum Rhizopoda.

Trichostomatia
Subclass in phylum Ciliophora.

Trilosporidae
Family in phylum Myxozoa.

Trimosinidae
Family in phylum Granuloreticulosa.

Trochamminacea
Superfamily in phylum Granuloreticulosa.

Trochamminidae
Family in phylum Granuloreticulosa.

Trocholonidae
Family in phylum Granuloreticulosa.

Trypanochloridaceae
Family in phylum Xanthophyta.

Trypanosomatidae
Family in phylum Zoomastigina.

Trypanosomatina
Suborder in phylum Zoomastigina.

Tuberitinidae
Family in phylum Granuloreticulosa.

Tubulina
Suborder in phylum Rhizopoda.

Turrilinacea
Superfamily in phylum Granuloreticulosa.

Turrilinidae
Family in phylum Granuloreticulosa.

Tuscaroridae
Family in phylum Actinopoda.

Tuzetiidae
Family in phylum Microspora.

U

Udoteaceae
Family in phylum Chlorophyta.

Ulotrichaceae
Family in phylum Chlorophyta.

Ulotrichales
Order in phylum Chlorophyta.

Ulvaceae
Family in phylum Chlorophyta.

Ulvales
Order in phylum Chlorophyta.

Ulvellaceae
Family in phylum Chlorophyta.

Ulvophyceae
Green seaweeds. Class in phylum Chlorophyta containing predominantly marine, sessile algae with multicellular or coenocytic, walled growing cells. Heteromorphic life history with multicellular or reduced diploid sporophyte. Mitosis with centrioles, cytokinesis by an in-growing cleavage furrow. Fertilization followed by zygote formation occurs when bi- or quadrimastigote isogamous cells fuse.

Unikaryonidae
Family in phylum Microspora.

Uradiophoridae
Family in phylum Apicomplexa.

Urophlyctaceae
Family in phylum Chytridiomycota.

Urosporidae
Family in phylum Apicomplexa.

Urostylina
Suborder in phylum Ciliophora.

Uvigerinidae
Family in phylum Granuloreticulosa.

V

Vaginulinidae
Family in phylum Granuloreticulosa.

Vahlkampfidae
Family in phylum Rhizopoda.

Valoniaceae
Family in phylum Chlorophyta.

Valvulinellidae
Family in phylum Granuloreticulosa.

Valvulinidae
Family in phylum Granuloreticulosa.

Variisporina
Suborder in phylum Myxozoa.

Vaucheriaceae
Family in phylum Xanthophyta.

Vaucheriales
Order in phylum Xanthophyta.

Verbeekinidae
Family in phylum Granuloreticulosa.

Verneuilinacea
Superfamily in phylum Granuloreticulosa.

Verneuilinidae
Family in phylum Granuloreticulosa.

Verrucalvaceae
Family of phylum Oomycota.

Vestibuliferida
Order in phylum Ciliophora.

Victoriellidae
Family in phylum Granuloreticulosa.

Virgulinellidae
Family in phylum Granuloreticulosa.

Volvocaceae
Family in phylum Chlorophyta.

Volvocales
Order in phylum Chlorophyta.

W

Warnowiaceae
Family in phylum Dinomastigota.

Wenyonellidae
Family in phylum Apicomplexa.

X

Xanthophyceae
Only class in phylum Xanthophyta.

Xanthophyta
Phylum of primarily freshwater, yellow-green, heterokont mastigote algae. Coccoid unicells and multicellular descendants; double-membrane-bounded plastids contain chlorophylls *a* and *c*. Plastids, which store fat or oil, not starch, are surrounded by plastid endoplasmic reticulum. Reproduction by zoospores or their amastigote equivalent (hemiautospores). Sexual fusion of egg and sperm reported.

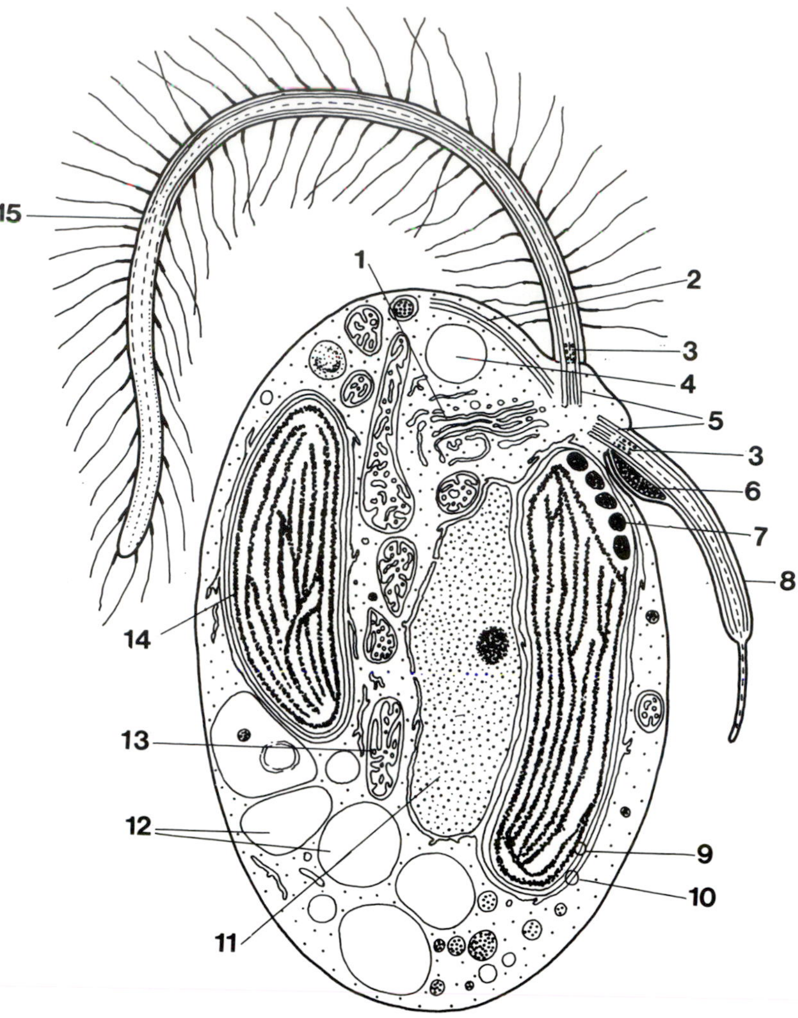

Xanthophyta
1=golgi; 2=microtubular undulipodial rootlets; 3=transitional helix; 4=contractile vacuole; 5=kinetosomes; 6=undulipodial swelling; 7=eyespot; 8=posterior undulipodium; 9=plastid envelope (single line); 10=plastid endoplasmic reticulum; 11=nucleus; 12=vacuoles; 13=mitochondrion; 14=plastid; 15=anterior undulipodium with mastigonemes.

Xenophyophora
Phylum of heterotrophic protoctists, all of which (except one group in shallow water) live in the abyssal marine benthos. Large ameboid organisms organized as plasmodia enclosed by a branched, tubelike organic cement. Tests patched from hard parts of skeletons, sponges, or foraminifera, radiolaria spicules, and mineral grains. Life history incompletely known. Illustration next page.

Y

Yamikovellidae
Family in phylum Apicomplexa.

Z

Zoomastigina

Phylum of diverse heterotrophic mastigotes. Undulipodiated organisms including solitary and colonial forms; freshwater or marine; free-living, symbiotrophic, or necrotrophic. Many lack mitochondria at all stages. See *Amebomastigota, Bicosoecids, Choanomastigotes, Diplomonadida, Kinetoplastida, Opalinata, Parabasalia, Proteromonadida, Pseudociliata, Pyrsonymphida, Retortamonadida.* Illustration below.

Zooxanthellaceae

Family in phylum Dinomastigota.

Zygnemataceae

Family in phylum Conjugaphyta.

Zygnematales

Order in phylum Conjugaphyta.

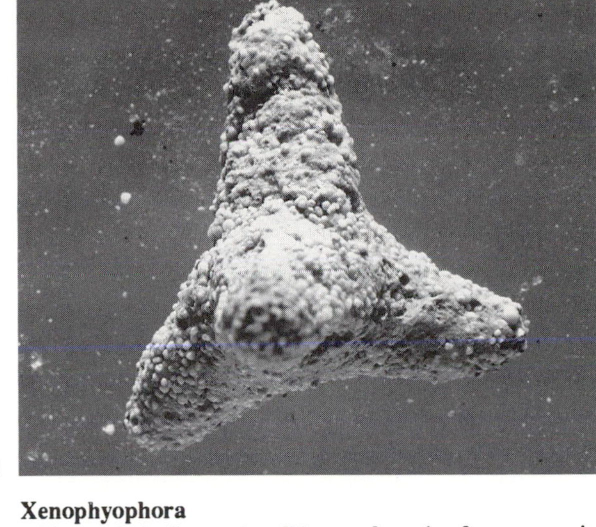

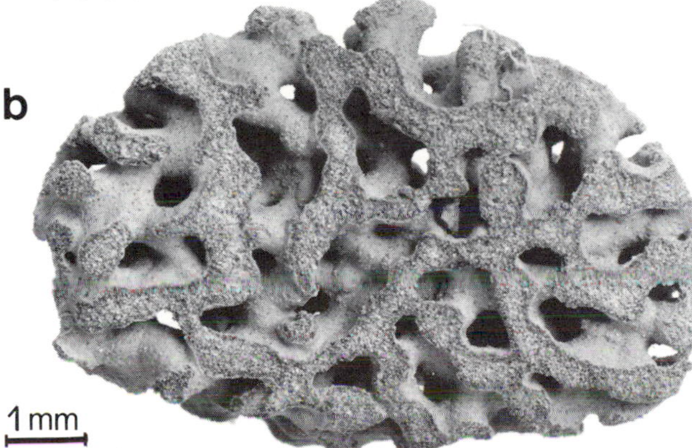

Xenophyophora
a. Test of *Galadheammina.* Distance from tip of one arm to tip of another is 18mm. b. Lower (flat) surface of test of *Reticulammina.*

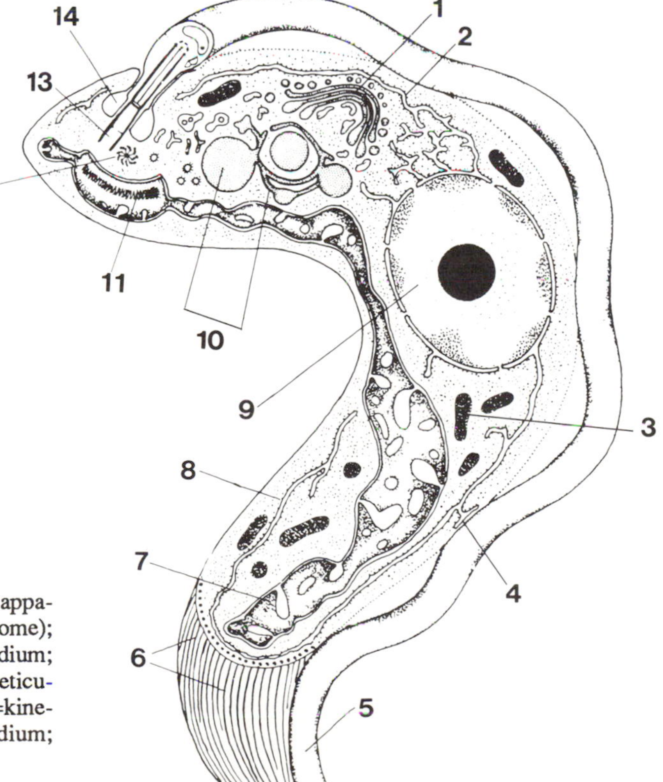

Zoomastigina
The trypanosomatid mastigote, *Trypanosoma congolese.* 1=golgi apparatus; 2=subtending endoplasmic reticulum; 3=microbody (glycosome); 4=undulipodium-associated endoplasmic reticulum; 5=undulipodium; 6=pellicular microtubules; 7=mitochondrion; 8=endoplasmic reticulum; 9=nucleus; 10=digestive smooth membrane network; 11=kinetoplast; 12=barren kinetosome; 13=kinetosome of undulipodium; 14=undulipodial pocket.

Bibliography of Glossaries

Bates, L., Jackson, J.A., eds.: *Dictionary of Geological Terms*, 3rd ed. Garden City, N.Y.: Anchor Press/Doubleday, 1984.

Blackmore, S., Tootill, E. eds.: *The Penguin Dictionary of Botany*. New York: Penguin Books, 1984.

Bold, H.C., Wynne, M.J.: *Introduction to the Algae: Structure and Reproduction*, 2nd ed. Englewood Cliffs, N.J.: Prentice-Hall, 1985.

Brown, R.W.: *Composition of Scientific Words*. Washington, D.C.: Smithsonian Institution Press, 1956.

Corliss, J.O.: *The Ciliated Protozoa: Characterization, Classification, and Guide to the Literature*, 2nd ed. Oxford-New York: Pergamon Press, 1979.

Corliss, J.O.: Problems in cytoterminology and nomenclature for the protists. In: *Advances in Culture Collections*, Vol. 1 (Huang, L.H., ed.). USFCC/ASM, pp. 23-37, 1991.

Dobell, C.: *Antony van Leeuwenhoek and His "Little Animals": Being some account of the father of protozoology and bacteriology and his multifarious discoveries in these disciplines*. New York: Russell and Russell, 1958.

Gary, M., McAfee, R., Jr., Wolf, C.L., eds.: *Glossary of Geology*. Washington, D.C.: American Geological Institute, 1974.

Holmes, S., ed.: *Henderson's Dictionary of Biological Terms*, 9th ed. New York: Van Nostrand Reinhold, 1979.

King, R.C., Stansfield, W.D.: *A Dictionary of Genetics*, 3rd ed. New York-Oxford: Oxford University Press, 1985.

Lee, J.J., Hutner, S.H., Bovee, E.C., eds.: *An Illustrated Guide to the Protozoa*. Lawrence, Kansas: Society of Protozoologists, 1985.

Lewis, K.B.: *A Key to Recent Genera of the Foraminiferida*. New Zealand Oceanographic Institute Memoirs No. 45, pp. 67-78, 1970.

Maclean, N.: *Dictionary of Genetics and Cell Biology*. New York: New York University Press, 1987.

Margulis, L., Corliss, J. O., Melkonian, M., Chapman, D. J., eds.: *Handbook of Protoctista*. Boston: Jones and Bartlett Publishers, 1990.

Margulis, L., Sagan, D.: *Origins of Sex*. New Haven-London: Yale University Press, 1985.

Martin, E.A., ed.: *Dictionary of Life Sciences*. New York: Pica Press, 1983.

Miyachi, S., Nakayama, O., Yokahama, Y., Hara, Y., Ohmori, M., Komagata, K., Sugawara, H., Ugawa, Y., eds.: *World Catalogue of Algae*, 2nd ed. Tokyo: Japan Scientific Societies Press, 1990.

Raven, P.H., Johnson, G.B.: *Biology*. St. Louis-Toronto-Santa Clara: Times Mirror/Mosby College Publishing, 1986.

Vávra, J., Sprague, V.: Glossary for the Microsporidia. In: *Biology of the Microsporidia, Comparative Pathobiology*, Vol. 1 (Bulla, L.A., Cheng, T.C., eds). New York-London: Plenum Press, 1976.

CLASSIFICATION

Table 4. Classes of the Phyla of the Kingdom Protoctista

Table 5. Informal Names of Protoctistan Phyla and Classes

Table 6A. Phyla of Protoctists

Table 6B. Classes of Phylum Zoomastigina

Table 6C. Classes of Phylum Chlorophyta

Table 7. Classification: Summary of Phyla and Lower Taxa

Classification

Given information by experts on various protist groups, we include classification as published except for the proviso that "botanical," "mycological," and "zoological" groups have been integrated and contradictions have been resolved. We offer the following tabulations fully aware of the impending revision based on new ultrastructural, genetic and molecular biological data. We anticipate refinement and reorganization by the International Society of Evolutionary Protistology (ISEP).

Table 4. Classes of the Phyla of the Kingdom Protoctista

I. No undulipodia; complex sexual cycles absent

Phylum	*Number of Classes*	*Names of Classes*
RHIZOPODA	2	Lobosea Filosea
HAPLOSPORIDIA	1	Haplosporea
PARAMYXEA	2	Paramyxidea Marteiliidea
MYXOZOA	2	Myxosporea Actinosporea
MICROSPORA[1]	2	Rudimicrosporea Microsporea

II. No undulipodia; complex sexual cycles present

Phylum	*Number of Classes*	*Names of Classes*
ACRASEA	1	Acrasids[5]
DICTYOSTELIDA	1	Dictyostelids[5]
RHODOPHYTA	1	Rhodophyceae
CONJUGAPHYTA	1	Conjugatophyceae

III. Reversible formation of undulipodia; complex sexual cycles absent

Phylum	*Number of Classes*	*Names of Classes*
XENOPHYOPHORA	2	Psamminida Stannominida
CRYPTOPHYTA (CRYPTOMONADS)	1	Cryptophyceae
GLAUCOCYSTOPHYTA (GLAUCOCYSTIDS)	1(or 3)	Glaucocystophyceae (Cyanophorales, Glaucocystales, Gloeochaetales)
KARYOBLASTEA[2]	1	Karyoblastea
ZOOMASTIGINA	11	Amebomastigota Bicosoecids (Bicoecids)[5] Choanomastigota Diplomonadida[3] Pseudociliata Kinetoplastida[3] Opalinata[4] Proteromonadida[4,10] Parabasalia (3 orders: Trichomonads,[5,10] Polymonads,[5,10] Hypermastigotes) Retortamonadida[3,10] Pyrsonymphida[10]
EUGLENIDA	1	Euglenophyceae
CHLORARACHNIDA	1	Chlorarachniophyceae
PRYMNESIOPHYTA	1	Prymnesiophyceae

Table 4 (*Continued*)

Phylum	*Number of Classes*	*Names of Classes*
Raphidophyta	1	Raphidophyceae
Eustigmatophyta	1	Eustigmatophyceae
Actinopoda	4	Polycystina Phaeodaria Heliozoa Acantharia
Hyphochytriomycota	1	Hyphochytrids[5]
Labyrinthulomycota	2	Labyrinthulids[5] Thraustochytrids[5]
Plasmodiophoromycota	1	Plasmodiophorids[5]
IV. Reversible formation of undulipodia; complex sexual cycles present		
Dinomastigota (Dinoflagellata)	2	Dinophyceae Syndiniophyceae
Chrysophyta	5	Chrysophyceae Synurophyceae Pedinellophyceae Dictyochophyceae Silicomastigota (Silicoflagellates)
Chytridiomycota	1	Chytridiomycetes
Plasmodial Slime Molds	2	Protostelids[5] Myxomycetes
Ciliophora[6]	8	Karyorelictea (PC) Spirotrichea (PC) Prostomatea (R) Litostomatea (R) Phyllopharyngea (R) Nassophorea (C) Oligohymenophorea (C) Colpodea (C)
Granuloreticulosa	2	Athalamea[7] Foraminiferea
Apicomplexa	3	Gregarinia Coccidia Hematozoa
Bacillariophyta	3	Coscinodiscophyceae Fragilariophyceae Bacillariophyceae
Chlorophyta[8]	4	Prasinophyceae Chlorophyceae Ulvophyceae Charophyceae

Table 4 ***(Continued)***

Phylum	*Number of Classes*	*Names of Classes*
OOMYCOTA	2	Saprolegniomycetidae[9] Peronosporomycetidae[9]
XANTHOPHYTA (TRIBOPHYTA)	1	Xanthophyceae
PHAEOPHYTA	1	Phaeophyceae
INCERTAE SEDIS	?	Ebridians, Ellobiopsids[5]

[1] Meiosis and fertilization reported in some genera, e.g., Canning, 1988.
[2] Shown conclusively to have nonmotile, often aberrant, undulipodia-like organelles during all stages of the life history investigated (Griffin, 1988; Seravin and Goodkov, 1987).
[3] Raised to class status although presented as orders in traditional zoological systematics.
[4] Or orders of class Slopalinata (Patterson, 1986)
[5] Names in vernacular given because taxonomy has always been presented assuming zoological or botanical rules. Formal systematics awaits official recognition of Kingdom Protoctista.
[6] Subphyla; PC = Postciliodesmatophora; R = Rhabdophora; C = Cyrtophora
[7] Group poorly known, sex not demonstrated.
[8] Twenty-eight orders of Chlorophyta classes, according to Melkonian, are listed below.
[9] Family terminology assuming that members belong to the Kingdom Fungi.
[10] Traditional "polymastigotes"

Orders of Chlorophyte Classes

Class Prasinophyceae
- Mamiellales
- Pseudoscourfieldiales
- Chlorodendrales
- Pyramimonadales

Class Chlorophyceae
- Dunaliellales
- Chlamydomonadales
- Volvocales
- Tetrasporales
- Chlorococcales
- Chlorosarcinales
- Sphaeropleales
- Microsporales
- Oedogoniales
- Cylindrocapsales
- Chaetophorales

Class Ulvophyceae
- Ulotrichales
- Ulvales
- Siphonocladales
- Dasycladales
- Caulerpales

Class Charophyceae
- Charales
- Chlorokybales
- Klebsormidiales
- Coleochaetales

Chlorophyte orders of uncertain affinities
- Pedinomonadales
- Microthamniales
- Prasiolales
- Trentepohliales

REFERENCES

Canning, E.U.: Nuclear division and chromosome cycle in microsporidia. *BioSystems* 21, 333-340 (1988).

Griffin, J.L.: Fine structure and taxomic position of the giant amoeboid flagellate *Pelomyxa palustris*. *Journal of Protozoology* 35, 300-315 (1988).

Patterson, D.J.: The fine structure of *Opalina ranarum* (family Opalinidae): Opalinid phylogeny and classification. *Protistologica* 21, 413-428 (1986).

Seravin, L.N., Goodkov, A.V.: The flagella of the freshwater amoeba *Pelomyxa palustrus*. *Tsitoligya* 29, 721-724 (1987) (in Russian).

Table 5. Informal Names of Protoctistan Phyla and Classes[1]

Formal Name	*Informal Name*	*Traditional Name*
1. RHIZOPODA	rhizopods	amebas
2. HAPLOSPORIDIA	haplosporidians	parasites[2]
3. PARAMYXEA	paramyxeans	parasites[2]
4. MYXOZOA	myxozoans	fish parasites[2]
5. MICROSPORA	microsporans	parasites[2]
6. ACRASEA	acrasids	cellular slime molds
7. DICTYOSTELIDA	dictyostelids	cellular slime molds
8. RHODOPHYTA	rhodophytes	red algae
9. CONJUGAPHYTA	gamophytes	conjugating green algae
10. XENOPHYOPHORA	xenophyophores	deep-sea protists
11. CRYPTOPHYTA	cryptomonads	flagellates,[3] phytoflagellates
12. GLAUCOCYSTOPHYTA	glaucocystids	algae, phytoflagellates
13. KARYOBLASTEA	karyoblasteans	giant amebas
14. ZOOMASTIGINA[3]	zoomastiginids	flagellates
a. Amebomastigota	amebomastigotes	ameboflagellates
b. Bicosoecids (Bicoecids)[4]	bicosoecids (bicoecids)	flagellates
c. Choanomastigota	choanomastigotes	collared mastigotes
d. Diplomonadida	diplomonads	flagellated parasites
e. Pseudociliata	pseudociliates	flagellates
f. Kinetoplastida	kinetoplastids	flagellated parasites
g. Opalinata	opalinids	flagellated parasites
h. Proteromonadida[4]	proteromonads	flagellated parasites
i. Parabasalia	parabasalians	flagellated parasites (polymastigotes, hypermastigotes)
j. Retortamonadida[4]	retortamonads	flagellated parasites
k. Pyrsonymphida	pyrsonymphids	flagellated parasites (polymastigotes)
15. EUGLENIDA	euglenids	flagellate algae, phytoflagellates
16. CHLORARACHNIDA	chlorarachnids	colored amebas
17. PRYMNESIOPHYTA	prymnesiophytes[5]	phytoflagellates
18. RAPHIDOPHYTA	raphidophytes	phytoflagellates
19. EUSTIGMATOPHYTA	eustigmatophytes	algae, phytoflagellates
20. ACTINOPODA	actinopods	radiolarians, heliozoans
a. Polycystina	polycystines	some radiolarians
b. Phaeodaria	phaeodarians	some radiolarians
c. Heliozoa	heliozoans	sun animalcules
d. Acantharia	acantharians	sun animalcules
21. HYPHOCHYTRIOMYCOTA	hyphochytrids	water molds, fungi
22. LABYRINTHULOMYCOTA	labyrinthulomycotes	water molds, fungi
a. Labyrinthulids[4]	slime nets	slime nets
b. Thraustochytrids[4]	water molds	water molds
23. PLASMODIOPHOROMYCOTA	plasmodiophorids	plant parasites, fungi
24. DINOMASTIGOTA	dinomastigotes	flagellate algae, plankton
25. CHRYSOPHYTA	chrysophytes	golden-yellow algae, plankton
26. CHYTRIDIOMYCOTA	chytridiomycotes	water molds, fungi

Table 5 (*Continued*)

Formal Name	*Informal Name*	*Traditional Name*
27. PLASMODIAL SLIME MOLDS	plasmodial slime molds	acellular slime molds
a. Myxomycota	myxomycotes	acellular slime molds
b. Protostelida	protostelids	acellular slime molds
28. CILIOPHORA	ciliates	ciliates, infusoria
29. GRANULORETICULOSA	granuloreticulosans	(includes all foraminiferans)
30. APICOMPLEXA	apicomplexans	sporozoan parasites[2]
31. BACILLARIOPHYTA	diatoms	algae, phytoplankton
32. CHLOROPHYTA	chlorophytes	green algae
a. Prasinophyceae	prasinophyceans	flagellates
b. Chlorophyceae	chlorophyceans	green algae
c. Charophyceae	charophyceans	pondweeds
d. Ulvophyceae	ulvophyceans	algae
33. OOMYCOTA	oomycotes	water molds, fungi
34. XANTHOPHYTA	xanthophytes	yellow-green algae
35. PHAEOPHYTA	phaeophytes	brown algae
36. INCERTAE SEDIS		
Ebridians	ebridians	plankton
Ellobiopsida	ellobiopsids	parasites

[1] For complete list of classes see Table 4.

[2] Traditionally called "sporozoa" in some literature.

[3] Although commonly named "flagellates," we discourage this term for reasons presented in the Introduction, sections 5 and 6 (p. xviii). All non-photosynthetic organisms listed here traditionally are called "protozoa."

[4] Names in vernacular given because taxonomy has always been presented assuming zoological or botanical rules. Formal systematics awaits official recognition of Kingdom Protoctista.

[5] Haptophytes, haptomonads, coccolithophorids, and other names applied to these marine algae.

CLASSIFICATION

Table 6A. Phyla of Protoctists

Table 6B. Classes of Phylum Zoomastigina

Table 6C. Classes of Phylum Chlorophyta

Table 7. Classification: Summary of Phyla and Lower Taxa

Table 6A. Phyla of Protoctists

Phylum	Morphology of Growing Organisms	Occurrence	Nutrition	Undulipodia	Mitochondria	Plastids/ Pigments
RHIZOPODA	Single cells, naked or enclosed by test.	Cosmopolitan in aquatic or terrestrial habitats.	Heterotrophic: phagotrophy, symbiotrophy.	Absent.	Present except in some *Entamoeba* sp.	Absent.
HAPLOSPORIDIA	Unicellular.	Symbiotrophic in freshwater and marine invertebrate animals; histozoic, coelozoic.	Heterotrophic: obligate symbiotrophy, histotrophy to necrotrophy.	Absent.	Absent(?)	Absent.
PARAMYXEA	Ameboid stem cell.	Symbiotrophic in marine invertebrate animals.	Heterotrophic: obligate symbiotrophy to necrotrophy.	Absent.	Light-matrix mitochondria present.	Absent.
MYXOZOA	Trophozoites often ameboid; may be pseudoplasmodia (one nucleus) or plasmodia (2-many nuclei); may assume appearance of cysts (Myxosporea); in Actinosporea trophozoite reduced to envelope cells of pansporoblast.	Symbiotrophic in animals.	Heterotrophic: obligate symbiotrophy; osmotrophy.	Absent.	Present.	Absent.

Reproduction	Propagules (Cysts)	Sexuality	Nuclear Features	Distinguishing Cytological Features	Presence in Fossil Record
Reproduction by cell division.	Produced by many but not all.	No sexuality confirmed.	Usu. uninucleate.	Pseudopods.	Testate species only.
Reproduction by sporulation; deposition of wall around multi-nucleated cells forms sporonts. Nuclear multiplication and increase in size followed by subdivision into uninucleate sporoblasts. Pairs of sporoblasts fuse, then karyogamy. Each cell becomes hourglass shaped and the nucleate half is engulfed by the anucleate half to form sporoplasm and epispore.	Spores.	No sexuality.	Uni- or multinucleate. Spores typically uninucleate. Karyokinesis with persistent nuclear envelope with spindle pole bodies at poles of mitotic apparatus.	Haplosporosomes, organelles of unknown function. Spherules at anterior of sporoplasm (may be modified golgi).	Absent.
Reproduction by sporulation resulting from series of internal cleavages within an ameboid stem cell. Offspring cells remaining within parent cell so that spores consist of several cells enclosed inside each other.	Spores.	No sexuality.		Centrioles with nine singlet microtubules. Haplosporosomes.	Absent.
Reproduction by multicellular spores containing 1-7 polar capsules with an extrusible filament of anchoring function to fix hatching spore that produces one to many infective sporoplasms of ameboid shape.	Spores.	Sexuality by autogamy; haploid spores (Myxosporea); fusion of gametes to form diploid spores (Actinosporea).	Acentric mitosis.	Lack centrioles. Polar capsules. Multicellular with distinct soma and generative (germen) constituents.	Absent.

Phylum	Morphology of Growing Organisms	Occurrence	Nutrition	Undulipodia	Mitochondria	Plastids/ Pigments
Microspora	Unicellular, induce single cell tumors.	Intracellular symbionts of protoctists, vertebrates and invertebrates.	Heterotrophic: osmotrophic, obligate symbiotrophy to necrotrophy.	Absent.	Absent.	Absent.
Acrasea	Ameboid organisms that aggregate to form pseudoplasmodia that develop into sorocarps.	Dead plant parts, soil, or dung.	Heterotrophic: phagotrophy.	In *Pocheina flagellata*, each spore or stalk cell may yield 2 bimastigote cells.	Present; platelike or tubular cristae.	Absent.
Dictyostelida	Ameboid. Pseudoplasmodium produced by aggregating myxamebas giving rise to multicellular sorocarp.	Soil.	Heterotrophic: phagotrophy (bactivorous).	Absent.	Present; tubular cristae.	Absent.
Rhodophyta	Morphologically diverse: unicells, uni- or multiseriate filaments, large pseudoparenchymatous, branched or unbranched cylindrical to bladelike thalli, including crustose and erect forms, some calcified.	Primarily marine, littoral and benthic, some restricted to freshwater habitats, others terrestrial. Some symbiotrophic on other red algae or other organisms.	Phototrophic. Some symbiotrophic.	Absent.	Present.	Rhodoplasts containing chlorophyll *a*, α and ß carotene, lutein, zeaxanthin and phycobilins (allophycocyanin, phycocyanin and phycoerythrin).

Reproduction	Propagules (Cysts)	Sexuality	Nuclear Features	Distinguishing Cytological Features	Presence in Fossil Record
Binary or multiple fission; reproduction by merogony. Sporulation: sporogony and spore morphogenesis; spores with long, coiled everting polar tubes (polar filaments) inoculate membrane-bounded nucleocytoplasm into host.	Spores.	Sexuality reported in some genera. Meiosis at onset of sporogony.	Nuclei isolated or in diplokaryon. Closed mitosis.	Polar tubes. Small ribosomes, rRNA with sizes typical of prokaryotes.	Absent.
Reproduction by spore dispersion in sorocarps.	Microcysts. Sphaerocysts in *Copromyxa protea*; spores; sorocysts.	Sexuality unknown.	Uninucleate; single central nucleolus.		Absent.
Reproduction by spores.	Spores (cysts): thick-walled and variable in size; germinate into amebas. Microcysts, macrocysts - may result from sexual fusion.	Sexuality involves fusion of myxamebas (heterothallic or homothallic) followed by macrocyst formation. Evidence for meiosis.	Persistent nuclear envelope, peripheral nucleoli; intranuclear spindle; extranuclear spindle pole bodies.		Absent.
Florideophycidae: complex life history composed of alternation of 2 free-living and independent generations (gametophyte and tetrasporophyte) and a third generation, the carposporophyte, that occurs on the female gametophyte. Apomictic and apogamic life histories known. Bangiophycidae: mostly reproduction by simple division or release of naked spores (monospores).	Absent.	Sexuality in Florideophycidae and in some Bangiophycidae.	Mitotic division: nuclear envelope intact with polar fenestrae present, intranuclear spindle, nucleus-associated organelle (NAO) present at each pole, thought to be MTOC.	Pit plugs (pit connections). Phycobilisomes.	Florideophycidae meager, date to Cretaceous. Corallinales dating to late Jurassic or mid-Cretaceous. Solenoporaceae from Cambrian to Paleocene. Bangiophycidae limited; *Bangia* fossils from late Vendian, Greenland, 750 mya.

Phylum	Morphology of Growing Organisms	Occurrence	Nutrition	Undulipodia	Mitochondria	Plastids/ Pigments
CONJUGAPHYTA (=order Zygnematales in phylum Chlorophyta)	Filamentous with cell walls lacking median incision or pores (Zygnemataceae) and unicellular (desmids). Placoderm desmids also filamentous or colonial.	Almost entirely freshwater in a wide variety of habitats including some extreme environments.	Phototrophic.	Absent.	Present.	Chloroplasts containing chlorophylls *a* and *b*, ß and γ carotenes, several xanthophylls.
XENOPHYO-PHORA	Large plasmodium enclosed by branched tube system made of transparent, cementlike organic substance.	Marine benthic in deep sea on bathyal, abyssal, and hadal sediments.	Heterotrophic; particle collecting via pseudopodia.	Two present in gametes.	?	Absent.
CRYPTOPHYTA (Cryptomonads)	Unicellular, flattened, motile; some nonmotile palmelloid forms that often form colonies.	Cosmopolitan in marine and freshwater habitats; most prominent in oligotrophic, temperate, and northern waters. May be symbiotic in other protoctists.	Phototrophic.	Two, slightly unequal with 2 rows of masti-gonemes on longer and one row on shorter. Fine fibrils and scales may also occur on surface.	Present; platelike cristae. In cryptophyte symbionts of *Mesodinium rubrum*, mitochondria with tubular cristae in plastid-mitochondrial complexes (PMCs).	Plastids (of different colors). Chlorophylls *a* and *c* and phycobilins as accessory pigments, carotene, and several xanthophylls.

Reproduction	Propagules (Cysts)	Sexuality	Nuclear Features	Distinguishing Cytological Features	Presence in Fossil Record
Fragmentation, cell division, akinetes, aplanospores, partheno-spores.	Zygospore formed following fertilization.	Sexuality by conjugation involving fusion of ameboid gametes to form thick-walled zygospore. Zygotic meiosis. Haplobiontic.	During mitosis, stainable material from nucleolus called nucleolar substance. Complex interphase nucleolus; absence of localized centromere on chromo-some of some species.	Cell wall with three layers.	Both zygo-spores and trophic cells present. Desmids dating to middle Devonian. Oldest zygnematacean from Carbonif-erous.
Incompletely known.	Not known.	Gametogamy occurs; gametes with 2 unduli-podia or ame-boid.	May be heterokary-otic.	Cytoplasm contains huge numbers of barite crystals (granellae); tests composed of foreign matter (xenophyae).	Suggested that *Paleodictyon* is fossil xeno-phyophore; if so, Ordovician to present.
Division of motile cells.	Thick-walled cysts in some.	Sexuality not known; fusion, interpreted to be sexuality, seen in symbiotic forms only.		Ejectosomes in gullet/furrow system. Nucleomorph in periplastidial compartment.	Absent.

Phylum	Morphology of Growing Organisms	Occurrence	Nutrition	Undulipodia	Mitochondria	Plastids/ Pigments
Glaucocystophyta	Unicellular mastigotes, nonmotile unicellular and colonial organisms with or without persistent contractile vacuoles.	All freshwater, living in plankton or benthos of lakes, ponds, or ditches.	Phototrophic.	Present in some members; two undulipodia both with mastigonemes. One in direction of swimming, one lateral. Kinetid with 2 or 4 multilayered structures.	Present; flattened cristae.	Cyanelles, organelles that are modified cyanobacterial symbionts. Chlorophyll *a*, ß-carotene, zeaxanthin, ß-cryptoxanthin, allophycocyanin, phycocyanin.
Karyoblastea (*Pelomyxa palustris*)	Giant, free-living ameba.	In mud or sand in many freshwater environments in Northern hemisphere. Usu. in stagnant or near-stagnant water.	Heterotrophic: phagotrophy (algivorous).	Nonmotile cilialike organelles present with extreme variations in axonemal morphology not [9(2)+2].	Absent.	Absent.
Zoomastigina (See Table 6B)	Unicellular mastigotes. Amebomastigotes with ameboid stage.	Free-living (amebomastigotes, bicosoecids, choanomastigotes, pseudociliata, some diplomonads, kinetoplastids); soil, marine, or freshwater. Symbionts (opalinids, proteromonads, parabasalia, retortamonads, pyrsonymphids) or parasites (some kinetoplastids and some diplomonads) of animals.	Heterotrophic: osmotrophic; symbiotrophic.	Formed reversibly in amebomastigotes. May be part of karyomastigont system (e.g., diplomonads); have paraxial rod (kinetoplastids); or be present in rows (opalinids, pseudociliates).	Lacking in some.	Absent.

Reproduction	Propagules (Cysts)	Sexuality	Nuclear Features	Distinguishing Cytological Features	Presence in Fossil Record
Mastigotes reproduce asexually by longitudinal binary fission; autospore formation.	?	Sexuality unknown.	Open spindle during mitosis. Centrioles and phycoplasts absent.	Cyanelles.	Unknown.
Plasmotomy.	Formed as overwintering forms.	Sexuality unknown.	Multinucleate. May be nuclear budding. Lack centrioles, centromeres, other MTOCs, chromosome coiling cycle. Perinuclear bacteria.	Lack golgi bodies but contain several types of endosymbiotic and parasitic organisms; three types of endosymbiotic bacteria: one Gram-positive, one Gram-negative, one Gram-variable; two are methanogens.	Absent.
Mastigote binary fission. Some choanomastigotes form mastigote swarmers; pseudociliates and opalinids undergo palintomy.	Formed by most.	Sexuality documented in some opalinids, in hypermastigotes and possibly *Mixotricha* (parabasalians), and in *Notila*, *Saccinobaculus*, and *Oxymonas* (pyrsonymphids).		Some lack golgi. Axostyle, hydrogenosomes present in some.	No direct record. Inferred from presence of fossil insects (wood-eating cockroaches, etc.).

Phylum	Morphology of Growing Organisms	Occurrence	Nutrition	Undulipodia	Mitochondria	Plastids/ Pigments
EUGLENIDA	Unicellular, green or colorless mastigotes. Two genera sessile.	Freshwater, esp. areas rich in decaying organic matter or in marine/brackish areas.	Colorless forms phagotrophic or osmotrophic. One-third phototrophic.	Usu. two; mastigonemate anterior one propelling cell, posterior trailing as a ventral skid or nonemergent.	Present; flattened cristae.	Chloroplasts with chlorophylls *a* and *b*; ß-carotene and several xanthophylls.
CHLORARACHNIDA (*Chlorarachnion reptans*)	Ameboid plasmodium with individual cells linked by a network of reticulopodia.	Known only from enrichment cultures from marine habitats.	Phototrophic; ameboid cells phagotrophic.	One in zoospores. Has hair point and bears fine lateral hairs. Appears to be wrapped around cell body during swimming.	Present; tubular cristae.	Chloroplasts with chlorophylls *a* and *b*.
PRYMNESIOPHYTA **(Haptophyta)**	Mostly unicellular, palmelloid, or coccoid but a few form colonies or short filaments.	Planktonic members found worldwide. A few species euryhaline. Occur rarely in fresh water. Symbiotic forms in 3 species of Acantharia and another form in solitary Radiolaria.	Phototrophic.	Two; equal or subequal nonmastigonemate; homo- or heterodynamic. Longer undulipodium of Pavlovales may have fine hairs or small, dense knoblike scales.	Present.	Plastids with chlorophylls *a*, c_1 and c_2. ß-carotene, diatoxanthin, diadinoxanthin, fucoxanthin.

Reproduction	Propagules (Cysts)	Sexuality	Nuclear Features	Distinguishing Cytological Features	Presence in Fossil Record
Binary fission. Sporadic meiosis occurs in some euglenids, believed to result from nonsexual autogamy.	May be formed under unfavorable environmental conditions.	Well-documented report of a complete sexual cycle in *Scytomonas*.	Uninucleate with prominent nucleolus. Closed mitosis, intranuclear spindle. Chromosomes remain condensed during interphase.	Trichocysts in some phagotrophic euglenoids and *Isonema*. Metaboly. In phototrophs a paracrystalline paraflagellar body attached to the paraflagellar rod of the anterior undulipodium.	Sparse; dating to middle Eocene. Freshwater forms do not fossilize well.
Zoospores. Walled stage formed that appears to give rise directly to ameboid stage and to zoospores via tetrad division. No sexuality known.	Walled structures formed: see Reproduction.	No sexuality known.	Each cell has a single central nucleus.	Lack photoreceptor apparatus but contain periplastidial compartment. Vesicles resembling extrusomes, but never seen to discharge.	Absent.
Asexual by binary fission; may be alternation between nonmotile form and motile (swarmer) form.	Few reports. In *Prymnesium*, cysts formed composed of layers of scales with siliceous material on the distal surfaces.	No sexuality documented.	Nuclear division open; spindle initiated outside nucleus and forms wide spindle with ill-defined poles.	In many, haptonemata; may be reduced or vestigial. Stigma consisting of single layer of lipid globules in some members of Pavlovales. Cells often covered with scales of varying degrees of complexity formed intracellularly. May be unmineralized or calcified (coccoliths).	Abundant. Included with "calcareous nanofossils." First found in abundance in early Jurassic. Occur in calcareous clays, marls, chalks, limestones, and occasionally coarser sediments.

Phylum	Morphology of Growing Organisms	Occurrence	Nutrition	Undulipodia	Mitochondria	Plastids/ Pigments
RAPHIDOPHYTA (Chloromonads)	Unicellular motile (mastigote) or palmelloid forms.	Freshwater or marine, planktonic. Marine-open sea or brackish water.	Phototrophic.	Heterokont: forward undulipodium bears tubular mastigonemes and both arise from a shallow pit at or near the apex of the cell.	Present; tubular cristae.	Chromoplasts with chlorophylls *a* and *c*. ß-carotene, lutein monoepoxide, antheraxanthin and xanthophyll that is trollixanthin-like. Fucoxanthin in some marine genera.
EUSTIGMATOPHYTA	Coccoid unicells: single, in pairs, or in colonies.	Marine and freshwater.	Phototrophic.	Most have single anteriorly-located undulipodium with mastigonemes. Two species have pair of unequal undulipodia, longer with mastigonemes.	Present.	Chloroplasts with chlorophyll *a*, ß-carotene, violaxanthin, vaucheriaxanthin.

Reproduction	Propagules (Cysts)	Sexuality	Nuclear Features	Distinguishing Cytological Features	Presence in Fossil Record
Binary fission.	?	Sexuality unknown.	Closed mitosis; extranuclear spindle enters through gaps in nuclear envelope. 97±2 chromosomes in *Vacuolaria virescens*; 65-75 in *Gonyostomum semen*.	Pyrenoids present only in plastids of some marine species. No eye-spots. Some contain organelles resembling microbodies. Large golgi over anterior surface of nucleus. Mucocysts and trichocysts in many species.	Absent.
Autospores, binary fission. Zoospores in some species.	?	Sexuality unknown.		Prominent red eyespot composed of osmiophilic globules.	Absent.

Phylum	Morphology of Growing Organisms	Occurrence	Nutrition	Undulipodia	Mitochondria	Plastids/ Pigments
ACTINOPODA: Polycystina (Radiolaria), Phaeodaria	Single cells with long processes called axopods. May form colonies.	Marine only, pelagic.	Heterotrophic; predaceous on nanoplankton, mastigotes, diatoms, and copepods.	Polycystina: bimastigote "spores" produced. One undulipodium emergent, other intracytoplasmic forming helical coil around body. Phaeodaria: polynucleated ameboids formed as result of successive binary fissions and these later develop 2 undulipodia.	Present.	None or algal symbionts (dinomastigote zooxanthellae).
ACTINOPODA: Heliozoa	Spherical unicells having naked, organic, or siliceous coated bodies from which axopods radiate.	Freshwater or marine. Generally in shallow water just above sediment-water interface.	Heterotrophic (predaceous).	Some species have one or more undulipodia in addition to axopods. May be undulipodiated gametes.	Present.	None or may contain symbiotic algae.

Reproduction	Propagules (Cysts)	Sexuality	Nuclear Features	Distinguishing Cytological Features	Presence in Fossil Record
Polycystina: binary fission, multiple fission ("sporogenesis"), or budding. No sexuality known. Phaeodaria: binary fission of multinucleate cell. Karyokinesis may be meiotic.	?	No sexuality known.	Polycystina: large nucleoli; during mitosis chromosomes form "parent plate" within which they are oriented parallel to their future direction of movement. Polyploidy. Phaeodaria: large nuclei. Polyploidy. During formation of offspring nuclei, structure formed composed of central substance from which irregular and anastomosed masses radiate, probably composed of DNA.	Polycystina: regularly perforated silica skeleton with radial axopods emerging among fine, ramified pseudopods. Phaeodaria: sometimes lack skeleton; when present, consists of isolated pieces of or numerous hollow tubes. Phaeodium in ectoplasm.	Late Precambrian/ early Paleozoic. Extensive deposits on ocean floor called radiolarian oozes which can form radiolarites and eventually chert.
Binary or multiple fission; budding.	Gametogenesis in gamontocyst. Resting cyst formed under unfavorable conditions.	Autogamy (pedogamy) in *Actinophrys sol* involving progamic fission, meiosis, and isogamous fertilization and resting cyst formation.	Growing stage mono- or multinucleate. Nuclear envelope remains intact through most of karyokinesis, disappearing at telophase.	Cell naked or covered with cell coat devoid of particular geometrical substructures in which organic or siliceous dispersed outer skeleton is embedded. Extrusomes of various types scattered in cytoplasm.	Absent.

Phylum	Morphology of Growing Organisms	Occurrence	Nutrition	Undulipodia	Mitochondria	Plastids/ Pigments
ACTINOPODA: Acantharia	Unicellular with skeletons of strontium sulfate.	Except for one sessile genus, exclusively marine plank-tonic.	Heterotrophic: phagotrophic.	Bimastigotes formed by changes in environmental conditions following rapid nuclear divisions. In *Haliom-matidium* endogenous karyokineses followed by fragmentation of cytoplasm and oval bimastigotes shed.	Present.	None or may contain symbiotic algae.
HYPHOCHYTRIO-MYCOTA	Polycentric thallus.	Soil or water, freshwater or marine.	Heterotrophic: osmotrophy or necrotrophy.	Zoospores with single anteriorly-directed mastigonemate undulipodium; second centriole that does not become kineto-some.	Present; tubular cristae.	Absent.
LABYRINTHULO-MYCOTA	Cells in wall-less ectoplasmic networks.	Cosmopolitan, primarily found in marine and estuarine environments, usu. associated with benthic algae, marine vascular plants, and detrital sediments.	Heterotrophic. Saprotrophic to weakly necrotrophic.	Zoospores heterokont bimastigotes. Anterior undulipodium with mastigonemes and almost twice as long as posterior, whiplash undulipodium.	Present; tubular cristae.	Absent.

Reproduction	Propagules (Cysts)	Sexuality	Nuclear Features	Distinguishing Cytological Features	Presence in Fossil Record
Multinucleate growing cells give rise to uninucleate bimastigotes which may or may not be associated with cyst transformation.	Formed in association with cell division.	Sexuality unknown.	All multinucleate except *Haliommatidium* which has one large nucleus and which is coated on its inner surface with a thick microfibrillar network, the fibrous lamina. Mitosis with persistent nuclear envelope. MTOC called the spindle pole body found in some nuclear pores.	Skeletons of strontium sulfate that exhibit strictly defined configuration of 10 diametral or 20 radial spicules, a cell body covered with an outer pellicle, the periplasmic cortex, and possession of numerous myonemes. May contain lithosomes.	Absent.
Formation of zoospores from multinucleate cytoplasm followed by formation of cleavage vesicles.	Zoospores form cysts which germinate to give rise to thallus.	Sexuality not well documented in any member; described in *Anisolpidium ectocarpii*.		Lack nuclear cap and microbody-lipid complex.	Absent.
Aggregation to form sporangia (sori) containing cysts within a common wall. Sporocytes divide to produce heterokont bimastigotes in *Labyrinthula vitellina* and *L. algeriensis*. In thraustochytrids, enlargement of cells which develop into sori and then release spores as result of mitotic divisions.	Formed in sporangia and give rise to zoospores in Labyrinthuliidae.	In *L. vitellina*, sexual reproduction leads to formation of 8 zoospores (meiospores). No sexual stages known in thraustochytrids.	Nuclei with prominent central nucleolus.	Cells with specialized organelles at surface called sagenogens. Thin wall scales produced in golgi. Pigmented eyespot in zoospores of Labyrinthuliidae.	Absent.

Phylum	Morphology of Growing Organisms	Occurrence	Nutrition	Undulipodia	Mitochondria	Plastids/ Pigments
Plasmodiophoromycota	Thalli composed of multinucleate, unwalled protoplasts (=plasmodia).	Soil or freshwater; obligate symbiotrophs.	Heterotrophic: osmotrophy. Symbiotrophic on plants, algae, other aquatic protoctists, and fungi. Some necrotrophic.	Zoospores bear 2 anteriorly directed whiplash undulipodia of unequal length.	Present.	Absent.
Dinomastigota (Dinoflagellata)	Walled or naked unicells.	Marine and freshwater. Intrazoic as zooxanthellae in a large range of protist and invertebrate hosts.	Roughly half are phototrophic. Nonphotosynthetic forms are phagotrophic or osmotrophic.	Typically bimastigote with one ribbonlike undulipodium, the transverse, wrapped around the cell, and one that beats posteriorly, the longitudinal. Arise from the side and lie in surface grooves.	Present; tubular cristae.	Plastids containing chlorophylls *a* and c_2, peridinin, ß-carotene, diadinoxanthin, dinoxanthin. May contain symbiotic cyanobacteria (phaeosomes), cryptomonads, or diatom/ chrysophyte-like symbionts.

Reproduction	Propagules (Cysts)	Sexuality	Nuclear Features	Distinguishing Cytological Features	Presence in Fossil Record
Plasmodia develop into sporangia giving rise to zoospores (secondary zoospores) or cystosori- which form resting bodies, cysts, that liberate zoospores (=primary zoospores). Meiosis occurs in some taxa just prior to the formation of thick-walled cysts.	Cysts produced either singly or in aggregations that develop from cyst-forming (cystogenous) plasmodia and liberate zoospores. Term not for cysts of zoospores.	Existence and timing of karyogamy not documented.	Cruciform division.	Specialized cell process for penetration of host: encysted zoospore forms approsorium at end of germ tube. Extracellular infection apparatus consists of the rohr, the schlauch, and the stachel.	Absent.
Cell division. Some blastodinialean parasites have specialized division called palisporogenesis.	Fewer than 10% of living forms known to form cysts. Form resting cysts (hypnozygote, resting spore), temporary cysts (pellicle cysts, ecdysal cyst), trophic cysts, digestion cysts.	Most haploid with post-zygotic meiosis. Iso- or anisogamic fusion forms tri- or quadrimastigote planozygote which forms nonmotile resting cyst (hypnospore). Excystment may precede or be followed by meiosis.	Dinokaryon. Prominent nucleolus also persistent. Histone proteins reduced, altered relative to other protoctists.	Thecate cells with cellulosic plates in patterns that form basis of taxonomy. Extrusomes including trichocysts, mucocysts, and in some genera, nematocysts. Taeniocysts in *Polykrikos schwartzii*. Luminescence associated with cytoplasmic bodies. Specialized vacuoles called pusules.	Earliest form assigned is from Silurian with many undoubted fossils from late Triassic onward. Acritarchs, some of which may be dinomastigotes, date to Precambrian.

Phylum	Morphology of Growing Organisms	Occurrence	Nutrition	Undulipodia	Mitochondria	Plastids/ Pigments
Chrysophyta	Single cells or colonies. Generally mastigotes, a few filamentous; may be ameboid.	Primarily planktonic in fresh water. Some epibiotic or neustonic, a few benthic. A few marine.	Phototrophic. Colorless forms osmotrophic or phagotrophic.	Swarmer (basic morphological type) heterokont with short undulipodium reduced and/or transformed into photoreceptor in more advanced forms.	Present; tubular cristae.	Plastids containing chlorophylls *a* and *c*, fucoxanthin.
Chytridiomycota	Thallus consisting of single cell or multicellular. Some filamentous.	Aquatic environments and soil in a wide range of geographical areas.	Heterotrophic with absorptive nutrition. Some obligate parasites on plants, fungi, or insects. Saprobic, necrotrophic.	Usu. unimastigote zoospores but some polymastigote. Undulipodia whiplash type and lack scales or mastigonemes. Have second kinetosome called nonfunctional centriole.	Present; vermiform. Absent in some anaerobic symbionts.	Absent.

Reproduction	Propagules (Cysts)	Sexuality	Nuclear Features	Distinguishing Cytological Features	Presence in Fossil Record
Cytokinesis. In some coccoid and filamentous forms, cell divides by autospore formation.	Stomatocysts (=statospores) with silicified walls. May be smooth or ornamented.	In small loricate monads, undifferentiated cells act as gametes, fuse apically and produce a globular zygote. In colonial species of *Dinobryon*, autogamy or gametic fusion to form zygotic cysts. In *Synura* and *Mallomonas*, normal scale-bearing cells act as gametes, with posterior fusion.	During mitosis, rhizoplasts from 2 kinetosomes act as poles for the organization of spindle microtubules.	Chrysolaminarin as storage. Eyespot often present.	Cysts date to upper Cretaceous. Silicoflagellates first appear in middle Cretaceous and reached highest development in Miocene. Used in biostratigraphy and as paleoecological indicators.
Spores borne in zoosporangia.	?	Zoospores can act as gametes and fuse. In other types, anisogametes fuse and in others zoospore fuses with developing sporangium which then becomes thick-walled. Karyogamy following fusion of rhizoids also known. In Monoblepharidales, sexual reproduction oogamous.	Most species haploid, but *Allomyces* has distinct haploid and diploid phases. Nucleus may be associated with (fixed to) kinetosome by rootlet microtubules or a fibrillar rhizoplast.	Rumposome present in orders Chytridiales, Monoblepharidales. Microbody-lipid complex. Nuclear cap in Monoblepharidales. Side body complex in Blastocladiales.	Rarely documented. Date to Pennsylvanian period but probably occurred earlier.

Phylum	Morphology of Growing Organisms	Occurrence	Nutrition	Undulipodia	Mitochondria	Plastids/ Pigments
Plasmodial Slime Molds: **Myxomycota**	Plasmodium: unwalled, multi-nucleate ameboid body.	Cosmopolitan in forest environ-ments. Generally in temperate and tropical forests.	Heterotrophic: phagotrophy.	Swarm cells with 2 undulipodia.	Present.	Absent.
Plasmodial Slime Molds: **Protostelida**	Ameboid with filose pseudopodia.	Aerial portions of dead plants and bark of living trees, soil, dung, and rotting wood.	Heterotrophic: phagotrophy. Feed on bacteria, yeasts, and spores of filamentous fungi.	Unimastigote. Kinetids with one to two kineto-somes.	Present; tubular or vesicular cristae.	Absent.
Ciliophora	Unicells.	Most free-living; found in lakes, ponds, estuaries, saltmarshes, and oceans; terrestrial soils; and as symbionts with a variety of animals; some histophagous.	Heterotrophic: mostly phagotrophic, ingesting particulate matter and/or prey in food vacuoles. *Tetrahymena vorax* alternates between bactivorous and ciliovorous forms.	Cells typically with files of cilia known as kineties on cell surface. Kinetids form basis of tax-onomy.	Present.	Absent or may contain algal symbionts or foreign plastids.

Reproduction	Propagules (Cysts)	Sexuality	Nuclear Features	Distinguishing Cytological Features	Presence in Fossil Record
Reproductive phase haploid. Formation of sporophores which contain spores from which emerge ameboid protoplasts (myxamebas) or undulipodiated swarm cells. Myxamebas may divide mitotically. Small plasmodia may also coalesce to form large plasmodia.	Microcysts may be formed from myxamebas. Sclerotia (with macrocysts) formed directly from plasmodia and are resistant.	Both myxamebas and swarmers can be gametes and fuse to form an ameboid zygote, which divides to form a plasmodium.	Plasmodia with synchronous division of nuclei. Intranuclear, acentriolar type of nuclear division. Closed mitosis except in myxamebas. Variable chromosome number, even within a species.	Reversible protoplasmic streaming in plasmodia of many species. Sporophores may contain calcium deposits.	Undocumented.
Spores borne in sporocarps. Also may form mastigotes in zoocyst which give rise to plasmodium that in turn cleaves into prespore cells.	In some species zoocysts, giving rise to mastigotes, e.g., *Ceratiomyxella tahitiensis*.	Meiosis in spores.	Mitosis with open centric spindle.		Absent.
Binary fission.	In absence of nutrients, many species encyst; may or may not withstand desiccation.	Sexuality involves conjugation between compatible mating types, or autogamy.	Nuclear dualism: macronucleus is physiologically active, containing thousands of copies of genes; diploid micronucleus is germ nucleus whose meiotic products are exchanged during conjugation.	Extrusomes, including mucocysts and toxicysts. Cell mouth, cytostome.	Poorly represented. Loricas of tintinnids date to Ordovician; peak of diversity in Jurassic and Cretaceous. Chitinozoa known from Proterozoic sediments, may be tintinnid cysts.

Phylum	Morphology of Growing Organisms	Occurrence	Nutrition	Undulipodia	Mitochondria	Plastids/ Pigments
GRANULO-RETICULOSA **(Foraminifera)**	Unicellular ameboid organisms with granular reticulopods; all enclosed in tests except class Athalamea.	Ubiquitous in marine habitats.	Heterotrophic; feeding on algae and bacteria.	Mastigote gametes produced by some species; bimastigote and unimastigote gametes reported.	Present; tubular crisate.	Absent or containing endosymbiotic algae, incl. diatoms, dinomastigotes, and unicellular red and green algae (some shallow benthic species).
APICOMPLEXA	Single cells with apical complex.	All parasites of invertebrate or vertebrate animals.	Heterotrophic: obligate symbiotrophy.	Zoites motile. In male gametes, axoneme that grows from persistent polar kinetosome. Undulipodia more or less developed among gregarines. Gregarines and coccidia have kinetosomes made of single microtubules; hematozoans bear standard kinetosomes.	Present.	Absent.

Reproduction	Propagules (Cysts)	Sexuality	Nuclear Features	Distinguishing Cytological Features	Presence in Fossil Record
Multiple fission (schizogony), binary fission, budding or cytotomy, or fragmentation.	*Rosalina bulloides* encysts during gamete formation.	"Classical" life cycle consisting of regular alternation of generations between sexual ("A form" = gamonts) and asexual ("B form" = agamonts) phase; possibly successive asexual generations. Sexual phase haploid and uninucleate; asexual phase diploid and multinucleate. May be autogamy or gamontogamy.	Heterokaryosis in some, e.g. *Rotaliella* spp. with one large somatic and three generative nuclei per individual. Chromosomes and division features vary greatly among foraminifera.	Granular pseudopods that form anastomosing reticulate networks with characteristic two-way streaming. Cell body of most foraminifera enclosed by a test which may be organic, agglutinated, or calcareous.	May have evolved in Proterozoic, but earliest fossil forams date from Cambrian. Extensive record of Cambrian and Ordovician single chambered fossil tests.
Merogony or endogeny.	Sporocysts and oocysts produced which derive directly or indirectly from zygote and which may be released from animal in feces or after death.	Life cycle consisting of 2 or 3 phases: gamogony, sporogony, and in some, merogony. Isogamy in gregarines; oogamy in coccidia and hematozoans. Zygotic meiosis. Sporogony leads to production of haploid sporozoites by the zygote.	Merogonic mitoses with centrocone. Centrioles made of 9 singlet microtubules present on top of centrocones except in hematozoans in which centrioles are not differentiated.	Zoites with apical complex.	Absent.

Phylum	Morphology of Growing Organisms	Occurrence	Nutrition	Undulipodia	Mitochondria	Plastids/ Pigments
BACILLARIOPHYTA (Diatoms)	Unicellular, silica-walled algae; may form colonies.	In all aquatic habitats; freshwater, marine, soil.	Phototrophic.	Only in gametes in species with anisogamous gametes (as in some centric genera).	Present.	Plastids containing chlorophylls *a* and *c*, fucoxanthin, and minor pigments, e.g., diatoxanthin and diadinoxanthin.
CHLOROPHYTA (See Table 6c)	Diverse morphologies from mastigotes to complex multicellular thalli.	Freshwater or marine, soil, terrestrial, subaerial. Some found in extreme environments. May be symbiotic in other protists, fungi (lichens), or animals.	Phototrophic.	Mastigotes as spores or gametes.	Present.	Chloroplasts containing chlorophylls *a* and *b*, accessory pigments.

Reproduction	Propagules (Cysts)	Sexuality	Nuclear Features	Distinguishing Cytological Features	Presence in Fossil Record
Cell division.	?	Sexual reproduction by auxospore formation. Isogamous (pennate) or anisogamous with motile gametes. Meiosis in gametogenesis and fertilization followed by development of zygote (=auxospore).	Dense granular material associated with nucleus during interphase in some species. MTOC breaks down at prophase and is replaced by complex and highly ordered spindle.	Siliceous walls with ornate pores, thickenings or spines, consisting of 2 valves. Valves may be centric or pennate.	Earliest whole diatom valve from lower Cretaceous with increasing abundance toward Tertiary. Freshwater forms thought to occur later. Absence from earlier strata due to conversion of silica to porcelanite and later to chert.
Thallus fragmentation; formation by many of akinetes, spores, autospores.	Spores, zygospores	Sexual reproduction may be isogamous, anisogamous, or oogamous. Many haplobiontic, but greater number diplobiontic and exhibit alternation of iso- or heteromorphic generations. Both mastigote and amastigote gametes.			Phycomata of Prasinophyceae in sediments from Precambrian to Recent. Calcified members of Dasycladales and some members of Caulerpales dating to Precambrian. Oospores of Charales (gyrogonites) dating to lower Devonian or even upper Silurian.

Phylum	Morphology of Growing Organisms	Occurrence	Nutrition	Undulipodia	Mitochondria	Plastids/ Pigments
Oomycota	Uninucleate or coenocytic protoplasts bounded by cell walls. Thalli that may be filamentous or monocentric.	Most species freshwater or terrestrial.	Heterotrophic: facultative or obligate symbiotrophs. Osmotrophs.	Present in zoospores. Heterokont with anterior bearing 2 rows of mastigonemes and posterior often coated with fine flexuous hairs.	Present; tubular cristae except in *Aqualinderella*.	Absent.
Xanthophyta (Tribophyta)	Coccoid unicells, some filamentous and siphoneous species.	Primarily freshwater.	Phototrophic, mixotrophic.	In zoospores, gametes. Spermatozoids and zoospores heterokont: shorter anterior undulipodium with lateral hairs; long, posterior one lacking hairs. In synzoospores, both undulipodia smooth.	Present; tubular cristae.	Plastids containing chlorophylls *a* and *c* (c_1 and c_2), ß-carotene, diatoxanthin, diadinoxanthin, heteroxanthin, vaucheriaxanthin ester.

Reproduction	Propagules (Cysts)	Sexuality	Nuclear Features	Distinguishing Cytological Features	Presence in Fossil Record
Protoplasmic cleavage within a sporangium, resulting in the production of naked mastigotes (zoospores).	Spores encyst and can, in some species, give rise to zoospores.	Life histories with sexual and asexual phases. Sexuality involving production of multinucleate antheridia and archegonia which develop in contact with each other. Haploid gametic nuclei produced in antheridium transferred directly to oogonial cavity and protoplasmic separation defines up to 40 oospheres which then form a wall, becoming thick-walled and resistant.	Nuclear membrane persists through mitosis and meiosis.	Dense body vesicle system.	Equivocal evidence.
Cell division, endogenous cysts in monadoid and rhizopodial forms; palmelloid forms reproduce by zoospores, hemiautospores, or aplanospores. Coccoid forms reproduce asexually by autospores or zoospores. Synzoospores resulting from incomplete cleavage.	Resting stages seen rarely with exception of oospores of *Vaucheria*. Endogenously produced cysts in monadoid and rhizopodial forms. Akinetes produced by coccoid and filamentous forms. Particular type of aplanospore described as resting stage.	Sexuality rare; appears to be confined to coenocytic genera *Botrydium* and *Vaucheria*.	17 chromosomes in 4 species of *Tribonema*.	Plastid endoplasmic reticulum enclosing periplastidial compartment in which the periplastidial network (a membranous reticulum) exists.	None positively identified.

Phylum	Morphology of Growing Organisms	Occurrence	Nutrition	Undulipodia	Mitochondria	Plastids/ Pigments
Phaeophyta	Multicellular organisms ranging from microscopic filaments to macroscopic fleshy thalli.	Almost exclusively marine; intertidal and subtidal.	Phototrophic.	Zoospores, gametes heterokont: 2 unequal undulipodia inserted laterally, anterior bearing mastigonemes and posterior lacking.	Present; flattened cristae.	Phaeoplasts containing chlorophylls *a*, *c*, and c_1, fucoxanthin, carotene, violaxanthin.
Incertae Sedis: **Ellobiopsids**	Large members arborescent.	Parasites of euphasids, mysids, amphipods, shrimp, and copepods.	Heterotrophic.	Present in zoospores. Lack mastigonemes and axial rod.	Present; tubular cristae.	Absent.

Reproduction	Propagules (Cysts)	Sexuality	Nuclear Features	Distinguishing Cytological Features	Presence in Fossil Record
Production of zoospores in sporangia on gametophyte or sporophyte.	?	Alternation of generations. Gametes produced in gametangia on gametophytic thalli. Fertilization isogamous to oogamous. Diploid zygote develops into sporophyte generation. Meiosis in sporangia on sporophyte thalli, leading to formation of meiospores.		Physodes in cell vacuole.	Rare. Identifiable forms date to Tertiary.
In one genus studied, gonomeres segment to form prespores that develop into mastigotes.	Spores produced by gonomere faceting.	Sexuality unknown.	Large irregular nuclei with one or more nucleoli. Unusual centriolar complexes comprised of a dense fusiform body and a pair of centrioles carried at each pole of the fusiform body. Gonomeres have small spherical nuclei without nucleoli; fusiform bodies absent and centrioles not observed.	Trophomere bounded by well-defined pellicle occasionally interrupted by flask-shaped organelles, resembling mucocysts.	?

Phylum	Morphology of Growing Organisms	Occurrence	Nutrition	Undulipodia	Mitochondria	Plastids/ Pigments
Incertae Sedis: **Ebridians**	Zoomastigotes with basketlike internal skeleton.	Confined to coastal marine environments.	Heterotrophic, feeding on phytoplankton, esp. diatoms.	Two; arise from near anterior end, apparently unequal and lacking hairs.	Present.	Absent.

Reproduction	Propagules (Cysts)	Sexuality	Nuclear Features	Distinguishing Cytological Features	Presence in Fossil Record
Simple fission.	Some fossil forms produce a chamberlike body— the lorica, which has been suggested to possibly correspond to a cyst stage.	Sexuality unknown.	Prominent nucleus toward cell anterior. In *Hermesium*, nucleus has granular appearance likened to that of dinomastigotes. In *Ebria*, the granulation is finer; prominent nucleolus.	Numerous refringent droplets of oil may be present. Skeleton, composed of silica rods branching in a regular way, entirely surrounded by cell.	Plentiful, dating to Cenozoic. Most diverse in Miocene.

Table 6B. Classes of Phylum Zoomastigina

Class	Morphology of Growing Organisms	Occurrence	Nutrition	Undulipodia	Mitochondria	Plastids/ Pigments
ZOOMASTIGINA: **Amebomastigota**	Unicellular mastigotes alternating with monopodial, uninucleate amebas.	Cosmopolitan; soil, freshwater, rarely marine.	Heterotrophic.	Reversibly formed.	Present.	Absent.
ZOOMASTIGINA: **Bicosoecids (Bicoecids)**	Solitary or colonial unicellular organisms with 2 undulipodia and a vaselike lorica of chitin.	Planktonic freshwater or marine.	Heterotrophic.	Two dissimilar undulipodia originating near anterior of cell. One with rows of mastigonemes. Tinsel undulipodium used for food collection; other serves as an attachment inside the lorica.	Present.	Absent.
ZOOMASTIGINA: **Choanomastigota**	Unimastigotes or colonial organisms with collars of tentacles and with either organic (thecae) or siliceous (loricae) structures enveloping cells.	Cosmopolitan in brackish or nearshore marine water. Planktonic, neustonic, epibiotic.	Heterotrophic. Phagotrophy (bactivory); osmotrophy, pinocytosis.	One present.	Present.	Absent.

Reproduction	Propagules (Cysts)	Sexuality	Nuclear Features	Distinguishing Cytological Features	Presence in Fossil Record
Cell division in ameboid stage.	Formed.	Sexuality not known.	Promitosis.		Not observed, although cysts may be present.
Little known.	?	Sexuality not reported.			Not known.
Longitudinal fission in most species. Some sedentary species reproduce by a mastigote swarmer.	?	?		Tentacle collars, vertical undulipodium.	Not known.

Class	Morphology of Growing Organisms	Occurrence	Nutrition	Undulipodia	Mitochondria	Plastids/ Pigments
ZOOMASTIGINA: Diplomonadida	Unicellular often with binary axial symmetry of body, "diplozoic forms" or without (monozoic forms).	Species of *Trepomonas* and *Hexamita* free-living in mesosaprobic or polysaprobic freshwater habitats and the sea. Most commensals or parasites in a variety of animals, usu. in alimentary tract.	Heterotrophic. Endocytosis through cytostome in bactivorous genera. Pinocytosis in non-phagotrophic genera.	One or two karyomastigont systems with 4 undulipodia each. One undulipodium typically recurrent and associated with cytostome or forms intracellular axis of cell.	Absent.	Absent.
ZOOMASTIGINA: Pseudociliata	Unicellular protists, slightly flattened dorsoventrally, with undulipodia in several rows primarily on the ventral surface.	Marine benthic.	Heterotrophic. Holozoic, feeding on diatoms, small mastigotes, and bacteria.	Undulipodia in rows, primarily on the ventral surface.	Present.	Absent.

Reproduction	Propagules (Cysts)	Sexuality	Nuclear Features	Distinguishing Cytological Features	Presence in Fossil Record
Binary fission of trophozoite.	Form oval quadrinucleate cysts and transmit endozoic species from one host to another. Cysts not known in free-living forms.	Sexuality unknown.	One or two nuclei, each associated in a karyo-mastigont system with 4 unduli-podia. Nu-clear di-vision is semi-open type and synchronous in 2 nuclei.	Lack golgi and axostyles.	Absent.
Cell division within a cyst by a kind of palintomy.	Form in which reproduction occurs.	No sexuality known.	Cells possessing 2-16 homo-karyotic nuclei, each with a single large endosome. Closed mitosis; acentric with an intra-nuclear spindle.	Mucocysts.	Absent.

Class	Morphology of Growing Organisms	Occurrence	Nutrition	Undulipodia	Mitochondria	Plastids/ Pigments
Zoomastigina: **Kinetoplastida**	Small colorless mastigotes with one or two undulipodia and massed mitochondrial DNA that forms a stainable structure, the kinetoplast, within the single mitochondrion.	Many of the bodonids are free-living, others epizoic or parasites of fish and other aquatic organisms. All trypanosomatids are parasitic, occurring in many animals and some plants.	Heterotrophic. Bodonids usu. phagotrophic, ingesting food through a cytopharynx. Trypanosomatids are all symbiotrophic to necrotrophic.	One or two, each typically possessing a paraxial rod and arising from an undulipodial pit or pocket.	Single mitochondrion typically extends the length of body and contains prominent DNA kinetoplast, usu. located close to the kinetosomes that insert on or near the mitochondrial outer membrane. Kinetoplast DNA consists of 25-50 maxicircles and 5,000-10,000 minicircles.	Absent.
Zoomastigina: **Opalinata**	Colorless, single-celled protists with a typically heavy covering of short undulipodia arranged in diagonal rows.	All symbionts in the posterior end of the digestive tract of cold-blooded vertebrates. Many found in tail-less amphibians.	Heterotrophic. Mouthless, feeding mainly by micropinocytosis.	Diagonal rows of short undulipodia in *Opalina*, somatic undulipodia appear to originate at or on the falx, somatic kinetosomes in each row interconnected by electron-dense connective structure of fibrous nature.	Present.	Absent.
Zoomastigina: **Proteromonadida**	Small, unicellular colorless heterokont mastigotes.	Posterior intestinal tract of many amphibians, reptiles, and mammals, esp. rodents.	Heterotrophic. Symbiotrophic.	Two unequal undulipodia. Mastigonemes absent.	Present.	Absent.

Reproduction	Propagules (Cysts)	Sexuality	Nuclear Features	Distinguishing Cytological Features	Presence in Fossil Record
Binary fission.	Common in free-living forms; rare in parasitic forms.	Sexual reproduction not demonstrated, but inferred in some species.	Single vesicular nucleus with a prominent nucleolus. Nuclear division with an intranuclear spindle, lacking polar structures.	Cells contain glycosomes, microbodylike organelles.	Absent.
Symmetrogenic binary fission, palintomy.	Serve as infective form, surviving the environment. May also be second round of cyst formation when host undergoes maturation. Zygocysts.	Sexual reproduction by complete fusion of (presumed) haploid uninucleate anisogametes; meiosis in gamete formation.	Two or many homokaryotic nuclei with distinct nucleoli. Mitoses acentric and closed with intranuclear spindle.	Falx.	Absent.
Longitudinal cell division. Trophocysts in which multiple fission occurs reported in *Proteromonas lacertae-viridis*.	Both *Proteromonas* and *Karotomorpha* form resistant cysts and multiplicative cysts have been seen in one species of *Proteromonas* (see Reproduction).	Sexuality unknown.	Single nucleus. Closed mitosis; intranuclear spindle.	In *Proteromonas*, filamentous coat of tubular hairs, somatonemes, covers posterior of cells.	Absent.

Class	Morphology of Growing Organisms	Occurrence	Nutrition	Undulipodia	Mitochondria	Plastids/ Pigments
Zoomastigina: Parabasalia (Trichomonads, calonymphids, hypermastigotes)	Single-celled, uni- and multinucleate, colorless protists with one or more parabasal bodies, with 3 to more than 10,000 undulipodia.	Symbiotic in digestive systems of some insects and in respiratory, digestive, and reproductive systems of mammals and birds.	Heterotrophic. Symbiotrophic.	Few to several thousand; may have specialized rootlets, cresta, or motile costa. Karyomastigont systems.	Absent.	Absent.
Zoomastigina: Retortamonadida	Small mastigotes with twisted cell bodies bearing cytostome in which a trailing undulipodium beats.	Symbiotrophs in intestines of animals.	Heterotrophic. Symbiotrophic.	Two in *Retortamonas*; four in *Chilomastix*; in both species, one recurrent, propelling food into cytostome.	Absent.	Absent.
Zoomastigina: Pyrsonymphida	Single-celled, colorless mastigotes.	Hindguts of wood-eating cockroaches and termites.	Heterotrophic.	Four, 8, or 12 undulipodia wrap around body in left-handed spiral and form undulating membrane with the cell surface.	Absent.	Absent.

Reproduction	Propagules (Cysts)	Sexuality	Nuclear Features	Distinguishing Cytological Features	Presence in Fossil Record
Binary fission.	Zygocyst may be formed in sexually-reproducing forms.	Sexual reproduction known in hypermastigotes and possibly *Mixotricha*.	Mitosis and meiosis in closed nucleus; extranuclear spindle and elaborated centrioles (= atractophore). In those with more than one nucleus, parabasal bodies, undulipodia, and nuclei in karyomastigont system.	One or more parabasal body. Possess hydrogenosomes. Have axostyle. In hypermastigotes and Trichomonadidae, a microtubular structure, the pelta, is attached to the axostyle. Undulating membranes in some.	Absent.
Longitudinal binary fission.	Resistant cysts formed that serve in host-to-host transmission.	Sexuality unknown.	One nucleus. Closed mitosis; intranuclear spindle.	Lack golgi.	Absent.
Binary fission.	?	Sexual reproduction in *Notila*, *Saccinobaculus*, and *Oxymonas*. Meiotic sex; autogamy or fertilization.		Lack parabasal bodies. Have motile axostyle that extends length of organism. Mitochondria and hydrogenosomes not known.	Absent.

Table 6c. Classes of Phylum Chlorophyta

Class/Order	Morphology of Growing Organisms	Occurrence	Nutrition	Undulipodia	Mitochondria	Plastids/ Pigments
CHLOROPHYTA: Class Prasinophyceae	Motile, unicellular chlorophytes covered with nonmineralized organic scales.	Most taxa marine, also in brackish and freshwater habitats. Common in temperate and cold regions.	Phototrophic.	One, 2, or 4 undulipodia. Originate from undulipodial groove. Tubular undulipodial hairs in 2 opposite rows except in *Mesostigma*.	Present.	Chloroplasts containing chlorophylls *a* and *b*, accessory pigments.
CHLOROPHYTA: Class Chlorophyceae	Unicells or colonial mastigotes; organisms with tetrasporal, coccoid, sarcinoid, filamentous, or parenchymatous organization.	Mostly freshwater; may be found in soil, snow, or ice. Some marine (Dunaliellales). May be symbionts (esp. *Chlorella*) in other protoctists, fungi, or animals.	Phototrophic. A few heterotrophic genera.	When present, inserted apically. Pseudoflagella (undulipodia) in Tetrasporales, undulipodia absent in Chlorococcales.	Present.	Chloroplasts containing chlorophylls *a* and *b*, accessory pigments.
CHLOROPHYTA: Class Ulvophyceae	Multicellular or coenocytic thalli: sarcinoid to bladelike to siphonous except for about 6 unicellular, coccoid species. Mostly macroscopic "seaweeds."	Predominantly marine, tropical. Ulotrichales well-represented in temperate regions. Some endobiotic, living in cell walls of other algae or vascular plants or in the nonliving strata produced by animals.	Phototrophic.	Bi- or quadrimastigote zoospores. Gametes usu. bimastigote. Kinetids with cruciate arrangement.	Present.	Chloroplasts containing chlorophylls *a* and *b*, accessory pigments.

Reproduction	Propagules (Cysts)	Sexuality	Nuclear Features	Distinguishing Cytological Features	Presence in Fossil Record
Cell division to form usu. 2 offspring cells.	Formed by *Pterosperma* and *Pachysphaera*.	No sexuality known.	All have persistent telophase spindle during mitosis except Chlorodendrales, in which spindle collapses during telophase and a phycoplast develops.	Golgi bodies in parabasal position. Muciferous bodies or trichocysts often present. Body covered with scales that may fuse to form a theca.	Phycomata are fossilizable and are found from Precambrian to Recent. *Tasmanites*, which resembles the phycoma stage of modern prasinophytes, found as early as Ordovician.
Cell division, autospores or autocolonies, aplanospores, zoospores, fragmentation.	May be formed as products of gametic fusion. Zygote may be thick-walled resting stage, the zygospore.	Sexual reproduction isogamous to oogamous. No sexuality known in Microsporales and rare in Chlorococcales.	Cell division with collapsing telophase spindle and phycoplast.	Chloroplast with pyrenoid; some have eyespot. With or without contractile vacuole. Bristles present in some members of Chlorococcales. Setae, cell wall, or plasmodesmata may be present in some.	Presumptive members identified in Precambrian. Colonies of *Botryococcus braunii* known from Carboniferous. Thought to be responsible for certain types of coals and petroleum.
Zoospores.	Not produced.	Life histories with alternation of iso- or heteromorphic generations. Isogamy to oogamy.	Mitotic spindle usu. closed and centric. Cells uni- or multinucleate.	Terminal cap and proximal sheath associated with kinetosomes in some. Calcium carbonate in cell walls of Dasycladales and Caulerpales. Pyrenoids may be present. Heteroplastidy in Caulerpales.	Calcified members of Dasycladales and Caulerpales found dating to Precambrian and fossils assignable to extant genera date to Cambrian.

Class/Order	Morphology of Growing Organisms	Occurrence	Nutrition	Undulipodia	Mitochondria	Plastids/ Pigments
CHLOROPHYTA: Class Charophyceae	Sarcinoid thallus, branched or unbranched filaments.	Terrestrial or aquatic.	Phototrophic.	Bimastigote zoospores with multilayered structures.	Present.	Chloroplasts containing chlorophylls *a* and *b*, accessory pigments.
CHLOROPHYTA: Order Pedinomo-nadales	Small, naked unimastigotes.	Mostly in freshwater, some in saline lakes. Three taxa are marine and two are symbionts, one in radiolar-ians, one in dinomastigotes.	Phototrophic.	One; second nonfunctional kinetosome present.	Present.	Chloroplasts containing chlorophylls *a* and *b*, accessory pigments.
CHLOROPHYTA: Order Microthamniales	Coccoid, sarcinoid, or filamentous.	Exclusively freshwater; subaerial or lichen phycobionts.	Phototrophic.	Two in zoospores.	Present.	Chloroplasts containing chlorophylls *a* and *b*, accessory pigments.

Reproduction	Propagules (Cysts)	Sexuality	Nuclear Features	Distinguishing Cytological Features	Presence in Fossil Record
Bimastigote zoospores in some, fragmentation, bulbils (Charales).	Zygote forms resistent oospore. Charales form bulbils, which are overwintering forms.	Sexuality not documented in unspecialized orders except in Coleochaetales which are oogamous with bimastigote sperm; haplobiontic. Charales monoecious or dioecious; may be autogamous. Form antheridium and oogonium. Only diploid stage is oospore.	Persistent interzonal spindle. Phragmoplast and cell plate in Coleochaetales and Charales. Polyploidy common in Charales.	Multilayered structure associated with kinetids. Swarmers (zoospores) covered with many square scales. Sheathed hair cells in some. Calcium carbonate deposited by some Charales.	The late Silurian- early Devonian fossil *Parka decipiens* closely resembles *Coleochaete* in morphology and habitat. Oospores of Charales preserved as gyrogonites, which date to lower Devonian or possibly upper Silurian.
Division into 2 offspring.	Formed.	Sexuality unknown.	Persistent telophase spindle.	Eyespot present.	Absent.
Naked bimastigote zoospores; symbiotic taxa frequently reproduce by autospores or aplanospores.	Not formed.	Sexuality unknown.	Closed metacentric spindle during mitosis. Centripetal furrow associated with a phycoplast during cytokinesis.	Eyespot in some.	Absent.

Class/Order	Morphology of Growing Organisms	Occurrence	Nutrition	Undulipodia	Mitochondria	Plastids/ Pigments
CHLOROPHYTA: **Order Prasiolales**	Sarcinoid, branched filaments, monostromatic blades.	Freshwater or marine; often in upper intertidal zone. Some subaerial. Some *Prasiola* sp. associated with ascomycetous fungi.	Phototrophic.	Present in gametes. When sperm fuse with egg, one undulipodium absorbed by egg and a posteriorly unimastigote planozygote results.	Present.	Chloroplasts containing chlorophylls *a* and *b*, accessory pigments.
CHLOROPHYTA: **Order Trentepohliales**	Microscopic filamentous.	Subaerial habitats on surfaces of rocks and trees. Parasites on leaves, stems, and fruits of tropical and subtropical plants; one genus is lichen phycobiont.	Phototrophic.	Two in gametes; four in zoospores. Multilayered structure associated with rootlets. Conical terminal caps.	Present.	Chloroplasts containing chlorophylls *a* and *b*, accessory pigments. Accumulate large amounts of ß-carotene.

Reproduction	Propagules (Cysts)	Sexuality	Nuclear Features	Distinguishing Cytological Features	Presence in Fossil Record
Aplanospores formed in packets.	?Can survive long periods of dryness.	Oogamous sexual reproduction known in a few species in which a bimastigote sperm fuses with egg and forms a unimastigote planozygote.	Uninucleate.	Single chloroplast with pyrenoid.	Absent.
Form quadrimastigote zoospores.	Sporopolleninlike outer wall. Zoosporangia borne on sporangiophores may be detached and serve as akinetes.	Sexual life history with an alternation of iso- or heteromorphic phases. Bimastigote isogametes.	Mitosis closed and centric with interzonal spindle present as a distinct membrane-enclosed bundle at telophase.	Several chloroplasts without pyrenoids. Plasmodesmata forming "pit field" in center of wall. Accumulate ß-carotene, giving cells bright orange color.	Present dating to Eocene.

Table 7. Classification: Summary of Phyla and Lower Taxa*

I. Phyla in which members lack undulipodia at all stages and lack complex sexual life cycles.

Classification	Examples of Genera
PHYLUM RHIZOPODA	
Class Lobosea	
Subclass Gymnamoebia	
Order Amoebida	
Suborder Tubulina	
Family Amoebidae	*Amoeba, Chaos*
Family Hartmannellidae	*Hartmannella*
Family Entamoebidae	*Endolimax, Entamoeba*
Family Vahlkampfidae	*Vahlkampfia*
Suborder Thecina	
Family Thecamoebidae	*Thecamoeba, Vannella*
Suborder Flabellina	
Family Flabellulidae	*Flabellula*
Suborder Conopodina	
Family Paramoebidae	*Mayorella, Paramoeba*
Suborder Acanthopodina	
Family Acanthamoebidae	*Acanthamoeba*
Family Echinamoebidae	*Echinamoeba*

Incertae sedis:

Pneumocystis spp. Organism with ameboid trophic and cystic stages in life cycle; parasitic in mammalian lung. Ribosomal RNA data suggests it is related to ascomycetous fungi.

Pansporella spp. Ameboid organism with multiple fission-trophic and cystic stages in life cycle; symbiotrophic in crustaceans (e.g., *Daphnia magna, D. pulex*); Chatton, 1925.

Classification	Examples of Genera
Subclass Testacealobosa	
Order Arcellinidae	
Family Arcellidae	*Arcella*
Family Centropyxidae	*Centropyxis*
Family Difflugiidae	*Difflugia*
Order Trichosida	*Trichosphaerium*
Class Filosea	
Order Aconchulinida	*Nuclearia*
Order Testaceafilosida	
Family Euglyphidae	*Euglypha*
PHYLUM HAPLOSPORIDIA	
Class Haplosporea	
Order Haplosporida	
Family Haplosporidiidae	*Haplosporidium, Minchinia, Urosporidium*
PHYLUM PARAMYXEA	
Class Paramyxidea	*Paramyxa*
Class Marteiliidea	*Marteilia, Paramarteilia*

* Live taxa only unless otherwise noted.

Taxon	Genera
PHYLUM MYXOZOA	
Class Myxosporea	
Order Bivalvulida	
Suborder Sphaeromyxina	
Family Sphaeromyxidae	*Sphaeromyxa*
Suborder Variisporina	
Family Myxidiidae	*Coccomyxa, Myxidium, Zschokkella*
Family Ortholineidae	*Neomyxobolus, Ortholinea*
Family Sinuolineidae	*Bipteria, Davisia, Myxoproteus, Schulmania, Sinuolinea*
Family Fabesporidae	*Fabespora*
Family Ceratomyxidae	*Ceratomyxa, Leptotheca*
Family Sphaerosporidae	*Hoferellus, Myxobilatus, Palliatus, Sphaerospora, Wardia*
Family Chloromyxidae	*Agarella, Caudomyxum, Chloromyxum*
Family Auerbachiidae	*Auerbachia, Globospora*
Family Alatosporidae	*Alatospora, Pseudoalatospora*
Family Parvicapsulidae	*Neoparvicapsula, Parvicapsula*
Suborder Platysporina	
Family Myxobolidae	*Dicauda, Henneguya, Lomosporus, Myxobolus (= Myxosoma), Neohenneguya, Phlogospora, Spirosuturia, Thelohanellus, Trigonosporus, Unicauda*
Order Multivalvulida	
Family Trilosporidae	*Trilospora, Unicapsula*
Family Kudoidae	*Kudoa*
Family Pentacapsulidae	*Pentacapsula*
Family Hexacapsulidae	*Hexacapsula*
Family Septemcapsulidae	*Septemcapsula*
Class Actinosporea	
Order Actinomyxida	
Family Tetractinomyxidae	*Tetractinomyxon*
Family Sphaeractinomyxidae	*Neoactinomyxon, Sphaeractinomyxon*
Family Triactinomyxidae	*Aurantiactinomyxon, Echinactinomyxon, Guyenotia, Raabeia, Triactinomyxon*
Family Synactinomyxidae	*Antonactinomyxon, Siedleckiella, Synactinomyxon*
Family Hexactinomyxidae	*Hexactinomyxon*
Family Ormieractinomyxidae	*Ormieractinomyxon*

Taxon	Genera
PHYLUM MICROSPORA	
Class Rudimicrosporea	
Order Metchnikovellida	
Family Metchnikovellidae	*Amphiamblys, Amphiacantha, Metchnikovella (Caulleryetta* or *Microsporidiopsis)*
Class Microsporea	
Order Minisporida	
Family Chytridiopsidae	*Chytridiopsis (Chytridioides), Steinhausia*
Family Hesseidae	*Hessea*
Family Burkeidae	*Burkea*
Family Buxtehudeidae	*Buxtehudea*

Taxon	Genera
Order Microsporida	
Suborder Pansporoblastina	
Family Glugeidae	*Baculea, Glugea, Loma, Mitoplistophora, Pleistophora (= Plistophora), Vavraia*
Family Pseudopleistophoridae	*Octosporea, Pseudopleistophora*
Family Duboscqiidae	*Duboscqia, Trichoduboscqia*
Family Thelohaniidae	*Agmasoma, Bohuslavia, Chapmanium, Cryptosporina, Helmichia, Heterosporis, Inodosporus, Ormieresia, Pegmatheca, Pilosporella, Systenostrema, Thelohania (= Orthothelohania), Toxoglugea (Toxonema, Spiroglugea, Toxospora, Spirospora,* or *Spirillonema)*
Family Burenellidae	*Burenella, Vairimorpha*
Family Amblyosporidae	*Amblyospora, Hyalinocysta, Parathelohania*
Family Culicosporidae	*Culicospora, Hazardia*
Family Gurleyidae	*Episeptum, Gurleya (Marssoniella), Norlevinea, Pyrotheca, Stempellia*
Family Telomyxidae	*Berwaldia, Issia (?), Neoperezia, Telomyxa*
Family Tuzetiidae	*Alfvenia, Janacekia, Nelliemelba, Tuzetia*
Suborder Apansporoblastina	
Family Unikaryonidae	*Encephalitozoon, Enterocytozoon, Microgemma, Nosemoides (Pleistosporidium), Orthosoma, Tetramicra, Unikaryon (Oligosporidium)*
Family Spraguidae	*Spraguea*
Family Pereziidae	*Ameson, Perezia*
Family Cougourdellidae	*Cougourdella*
Family Caudosporidae	*Caudospora, Culicosporella, Golbergia, Weiseria*
Family Nosematidae	*Hirsutosporos, Ichthyosporidium, Nosema*
Family Mrazekiidae	*Jirovecia, Mrazekia*
Incertae sedis: unclassified microspora	*Microsporidium* (collective group name)

II. Phyla in which members lack undulipodia at all stages of their life cycles and display complex sexual life cycles.

Taxon	Genera
PHYLUM ACRASEA	
Class Acrasea	
Order Acrasida	
Family Acrasidae	*Acrasis, Pocheina*
Family Copromyxidae	*Copromyxa, Copromyxella*
Family Guttulinopsidae	*Guttulinopsis*
Family Fonticulidae*	*Fonticula*

Taxon	Genera
PHYLUM DICTYOSTELIDA	
Class Dictyostelids	
Family Dictyosteliaceae	*Dictyostelium, Polysphondylium*
Family Acytosteliaceae	*Acytostelium*

* Family of uncertain affinity.

Taxon	Genera
PHYLUM RHODOPHYTA	
Class Rhodophyceae	
Subclass Bangiophycidae	
Order Porphyridiales	
Family Porphyridiaceae	*Bangiopsis, Chroodactylum, Chroothece, Colacodictyon, Porphyridium, Rhodella, Rhodosorus, Stylonema, Vanhoffenia*
Family Phragmonemataceae	*Cyanidium, Cyanoderma, Flintiella, Goniotrichopsis, Kneuckeria, Kyliniella, Neevea, Phragmonema, Rhodospora*
Order Compsopogonales	
Family Compsopogonaceae	*Compsopogon, Compsopogonopsis*
Family Erythropeltidaceae	*Erythrocladia, Erythrotrichia, Erythrotrichopeltis, Membranella, Porphyropsis, Smithora*
Family Boldiaceae	*Boldia*
Order Bangiales	
Family Bangiaceae	*Bangia, Porphyra*
Order Rhodochaetales	
Family Rhodochaetaceae	*Rhodochaete*
Subclass Florideophycidae	
Order Nemaliales	
Order Acrochaetiales	
Order Batrachospermales	
Order Gelidiales	
Order Bonnemaisoniales	
Order Corallinales	
Order Hildenbrandiales	
Order Gigartinales	
Order Rhodymeniales	
Order Palmariales	
Order Ceramiales	

Taxon	Genera
PHYLUM CONJUGAPHYTA	
Class Conjugatophyceae	
Order Zygnematales	
Family Mesotaeniaceae	*Ancylonema, Cylindrocystis, Mesotaenium, Netrium, Roya, Spirotaenia*
Family Desmidiaceae	*Actinotaenium, Allorgeia, Amscottia, Arthrodesmus, Bambusina, Closterium, Cosmarium, Cosmocladium, Desmidium, Docidium, Euastridium, Euastrum, Genicularia, Gonatozygon, Groenbladia, Hyalotheca, Ichthyocercus, Ichthyodontum, Micrasterias, Onychonema, Oocardium, Penium, Phymatodocis, Pleurotaenium, Prescottiella, Sphaerozosma, Spinoclosterium, Spinocosmarium, Spondylosium, Staurastrum, Staurodesmus, Streptonema, Teilingia, Tetmemorus, Triplastrum, Triploceras, Xanthidium*
Family Zygnemataceae	*Debarya, Hallasia, Mougeotia, Mougeotiopsis, Pleurodiscus, Sirocladium, Sirogonium, Spirogyra, Temnogametum, Zygnema, Zygnemopsis, Zygogonium*

III. Phyla in which members display reversible formation of undulipodia and lack complex sexual life cycles.

Taxon	Genera
PHYLUM XENOPHYOPHORA	
Class Psamminida	
Family Psammettidae	*Maudammina, Psammetta*
Family Psamminidae	*Cerelpemma, Galatheammina, Psammina, Reticulammina, Semipsammina*
Family Syringamminidae	*Aschemonella, Ocultammina, Syringammina*
Family Cerelasmidae	*Cerelasma*
Class Stannomida	
Family Stannomidae	*Stannoma, Stannophyllum*
PHYLUM CRYPTOPHYTA	
Class Cryptophyceae	*Chilomonas, Chroomonas, Cryptomonas, Hemiselmis, Pyrenomonas, Rhodomonas*
PHYLUM GLAUCOCYSTOPHYTA	
Class Glaucocystophyceae	
Order Cyanophorales	
Family Cyanophoraceae	*Cyanophora*
Order Gloeochaetales	
Family Glaucosphaeraceae	*Glaucosphaera*
Family Gloeochaetaceae	*Gloeochaete*
Order Glaucocystales	
Family Glaucocystaceae	*Glaucocystis*
Coccoid endocyanomes of uncertain affiliation with the Glaucocystales (genera et species inquirendae)	*Archaeopsis, Glaucocystopsis*
Mastigotes of uncertain affiliation with the Cyanophorales (genera et species inquirendae)	*Peliaina, Strobilomonas*
A capsalean endocyanome of uncertain affiliation with the Glaucosphaeraceae (genera et species inquirendae)	*Cyanoptyche, Chalarodora*
PHYLUM KARYOBLASTEA	
Class Karyoblastea	*Pelomyxa*
PHYLUM ZOOMASTIGINA	
Class Amebomastigota	*Adelphamoeba, Heteroamoeba, Naegleria, Paratetramitus, Protonaegleria (Willaertia), Tetramitus*
Class Bicosoecids (Bicoecids)*	*Bicosoeca (Bicoeca), Pseudobodo*
Class Choanomastigotes	
Family Salpingoecidae	*Choanoeca, Diploeca, Salpingoeca*
Family Codonosigidae	*Codosiga, Desmorella, Monosiga*
Family Acanthoecidae	*Bicosta, Diaphanoeca, Parvicorbicula*
Class Diplomonadida	
Family Enteromonadidae	*Caviomonas, Enteromonas, Trimitus*
Family Hexamitidae	
Subfamily Hexamitinae	*Hexamita, Spironucleus, Trepomonas*
Subfamily Giardiinae	*Giardia, Octomitus*

* See Corliss note p. xxx.

Taxon	Genera
Class Pseudociliata	*Stephanopogon*
Class Kinetoplastida	
Order Bodonina	
Family Bodonidae	*Bodo, Cephalothamnium, Cryptobia (Trypanoplasma), Dimastigella, Ichthyobodo (Costia), Procryptobia, Rhynchomonas*
Order Trypanosomatina	
Family Trypanosomatidae	*Blastocrithidia, Crithidia, Endotrypanum, Herpetomonas, Leishmania, Leptomonas, Phytomonas, Rhynchoidomonas, Trypanosoma*
Class Opalinata	
Order Opalinida	
Family Opalinidae	*Cepedea, Opalina, Protoopalina, Zelleriella*
Class Proteromonadida	
Family Proteromonadidae	*Karotomorpha, Proteromonas*
Class Parabasalia	
Order Trichomonads	
Family Monocercomonae	*Hexamastix, Histomonas, Monocercomonas*
Family Devoscoviidae	*Devoscovinia, Metadevoscovinia*
Family Trichomonadidae	*Mixotricha, Trichomitus, Trichomonas*
Order Polymonads	
Family Calonymphidae	*Calonympha, Snyderella*
Order Hypermastigotes	
Family Holomastigotidae	*Holomastigotoides*
Family Lophomonadidae	*Lophomonas*
Family Hoplonymphidae	*Barbulanympha*
Family Staurojoenidae	*Staurojoenia*
Family Kofoidiidae	
Family Trichonymphidae	*Trichonympha*
Family Joeniidae	*Joenia*
Family Rhizonymphidae	*Rhizonympha*
Family Spirotrichonymphidae	*Spirotrichonympha*
Family Eucomonymphidae	*Eucomonympha*
Family Teratonymphidae	*Teratonympha*
Family Spirotrichosomidae	*Spirotrichosoma*
Class Retortamonadida	
Family Retortamonadidae	*Chilomastix, Retortamonas*
Class Pyrsonymphida	
Family Pyrsonymphidae	*Notila, Oxymonas, Pyrsonympha, Saccinobaculus*

Taxon	Genera
PHYLUM EUGLENIDA	
Class Euglenophyceae	
Order Eutreptiales	*Distigmopsis, Distigma, Eutreptia, Eutreptiella*
Order Euglenales	*Ascoglena, Astasia, Colacium, Cyclidiopsis, Euglena, Euglenopsis, Hyalophacus, Khawkinea, Klebsiella, Lepocinclis, Phacus, Strombomonas, Trachelomonas*
Order Rhabdomonadales	*Gyropaigne, Menoidium, Parmidium, Rhabdomonas, Rhabdospira*
Order Sphenomonadales	*Anisonema, Atraktomonas, Calycimonas, Notosolenus, Petalomonas, Sphenomonas, Tropidoscyphus*
Order Heteronematales	*Dinema, Entosiphon, Heteronema, Peranema, Peranemopsis, Urceolus*
Order Euglenamorphales	*Euglenamorpha, Hegneria*

Taxon	Genera
PHYLUM CHLORARACHNIDA	
Class Chorarachniophyceae	*Chorarachnion*
PHYLUM PRYMNESIOPHYTA	
Class Prymnesiophyceae	
Order Isochrysidales	*Chrysotila, Cricosphaera, Dicrateria, Emiliania, Imantonia, Isochrysis, Ochrosphaera, Pleurochrysis*
Order Coccosphaerales	*Acanthoica, Braarudosphaera, Calyptrosphaera, Coccolithus, Corisphaera, Crenalithus, Cyclolithella, Discosphaera, Helicosphaera, Laminolithus, Rhabdosphaera, Syracosphaera, Umbellosphaera, Umbilicosphaera*
Order Prymnesiales	*Chrysochromulina, Corymbellus, Phaeocystis, Platychrysis, Prymnesium*
Order Pavlovales	*Diacronema, Pavlova*
PHYLUM RAPHIDOPHYTA	
Class Raphidophyceae	*Chattonella, Gonyostomum, Merotricha, Vacuolaria*
PHYLUM EUSTIGMATOPHYTA	
Class Eustigmatophyceae	
Order Eustigmatales (= Pseudocharaciopsidales)	
Family Eustigmataceae	*Eustigmatos, Vischeria*
Family Pseudocharaciopsidaceae	*Pseudocharaciopsis*
Family Chlorobotryaceae	*Chlorobotrys*
Family Monodopsidaceae	*Monodopsis, Nannochloropsis*
PHYLUM ACTINOPODA	
Class Polycystina	
Order Spumellarida	
Suborder Sphaerocollina	
Superfamily Exoaxoplastidiata	
Family Thalassicollidae	*Sphaerocolla, Thalassicolla, Thalassoxanthium*
Family Thalassophysidae	*Thalassophysa*
Family Polycyttaria	
Superfamily Exocryptoaxoplastidiata	
Family Physematidae	*Physematium, Thalassolampe*
Superfamily Centroaxoplastidiata	
Family Centrocollidae	*Centrocolla*
Suborder Sphaerellarina	
Superfamily Cryptoaxoplastidiata	*Arachnosphaera, Diplosphaera, Halosphaera*
Superfamily Centroaxoplastidiata	*Rhizosphaera, Spongosphaera*
Superfamily Periaxoplastidiata	*Cenosphaera, Tetrapetalon*
Families whose cytological fine structure has not been determined:	
Family Prunoidae	*Cenellipsis, Panartus, Spongoliva*
Family Discoidae	*Astrophacus, Euchitonia?, Triolena*
Family Larcoidae	*Monozonium, Streblopyle, Tetrapyle*

Taxon	Genera
Order Nassellarida	
Superfamily Proaxoplastidiata	
Family Cyrtoidea	*Cyrtocalpis*
Family Botryoidea	*Botryopera*
Superfamily Apoaxoplastidiata	
Family Plectoidea	*Plagiocarpa, Plagonidium*
Family Stephoidea	*Archicircus, Cartina*
Family Spyroidea	*Anthospyris*
Class Phaeodaria	
Order Phaeogymnocellida	
Family Phaeosphaeridae	*Phaeosphaera*
Family Phaeodinidae	*Phaeodina*
Family Atlanticellidae*	*Gymnocella, Halocella, Lobocella, Miracella, Planktonetta*
Order Phaeocystida	
Family Aulacanthidae	*Aulacantha, Auloceros, Aulodendron, Aulographis, Aulographonium, Aulokleptes, Aulospathis*
Family Astracanthidae	*Astracantha*
Order Phaeosphaerida	
Family Aulosphaeridae	*Aulastrum, Auloscena, Aulosphaera, Aulotractus*
Family Cannosphaeridae*	*Coelacantha*
Family Sagosphaeridae	*Sagenoscena, Sagonoarium, Sagoscena*
Order Phaeocalpida	
Family Castanellidae	*Castanarium, Castanea, Castanella, Castanidium, Castanissa, Castanura, Circocastanea*
Family Circoporidae*	*Circospathis, Haeckeliana*
Family Tuscaroridae	*Tuscaretta, Tuscarilla, Tuscarora*
Family Porospathidae	*Porospathis*
Family Polypyramidae	*Polypyramis*
Order Phaeogromida	
Family Challengeriidae	*Cadium, Challengeria, Protocystis*
Family Medusettidae	*Cortinetta, Euphysetta, Gazelletta, Gorgonetta, Medusetta, Polypetta*
Order Phaeoconchia	
Family Concharidae	*Conchidium*
Order Phaeodendrida	
Family Coelodendriae	*Coelodendrum, Coeloplegma*
Class Heliozoa	
Order Cryptaxohelida	
Suborder Actinophryida	
Family Actinophryidae	*Actinophrys, Actinosphaerium, Echinosphaerium*
Suborder Desmothoracida	
Family Clathrulinidae	*Clathrulina, Hedriocystis*
Suborder Ciliophryida	
Family Ciliophryidae	*Ciliophrys*
Suborder Taxopodida	
Family Sticholonchidae	*Sticholonche*
Order Phaneraxohelida	
Suborder Axoplasthelida	
Family Gymnidae	*Actinocoryne, Gymnosphaera*
Family Hedraiophryidae	*Hedraiophrys*

* Absence of type-genera renders family names suspect. We respect the judgment of the authors recognizing the tentative nature of all higher protoctistan taxa.

Taxon	Genera
Suborder Centroplasthelida	
Family Heterophryidae	*Cienkowskya, Heterophrys*
Family Acanthocystidae	*Acanthocystis*
Family Raphidiophryidae	*Raphidiophrys*
Suborder Endonucleoaxoplasthelida	
Family Dimorphyidae	*Dimorpha*
Suborder Exonucleoaxoplasthelida	
Family Tetradimorphyidae	*Tetradimorpha*
Class Acantharia	
Order Holacanthida	
Family Acanthochiasmidae	*Acanthochiasma, Acanthocyrtha*
Family Acanthoplegmidae	*Acanthocolla, Acanthoplegma, Acanthospira*
Order Symphiacanthida	
Family Astrolithidae	*Acantholithium, Astrolithium, Astrolonche, Heliolithium*
Family Amphilithidae	*Amphibelone, Amphilithium*
Family Pseudolithidae	*Dicranophora (Dipelicophora), Pseudolithium*
Family Haliommatidae	*Haliommatidium*
Order Chaunacanthida	
Family Gigartaconidae	*Amphiacon, Gigartacon, Heteracon*
Family Conaconidae	*Conacon*
Family Stauraconidae	*Stauracon*
Order Arthracanthida	
Suborder Sphaenacantha	
Family Acanthometridae	*Acanthometra, Amphilonche, Tetralonche*
Family Lithopteridae	*Lithoptera*
Family Dorataspidae	*Dorataspis, Pleuraspis, Stauraspis*
Family Hexalaspidae	*Coleaspis, Hexaconas, Hexalaspis*
Family Pharactopeltidae	*Pharactopelta*
Family Diploconidae	*Diploconus*
Family Nivalidae	
Suborder Phyllacantha	
Family Phyllostauridae	*Acanthostaurus, Phyllostaurus, Zygostaurus*
Family Stauracanthidae	*Pristacantha, Stauracantha, Xiphacantha*
Family Dictyacanthidae	*Dictyacantha*
PHYLUM HYPHOCHYTRIOMYCOTA	
Class Hyphochytrids	
Family Anisolpidiaceae	*Anisolpidium*
Family Rhizidiomycetaceae	*Elina?, Latrostium, Rhizidiomyces*
Family Hyphochytriaceae	*Hyphochytrium*
PHYLUM LABYRINTHULOMYCOTA	
Class Labyrinthulea	
Order Labyrinthulida	
Family Labyrinthuliidae	*Labyrinthula*
Family Thraustochytriidae	*Aplanochytrium, Japonochytrium, Labyrinthuloides, Schizochytrium, Thraustochytrium, Ulkenia*
PHYLUM PLASMODIOPHOROMYCOTA	
Class Plasmodiophoromycetes	
Order Plasmodiophorales	
Family Plasmodiophoraceae	*Ligniera, Membranosorus, Octomyxa, Plasmodiophora, Polymyxa, Sorodiscus, Sorosphaera, Spongospora, Tetramyxa, Woronina*

IV. Phyla in which members display reversible formation of undulipodia and complex sexual life cycles.

PHYLUM DINOMASTIGOTA

Taxon	Genera
Order Desmomonadales	
Family Adinomonadaceae	*Adinomonas*
Family Desmomonadaceae	*Desmomastix, Haplodinium, Pleromonas*
Order Desmocapsales	
Family Desmocapsaceae	*Desmocapsa*
Order Prorocentrales	
Family Prorocentraceae	*Mesoporus, Prorocentrum*
Order Dinophysiales	
Family Dinophysiaceae	*Citharistes, Dinofurcula, Dinophysis, Histioneis, Histiophysis, Latifascia, Metadinophysis, Metaphalacroma, Ornithocercus, Parahistioneis, Proheteroschisma, Pseudophalacroma, Sinophysis, Thaumatodinium*
Family Amphisoleniaceae	*Amphisolenia, Triposolenia*
Family Oxyphysaceae	*Oxyphysis*
Order Gonyaulacales	
Family Ceratiaceae	*Ceratium*
Family Ceratocoryaceae	*Ceratocorys*
Family Cladopyxidaceae	*Cladopyxis, Micracanthidinium, Palaeophalacroma, Sinodinium*
Family Crypthecodiniaceae	*Crypthecodinium*
Family Gonyaulacaceae	*Alexandrium (= Gessnerium), Amphidoma, Gonyaulax, Protoceratium, Protogonyaulax, Pyrodinium, Spiraulax*
Family Heterodiniaceae	*Dolichodinium, Heterodinium*
Family Ostreopsidaceae	*Coolia, Gambierdiscus, Ostreopsis*
Family Oxytoxaceae	*Centrodinium, Oxytoxum, Pavillardinium*
Family Pyrophacaceae	*Fragilidium, Helgolandinium, Pyrophacus*
Family Triadiniaceae	*Triadinium (= Heteraulacus)*
Order Pyrocystales	
Family Pyrocystaceae	*Pyrocystis*
Order Peridiniales	
Family Peridiniaceae	*Apsteinia, Chalubinskia, Dinosphaera, Diplopsalis, Diplopsalopsis, Dissodium, Ensiculifera, Glenodiniopsis, Glenodinium (= Sphaerodinium), Gotoius, Heterocapsa* (incl. *Cachonina*), *Oblea, Peridiniopsis, Peridinium, Protoperidinium, Thompsodinium, Zygabikodinium*
Family Podolampaceae	*Blepharocysta, Podolampas*
Family Thecadiniaceae	*Amphidiniopsis, Roscoffia, Thecadinium*
Order Thoracosphaerales	
Family Thoracosphaeraceae	*Thoracosphaera*
Order Gymnodiniales	
Family Gymnodiniaceae	*Amphidinium, Cochlodinium, Gymnodinium, Gyrodinium, Katodinium, Torodinium, Woloszynskia*
Family Lophodiniaceae	*Lophodinium*
Family Polykrikaceae	*Pheopolykrikos, Polykrikos*
Family Warnowiaceae	*Erythropsidinium, Greuetodinium (= Leucopsis), Nematodinium, Nematopsides, Proterythropsis, Protopsis, Warnowia*
Family Zooxanthellaceae	*Endodinium, Symbiodinium, Zooxanthella*

Taxon	Genera
Order Kolkwitziellales	
Family Brachydiniaceae	*Asterodinium, Brachydinium*
Family Kolkwitziellaceae	*Balechina, Berghiella, Kolkwitziella, Lophodinium, Ptychodiscus, Sclerodinium*
Order Actiniscales	
Family Actiniscaceae	*Achradina, Actiniscus, Dicroerisma, Monaster, Plectodinium*
Order Noctilucales	
Family Kofoidiniaceae	*Kofoidinium, Pomatodinium, Spatulodinium*
Family Leptodiscaceae	*Abedinium (= Leptophyllus), Cachonodinium (= Leptodinium), Craspedotella, Cymbodinium, Leptodiscus, Petalodinium, Scaphodinium*
Family Noctilucaceae	*Noctiluca, Pavillardia, Pronoctiluca*
Order Phytodiniales	
Family Gloeodiniaceae	*Gloedinium, Rufusiella*
Family Phytodiniaceae	*Cystodinedria, Cystodinium* (incl. *Dinococcus?*), *Dinastridium* (incl. *Bourrellyella*), *Dinopodiella, Hypnodinium, Manchudinium, Paulsenella, Phytodinedria, Phytodinium, Rhizodinium, Stylodinium, Tetradinium*
Order Blastodiniales	
Family Apodiniaceae	*Apodinium*
Family Blastodiniaceae	*Blastodinium*
Family Cachonellaceae	*Actinodinium, Cachonella*
Family Haplozoaceae	*Haplozoon*
Family Oodiniaceae	*Amylodinium, Dissodinium, Oodinium*
Family Protoodiniaceae	*Crepidoodinium, Piscinoodinium, Protoodinium*
Order Chytriodiniales	
Family Chytriodiniaceae	*Chytriodinium, Myxodinium, Sporodinium*
Order Syndiniales	
Family Amoebophryaceae	*Amoebophrya*
Family Coccidiniaceae	*Coccidinium*
Family Duboscquellaceae	*Dogelodinium, Duboscquella, Keppenodinium*
Family Sphaeriparaceae	*Sphaeripara*
Family Syndiniaceae	*Hematodinium, Ichthyodinium, Merodinium, Solenodinium, Syndinium, Trypanodinium*
Order Oxyrrhinales	
Family Oxyrrhinaceae	*Oxyrrhis*

Incertae sedis

Adenoides	*Dinamoeba*	*Pseliodinium*
Archaeosphaerodiniopsis	*Filodinium*	*Spiromonas*
Amphilothus	*Lissodinium*	*Thaurilens*
Ceratoperidinium	*Protaspis*	

PHYLUM CHRYSOPHYTA

Taxon	Genera
Class Chrysophyceae	
Order Ochromonadales	
Family Ochromonadaceae	*Ochromonas*
Family Chromulinaceae*	*Chromulina*
Family Dinobryaceae	*Dinobryon*
Family Chrysococcaceae*	*Chrysococcus*
Family Paraphysomonadaceae	*Paraphysomonas*

* Family distinguishable from preceding family only on the basis of number of undulipodia.

Taxon	Genera
Order Mallomonadales	
Family Mallomonadaceae	*Mallomonas*
Order Chrysamoebales	
Family Rhizochrysidaceae	*Rhizochromulina*
Family Chrysamoebaceae*	*Chrysamoeba*
Family Stylococcaceae	
Family Kybotiaceae*	*Kybotion*
Family Myxochrysidaceae	*Myxochrysis*
Order Chrysocapsales	
Family Chrysocapsaceae	
Family Chrysosaccaceae*	
Family Naegeliellaceae	
Family Chrysochaetaceae*	
Family Hydruraceae	*Hydrurus*
Order Chrysosphaerales	
Family Chrysosphaeraceae	
Family Chrysapiaceae*	
Family Stichogloeaceae	
Family Aurosphaeraceae	
Order Phaeothamniales	
Family Sphaeridiothricaceae	*Sphaleromantis*
Family Phaeoplacaceae	
Family Phaeothamniaceae	
Family Phaeodermatiaceae	*Phaeodermatium*
Order Sarcinochrysidales	
Family Sarcinochrysidaceae	
Family Chrysomeridaceae	
Family Phaeosacciaceae	*Phaeosaccion*
Family Nematochrysidaceae	
Class Pedinellophyceae	
Order Pedinellales	
Family Pedinellaceae	*Pedinella*
Class Dictyochophyceae (silicoflagellates)	
Order Dictyochales	
Family Dictyochaceae	*Dictyocha*

PHYLUM CHYTRIDIOMYCOTA

Taxon	Genera
Class Chytridiomycetes	
Order Chytridiales	
Family Chytridiaceae	*Chytridium, Chytriomyces, Polyphagus, Rhizoclosmatium, Rhizophydium*
Family Harpochytriaceae	*Harpochytrium*
Family Endochytriaceae	*Endochytrium*
Family Synchytriaceae	*Synchytrium*
Family Cladochytriaceae	*Cladochytrium, Nowkowskiella*
Order Spizellomycetales	
Family Spizellomycetaceae	*Gaertneriomyces, Karlingia, Kochiomyces, Spizellomyces, Triparticalcar*
Family Caulochytriaceae	*Caulochytrium*
Family Neocallimasticaceae	*Neocallimastix*
Family Olpidiaceae	*Entophlyctis, Olpidium, ?Rozella*
Family Urophlyctaceae	*Urophlyctis*

* Family distinguishable from preceding family only on the basis of number of undulipodia.

Taxon	Genera
Order Monoblepharidales	
Family Monoblepharidaceae	*Monoblepharella*
Family Gonapodyaceae	
Family Oedogoniomycetaceae	*Oedogoniomyces*
Order Blastocladiales	
Family Catenariaceae	*Catenaria*
Family Physodermataceae	*Physoderma*
Family Coelomomycetaceae	*Callimastix, Coelomomyces, Coelomycidium*
Family Blastocladiaceae	*Allomyces, Blastocladiella*
PHYLUM PLASMODIAL SLIME MOLDS	
Class Myxomycetes	
Subclass Ceratiomyomycetidae*	
Order Ceratiomyxales	
Family Ceratiomyxaceae	
Subclass Myxogastromycetidae	
Order Echinosteliales	
Family Clastodermataceae	
Family Echinosteliaceae	*Echinostelium*
Order Liceales	
Family Cribrariaceae	*Dictydium*
Family Enteridiaceae	*Enteridium*
Family Liceaceae	
Order Physarales	
Family Didymiaceae	*Didymium*
Family Elaeomyxaceae	
Family Physaraceae	*Fuligo, Physarum*
Order Trichiales	
Family Dianemaceae	
Family Trichiaceae	*Arcyria, Hemitrichia*
Subclass Stemonitomycetidae	
Order Stemonitales	
Family Schnellaceae	
Family Stemonitaceae	*Comatricha, Diachea, Stemonitis*
Class Protostelida	
Family Cavosteliidae	*Cavostelium, Ceratiomyxella, Clastostelium, "Echinostelium," Planoprotostelium, Protosporangium*
Family Ceratiomyxidae	*Ceratiomyxa**
Family Protosteliidae	*Protostelium, Microglomus, Nematostelium, Protosteliopsis, Schizoplasmodiopsis, Schizoplasmodium*
Family Echinosteliopsidae	*Echinosteliopsis*
PHYLUM CILIOPHORA**	
SUBPHYLUM POSTCILIODESMATOPHORA	
Class Karyorelictea	
Order Protostomatida	
Family Trachelocercidae	*Trachelocerca, Tracheloraphis*
Family Kentrophoridae	*Kentrophoros, Trachelonema*
Order Loxodida	
Family Loxodidae	*Loxodes, Remanella*
Family Cryptopharyngidae	*Cryptopharynx*

* Classification of ceratiomyxids unresolved.
** This table does not include all families in the phylum Ciliophora. Refer to Corliss (1979) for a more detailed listing.

Taxon	Genera
Order Protoheterotrichida	
Family Geleiidae	*Avelia, Geleia*
Order Protocruziida	
Family Protocruziidae	*Protocruzia*
Class Spirotrichea	
Subclass Heterotrichia	
Order Heterotrichida	
Suborder Heterotrichina	
Family Blepharismidae	*Anigsteinia, Blepharisma, Parablepharisma, Pseudoblepharisma*
Family Climacostomidae	*Climacostomum, Copemetopus, Fabrea*
Family Condylostomatidae	*Condylostoma*
Family Spirostomidae	*Gruberia, Spirostomum*
Family Stentoridae	*Stentor*
Suborder Coliphorina	
Family Folliculinidae	*Folliculina, Magnifolliculina, Metafolliculina, Parafolliculina, Pebrilla*
Order Clevelandellida	
Family Clevelandellidae	*Clevelandella*
Family Nyctotheridae	*Metanyctotherus, Nyctotherus, Nyctotheroides, Pronyctotherus*
Family Sicuophoridae	*Metasicuophora, Parasicuophora, Prosicuophora, Sicuophora*
Order Plagiotomida	
Family Plagiotomidae	*Plagiotoma*
Order Armophorida	
Family Caenomorphidae	*Caenomorpha, Cirranter, Ludio*
Family Metopidae	*Bothrostoma, Brachonella, Eometopus, Metopus, Palmarella, Parametopus, Tesnospira*
Order Phacodiniida	
Family Phacodiniidae	*Phacodinium, Transitella*
Order Licnophorida	
Family Licnophoridae	*Licnophora*
Order Odontostomatida	
Family Epalxellidae	*Atopodinium, Epalxella, Saprodinium*
Family Discomorphellidae	*Discomorphella*
Family Mylestomatidae	*Mylestoma*
Incertae sedis	
Order Peritromida	
Family Peritromidae	*Peritromus*
Subclass Choreotrichia	
Order Choreotrichida	
Suborder Tintinnina	
Family Ascampbelliellidae	*Acanthostomella, Ascampbelliella*
Family Codonellidae	*Codonaria, Codonella, Tintinnopsis*
Family Codonellopsidae	*Codonellopsis, Stenosemella*
Family Cyttarocylididae	*Cyttarocylis, Petalotricha*
Family Dictyocystidae	*Dictyocystis*
Family Epiplocylididae	*Epipocylis*
Family Metacylididae	*Climacocylis, Coxliella, Helicostomella, Metacylis*
Family Ptychocylididae	*Favella, Poroecus, Ptychocylis*
Family Rhabdonellidae	*Protorhabdonella, Rhabdonella, Rhabdonellopsis*
Family Tintinnidae	*Amphorellopsis, Amphorides, Eutintinnus, Salpingacantha, Salpingella, Steenstrupiella, Tintinnus*

Taxon	Genera
Family Tintinnidiidae	*Tintinnidium*
Family Undellidae	*Proplectella, Undella*
Family Xystonellidae	*Parafavella, Parundella, Xystonella*
Suborder Strombidinopsina	
Family Strombidinopsidae	*Strombidinopsis*
Suborder Strobilidiina	
Family Strobilidiidae	*Strobilidium*
Order Oligotrichida	
Family Halteriidae	*Halteria*
Family Strombidiidae	*Laboea, Strombidium, Tontonia*
Subclass Stichotrichia	
Order Stichotrichida	
Suborder Stichotrichina	
Family Amphisiellidae	*Amphisiella, Eschaneustyla, Kahliella, Onychodromopsis, Onychodromus, Paraurostyla*
Family Chaetospiridae	*Chaetospira*
Family Cladotrichidae	*Cladotricha, Engelmanniella, Lamtostyla, Perisincirra, Uroleptoides*
Family Gonostomatidae	*Gonostomum, Trachelochaeta, Wallackia*
Family Keronidae	*Kerona, Keronopsis*
Family Psilotrichidae	*Caryotricha, Kiitricha, Psilotricha*
Family Spirofilidae	*Atractos, Hypotrichidium, Spiretella, Stichotricha, Urostrongylum*
Family Strongylidiidae	*Strongylidium*
Suborder Urostylina	
Family Pseudokeronopsidae	*Pseudokeronopsis, Thigmokeronopsis*
Family Pseudourostylidae	*Pseudourostyla*
Family Urostylidae	*Bakuella, Holosticha, Uroleptus, Urostyla*
Suborder Sporadotrichina	
Family Oxytrichidae	*Ancystropodium, Gastrostyla, Histriculus, Laurentiella, Oxytricha, Parastylonychia, Pleurotricha, Stylonychia, Tachysoma*
Family Trachelostylidae	*Psammomitra, Trachelostyla, Urosoma, Urosomoida*
SUBPHYLUM RHABDOPHORA	
Class Prostomatea	
Order Prostomatida	
Family Holophryidae	*Bursellopsis, Holophrya*
Family Metacystidae	*Metacystis, Pelatractus, Vasicola*
Order Prorodontida	
Family Balanionidae	*Balanion*
Family Colepidae	*Coleps, Nolandia, Plagiopogon, Tiarina*
Family Placidae	*Placus, Spathidiopsis*
Family Plagiocampidae	*Plagiocampa*
Family Prorodontidae	*Mimeticus, Prorodon, Pseudoprorodon*
Family Urotrichidae	*Patschia, Rhagadostoma, Urotricha*
Class Litostomatea	
Subclass Haptoria	
Order Haptorida	
Family Didiniidae	*Didinium, Monodinium*
Family Enchelyidae	*Enchelyodon, Enchelys, Homalozoon*
Family Lacrymariidae	*Lacrymaria*
Family Mesodiniidae	*Askenasia, Mesodinium, Myrionecta*
Family Spathidiidae	*Bryophyllum, Myriokaryon, Spathidium*
Family Trachelophyllidae	*Chaenea, Lagynophrya, Trachelophyllum*

Order Pleurostomatida	
Family Amphileptidae	*Amphileptus, Litonotus, Loxophyllum*
Order Pharyngophorida	
Family Actinobolinidae	*Actinobolina*
Family Helicoprorodontidae	*Helicoprorodon*
Family Tracheliidae	*Dileptus, Paradileptus, Teuthophrys, Trachelius*
Subclass Trichostomatia	
Order Vestibuliferida	
Family Balantidiidae	*Balantidium*
Family Isotrichidae	*Dasytricha, Isotricha*
Family Paraisotrichidae	*Paraisotricha*
Order Entodiniomorphida	
Suborder Archistomatina	
Family Buetschliidae	*Alloiozona, Didesmis, Polymorphella*
Suborder Blepharocorythina	
Family Blepharocorythidae	*Blepharocorys, Circodinium, Ochoterenaia, Raabena*
Suborder Entodiniomorphina	
Family Cycloposthiidae	*Cycloposthium, Triplumaria*
Family Ophryoscolecidae	*Enoploplastron, Entodinium, Eremoplastron, Eudiplodinium, Ophryoscolex, Ostracodinium*
Family Spirodiniidae	*Cochliatoxum*
SUBPHYLUM CYRTOPHORA	
Class Phyllopharyngea	
Subclass Phyllopharyngia	
Order Cyrtophorida	
Suborder Chlamydodontina	
Family Chilodonellidae	*Alinostoma, Chilodonella, Phascolodon, Thigmogaster, Trithigmostoma*
Family Chitonellidae	*Chitonella*
Family Chlamydodontidae	*Chlamydodon, Cyrtophoron*
Family Lynchellidae	*Atopochilodon, Chlamydonella, Coeloperix, Gastronauta, Lynchella*
Suborder Dysteriina	
Family Dysteriidae	*Dysteria, Hartmannulopsis, Microdysteria, Trochilia*
Family Hartmannulidae	*Aegyriana, Brooklynella, Hartmannula, Microxysma, Orthotrochilia, Trichopodiella, Trochilioides*
Family Plesiotrichopidae	*Atelepithites, Pithites, Plesiotrichopus, Trochochilodon*
Suborder Hypocomatina	
Family Hypocomidae	*Crateristoma, Hypocoma, Parahypocoma*
Order Rhynchodida	
Suborder Rhynchodina	
Family Sphenophryidae	*Sphenophrya*
Suborder Ancistrocomina	
Family Ancistrocomidae	*Ancistrocoma, Colligocineta, Crebricoma, Heterocinetopsis, Hypocomatidium, Hypocomella, Hypocomides, Ignotocoma, Insignicoma, Raabella*
Subclass Chonotrichia	
Order Exogemmida	
Family Chilodochonidae	*Chilodochona, Vasichona*
Family Filichonidae	*Aurichona, Filichona*
Family Heliochonidae	*Heliochona, Heterochona*
Family Lobochonidae	*Eleutherochona, Lobochona, Oenophorachona, Toxochona*

Taxon	Genera
Family Phyllochonidae	*Phyllochona*
Family Spirochonidae	*Cavichona, Serpentichona, Spirochona*
Order Cryptogemmida	
Family Actinichonidae	*Actinichona, Carinichona, Cristichona, Rhizochona*
Family Echinichonidae	*Coronochona, Echinichona, Eurychona*
Family Inversochonidae	*Ceratochona, Chonosaurus, Inversochona, Kentrochona, Pleochona*
Family Isochonidae	*Isochona, Trichochona*
Family Stylochonidae	*Armichona, Dentichona, Eriochona, Stylochona*
Subclass Suctoria	
Order Exogenida	
Family Ephelotidae	*Ephelota*
Family Ophryodendridae	*Loricodendron, Ophryodendron*
Family Parapodophryidae	*Parapodophrya*
Family Podophryidae	*Corynophrya, Kystopus, Podophrya, Sphaerophrya*
Family Rhabdophryidae	*Dendrosomides, Trophogemma*
Family Spelaeophryidae	*Spelaeophrya*
Family Tachyblastonidae	*Tachyblaston*
Family Thecacinetidae	*Thecacineta*
Family Urnulidae	*Metacineta, Paracineta, Urnula*
Order Endogenida	
Family Acinetidae	*Acineta, Acinetopsis, Loricophrya*
Family Dendrosomatidae	*Dendrosoma, Lernaeophrya, Platophrya, Trichophrya*
Family Endosphaeridae	*Endosphaera*
Family Tokophryidae	*Choanophrya, Hypophrya, Multifasciculatum, Pottsiocles, Pseudogemma, Rhyncheta, Tokophrya*
Order Evaginogenida	
Family Cyathodiniidae	*Cyathodinium*
Family Dendrocometidae	*Cometodendron, Dendrocometes, Dendrocometides, Stylocometes*
Family Discophryidae	*Anarma, Discophrya, Periacineta, Prodiscophrya, Squalorophrya, Testudinicola*
Family Heliophryidae	*Heliophrya*
Class Nassophorea	
Subclass Nassophoria	
Order Synhymeniida	
Family Orthodonellidae	*Orthodonella, Zosterodasys*
Family Nassulopsidae	*Nassulopsis, Phasmatopsis*
Family Scaphidiodontidae	*Chilodontopsis, Schaphidiodon*
Family Synhymeniidae	*Synhymenia*
Order Nassulida	
Suborder Nassulina	
Family Nassulidae	*Nassula*
Suborder Parahymenostomatina	
Family Furgasoniidae	*Furgasonia*
Family Paranassulidae	*Enneameron, Paranassula*
Order Microthoracida	
Family Discotrichidae	*Discotricha*
Family Microthoracidae	*Drepanomonas, Kreyella, Microthorax*
Order Propeniculida	
Family Leptopharyngidae	*Leptopharynx, Pseudomicrothorax*
Order Peniculida	
Suborder Frontoniina	
Family Clathrostomatidae	*Clathrostoma*
Family Frontoniidae	*Didieria, Frontonia, Paraclathrostoma, Wenrichia*
Family Lembadionidae	*Lembadion*

Taxon	Genera
Family Maritujidae	*Marituja*
Family Stokesiidae	*Stokesia*
Suborder Parameciina	
Family Parameciidae	*Paramecium*
Family Neobursaridiidae	*Neobursaridium*
Family Urocentridae	*Urocentrum*
Subclass Hypotrichia	
Order Euplotida	
Suborder Euplotina	
Family Aspidiscidae	*Aspidisca, Euplotaspis*
Family Euplotidae	*Certesia, Diophrys, Euplotes*
Family Gastrocirrhidae	*Cytharoides, Euplotidium, Gastrocirrhus*
Family Uronychiidae	*Uronychia*
Suborder Discocephalina	
Family Discocephalidae	*Discocephalus, Marginotricha, Prodiscocephalus, Psammocephalus*
Family Erionellidae	*Erionella*
Class Oligohymenophorea	
Subclass Hymenostomatia	
Order Hymenostomatida	
Suborder Tetrahymenina	
Family Curimostomatidae	*Curimostoma*
Family Glaucomidae	*Epenardia, Espejoia, Glaucoma, Glaucomella, Jaocorlissia*
Family Tetrahymenidae	*Deltopylum, Lambornella, Tetrahymena*
Family Turaniellidae	*Colpidium, Turaniella*
Suborder Ophryoglenina	
Family Ichthyophthiriidae	*Ichthyophthirius*
Family Ophryoglenidae	*Ophryoglena*
Order Scuticociliatida	
Suborder Philasterina	
Family Cinetochilidae	*Cinetochilum, Sphenostomella*
Family Cohnilembidae	*Cohnilembus*
Family Cryptochilidae	*Biggaria*
Family Entodiscidae	*Entodiscus*
Family Entorhipidiidae	*Entorhipidium*
Family Loxocephalidae	*Balanonema, Cardiostomatella, Dexiotricha, Loxocephalus, Paradexiotricha, Paraloxocephalus, Paratetrahymena*
Family Paralembidae	*Magnalembus, Mesolembus, Ovolembus, Paralembus*
Family Paranophryidae	*Anophryoides, Mesanophrys, Metanophrys, Mugardia, Paranophrys*
Family Parauronematidae	*Glauconema, Miamiensis, Parauronema, Potomacus*
Family Philasteridae	*Philaster, Philasterides*
Family Pseudocohnilembidae	*Pseudocohnilembus*
Family Thyrophylacidae	*Plagiopyliella, Thyrophylax*
Family Thigmophryidae	*Thigmophrya*
Family Uronematidae	*Homologastra, Urocyclon, Uronema, Uropedalium*
Family Urozonidae	*Urozona*
Suborder Pleuronematina	
Family Calyptotrichidae	*Calyptotricha*
Family Conchophthiridae	*Conchophthirus*
Family Ctedectomatidae	*Compsosomella, Ctedectoma, Hippocomos, Paractedectoma*
Family Cyclidiidae	*Cristigera, Cyclidium, Echinocyclidium, Paracyclidium, Pseudocyclidium*

Taxon	Genera
Family Dragescoidae	*Dragescoa*
Family Histiobalantiidae	*Histiobalantium*
Family Peniculistomatidae	*Echinosociella, Peniculistoma*
Family Pleuronematidae	*Pleurocoptes, Pleuronema, Schizocalyptra*
Family Thigmocomidae	*Thigmocoma*
Suborder Thigmotrichina	
Family Ancistridae	*Ancistrum, Ancistrumina, Protophrya*
Family Hemispeiridae	*Ancistrospira, Boveria, Hemispeira, Plagiospira, Proboveria*
Family Hysterocinetidae	*Hysterocineta, Kozloffia, Ptychostomum*
Subclass Peritrichia	
Order Sessilida	
Family Astylozoidae	*Astylozoon, Hastatella*
Family Ellobiophryidae	*Ellobiophrya*
Family Epistylididae	*Campanella, Epistylis, Heteropolaria, Nidula, Rhabdostyla*
Family Lagenophryidae	*Circolagenophrys*
Family Operculariidae	*Ampullaster, Opercularia, Orbopercularia, Propyxidium*
Family Ophrydiidae	*Ophrydium*
Family Opisthonectidae	*Opisthonecta, Telotrochidium*
Family Rovinjellidae	*Rovinjella*
Family Scyphidiidae	*Ambiphrya, Apiosoma, Scyphidia*
Family Termitophryidae	*Termitophrya*
Family Vaginicolidae	*Cothurnia, Platycola, Pyxicola, Thuricola, Vaginicola*
Family Vorticellidae	*Carchesium, Vorticella*
Family Zoothamniidae	*Craspedomyoschiston, Intranstylum, Zoothamnium*
Order Mobilida	
Family Leiotrochidae	*Leiotrocha*
Family Polycyclidae	*Polycycla*
Family Trichodinidae	*Trichodina, Vauchomia*
Family Trichodinopsidae	*Trichodinopsis*
Family Urceolariidae	*Urceolaria*
Subclass Astomatia	
Order Astomatida	
Family Anoplophryidae	*Almophrya, Anoplophrya*
Family Buetschliellidae	*Buetschliella*
Family Clausilocolidae	*Clausilocola, Haptophryopsis*
Family Contophryidae	*Contophrya, Dicontophrya*
Family Haptophryidae	*Cepedietta, Haptophrya, Steinella*
Family Hoplitophryidae	*Buetschliellopsis, Delphyella, Hoplitophrya, Juxtaradiophrya, Radiophryoides*
Family Intoshellinidae	*Intoshellina*
Family Maupasellidae	*Maupasiella*
Family Radiophryidae	*Acanthodiophrya, Helella, Metaracoelophrya, Metaradiophrya, Mrazekiella, Radiophrya*
Incertae sedis	
Family Archiastomatidae	*Archiastoma*
Subclass Apostomatia	
Order Apostomatida	
Family Colliniidae	*Metacollinia*
Family Foettingeriidae	*Gymnodinioides, Hyalophysa, Phoretophrya, Phtorophyra, Spirophyra*

Taxon	Genera
Order Astomatophorida	
Family Opalinopsidae	*Chromidina*
Order Pilisuctorida	
Family Conidophryidae	*Ascophrys, Askoella, Condiophrys*
Subclass Plagiopylia	
Order Plagiopylida	
Family Plagiopylidae	*Lechriopyla, Plagiopyla, Pseudoplagiopyla*
Family Sonderiidae	*Parasonderia, Sonderia*
Class Colpodea	
Order Cyrtolophosidida	
Family Cyrtolophosididae	*Cyrtolophosis*
Family Grossglockneriidae	*Grossglockneria*
Family Woodruffiidae	*Enigmostoma, Platyophrya, Rostrophrya, Woodruffia*
Order Bryophryida	
Family Bryophryidae	*Bryophrya, Puytoraciella*
Order Colpodida	
Family Colpodidae	*Bresslaua, Colpoda, Tillina*
Family Marynidae	*Maryna, Mycterothrix*
Order Bursariomorphida	
Family Bursariidae	*Bursaria*

PHYLUM GRANULORETICULOSA

Taxon	Genera
Class Athalamea	
Order Biomyxida	
Class Foraminiferea	
Order Allogromida	
Suborder Lagyniina	
Family Lagynidae	*Apogromia, Belaria, Lagynis*
Family Maylisoriidae*	
Family Allogromiidae	*Allogromia, Chitinosaccus, Diplogromia*
Family Hospitellidae	*Hospitella, Placopsilinellaia, Thalamophaga*
Family Phthanotrochidae	*Phthanotrochus*
Order Textulariida	
Superfamily Astrorhizacea	
Family Astrorhizidae	*Astrorhiza, Pelosina, Radicula*
Family Bathysiphonidae	*Astrorhizinulla, Bathysiphon, Rhabdanminella*
Family Rhizamminidae	*Marsipella, Oculosiphon, Rhabdammina*
Family Dryorhizopsidae	*Sagenina*
Family Silicotubidae*	
Family Hippocrepinellidae	*Hippocrepinella*
Family Schizamminidae	*Jullienella, Schizammina*
Family Psammosphaeridae	*Psammophax, Psammosphaera, Sorosphaera*
Family Saccamminidae	*Brachysiphon, Ovammina, Saccammina*
Family Polysaccamminidae	*Goatapitigba, Polysaccammina*
Family Hemisphaeramminidae	*Ammopemphix, Hemisphaerammina, Ividia*
Family Diffusilinidae	*Diffusilina*
Superfamily Komokiacea	
Family Komokiidae	*Ipoa, Komokia, Lana*
Family Baculellidae	*Baculella, Edgertonia*
Superfamily Hyperamminacea	
Family Hyperamminidae	*Botellina, Hyperammina, Saccorhiza*

* In the phylum Granuloreticulosa, an asterisk indicates an extinct group.

Family Hyperamminoididae*	
Family Notodendrodidae	*Notodendrodes*
Superfamily Ammodiscacea	
Family Ammodiscidae	*Psammonyx*
Superfamily Rzehakinacea	
Family Rzehakinidae	*Ammoflintina, Birsteiniolla, Miliammina*
Superfamily Hormosinacea	
Family Aschemonellidae	
Family Hormosinidae	*Hormosinella, Hormosinoides, Kalamopsis, Nodulina*
Family Thomasinellidae	*Protoschista*
Family Dusenburyinidae	*Dusenburyina*
Family Cribratinidae*	
Superfamily Lituolacea	
Family Oxinoxisidae*	
Family Haplophragmoididae	*Buzasina, Gobbettia, Haplophragmoides*
Family Discamminidae	*Ammoscalaria, Discammina, Glaphyrammina*
Family Sphaeramminidae	*Ammosphaerulina, Canepaia, Sphaerammina*
Family Lituotubidae	*Lituotuba, Trochamminoides*
Family Lituolidae	*Ammobaculites, Ammotium, Eratidus*
Family Ammosphaeroidinidae	*Adercotryuma, Ammosphaeroidina, Cystammina*
Family Haplophragmiidae*	
Family Nezzazatidae*	
Family Barkerinidae*	
Family Placopsilinidae	*Ammocibicides, Ammocibicoides, Placopsilina*
Superfamily Coscinophragmatacea	
Family Lituoliporidae*	
Family Biokovinidae*	
Family Haddoniidae	*Haddonia*
Family Coscinophragmatidae	*Ammotrochoides, Bdelloidina*
Superfamily Loftusiacea	
Family Cyclamminidae	
Family Spirocyclinidae*	
Family Loftusiidae*	
Superfamily Spiroplectamminacea	
Family Spiroplectamminidae	*Orectostomina, Spiroplectammina, Spiroplectinella*
Superfamily Trochamminacea	
Family Trochamminidae	*Ammoglobigerina, Asarotammina, Paratrochammina*
Family Remaneicidae	*Asteroparatrochammina, Asterotrochammina*
Superfamily Verneuilinacea	
Family Textulariopsidae*	
Family Verneuilinidae	*Barbourinella, Gaudryina, Latentoverneuilina*
Family Plectorecurvoididae*	
Family Pseudobolivinidae	*Lacroixina, Parvigenerina, Pseudobolivina*
Family Nouriidae	*Nouria*
Superfamily Tawitawiacea	
Family Tawitawiidae	
Superfamily Ataxophragmiacea	
Family Ataxophragmiidae*	
Family Dorothiidae	
Family Eggerellidae	*Bannerella*
Family Globotextulariidae	*Globotextularia, Rhumblerella, Tetrataxiella*
Family Textulariellidae	*Guppyella, Hagenowinoides, Textulariella*
Family Cuneolinidae*	
Family Dicyclinidae*	
Family Coskinolinidae*	
Family Pfenderinidae*	

Family Coskinolinidae*
Superfamily Textulariacea
Family Textulariidae — *Bigenerina, Sahulia, Tetragonostomina*
Family Valvulinidae — *Clavulina, Cribrobulimina, Goesella*
Family Chrysalidinidae*
Superfamily Cyclolinacea*
Family Cyclolinidae
Family Orbitopsellidae
Superfamily Orbitolinacea*
Family Dictyopsellidae
Family Orbitolinidae
Order Fusulinida*
Suborder Parathuramminiina
Superfamily Parathuramminacea
Family Parathuramminidae
Superfamily Earlandiacea
Family Earlandiidae
Family Pseudoammodiscidae
Superfamily Archaediscacea
Family Archaediscidae
Family Lasiodiscidae
Superfamily Moravamminacea
Family Caligellidae
Family Moravamminidae
Family Paratikhinellidae
Superfamily Nodosinellacea
Family Earlandinitidae
Family Tuberitinidae
Family Nodosinellidae
Superfamily Geinitzinacea
Family Geinitzinidae
Family Pachyphloiidae
Superfamily Colaniellacea
Family Colaniellidae
Superfamily Ptychocladiacea
Family Ptychocladiidae
Superfamily Palaeotextulariacea
Family Semitextulariidae
Family Palaeotextulariidae
Family Biseriamminidae
Superfamily Tournayellacea
Family Tournayellidae
Family Palaeospiroplectamminidae
Superfamily Endothyracea
Family Endothyridae
Family Dariopsidae
Family Bradyinidae
Family Loeblichiidae
Family Eocristellariidae
Superfamily Tetrataxacea
Family Pseudotaxidae
Family Tetrataxidae
Family Valvulinellidae
Family Abadehellidae
Superfamily Fusulinacea
Family Ozawainellidae
Family Staffellidae

Taxon	Genera
Family Schubertellidae	
Family Schwagerinidae	
Family Fusulinidae	
Family Pseudoendothyridae	
Family Verbeekinidae	
Family Neoschwagerinidae	
Order Involutinida	
Family Involutinidae*	
Family Trocholinidae	
Order Miliolida	
Superfamily Squamulinacea	
Family Squamulinidae	*Squamulina*
Superfamily Cornuspiracea	
Family Cornuspiridae	*Cornuspirella, Cornuspira, Cornuspiroides*
Family Fischerinidae	*Fischerina, Planispirina, Planispirinella*
Family Nubeculariidae	*Calcituba, Cornuspiramia, Nubeculina*
Family Ophthalmidiidae	*Cornuloculina, Edentostomina, Opthalmina*
Family Discospirinidae	*Discospirina*
Superfamily Miliolacea	
Family Miliolidae	*Rupertianella*
Family Riveroinidae	*Pseudohauerinella, Riveroina*
Family Fabulariidae*	
Superfamily Soritacea	
Family Milioliporidae*	
Family Peneroplidae	*Coscinospira, Dendritina, Peneroplis*
Family Meandropsinidae*	
Family Rhapydioninidae	*Crateritina, Ripacubaria*
Family Soritidae	*Androsina, Archaias, Sorites*
Family Keramosphaeridae	*Keramosphaera*
Family Alveolinidae	*Alveolinella, Borelis, Flosculinella*
Order Silicoloculinida	
Family Silicoloculinidae	*Miliammellus*
Order Spirillinida	
Family Spirillinidae	*Mychostomina, Sejunctella, Spirillina*
Family Patellinidae	*Heteropatellina, Patellina, Patellinoides*
Family Placentulinidae	*Ashbrookia, Patellinella, Subpatellinella*
Order Lagenida	
Superfamily Robuloidacea*	
Family Syzraniidae	
Family Robuloididae	
Family Partisaniidae	
Superfamily Nodosariacea	
Family Nodosariidae	*Botuloides, Dentalina, Nodosaria*
Family Vaginulinidae	*Dimorphina, Lenticulina, Marginulinopsis*
Family Polymorphinidae	*Globulina, Gultulina, Pyrulina*
Family Glandulinidae	*Esosyrinx, Globulotuba, Globulotuboides*
Order Robertinida	
Superfamily Duostominacea*	
Family Duostominidae	
Family Asymmetrinidae	
Family Oberhauserellidae	
Superfamily Robertinacea	
Family Conorboididae*	
Family Ceratobuliminidac	*Ceratobulimina, Lamarckina, Rubratella*
Family Robertinidae	*Alliatina, Geminospira, Robertina*
Family Epistominidae	*Hoeglundina*

Taxon	Genera
Family Mississippinidae	*Mississippina, Stomatorbina*
Order Globigerinida	
Superfamily Heterohelicacea	
Family Heterohelicidae*	
Family Chiloguembelinidae	*Laterostomella*
Superfamily Planomalinacea*	
Family Globigerinelloididae	
Family Planomalinidae	
Family Schackoinidae	
Superfamily Rotaliporacea*	
Family Hedbergellidae	
Family Globuligerinidae	
Family Favusellidae	
Family Rotaliporidae	
Superfamily Globotruncanacea*	
Family Globotruncanidae	
Family Rugoglobigerinidae	
Superfamily Hantkeninacea*	
Family Globanomalinidae	
Family Hantkeninidae	
Family Cassigerinellidae	
Superfamily Globigerinacea	
Family Globigerinidae	*Globigerina, Globigerinella, Globigerinoides*
Family Candeinidae	*Candeina, Tenuitella*
Family Hastigerinidae	*Hastigerina, Hastigerinopsis, Orcadia*
Superfamily Globorotaliacea	
Family Eoglobigerinidae*	
Family Globorotaliidae	*Berggrenia, Clavatorella, Globorotalia*
Family Pulleniatinidae	*Pulleniatina*
Family Globigerinitidae	*Tinophodella*
Family Catapsydracidae	
Order Rotaliida	
Superfamily Turrilinacea	
Family Turrilinidae*	
Family Trimosinidae	*Mimosina, Trimosina*
Family Pavoninidae	*Bifarinella, Pavonina*
Family Sphaeroidinidae	*Sphaeroidina*
Superfamily Eouvigerinacea	
Family Eouvigerinidae*	
Family Bolivinidae	*Bolivina, Bolivinellina, Brizalina*
Family Bolivinoididae*	
Family Islandiellidae	
Family Elhasaellidae	
Family Stilostomellidae	*Nodogenerina, Orthomorphina, Siphonodosaria*
Family Lacosteinidae*	
Superfamily Buliminacea	
Family Praebuliminidae*	
Family Bolivinitidae	*Abditodentrix, Bolivinita*
Family Stainforthiidae	*Cassidelina, Hopkinsina, Stainforthia*
Family Buliminidae	*Bulimina, Globobulimina, Praeglobobulimina*
Family Buliminellidae	*Buliminella*
Family Reussellidae	*Acostina, Reysella, Valvobifarina*
Family Siphogenerinoididae	*Euloxostomum, Hopkinsinella, Loxostomina*
Family Uvigerinidae	*Euuvigerina, Uvigerina, Neouvigerina*
Family Millettiidae	*Millettia*

Superfamily Fursenkoinacea	
Family Fursenkoinidae	*Fursenkoina, Coryphostoma*
Family Virgulinellidae	*Virgulinella*
Superfamily Cassidulinacea	
Family Cassidulinidae	*Cassidulina, Cassidulinoides, Evolvocassidulina*
Family Loxostomatidae	
Family Bolivinellidae	*Bolivinella*
Family Annulopatellinidae	*Annulopatellina*
Superfamily Pleurostomellacea	
Family Pleurostomellidae	*Ellipsobulimina, Nodosarella, Pleurostomella*
Superfamily Delosinacea	
Family Caucasinidae	*Francesita*
Family Tremachoridae*	
Family Delosinidae	*Delosina, Neodelosina*
Superfamily Discorbacea	
Family Conorbinidae*	
Family Bagginidae	*Baggina, Cancris, Natlandia*
Family Eponididae	*Alabaminella, Donsissonia, Eponides*
Family Bueningiidae	*Bueningia*
Family Pegidiidae	*Pegidia, Siphonidia, Sphaeridia*
Family Discorbidae	*Disconorbis, Discorbis, Obitina*
Family Pannellainidae	*Pannellaina*
Family Bronnimanniidae	*Bronnimannia*
Family Rotaliellidae	*Metarotaliella, Rotaliella*
Superfamily Glabratellacea	
Family Glabratellidae	*Angulodiscorbis, Conorbella, Crosbyia*
Family Buliminoidiae	*Buliminoides, Fredsmithia*
Superfamily Siphoninacea	
Family Siphoninidae	*Siphonina, Siphoninella, Siphoninoides*
Family Pseudoparrellidae	*Alabaminoides, Alexanderina, Ambitropus*
Family Planulinoididae	*Planulinoides*
Family Discorbinellidae	*Colonimilesia, Discorbinella, Discorbitina*
Superfamily Planorbulinacea	
Family Planulinidae	*Hyalinea, Planulina*
Family Cibicididae	*Cibicides, Cibicicoides, Discorbia*
Family Planorbulinidae	*Caribeanella, Planorbulina, Planorbulinella*
Family Cymbaloporidae	*Cymbaloporella, Cymbaloporelta, Milleltiana*
Family Victoriellidae	*Biarritzina, Carpenteria, Rupertina*
Superfamily Acervulinacea	
Family Acervulinidae	*Acervulina, Discogypsina, Gypsina*
Family Homotrematidae	*Homotrema, Miniacina, Sporadotrema*
Superfamily Asterigerinacea	
Family Epistomariidae	*Asanonella, Eponidella, Hildemannia*
Family Alfredinidae	*Epistomaroides, Mullinia*
Family Asterigerinatidae	*Asterigerinata, Biasterigerina, Dublinia*
Family Asterigerinidae	*Asterigerina*
Family Amphisteginidae	*Amphistegina*
Superfamily Nonionacea	
Family Nonionidae	*Evolutononion, Haynesina, Nonion*
Family Melonidae	
Family Spirotectinidae	*Spirotectina*
Family Almaenidae	*Anomalinella*
Superfamily Chilostomellacea	
Family Quadrimorphinidae	*Quadrimorphina*
Family Chilostomellidae	*Allomorphina, Chilostomella*
Family Globorotalitidae*	

Taxon	Genera
Family Parrelloididae	*Cibicidoides, Parrelloides*
Family Alabaminidae	*Alabamina, Svratkina*
Family Oridorsalidae	*Oridorsalis, Schwantzia*
Family Osangulariidae	*Osangularia*
Family Gavelinellidae	*Gyroidella, Gyroidina, Gyroidinoides*
Family Karreriddae	*Karreria*
Family Coleitidae*	
Family Trichohyalidae	*Buccella, Neobuccella, Trichohyalus*
Superfamily Orbitoidacea*	
Family Linderinidae	
Family Orbitoididae	
Family Lepidocyclinidae	
Superfamily Rotaliacea	
Family Rotaliidae	*Ammonia, Pararotalia, Rotalidium*
Family Calcarinidae	*Baculogypsina, Calcarina, Schulumbergerella*
Family Elphidiidae	*Cribroelphidium, Elphidium, Ozawaia*
Family Lepidorbitoididae*	
Family Miogypsinidae*	
Superfamily Nummulitacea	
Family Nummulitidae	*Assilina, Cycloclypeus, Nummulites*
Family Pseudorbitoididae*	
Family Discocyclinidae*	
Family Orbitoclypeidae	
Order Carterinida	
Family Carterinidae	*Carterina*

PHYLUM APICOMPLEXA

Taxon	Genera
Class Gregarinia	
Order Blastogregarinida	
Order Archigregarinida	
Family Selenidiidae	*Heterospora, Selenidium, Selenocystis*
Family Merogregarinidae	
Order Eugregarinida	
Family Lecudinidae	*Illivina, Kofoidina, Lecudina*
Family Ganymedidae	*Ganymedes*
Family Uradiophoridae	
Family Cephaloidophoridae	*Cephaloidophora*
Family Cephalolobidae	*Cephalolobus*
Family Porosporidae	*Nematopsis*
Family Thalicolidae	
Family Urosporidae	*Urospora*
Family Monocystidae	*Monocystis*
Family Stenophoridae	*Stenophora*
Family Monoductidae	
Family Dactylophoridae	*Echinomera*
Family Gregarinidae	*Gregarina*
Family Hirmocystidae	*Hirmocystis*
Family Enterocystidae	*Enterocystis*
Family Stylocephalidae	*Stylocephalus*
Family Actinocephalidae	*Actinocephalus*
Family Diplocystidae	*Diplocystis*
Order Neogregarinida	
Class Coccidia	
Order Coelotrophiida	
Family Coelotrophiidae	

Taxon	Genera
Family Angeiocystidae	
Family Myriosporidae	
Family Mackinnoniidae	
Family Eleutheroschizonidae	
Order Adeleida	
Family Adeleidae	*Adelea, Adelina, Klossia*
Family Klossiidae	*Klossiella?*
Family Legerellidae	*Legerella*
Family Dobelliidae	*Dobellia*
Family Haemogregarinidae	*Haemogregarina, Hepatozoon, Karyolysus*
Order Eimeriida	
Family Cryptosporidae	*Cryptosporidium*
Family Mantonellidae	
Family Cyclosporidae	
Family Pfeifferinellidae	*Pfeifferinella*
Family Caryosporidae	
Family Diplosporidae	
Family Eimeriidae	*Eimeria, Mantonella, Tyzzeria*
Family Dorisiellidae	
Family Wenyonellidae	
Family Caryotrophidae	
Family Lankesterellidae	*Atoxoplasma, Lankesterella, Schellackia*
Family Yamikovellidae	
Family Angeiocystidae	
Family Psyedoklossiidae	
Family Myriosporidae	
Family Merocystidae	
Family Aggregatidae	
Family Sarcocystidae	*Arthrocystis, Frenkelia, Sarcocystis*
Class Hematozoa	
Order Haemosporida	*Haemoproteus, Hepatocystis, Leucocytozoon, Plasmodium*
Order Piroplasmida	*Babesia, Theileria*
PHYLUM BACILLARIOPHYTA	
Class Coscinopiscophyceae	
Subclass Thalassiosirophycidae	
Order Thalassiosirales	
Family Thalassiosiraceae	*Bacteriosira, Mindiscus, Planktoniella, Porosira, Thalassiosira*
Family Sceletonemataceae	*Sceletonema, Schroederella*
Family Cyclotellaceae	*Cyclostephanos, Cyclotella, Stephanodiscus*
Family Lauderiaceae	*Lauderia*
Subclass Coscinodiscophycidae	
Order Chrysanthemodiscales	
Family Chrysanthemodiscaceae	*Chrysanthemodiscus*
Order Melosirales	
Family Melosiraceae	*Melosira*
Family Stephanopyxidaceae	*Stephanopyxis*
Family Endictyaceae	*Endictya*
Family Hyalodiscaceae	*Druridgea, Hyalodiscus, Podosira*
Order Paraliales	
Family Paraliaceae	*Paralia*

Taxon	Genera
Order Aulacosirales	
Family Aulacosiraceae	*Aulacosira*
Order Coscinodiscales	
Family Coscinodiscaceae	*Coscinodiscus, Palmeria, Symbolophora*
Family Aulacodiscaceae	*Aulacodiscus*
Family Gossleriellaceae	*Gossleriella*
Family Hemidiscaceae	*Actinocyclus, Hemidiscus, Roperia*
Family Actinoptychaceae	*Actinoptychus*
Order Ethmodiscales	
Family Ethmodiscaceae	*Ethmodiscus*
Order Stictocyclales	
Family Stictocyclaceae	*Stictocyclus*
Order Asterolamprales	
Family Asterolampraceae	*Asterolampra, Asteromphalus*
Order Arachnoidiscales	
Family Arachnoidiscaceae	*Arachnoidiscus*
Subclass Biddulphiophycidae	
Order Eupodiscales	
Family Eupodiscaceae	*Auliscus, Cerataulus, Eupodiscus, Odontella, Pleurosira, Pseudoauliscus, Triceratium*
Order Biddulphiales	
Family Biddulphiaceae	*Biddulphia, Hydrosera, Isthmia, Trigonium, Terpsinoe*
Order Hemiaulales	
Family Hemiaulaceae	*Baxteriopsis, Cerataulina, Climacodium, Eucampia, Hemiaulus, Pseudorutilaria, Riedelta, Trinacria*
Order Anaulales	
Family Anaulaceae	*Anaulus, Eunotogramma, Porpeia*
Subclass Lithodesmiophycidae	
Order Lithodesmiales	
Family Lithodesmiaceae	*Ditylum, Lithodesmium*
Family Bellerocheaceae	*Bellerochea, Neostreptotheca, Streptotheca*
Subclass Corethronophycidae	
Order Corethrales	
Family Corethraceae	*Corethron*
Subclass Cymatosirophycidae	
Order Cymatosirales	
Family Cymatosiraceae	*Arcocellulus, Brockmaniella, Campylosira, Cymatosira, Extubocellulus, Leyanella, Minutocellus, Papiliocellulus, Plagiogrammopsis*
Subclass Rhizosoleniophycidae	
Order Rhizosoleniales	
Family Rhizosoleniaceae	*Dactyliosolen, Gladius, Guinardia, Gyrodiscus, Mastogonia, Pyrgupyxis, Pyxilla, Rhizosolenia*
Family Acanthoceraceae	*Acanthoceros*
Family Attheyaceae	*Attheya*
Subclass Chaetocerophycidae	
Order Chaetocerotales	
Family Chaetocerotaceae	*Bacteriastrum, Chaetoceros*
Order Leptocylindrales	
Family Leptocylindraceae	*Leptocylindrus*
Subclass Fragilariophycidae	
Order Fragilariales	
Family Fragilariaceae	*Asterionella, Asterionellopsis, Centronella, Ceratoneis, Diatoma, Fragilaria, Meridion, Opephora, Synedra*

Taxon	Genera
Family Tabellariaceae	*Tabellaria, Tetracyclus*
Family Licmophoraceae	*Licmophora*
Order Rhaphoneidales	
Family Rhaphoneidaceae	*Delphineis, Diplomenora, Neodelphineis, Perissonoe, Rhaphoneis, Sceptroneis*
Family Psammodiscaceae	*Psammodiscus*
Order Ardissoniales	
Family Ardissoniaceae	*Ardissonia*
Order Toxariales	
Family Toxariaceae	*Toxarium*
Order Thalassionematales	
Family Thalassionemataceae	*Thalassionema*
Order Rhabdonematales	
Family Rhabdonemataceae	*Cyclophora, Grammatophora, Rhabdonema, Striatella, Tessella*
Order Climacospheniales	
Family Climacospheniaceae	*Climacosphenia*
Order Entopylales	
Family Entopylaceae	*Entopyla, Gephyria*
Order Plagiogrammales	
Family Plagiogrammaceae	*Dimerogramma, Glyphodesmis, Plagiogramma*
Order Pseudohimantidiales	
Family Pseudohimantidiaceae	*Protoraphis, Pseudohimantidium*
Class Bacillariophyceae	
Subclass Eunotiophycidae	
Order Eunotiales	
Family Eunotiaceae	*Actinella, Desmogonium, Eunotia, Semiorbis*
Family Peroniaceae	*Peronia*
Subclass Bacillariophycidae	
Order Lyrellales	
Family Lyrellaceae	*Lyrella, Petroneis*
Order Naviculales	
Suborder Sellaphorineae	
Family Sellaphoraceae	*Sellaphora*
Family Pinnulariaceae	*Dimidiata, Pinnularia, Oestrupia*
Family Berkeleyaceae	*Berkeleya, Climaconeis*
Family Psycheneidaceae	*Psycheneis*
Family Scolioneidaceae	*Scolioneis*
Suborder Neidiineae	
Family Amphipleuraceae	*Amphipleura, Frickea, Frustulia*
Family Brachysiraceae	*Brachysira*
Family Neidiaceae	*Neidium*
Family Diadesmidaceae	*Diadesmis*
Suborder Naviculineae	
Family Naviculaceae	*Cymatoneis, Haslea, Navicula, Rhoikoneis, Trachyneis*
Family Proschkiniaceae	*Proschkinia*
Family Pleurosigmataceae	*Donkinia, Gyrosigma, Pleurosigma, Rhoicosigma, Toxonidea*
Family Plagiotropidaceae	*Plagiotropis*
Family Stauroneidaceae	*Craticula, Stauroneis*
Suborder Diploneidineae	
Family Diploneidaceae	*Diploneis*
Family Progonoiaceae	*Progonoia*
Family Scoliotropidaceae	*Scoliopleura, Scoliotropis*
Order Mastogloiales	
Family Mastogloiaceae	*Mastogloia*

Taxon	Genera
Order Dictyoneidales	
Family Dictyoneidaceae	*Dictyoneis*
Order Cymbellales	
Family Cymbellaceae	*Brebissonia, Cymbella, Encyonema, Placoneis*
Family Gomphonemataceae	*Didymosphenia, Gomphoneis, Gomphonema*
Family Anomoeoneidaceae	*Anomoeoneis*
Family Rhoicospheniaceae	*Campylopyxis, Cuneolus, Rhoicosphenia*
Order Achnanthales	
Family Achnanthaceae	*Achnanthes*
Family Achnanthidiaceae	*Achnanthidium, Eucocconeis*
Family Cocconeidaceae	*Anorthoneis, Bennettella, Campyloneis, Cocconeis, Epipellis*
Order Amphorales	
Family Amphoraceae	*Amphora, Undatella*
Family Thalassiophysaceae	*Thalassiophysa*
Order Bacillariales	
Family Bacillariaceae	*Cymbellonitzschia, Denticula, Gomphonitzschia, Gomphotheca, Hantzschia, Nitzschia, Psammodictyon, Simonsenia, Tryblionella*
Family Perryaceae	*Perrya*
Order Epithemiales	
Family Epithemiaceae	*Epithemia, Rhopalodia*
Family Entomoneidaceae	*Entomoneis*
Family Auriculaceae	*Auricula*
Order Surireliales	
Family Surirellaceae	*Campylodiscus, Cymatopleura, Hydrosilicon, Peterodictyon, Plagiodiscus, Stenoterobia, Surirella*

PHYLUM CHLOROPHYTA

Taxon	Genera
Class Prasinophyceae	
Order Mamiellales	
Family Mamiellaceae	*Dolichomastix, Mamiella, Mantoniella*
Family Micromonadaceae (?)	*Micromonas*
Order Pseudoscourfieldiales	
Family Pseudoscourfieldiaceae	*Pseudoscourfieldia*
Family Nephroselmidaceae	*Nephroselmis*
Order Chlorodendrales	
Family Chlorodendraceae	*Scherffelia, Tetraselmis*
Order Pyramimonadales	
Family Pterospermataceae	*Pachysphaera, Pterosperma*
Family Pyramimonadaceae	*Halosphaera, Pyramimonas*
Family Mesostigmataceae	*Mesostigma*
Class Chlorophyceae	
Order Dunaliellales	
Order Chlamydomonadales	
Family Chlamydomonadaceae	*Brachiomonas, Chlamydomonas, Chlorogonium, Chloromonas, Lobomonas, Polytoma, Polytomella, Spermatozopsis*
Family Phacotaceae	*Dysmorphococcus, Phacotus, Pteromonas*
Family Haematococcaceae	*Haematococcus, Stephanosphaera*
Family Carteriaceae	*Carteria*
Order Volvocales	
Family Volvocaceae	*Eudorina, Gonium, Pandorina, Platydorina, Pleodorina, Volvox, Volvulina*

Taxon	Genera
Family Astrephomenaceae	*Astrephomene*
Family Pyrobotryaceae	*Chlorocorona, Pyrobotrys, Pascherina*
Order Tetrasporales	
Family Palmellopsidaceae	*Asterococcus, Palmellopsis, Tetrasporidium*
Family Characiochloridaceae	*Characiochloris, Chlorangiella, Chlorangiopsis, Pseudochlorangium, Stylosphaeridium*
Family Chaetochloridaceae	*Chaetochloris, Chlorangiochaete, Dicranochaete*
Family Tetrasporaceae	*Apiocystis, Fottiella, Paulschulzia, Tetraspora*
Family Nautococcaceae	*Actinochloris, Apiococcus, Hypnomonas, Nautococcopsis (?), Nautococcus*
Order Chlorococcales	
Family Coccomyxaceae	*Coccomyxa, Gloeocystis, Radiococcus, Sphaerocystis*
Family Chlorococcaceae	*Bracteacoccus, Chlorococcum, Dictyococcus, Neochloris, Spongiochloris, Spongiococcum*
Family Characiaceae (?)	*Ankyra, Characium, Hydrianum, Schroederia*
Family Chlorochytriaceae (?)	*Chlorochytrium*
Family Rhodochytriaceae (?)	*Rhodochytrium*
Family Micractiniaceae	*Golenkinia, Golenkiniopsis, Micractinium*
Family Hydrodictyaceae	*Euastropsis, Hydrodictyon, Pediastrum, Sorastrum*
Family Protosiphonaceae	*Protosiphon*
Family Characiosiphonaceae	*Characiosiphon*
Family Hormotilaceae (?)	*Hormotila, Hormotilopsis*
Family Palmodictyaceae	*Palmodictyon*
Family Oocystaceae	*Ankistrodesmus, Chlorella, Chodatella, Franceia, Lagerheimia, Monoraphidium, Oocystis, Selenastrum, Tetraedron*
Family Scenedesmaceae	*Actinastrum, Coelastrum, Crucigenia, Scenedesmus*
Family Eremosphaeraceae	*Eremosphaera*
Family Treubariaceae	*Desmatractum, Treubaria*
Family Dictyosphaeriaceae	*Dictyosphaerium, Dimorphococcus*
Family Botryococcaceae	*Botryococcus*
Family Heleochloridaceae (?)	*Heleochloris, Heleococcus*
Order Chlorosarcinales	*Chlorosarcina, Chlorosarcinopsis, Chlorosphaeropsis, Tetracystis*
Order Sphaeropleales	*Atractomorpha, Sphaeroplea*
Order Microsporales	*Microspora*
Order Oedogoniales	
Family Oedogoniaceae	*Bulbochaete, Oedocladium, Oedogonium*
Order Cylindrocapsales	*Cylindrocapsa*
Order Chaetophorales	
Family Chaetophoraceae	*Chaetophora, Draparnaldia, Draparnaldiopsis, Fritschiella, Stigeoclonium, Uronema*
Family Aphanochaetaceae	*Aphanochaete, Chaetonema (?), Thamniochaete (?)*
Family Schizomeridaceae	*Schizomeris*
Class Ulvophyceae	
Order Ulotrichales	
"*Eugomontia* group"	*Eugomontia, Pseudendoclonium sensu Vischer, Pseudopringsheimia sensu Perrot*
Family Monostromataceae	*Collinsiella, (Gemina?), Gomontia, Monostroma, Protomonostroma*
"*Gayralia* group"	*Capsosiphon, Gayralia, Trichosarcina*
Family Ulotrichaceae	*Ulothrix*
"*Chlorocystis* group"	*Chlorocystis, Halochlorococcum*
Family Acrosiphoniaceae	*Acrosiphonia, Chlorothrix, Spongomorpha, Urospora*

Taxon	Genera
Order Siphonocladales	
Family Chaetosiphonaceae	*Blastophysa, Chaetosiphon*
Family Arnoldiellaceae	*Arnoldiella, Basicladia, Dermatophyton*
Family Cladophoraceae	*Bryobesia, Chaetocladiella, Chaetomorpha, Chaetonella, Cladophora, Cladophorella, Cladostromia, Gemmiphora, Pithophora, Rhizoclonium*
Family Anadyomenaceae	*Anadyomene, Cystodictyon, Microdictyon, Willeella*
Family Siphonocladaceae	*Apjonnia, Boergesenia, Boodlea, Chamaedoris, Cladophoropsis, Pseudostruvea, Siphonocladus, Struvea*
Family Valoniaceae	*Dictyosphaeria, Ernodesmis, Valonia, Valoniopsis*
Order Ulvales	
Family Ulvaceae	*Chloropelta, Enteromorpha, Letterstedtia, Percursaria, Ulva, Ulvaria*
Family Ulvellaceae	*Acrochaete, Endophyton, Entocladia, Ochlochaete, Pilinella, Pringsheimiella, Pseudopringsheimia, Syncoryne, Tellamia, Ulvella*
"*Blidingia* group"	*Blidingia, Kornmannia, Pseudendoclonium*
"*Bolbocoleon* group"	*Acroblaste, Bolbocoleon*
"*Phaeophila* group"	*Phaeophila*
Order Dasycladales	
Family Dasycladaceae	*Acetabularia, Acicularia, Batophora, Bornetella, Chalmasia, Cymopohlia, Dasycladus, Halycoryne, Neomeris*
Order Caulerpales	
Family Ostreobiaceae	*Ostreobium*
Family Bryopsidaceae	*Bryopsidella, Bryopsis, Derbesia, Pedobesia, Pseudobryopsis*
Family Codiaceae	*Codium, Pseudocodium*
Family Udoteaceae	*Avrainvillea, Boodleopsis, Callipsygma, Chlorodesmis, Cladocephalus, Flabellaria, Halimeda, Johnson-sea-linkia, Penicillus, Pseudochlorodesmis, Rhipilia, Rhipiliopsis, Rhipocephalus, Tydemania, Udotea*
Family Caulerpaceae	*Caulerpa*
Family Dichotomosiphonaceae	*Dichotomosiphon*
Class Charophyceae	
Order Chlorokybales	*Chlorokybus atmophyticus*
Order Klebsormidiales	*Klebsormidium, Raphidonema, Stichococcus*
Order Coleochaetales	*Chaetosphaeridium, Coleochaete*
Order Charales	*Chara, Lamprothamnium, Lychnothamnus, Nitella, Nitellopsis, Tolypella*
Orders of uncertain affinity	
Order Pedinomonadales	*Pedinomonas*
Order Microthamniales	*Friedmannia, Microthamnion, Pleurastrum, Pseudotrebouxia, Trebouxia*
Order Prasiolales	
Family Prasiolaceae	*Prasiococcus, Prasiola, Prasiolopsis, Rosenvingiella*
Order Trentepohliales	
Family Trentepohliaceae	*Cephaleuros, Phycopeltis, Stomatochroon, Trentepohlia*

Taxon	Genera
PHYLUM OOMYCOTA*	
Class Peronosporomycetidae	
Order Leptomitales	
Family Leptomitaceae	*Apodachlya, Leptomitus, Plerogone*
Family Apodachlyellaceae	*Apodachlyella*
Order Rhipidiales	
Family Rhipidiaceae	*Aqualinderella, Araiospora, Mindeniella, Rhipidium, Sapromyces*
Order Sclerosporales	
Family Verrucalvaceae	*Sclerophthora, Verrucalvus*
Family Sclerosporaceae	*Peronosclerospora, Sclerospora*
Order Pythiales	
Family Leptolegniellaceae	*Aphanodictyon, Aphanomycopsis, Brevilegniella, Leptolegniella, Nematophthora*
Family Lagenidiaceae	*Lagena, Lagenidium, Myzocytium*
Family Pythiaceae	*Diasporangium, Peronophythora, Phytophthora, Pythiogeton, Pythium, Trachysphaera, Zoophagus*
Order Peronosporales	
Family Peronosporaceae	*Basidiophora, Bremia, Bremiella, Peronospora, Plasmopara, Pseudoperonospora*
Family Albuginaceae	*Albugo*
Class Saprolegniomycetidae	
Order Saprolegniales	*Achlya, Aphanomyces, Aplanopsis, Brevilegnia, Calyptralegnia, Dictyuchus, Geolegnia, Leptolegnia, Plectospira, Pythiopsis, Saprolegnia, Scoliolegnia, Sommerstorffia, Thraustotheca*

Taxon	Genera
PHYLUM XANTHOPHYTA	
Class Xanthophyceae	
Order Chloramoebales	
Family Chloramoebaceae	*Chloramoeba*
Order Rhizochloridales	
Family Rhizochloridaceae	*Rhizochloris*
Family Stipitococcaceae	*Rhizolekane, Stipitococcus, Stipitoporos*
Family Myxochloridaceae (Chlamydomyxaceae)	*Chlamydomyxa, Myxochloris*
Order Heterogloeales	
Family Heterogloeaceae	*Gloeochloris, Helminthogloea, Heterogloea*
Family Mallodendraceae	*Mallodendron*
Family Pleurochloridellaceae	*Pleurochloridella*
Family Characidiopsidaceae	*Characidiopsis*
Order Mischococcales	
Family Pleurochloridaceae	
Family Botrydiopsidaceae	
Family Botryochloridaceae	
Family Gloeobotrydaceae	
Family Gloeopodiaceae	*Gloeopodium*
Family Mischococcaceae	*Mischococcus*
Family Characiopsidaceae	*Characiopsis*
Family Chloropediaceae	*Chloropedia*
Family Trypanochloridaceae	*Trypanochloris*
Family Centritractaceae	*Bumilleriopsis, Centritractus*
Family Ophiocytaceae (Sciadiaceae)	*Ophiocytium*

* Some genera of uncertain affinity have been omitted.

Taxon	Genera
Order Tribonematales	
Family Neonemataceae	*Chadefaudiothrix, Neonema*
Family Tribonemataceae	*Brachynema, Bumilleria, Heterothrix, Heterotrichella, Tribonema*
Family Heterodendraceae	*Heterodendron*
Family Heteropediaceae	*Heterococcus*
Order Vaucheriales	
Family Botrydiaceae	*Botrydium*
Family Vaucheriaceae	*Asterosiphon, Vaucheria*

PHYLUM PHAEOPHYTA

Taxon	Genera
Class Phaeophyceae	
Order Ectocarpales	
Family Ectocarpaceae	*Acinetospora, Asteronema, Bachelotia, Bodanella, Climacosorus, Dermatocelis, Dichosporangium, Ectocarpus, Endodictyon, Entonema, Feldmannia, Geminocarpus, Giffordia, Gononema, Hamelella, Herponema, Kuckuckia, Kuetzingiella, Laminariocolax, Mikrosiphar, Phaeostroma, Pilayella, Pleurocladia, Spongonema, Streblonema, Waerniella, Zosterocarpus*
Family Sorocarpaceae	*Hummia, Polytretus, Sorocarpus*
Family Ralfsiaceae	*Basispora, Diplura, Endoplura, Hapalospongidion, Hapterophycus, Heribaudiella, Jonssonia, Lithoderma, Mesospora, Nemoderma, Petroderma, Porterinema, Pseudolithoderma, Ralfsia, Sorapion, Symphyocarpus*
Family Heterochordariaceae	*Analipus*
Family Myrionemataceae	*Chilionema, Compsonema, Hecatonema, Microspongium, Myrionema, Pleurocladia, Protectocarpus, Ulonema*
Family Choristocarpaceae	*Choristocarpus, Discosporangium*
Order Chordariales	
Family Elachistaceae	*Elachista, Halothrix, Herpodiscus, Leptonematella, Portphillipia*
Family Leathesiaceae	*Corynophlaea, Cylindrocarpus, Leathesia, Microcoryne, Myriactula*
Family Chordariaceae	*Caepidium, Chordaria, Cladosiphon, Eudesme, Halonema, Haplogloia, Levringia, Liebmannia, Mesogloia, Myelophycus, Myriocladia, Myriogloia, Papenfussiella, Polycerea, Pseudochorda, Saundersella, Sauvageaugloia, Sphaerotrichia, Stereocladon, Strepsithalia, Suringaria, Tinocladia*
Family Ishigeaceae	*Ishige*
Family Acrotrichaceae	*Acrothrix*
Family Spermatochnaceae	*Chordariopsis, Nemacystus, Spermatochnus, Stilophora, Stilopsis*
Family Splachnidiaceae	*Splachnidium*
Family Notheiaceae	*Notheia*
Order Scytosiphonales	
Family Scytosiphonaceae	*Chnoospora, Colpomenia, Endarachne, Hydroclathrus, Iyengaria, Petalonia, Rosenvingea, Scytosiphon*

Order Tilopteridales	
Family Tilopteridaceae	*Haplospora, Tilopteris*
Order Dictyosiphonales	
Family Myriotrichiaceae	*Leblondiella, Myriotrichia*
Family Giraudiaceae	*Giraudia*
Family Punctariaceae	*Adenocystis, Asperococcus, Corycus, Desmotrichum, Halorhipis, Litosiphon, Melanosiphon, Omphalophyllum, Phaeostrophion, Platysiphon, Pogotrichum, Punctaria, Soranthera, Utriculidium*
Family Buffhamiaceae	*Buffhamia*
Family Delamareaceae	*Delamarea*
Family Coelocladiaceae	*Coelocladia*
Family Striariaceae	*Cladothele, Isthmoplea, Stictyosiphon, Striaria*
Family Coilodesmaceae	*Akkisiphycus, Coilodesme*
Family Dictyosiphonaceae	*Dictyosiphon, Scytothamnus*
Order Cutleriales	
Family Cutleriaceae	*Cutleria, Microzonia, Zanardinia*
Order Syringodermatales	
Family Syringodermataceae	*Syringoderma*
Order Sphacelariales	
Family Sphacelariaceae	*Sphacelaria, Sphacella*
Family Stypocaulaceae	*Alethocladus, Halopteris, Phloiocaulon, Ptilopogon*
Family Cladostephaceae	*Cladostephus*
Order Dictyotales	
Family Dictyotaceae	*Chlanidophora, Dictyopteris, Dictyota, Dilophus, Distromium, Glossophora, Homoeostrichus, Lobospira, Pachydictyon, Padina, Padinopsis, Pockockiella, Spatoglossum, Stoechospermum, Stypopodium, Taonia, Zonaria*
Family Dictyotopsidaceae	*Dictyotopsis*
Order Sporochnales	
Family Sporochnaceae	*Bellotia, Carpomitra, Encyothalia, Nereia, Perisporochnus, Perithalia, Sporochnus*
Order Desmarestiales	
Family Arthrocladiaceae	*Arthrocladia*
Family Desmarestiaceae	*Desmarestia, Himanthothallus, Phaeurus*
Order Laminariales	
Family Chordaceae	*Chorda*
Family Laminariaceae	*Agarum, Arthrothamnus, Costaria, Costularia, Cymathere, Feditia, Hedophyllum, Kjellmaniella, Laminaria, Phyllariella, Pleurophycus, Saccorhiza, Streptophyllum, Thalassiophyllum*
Family Alariaceae	*Alaria, Ecklonia, Eckloniopsis, Egregia, Eisenia, Pleuropterum, Pterygophora, Undaria*
Family Lessoniaceae	*Dictyoneuropsis, Dictyoneurum, Lessonia, Lessoniopsis, Macrocystis, Nereocystis, Pelagophycus, Postelsia*
Order Acroseirales	
Family Acroseiraceae	*Acroseira*
Order Durvillaeales	
Family Durvillaeaceae	*Durvillaea*
Order Fucales	
Family Hormosiraceae	*Hormosira*

Family Cystoseiraceae	*Acrocarpia, Acystis, Bifurcaria, Bifurcariopsis, Carpoglossum, Caulocystis, Coccophora, Cystophora, Cystoseira, Halidrys, Hormophysa, Landsburgia, Myragropsis, Myriodesma, Scaberia, Stolonophora*
Family Himanthaliaceae	*Himanthalia*
Family Sargassaceae	*Anthophycus, Carpophyllum, Cladophyllum, Hizikia, Oerstedtia, Sargassum, Turbinaria*
Family Seirococcaceae	*Axillariella, Cystosphaera, Marginariella, Phyllospora, Scytothalia, Seirococcus*
Family Fucaceae	*Ascophyllum, Fucus, Hesperophycus, Pelvetia, Pelvetiopsis, Xiphophora*

Incertae sedis

ELLOBIOPSIDA	*Ellobiopsis, Thallassomyces, others?*
EBRIDIANS	*Ebria, Hermesinum*

REFERENCES

Chatton, E.: *Pansporella perplexa*, amoebien á spores protégées, parasite des Daphnies. Réflexions sur la biologie et la phylogénie des Protozoaires. *Ann. Sci. Nat. Zool.* 10e ser., VIII, 5-84 (1925).

Corliss, J.O.: *The Ciliated Protozoa. Characterization, Classification and Guide to the Literature*, 2nd ed. New York: Pergamon Press, 1979.

Gabrielson, P.W., Garbary, D.: Systematics of red algae (Rhodophyta). CRC *Critical Reviews in Plant Sciences* 3, 325-366 (1986).

Irvine, D.E.G., John, D.M., eds.: *The Systematics of the Green Algae*. London: Academic Press, 1984.

Kadlubowska, J.Z.: *Zygnemaceae*. Tom 12A. *Flora Slodkowodna Polski*. Krakow, Poland: Polska Academia Nauk Instytut Botaniki, 1972.

Leedale, G.F.: *Euglenoid Flagellates*. Englewood Cliffs, NJ: Prentice-Hall, 1967.

Loeblich, R., Tappan, H.: *Foraminiferal Genera and Their Classification*. New York: Van Nostrand Reinhold, 1988.

Margulis, L., Corliss, J. O., Melkonian, M., Chapman, D. J., eds.: *Handbook of Protoctista*. Boston: Jones and Bartlett Publishers, 1990.

Martin, G.W., Alexopoulos, C.J., Farr, M.L.: *The Genera of Myxomycetes*. Iowa City: University of Iowa Press, 1983.

McCracken, D.A., Nadakavukaren, M.J., Cain, J.R.: A biochemical and ultrastructural evaluation of the taxonomic position of *Glaucosphaera vacuolata* Korsh. *New Phytologist* 86, 39-44 (1980).

Mix, M.: Die Feinstruktur der Zellwände der Conjugaten und ihre systemische Bedeutung. *Beihefte zur Nova Hewigia* 42, 179-194 (1975).

Olive, L.S.: *The Mycetozoans*. New York: Academic Press, 1975.

Olive, L.S.: Eumycetozoa. In: *Synopsis and Classification of Living Organisms* (Parker, S.P., ed.), pp. 521-525. New York: McGraw-Hill, 1982.

Parke, M., Dixon, P.S.: Check-list of British marine algae, third revision. *Journal of the Marine Biological Association of the United Kingdom* 56, 527-594 (1976).

Prescott, G.W., Croasdale, H.T., Vinyard, W.C.: Desmidiales, Part 1. Saccodermae, Mesotaeniaceae. *North American Flora* II, Part 6. Bronx, NY: The New York Botanical Garden, 1972.

Prescott, G.W., Croasdale, H.T., Vinyard, W.C.: *A Synopsis of North American Desmids*. Part II. Desmidiaceae: Placodermae. Section 1. Lincoln: University of Nebraska Press, 1975.

Round, F.E., Crawford, R.M., Mann, D.G.: *The Diatoms: Biology and Morphology of the Genera*. Cambridge: Cambridge University Press, 1989.

Silva, P.C.: Chlorophyceae. In: *Synopsis and Classification of Living Organisms* (Parker, S.P., ed.), pp. 133-161. New York: McGraw-Hill, 1982.

Small, E.B., Lynn, D.H.: Phylum Ciliophora. In: *Illustrated Guide to the Protozoa* (Hutner, S.H., Lee, J.J., Bovee, E.C., eds.), pp. 393-575. Lawrence: Allen Press, 1985.

Figure Credits

HOP = *Handbook of Protoctista*, L. Margulis, J. O. Corliss, M. Melkonian, and D. J. Chapman, eds. Boston: Jones and Bartlett Publishers, 1990.
[*HOP*] = Refer to the *Handbook of Protoctista* for additional information about this figure.

Frontispiece
Figs. 1B, 2B, 4A, 5B courtesy of Jørgen Kristiansen. Fig. 3E courtesy of Ralph A. Lewin. Figs. 6A, 6B, 6E courtesy of Bruce Parker. Fig. 6C courtesy of L. V. Evans. Fig. 2C, 2E from the personal collection of Robert T. Wilce. Fig., p. iv, courtesy of Rhoda Honigberg. All other photographs from the personal collection of John O. Corliss.

General Glossary
Acanthopodium, from Schuster. In: *HOP*.
Acetabularian life history, from Floyd and O'Kelly. In *HOP*. [*HOP*.]
Acritarch, photo by Gonzalo Vidal.
Actinopod, redrawn from Cachon *et al.* In: *HOP*. Drawing by Steve Alexander.
Aethalia, redrawn from Frederick. In: *HOP*. Drawing by Sheila Manion-Artz.
Aggregation center, redrawn from Blanton. In: *HOP*. Drawing by Kathryn Delisle.
Anisokont, from L. Margulis and D. Sagan, *Origins of Sex*. New Haven, CT: Yale University Press, 1986. Drawing by Christie Lyons.
Antheridium, from Floyd and O'Kelly. In: *HOP*. [*HOP*.]
Atractophore, from L. Margulis and D. Sagan, *Origins of Sex*. New Haven, CT: Yale University Press, 1986. [*HOP*.] Drawing by Christie Lyons.
Autospore, redrawn from Kies and Kremer. In: *HOP*. Drawing by Kathryn Delisle.
Brevetoxin complex, from Taylor. In: *HOP*.
Capillitium, redrawn from Frederick. In: *HOP*. Drawing by Sheila Manion-Artz.
Centrocone, redrawn from Vivier and Desportes. In: *HOP*. Drawing by Sheila Manion-Artz.
Centroplast, from Febvre-Chevalier. In: *HOP*. Courtesy of C. F. Bardele.
Centrosphere, from Febvre-Chevalier. In: *HOP*.
Chromatoid bodies, from Schuster. In: *HOP*.
Chromosome, drawing by Lorraine Olendzenski.
Cladogram, from Floyd and O'Kelly. In: *HOP*.
Closed mitosis, drawing by Steve Alexander.
Coated vesicles, from D. J. Hibberd, *Botanical Journal of the Linnean Society* 72:55-80 (1976).
Coccidian life history, from H. Mehlhorn *et al.*, *Année Biologique* 18:97-120 (1979).
Coccolith, redrawn in part from Green *et al.* In: *HOP*. Based in part on information from Mark Leckie. Drawing by Kathryn Delisle.
Conceptacle, redrawn from Gabrielson *et al.* In: *HOP*. Drawing by Sheila Manion-Artz.
Conoid, from Vivier and Desportes. In: *HOP*.
Contractile vacuole, drawing by Christie Lyons.
Cortex, from Lynn and Small. In: *HOP*.
Cristae, drawing by Kathryn Delisle.
Curved vane assembly, redrawn from P. A. Kivic and P. L. Walne, *Origins of Life* 13:269-288 (1984). Drawing by Kathryn Delisle.
Cyst, drawing by Sheila Manion-Artz.
Cystosorus, from Dylewski. In: *HOP*.
Cytotomy, from Lee. In: *HOP*.
Dense body vesicle, from Dick. In: *HOP*.
Desmokont, from Taylor. In: *HOP*.
Dinokont, from Taylor. In: *HOP*.
Dinomastigote life history, from Taylor. In: *HOP*. [*HOP*.]
Elastic junctions, from Febvre. In: *HOP*.
Endocytosis, from Brugerolle and Mignot. In: *HOP*.
Endoplasmic reticulum, from Brugerolle and Mignot. In: *HOP*.
Epicone, from L. Margulis and K. V. Schwartz, *Five Kingdoms*. Copyright (c) 1988 by W. H. Freeman and Company. Reprinted with permission. Drawing by Robert Golder.
Epiplasm, from Lynn and Small. In: *HOP*.
Extrusome, from Taylor. In: *HOP*. [*HOP*.]
Eyespot, from D. J. Hibberd, *Botanical Journal of the Linnean Society* 72:55-80 (1976).
Falx, from Corliss. In: *HOP*. [*HOP*.]
Filament, from Round and Crawford. In: *HOP*.
Foraminiferan test, from Lee. In: *HOP*.
Frustule, from Round and Crawford. In: *HOP*.
Funis, from Kulda and Nohŷnkovà. In: *Parasitic Protozoa* (J. P. Kreier, ed.), Vol. 2, pp. 2-138. New York: Academic Press, 1978. [*HOP*.]
Gametangium, from Floyd and O'Kelly. In: *HOP*. [*HOP*.]
Gamete, redrawn from Lee. In: *HOP*. Drawing by Steve Alexander.
Genophore, from L. Margulis and D. Sagan, *Origins of Sex*. New Haven, CT: Yale University Press, 1986. Drawing by Steve Alexander.
Gonomere, from Whisler. In: *HOP*.
Habitat, drawing by Kathryn Delisle.
Haptonema, from Green *et al.* In: *HOP*.
Heterothallism, from Dick. In: *HOP*.
Hilum, reproduced from R. L. Blanton, *Journal of the Elisha Mitchell Scientific Society* 97:95-100 (1981) with copyright permission of the North Carolina Academy of Sciences.
Hypha, from Barr. In: *HOP*.
Hypovalve, from L. Margulis and K. V. Schwartz, *Five Kingdoms*. Copyright (c) 1988 by W. H. Freeman and Company. Reprinted with permission. Drawing by Emily Hoffman.
Interaxonemal substance, from Febvre-Chevalier. In: *HOP*.
Interphase, from P. Heywood, *Journal of Cell Science* 31:37-51 (1978).
Karyomastigont system, drawing by Barbara Dorritie.
Keel, from Walne and Kivic. In: *HOP*.
Kinete, from H. Mehlhorn *et al.*, *Année Biologique* 18:97-120 (1979).
Kinetid, from L. Margulis and D. Sagan, *Origins of Sex*. New Haven, CT: Yale University Press, 1986. Drawing by Laszlo Meszoly.
Kinetocyst, from Febvre-Chevalier. In: *HOP*. Courtesy of C. F. Bardele.
Kinetosomes, from L. Margulis and D. Sagan, *Origins of Sex*. New Haven, CT: Yale University Press, 1986. Drawing by Laszlo Meszoly.
Knob scales, from Green *et al.* In: *HOP*.
Lacunae, from Kies and Kremer. In: *HOP*.
Lamella, from K. R. Roberts *et al.*, *Journal of Phycology* 17:159-167 (1981).
Life cycle, redrawn by Kathryn Delisle from an original drawing by Laszlo Meszoly.
Ligula, from Round and Crawford. In: *HOP*.
Litholophus, from Febvre. In: *HOP*.
Luciferin, from Taylor. In: *HOP*.
Macrogamont, from J. Senaud *et al.*, *Protistologica* 16:241-257 (1980).
Macrostome, from Lynn and Small. In: *HOP*. [*HOP*.] Drawing by Steve Alexander.
Mastigoneme, drawing by Kathryn Delisle.
Mastigote division, from Brugerolle and Mignot. In: *HOP*.
Meiosis, from L. Margulis, *Early Life*. Boston: Jones and Bartlett Publishers, 1982.
Merogony, from Porchet-Henneré and Richard, *Protistologica* 7:227-259 (1971).
Microbody, from Febvre. In: *HOP*.
Microcyst, from Blanton. In: *HOP*. [*HOP*.]
Microgamete, from J. Senaud *et al.*, *Protistologica* 16:241-257 (1980).
Mitosis, redrawn from I. B. Raikov, *The Protozoan Nucleus*. Vienna: Springer-Verlag, 1982. Drawing by Kathryn Delisle.
Mitotic apparatus, drawing by Kathryn Delisle.
Monosporangium, from Gabrielson *et al.* In: *HOP*. Courtesy of M. Sommerfeld.
Müller's Law, from Febvre. In: *HOP*.
Multilayered structure, from Graham. In: *HOP*. [*HOP*.] Drawing by Steve Alexander.
Myonemes, from Febvre. In: *HOP*.
Nuclear cap, from Barr. In: *HOP*.
Ocellus, from Taylor. In: *HOP*. [*HOP*.]
Oospore, from Dick. In: *HOP*.
Oral region, from Lynn and Small. In: *HOP*.
Organelle, drawing by Kathryn Delisle.
Palmelloid, from Gillott. In: *HOP*.
Pansporoblast, from E. U. Canning *et al.*, *Systematic Parasitology* 5:147-159 (1983). Reprinted by permission of Kluwer Academic Publishers.
Papilla, from Barr. In: *HOP*.
Paraxial rod, from Vickerman. In: *HOP*.
Pedicel, courtesy of K. R. Buck.
Pellicle, from Walne and Kivic. In: *HOP*. [*HOP*.]
Periplastidial compartment, from Hibberd. In: *HOP*.
Perispicular cone, from Febvre. In: *HOP*.
Phragmoplast, from Grant. In: *HOP*. Courtesy of J. D. Pickett-Heaps.
Phylogeny, based on information from E. B. Small. Drawing by Kathryn Delisle.
Pit connections, from Gabrielson *et al.* In: *HOP*. Courtesy of B. J. Hymes and K. M. Cole.
Plasmodiocarp, from Frederick. In: *HOP*. [*HOP*.]
Plastid, based on information from P. W. Gabrielson, D. J. Hibberd, and J. Kristiansen. Drawing by Kathryn Delisle.
Plate formula, from Taylor. In: *HOP*. [*HOP*.]
Polykinetoplastic, from Vickerman. In: *HOP*. [*HOP*.]

Porelli, from Round and Crawford. In: *HOP*.
Prespore cell, redrawn from Spiegel. In: *HOP*. Drawing by Steve Alexander.
Procentriole, from Febvre. In: *HOP*.
Protoplast, from Porter. In: *HOP*. Drawing by Andrew J. Lampkin, III.
Proximal sheath, from Floyd and O'Kelly. In: *HOP*. [*HOP*.]
Pseudoparenchyma, from O'Kelly and Floyd. In: *HOP*. Drawing by Sheila Manion-Artz.
Raphe, from Round and Crawford. In: *HOP*.
Reservoir, from Walne and Kivic. In: *HOP*. [*HOP*.]
Residual body, from I. Desportes, *Annales des Sciences Naturelles, Zoologie et Biologie Animale*, 12ème Serie 17:215-228 (1975).
Rhizoplast, from Floyd and O'Kelly. In: *HOP*. [*HOP*.]
Rhizostyle, from Gillott. In: *HOP*.
Rimoportule, from Round and Crawford. In: *HOP*.
Sagenogen, from Porter. In: *HOP*. Drawing by Andrew J. Lampkin, III.
Semicell, from Hoshaw *et al*. In: *HOP*. Courtesy of J. D. Pickett-Heaps.
Somatonemes, from Brugerolle and Mignot. In: *HOP*.
Sorocarp, reproduced from M. C. Deasey and L. S. Olive, *Science* 213:561-563 (1981). Copyright 1981 by the AAAS. Courtesy of K. B. Raper.
Sorocyst, reproduced from K. B. Raper *et al.*, *American Journal of Botany* 65:1011-1026 (1978) with copyright permission of the Botanical Society of America. Courtesy of F. W. Spiegel.
Sphaerocyst, from F. W. Spiegel and L. S. Olive, *Mycologia* 70:843-847. Copyright 1978, The New York Botanical Garden. Reprinted with permission. [*HOP*.]
Spicule, from Febvre-Chevalier. In: *HOP*.
Spindle pole body, from Perkins. In: *HOP*.
Sporangium, from L. Margulis and K. V. Schwartz, *Five Kingdoms*. Copyright (c) 1988 by W. H. Freeman and Company. Reprinted with permission. Drawing by Laszlo Meszoly.
Sporelings, from Clayton. In: *HOP*.
Sporoblasts, from Lom. In: *HOP*.
Sporocyst, from L. Margulis and K. V. Schwartz, *Five Kingdoms*. Copyright (c) 1988 by W. H. Freeman and Company. Reprinted with permission. Drawing by Laszlo Meszoly.
Sporozoite, from I. Desportes, *Annales des Sciences Naturelles, Zoologie et Biologie Animale,* 12ème Serie 11:31-96 (1969).
Stachel, from Dylewski. In: *HOP*.
Stomatocyst, from Kristiansen. In: *HOP*.
Swarmer, drawing by Kathryn Delisle.
Synaptonemal complex, from Dylewski. In: *HOP*.
System I fiber, from Melkonian. In: *HOP*.
Taeniocyst, from Taylor. In: *HOP*. [*HOP*.]
Telophase, from L. Margulis and D. Sagan. *Origins of Sex*. New Haven, CT: Yale University Press, 1986.
Terminal cap, from Floyd and O'Kelly. In: *HOP*. [*HOP*.]
Thecal plate, from Taylor. In: *HOP*.
Thylakoid, from Kies and Kremer. In: *HOP*.
Tomont, from Lynn and Small. In: *HOP*. [*HOP*.] Drawing by Steve Alexander.
Transverse ribbon, from Lynn and Small. In: *HOP*. [*HOP*.]
Trophozoite, from Vivier and Desportes. In: *HOP*.
Trypomastigote, from Vickerman. In: *HOP*. [*HOP*.]
Undulipodial groove, from Melkonian. In: *HOP*. [*HOP*.]
Undulipodial hairs, from Melkonian. In: *HOP*.
Undulipodial rootlet, from M. A. Gillott and S. P. Gibbs, *Canadian Journal of Botany* 61:1964-1978 (1983).
Undulipodium, drawing by Kathryn Delisle.
Valve face, from Round and Crawford. In: *HOP*.
Velum, from Round and Crawford. In: *HOP*.

Organism Glossary

Acantharia, drawing by Kathryn Delisle.
Acrasea, from L. Margulis and D. Sagan, *Origins of Sex*. New Haven, CT: Yale University Press, 1986. Drawing by Laszlo Meszoly.
Actinopoda, from L. Margulis and K. V. Schwartz, *Five Kingdoms*. Copyright (c) 1988 by W. H. Freeman and Company. Reprinted with permission. Drawing by Laszlo Meszoly.
Amebomastigota, from Dyer. In: *HOP*. [*HOP*.]
Apicomplexa, drawing by Steve Alexander.
Bacillariophyta, from L. Margulis and D. Sagan, *Origins of Sex*. New Haven, CT: Yale University Press, 1986. Drawing by Emily Hoffman.
Bicosoecids, courtesy of J. P. Mignot.
Bodonidae, from G. Brugerolle *et al.*, *Protistologica* 15: 197-221 (1979).
Chlorarachnida, based on information from D. J. Hibberd. Drawing by Kathryn Delisle.
Chlorophyta, based on information from M. Melkonian. Drawing by Kathryn Delisle.
Choanomastigota, (a) courtesy of K. R. Buck; (b) from Buck. In: *HOP*. Drawing by Steve Alexander.
Chrysophyta, based on information from J. Kristiansen. Drawing by Kathryn Delisle.
Chytridiomycota, (a) from D. J. S. Barr. In: *HOP*; (b) based on information from D. J. S. Barr. Drawing by Kathryn Delisle.
Ciliophora, based on information from D. H. Lynn. Drawing by Kathryn Delisle.
Conjugaphyta, based on information from R. W. Hoshaw. Drawing by Kathryn Delisle.
Cryptophyta, based on information from M. Gillott. Drawing by Kathryn Delisle.
Dictyostelida, (a) drawing by Kathryn Delisle; (b) based on information from J. C. Cavender. Drawing by Kathryn Delisle.
Dinomastigota, from Taylor. In: *HOP*. [*HOP*.]
Diplomonadida, from Kulda and Nohŷnkovà. In: *Parasitic Protozoa* (J. P. Kreier, ed.), Vol. 2, pp. 2-138. New York: Academic Press, 1978.
Ebridians, from Taylor. In: *HOP*. [*HOP*.] All drawings by Kathryn Delisle.
Ellobiopsida, from Whisler. In: *HOP*.
Euglenida, from Walne and Kivic. In: *HOP*. [*HOP*.]
Eustigmatophyta, from Hibberd. In: *HOP*.
Florideophycidae, (a) based on information from P. W. Gabrielson. Drawing by Kathryn Delisle; (b) from Gabrielson *et al.* In: *HOP*. [*HOP*.]
Foraminifera, from M. McEnery and J. J. Lee, *Journal of Protozoology* 23:94-108 (1976).
Glaucocystophyta, based on information from L. Kies. Drawing by Kathryn Delisle.
Granuloreticulosa, based on information from M. Leckie and J. J. Lee. Drawing by Kathryn Delisle.
Haplosporidia, based on information from F. O. Perkins. Drawing by Kathryn Delisle.
Heliozoa, drawing by Kathryn Delisle.
Hyphochytriomycota, based on information from M. S. Fuller. Drawing by Kathryn Delisle.
Karyoblastea, (a) drawing by Christie Lyons; (b) drawing by Kathryn Delisle.
Kinetoplastida, from Vickerman. In: *HOP*. [*HOP*.]
Labyrinthulomycota, from Porter. In: *HOP*. Drawing by Andrew J. Lampkin, III.
Microspora, based on information from E. U. Canning. Drawing by Kathryn Delisle.
Myxomycota, from Frederick. In: *HOP*. Photo by Ray Simons.
Myxozoa, based on information from J. Lom. Drawing by Kathryn Delisle.
Oomycota, based on information from I. B. Heath. Drawing by Kathryn Delisle.
Opalinata, from Corliss. In: *HOP*. [*HOP*.]
Parabasalia, from L. Margulis amd K. V. Schwartz, *Five Kingdoms*. Copyright (c) 1988 by W. H. Freeman and Company. Reprinted with permission. Drawing by Robert Golder.
Paramyxea, from I. Desportes, *Origins of Life* 13:343-352 (1984). Reprinted by permission of Kluwer Academic Publishers.
Phaeophyta, (a) drawing by Christie Lyons; (b) based on information from M. N. Clayton. Drawing by Kathryn Delisle.
Plasmodial Slime Molds, (a) from Frederick. In: *HOP*. [*HOP*.] Drawing by Sheila Manion-Artz; (b) drawing by Kathryn Delisle.
Plasmodiophoromycota, (a) from Dylewski. In: *HOP*; (b) redrawn from D. J. S. Barr and P. M. E. Allan, *Canadian Journal of Botany* 60:2496-2504 (1982) with additional information from D. J. S. Barr. Drawing by Kathryn Delisle.
Proteromonadida, from Brugerolle and Mignot. In: *HOP*.
Protostelida, (a) from Spiegel. In: *HOP*; (b) drawing by Frederick Spiegel.
Prymnesiophyta, based on information from J. C. Green. Drawing by Kathryn Delisle.
Pseudociliata, from J. O. Corliss, *The Ciliated Protozoa*. Oxford, N.Y.: Pergamon Press, 1979.
Pyrsonymphida, from L. Margulis and D. Sagan, *Origins of Sex*. New Haven, CT: Yale University Press, 1986. Drawing by Christie Lyons.
Raphidophyta, based on information from P. Heywood. Drawing by Kathryn Delisle.
Retortamonadida, from Brugerolle and Mignot. In: *HOP*.
Rhizopoda, from Schuster. In: *HOP*. [*HOP*.]
Rhodophyta, from Gabrielson *et al.* In: *HOP*. Courtesy of M. Sommerfeld.
Xanthophyta, based on information from D. J. Hibberd. Drawing by Kathryn Delisle.
Xenophyophora, photos by Øle Tendal.
Zoomastigina, drawing by Keith Vickerman.